# Insects in Perspective

# INSECTS IN Perspective

MICHAEL D. ATKINS

*San Diego State University*

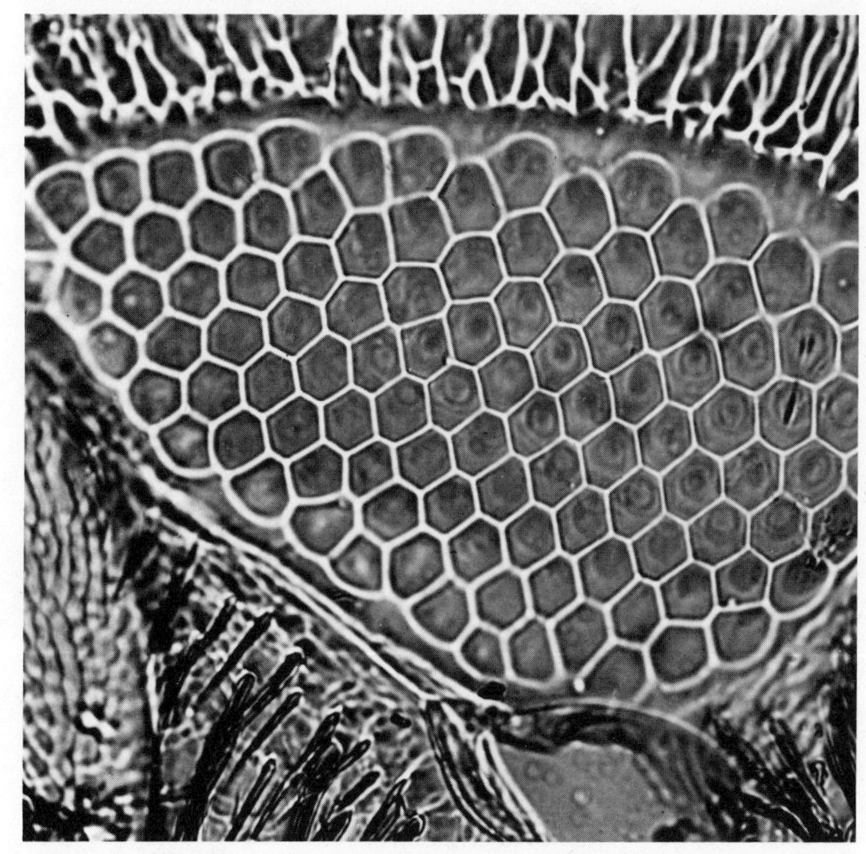

MACMILLAN PUBLISHING CO., INC.
NEW YORK
Collier Macmillan Publishers
LONDON

In memory of JOHN A. CHAPMAN
who first introduced me to the fascinating world of insects and thereby provided me with countless hours of enjoyment.





Printed in the United States of America



Macmillan Publishing Co., Inc.
866 Third Avenue, New York, New York 10022

Collier Macmillan Canada, Ltd.

Library of Congress Cataloging in Publication Data

Atkins, Michael D
Insects in perspective.

Includes bibliographies and index.
1. Insects. I. Title.
QL463.A87 595.7 77-705
ISBN 0-02-304500-0

Printing: 1 2 3 4 5 6 7 8 Year: 8 9 0 1 2 3 4

# Preface

I have never met an entomologist who did not thoroughly enjoy his occupation. The reason for this, I believe, lies in the fact that the study of insects presents a constant challenge and repeatedly rewards the investigator with the discovery of something he did not know before. Even though hundreds of thousands of insect species have been described over the centuries and a huge body of knowledge has already been amassed about them, new and exciting discoveries are being made every day.

The field of entomology is so dynamic that teaching even an introductory course presents a real challenge. The field is so broad and so many other disciplines come to bear on the study of insects that presenting a balanced overview can be difficult. If one is not careful, it is easy to become so involved with the basic information about morphology and classification that many of the really interesting aspects of insect biology are given little or no attention. The authors of introductory texts face a similar problem in that by the time they have discussed the structure, function, classification, and identification of insects, there is little space left within a book of reasonable length to devote to other aspects.

I faced this problem too, and decided there was no reason to publish another introductory book about insects unless it was rather different from the others available. Over the years I have found that students rarely request additional reading in support of laboratory exercises, probably because it is relatively easy to provide the resources necessary for a good laboratory learning experience. However, students often ask for reading material that expands on what I have covered in lecture.

My reason for writing this book, therefore, was to provide some up-to-date background reading for a broad range of lecture topics. From the outset, I prepared chapters and distributed them as background reading to students in classes of both majors and nonmajors in entomology. The comments have been very favorable, and I am confident that this text can serve satisfactorily in introductory entomology courses as well as courses about insects taken by nonmajors who wish to expand their general education through courses in the life sciences. I am hopeful that professional entomologists will find it interesting reading as well.

I have tried to organize the chapters into a logical sequence, but I have also tried to keep each chapter sufficiently self-contained that an instructor can develop his own arrangement of lectures. For this reason I have provided a bibliography at the end of each chapter rather than a single list of references at the end of the book.

My over-all objective, of course, has been to present a balanced view of the biology of insects, and thereby stimulate the reader's interest in these fascinating creatures. In Part I, I have tried to show how the development of entomology paralleled and perhaps influenced the development of science in general, to show the relationship of insects within the animal kingdom, and to illustrate how the evolution of insects has led to such a great diversity of kinds. Although it was never my intent to provide an aid to insect identification, the orders are presented in an appendix in the hope that the discussion of insect groups with which students are unfamiliar will be more meaningful. In Part II my objective was to cover the important aspects of structure and function while avoiding, as much as possible, tedious descriptions of anatomy. Consequently, I chose a process orientation and concentrated on the adaptive breakthroughs that have clearly played a major role in the insect success story. Part III is devoted to behavior, an aspect of insect biology so often treated inadequately in general texts. Once again, I based this discussion on a selected list of processes of both biological and practical importance. Finally, in Part IV I have examined the relationships of insects with other organisms and their environment. This is where both the beneficial and harmful roles of insects are reviewed, preparatory to a consideration of how we can reduce pest problems without damaging our natural heritage.

I am indebted to many people who have provided assistance and encouragement throughout this project, as well as to all the entomologists who made the basic discoveries that form the foundation of any book about insects. I am especially grateful to Woody Chapman, Biology Editor for Macmillan Publishing Co., Inc., who guided the project from conception to production, and to John A. Davidson and Fred Delcomyn, who offered valuable criticism of the entire manuscript.

I also thank T. J. Cohn, G. W. Cox, E. Huffman, R. E. Monroe, and C. E. Norland, who freely gave of their time for discussion and commented on selected chapters as I progressed. I am grateful for the assistance and corrections that they and others provided; the errors that remain are solely my responsibility.

I thank the following for generously allowing me to use original material or for providing one or more photographs: Max Badgley, Canadian Forestry Service, Thomas Eisner, W. G. Evans, L. A. Falcon, E. F. Knipling, Jason Lillegraven, Daniel Mahr, Neil Marshall, M. L. McManus, Gary Pitman, William Stephen, Vern Stern, United States Department of Agriculture, University of California, E. O. Wilson, and W. G. Wellington.

M. D. A.

# Contents



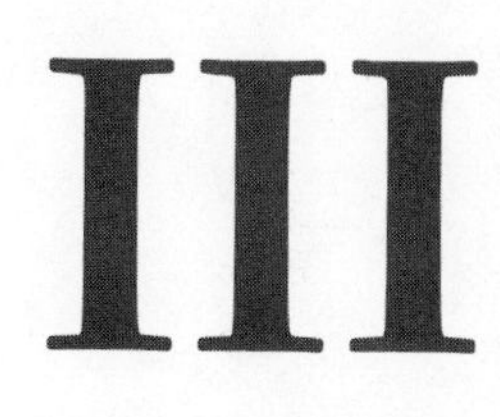

# Introduction

New discoveries of ancient humanoid remains have resulted in an ever earlier estimate of our origin. Some anthropologists now believe that man or his immediate ancestors have walked the face of the earth for perhaps two million years, or twice as long as was considered possible just a few years ago. But even if these new estimates are accurate, we are relative newcomers in the eyes of the insects. The insects as a group have their origin somewhere in the order of 350 million years ago, when the climate of the earth was generally warm and humid and a vegetation of giant horsetails, ferns, and club mosses extended northward into Norway and Sweden. The remains of winged insects do not appear in the fossil record until about 80 million years later. By that time, many of the orders in existence were indistinguishable from present day forms. Some orders did not really proliferate until much later when the flowering plants evolved, and others, such as the lice, did not appear until warm-blooded animals were available as hosts. But even the most recently evolved orders of insects have been here for at least as long as man.

By the time man came on the scene and was present in sufficient numbers to spread out from the cradle of his origin, the insects were already a highly integrated part of every terrestrial and fresh-water ecosystem. In fact, the insects had already had a profound influence on all of these systems and on the evolution of many of their plant and animal components. Man, then, was an outsider trying to break in, and to do so, he had to find his place among all of the long-time relationships that already existed as part of his environment.

Undoubtedly from the very beginning, insects that were ectoparasitic, such as the blood-feeding lice and flies, simply adopted man as another animal host and source of food. Many such insects had already established themselves as vectors of disease-causing organisms such as protozoans (sporozoans, trypanosomes) and filarial worms. Therefore, we can safely assume that primitive man was infected by insect-borne pathogens. Parasitologists have postulated that man's serious reaction to sleeping sickness, caused by a tsetse fly-transmitted trypanosome, is indicative of his recent relationship as a host for this protozoan parasite. Animals such as native cattle that become infected by these organisms apparently do not display the same severity of symptoms because they have had time to evolve a more tolerant relationship with them.

Humanoids with the simplest of life styles would have been producers of waste in the form of food scraps, excrement, and corpses, which would have been fed upon by many forms of insect scavengers and carrion feeders. In some cases, these early associations have persisted. Some species of insects, such as the house fly, have become so intimately associated with man-produced waste that they are only found in areas inhabited by man.

We do not know for certain how long man spent as a more or less free-living animal, but archeological evidence indicates that early in our existence we lived in family groups or small bands that had at least temporary home ranges with a living place such as a cave at its center. These earliest of dwellings provided the basis for some additional relationships between the insects and ourselves. Caves were the natural nesting and resting places for other animals as well as for man. Parasitic and scavenging insects, such as fleas, lice, and cockroaches, probably found man and his leavings just as much to their liking as they did the other cave occupants and their leavings.

As time went on, primitive man gradually changed from being strictly a hunter and gatherer to a storer of small surpluses and a possessor of some simple belongings such as hides used for clothing and sleeping pads. We need not stretch our imagination to visualize how insects such as booklice (Psocoptera) and small beetles (Coleoptera) that are common to bird nests might simply have moved to small caches of grain or other food placed in trees to keep the foodstuffs out of reach of ground-dwelling animals. Nor is it difficult to imagine how insects that fed on the skin and hair of carcasses would have become inhabitants of the hides used by man. As man moved about, carrying his simple possessions with him, he began to transport these new insect associates from place to place. From that time onward, man has had a continuous relationship with fleas, bedbugs, roaches, dermestid beetles, and other food pests that infest cereals and dried fruits.

Some time later, man began to devote less time to hunting and wandering in search of food and more time to the cultivation of food plants and the domestication of animals. This led to a new way of life which involved the settlement of tribes in villages adjacent to crop-growing areas. Rather than living in available caves, these early farmers began to build simple living structures from whatever material resources were available. The cultivation of food led to larger surpluses that had to be stored. The reduced need to hunt and gather provided more time to engage in the development of tools and in the manufacture of personal possessions. During this time, referred to as the Neolithic period, which flourished some 10,000 to 15,000 years ago, man's association with the insects expanded still further. The household pests brought from the caves flourished in the villages. The storage pests that had transferred from animal caches, as well as those which normally fed inside the seeds of plants now under cultivation, were provided with a continuous place in which to feed and reproduce. The cultivation of native plants provided conditions, not unlike modern monocultures, which were ideal for the build-up of larger populations of indigenous plant-feeding insects of great variety. Some of the wood used to build houses was almost certainly infested with termites and other wood borers that are now structural pests of some concern. On the other hand, the abundance of the harvest from these early farms was made possible by the beneficial insects that year after year pollinated the crop plants and improved the structure of the soil.

We need also to consider the impact of the early aggregation of larger numbers of people in villages and of the confinement of herds of domesticated animals. Both of those activities would have affected the populations of ectoparasites and blood-feeding insects by providing them with readily available hosts. The concentration of natural odors and the attractant chemicals present in accumulations of their urine and excrement made these corralled animals easy to locate. In addition, the creation of village watering holes and water impoundments for irrigation provided breeding sites for aquatic insects such as mosquitoes that transmit disease-causing organisms.

By about 5,000 B.C., people had begun to live and work in larger cities. Many of these cities were dependent upon the agricultural production of the surrounding countryside. A predictable outcome of this was an increased concern about insect problems in relation to the crops, as well as about the insect pests directly associated with urbanization. There was considerable commerce, which facilitated the transport of storage pests, livestock pests, and crop pests from place to place. Urbanization itself concentrated the production of human waste and increased the incidence of flies, lice, and the rodent reservoirs of insect-borne diseases such as typhus, typhoid, bubonic plague, and dysentary.

Clearly then, the ascent of man, which began long after the insects had established all of their fundamental ecological relationships, made inevitable a complex set of interactions that could only grow in both their beneficial and detrimental consequences as man altered his way of life. Perhaps we should not be surprised, therefore, that some of the very first elements of science, including the work of Aristotle and others, involved the observation of insects and insect-related problems. The relationship between man and the insects, as recent as it is, has provided the source of much concern, fascination, and delight. In the following pages, I hope to reveal some of the reasons for this wide range of attitudes.

# I

# Some Generalizations

# A little history

# 1

Entomology is the scientific study of insects and therefore contributes to the body of knowledge we refer to as science. This body of knowledge has been accumulated over a long period of time and has been influenced by the sociological and philosophical changes that have transpired. It is impossible therefore to separate the history of entomology from the history of science or the history of science from history in general. What follows then is merely an identification of some of the landmark events in the development of science in general and of entomology in particular.

In the early stages of human development, man was simply a part of natural history, another animal living in concert with all the other components of his environment. As a hunter and gatherer, and a little later, as a crude horticulturist, man had to learn certain fundamentals about the plants and animals he relied upon for food and shelter and about their relations with each other and the passing seasons. He gained the information he needed for survival through observation and trial and error, but made no attempt to gain experimental proof of cause-and-effect relationships. At first, most observed events were given supernatural explanations, and the observations and explanations were then passed from generation to generation by way of simple stories that often became more fiction than fact. Later, as man learned the cycles of plants and seeds and applied his knowledge to the cultivation of crops, he became, in a sense, a scientist, with a rather extensive understanding of nature and a basic curiosity about the weather and the influence of the passing seasons.

Man's technological advancement proceeded at different rates in different regions in relation to the climate, ecological diversity, and availability of material resources. The real scientific revolution did not occur until just a few hundred years before the birth of Christ. By this time, many of the fundamental technological advances connected with agriculture and construction had been made, and the earlier time-consuming activities related to the provision of food, shelter, and protection had been simplified considerably. The major problems of the day were war and pestilence, so it is not surprising that some of the earliest detailed "scientific" observations were concerned with insect pests and diseases. Large cities were repeatedly swept by maladies, sometimes unknowingly transmitted by insects. Crops were ravaged by insect pests, and the grains and fruits in storage vaults were decimated by insects and rodents.

The writings of some of the earliest scientists dwelt heavily upon the insects. Theophrastus (ca. 380–287 B.C.) catalogued many plant diseases and insect pests of crop plants. Aristotle (384–322 B.C.), considered the "Father of Zoology," established the basis for the scientific study of insects. He was a careful observer of natural phenomena who by the application of *deductive reasoning* (reasoning from the general to the particular) was able to gain considerable insight into the functioning of natural events. For example, he realized that insects had different stages in their life history and underwent metamorphosis, although he was not clear as to the interrelationships of all the stages, and thought that pupae were eggs.

A little later, the Roman authors, led by Pliny (A.D. 23–79), became the first true encyclopaedists and recorded voluminous information on agriculture, engineering, architecture, crop pests, and veterinary problems. Book eleven of *Historia Naturales* (A.D. 77) by Pliny was devoted entirely to insects.

Following the partition of the Roman Empire in 395, the West Roman Empire fell into chaos and the ancient scientific tradition was largely lost to western Europe. Such manuscripts and knowledge that survived were managed by the monastic schools, but the works of Aristotle and Theophrastus were lost. Fortunately, the East Roman Empire of Byzantium gathered the manuscripts of the ancients into its libraries where they were translated and copied. These formed the basis for the development of Arabic science over the next several hundred years.

In the West, scientific progress was almost nonexistent throughout the Medieval period. The church diverted the best minds from science to theological writing, and the divinity became the predominant theme of art, music, and literature. Throughout this entire period there were only a few significant works dealing with natural history. Insects were mentioned only in the encyclopaedic *Origines sive Etymologiae* by Isidorus of Seville (ca. 560–636) and in *De Universo* by Rhabanus Maurus (776–856) until late in the period when Albertus Magnus (1193–1280) re-established a link with the ancients by paraphrasing Aristotle in his *De Animalibus* (1255–1270). The last great Western work of the Middle Ages that treated entomological subject matter, and perhaps the most influential in that it appeared just prior to the Renaissance, was the *Ruralium Commodorum* (1304–1309), in which Petrus Crescenti of Italy discussed many insect pests and their control.

The Renaissance (1400–1600) was a period of intellectual awakening that encompassed a rediscovery of the Greek and Roman contributions to science, which had been retained largely in institutions of the Eastern Roman Empire and referred to extensively during the rise of Islam. The voyages that returned to Europe from the East brought new knowledge along with the old, and established a new sense of inquiry. Scientists again began to question freely the authority of the church and to lean less heavily on the discoveries and methods of Aristotle, Pliny, and the Scriptures. By the end of the Renaissance, the Modern Scientific Revolution had begun, stimulated in part by the development of a new scientific method that stressed *inductive reasoning* (reasoning from specific questions to general concepts), brought into prominence by Francis Bacon and Galileo. The scientific method made investigation more rigorous because it was based on the assumption that all of the components of the universe interact in a predictable way, and therefore it should be possible to develop an understanding of the laws governing these interactions. Scientists believed that such an understanding would make possible the prediction of certain events that in turn would be useful in the development of ways to modify or control them. (The basic steps in the application of the modern scientific method are described in the accompanying box.)

Every scientific investigation begins with the observation and description of a set of occurrences from which a pattern is established that permits or suggests the formulation of a specific hypothesis. The early advancement of science was therefore limited to the range of objects and phenomena that could be observed with the unassisted senses. In order to investigate the heavens and the microscopic world, technological advancements that would

**The Scientific Method**

The modern approach to scientific discovery involves the adherence to a sequence of logical steps. Each investigation begins with the organization of observed occurrences into a conceptual model. Such a model consists of a description of a process, based on experience, insight, and logic. The conceptual model forms the basis for one or more questions, which are phrased as either a positive statement called a *hypothesis* or a negative statement called a *null-hypothesis*.

An experiment is then designed to test the validity of the hypothesis or null-hypothesis. This involves the establishment of a set of tests in which the effect of treatments or conditions under investigation can be demonstrated clearly. This involves a comparison of treated groups with untreated (control) groups. The experiment must be replicated sufficiently to reduce the influence of anomalies as a result of drawing the samples from a narrow portion of the range of natural variation.

The data obtained from the experiment must then be analyzed statistically to demonstrate that the treatments resulted in effects that were different from those obtained from the controls owing to something other than chance.

The analyzed data must then be interpreted relative to the original hypothesis or null-hypothesis. If the data analysis supports the rejection of the null-hypothesis or the retention of the hypothesis, they stand as factual until such time as refuted by further experimentation. These facts are then incorporated into the formulation of more complete conceptual models that in turn lead to new hypotheses and further experimentation.

increase the power of critical observation were needed. The need for such improvements was obviously recognized, as indicated by the almost simultaneous invention of the microscope and the telescope by the Janssens (father and son) and by Galileo. The use of these magnifying devices ushered in the age of microscopy that led to the devlopment of a basic understanding of plant anatomy (Grew and Malpighi), insect anatomy (Malpighi and Swammerdam), and microbiology (Leeuwenhock).

Typical of the progress made during this period as a result of applying the scientific method and the new tools of science were the disproof by Francesco Redi (1626–1697) of the spontaneous generation of maggots from spoiled meat and his description of ectoparasitic lice. At about the same time, John Ray (1627–1705), a botanist with a basic interest in insects, published the first descriptions of insect life histories, including an accurate account of caterpillar metamorphosis. He also made observations on the parasitism of caterpillars, and in one of his articles stated:

> I shut up ten or so of these in a wooden box at the end of August 1658. They fed for a few days, and fixed themselves to the sides or lid of the box. Seven of them proved to be viviparous or vermiparous: from their backs and sides very many, from thirty to sixty apiece, wormlike animalcules broke out; they were white, glabrous, footless, and under the microscope transparent. As soon as they were born they began to spin silken cocoons, finished them in a couple of hours, and in early October came out as

Jan Swammerdam (1637-1680), the most famous of the early insect anatomists.

flies, black all over with reddish legs and long antennae, and about the size of a small ant. The three or four caterpillars which did not produce maggots after a long interval changed into angular and humped chrysalids which came out in mid-April as white butterflies.

He later wrote in his *Historia Insectorum,* 1710:

Whence these maggots arise is a great problem. I think that the ichneumon wasps prick these caterpillars with the hollow tube of their ovipositor and insert eggs into

Francesco Redi (1626-1698) contributed to the disproof of spontaneous generation through his study of maggots in spoiled meat.

their bodies: the maggots are hatched by the warmth of them, and feed there until they are full grown; then they gnaw through the skin, come out, and spin their cocoons.

The adoption of the scientific method and the use of the microscope opened the floodgates to an ever-increasing flow of information about the natural world in general and the insects in particular. This great expansion of knowledge, combined with the growing lists of plants and animals described at home and abroad, created the need for a refined system for organizing what was known. The development of such a system became the main thrust of many biologists during the eighteenth century. René Antoine Ferchault de Réaumur (1683–1757) produced the first entirely original compendium of entomology since Aristotle. Through his six-volume *Memoires pour server a l'histoire des insects* (1734–1742), he initiated modern entomology and produced the first well-illustrated classification. Réaumur also introduced behavior and ecology, and provided the first scientific rationale for insect control. At about the same time, a Swedish physician and botanist, Carl von Linné, later called Linnaeus (1707–1778), introduced the binomial system of nomenclature which reduced lengthy Latin descriptions to a two-part name. Linnaeus was clearly a superb observer with a sophisticated understanding of the natural world even though he subscribed to the popular belief of special creation. His ability to select strong diagnostic characters as the basis for classification, his strict adherence to the system he proposed, and his recognition as an authoritative scientist led to widespread acceptance of his work that culminated in the publication of *Systema Naturae*, 10th edition, in 1758.

After the publication of *Systema Naturae*, taxonomy was no longer cumbersome and rapidly gained popularity, particularly among the professional

René Antoine Ferchault de Réaumur (1683–1757) is considered to have laid the foundations of modern entomology.

Carl von Linné (Linnaeus) (1707–1778) introduced the modern system of binomial nomenclature.

upper class who engaged in natural history as a side interest or hobby. But a student of Linnaeus by the name of Johann Christian Fabricius (1745–1808) overshadowed all the earlier insect taxonomists. In his *Systema Entomologica* (1775) and subsequent volumes, he attempted to classify the insect fauna of the entire world; no entomologist has attempted to do so since.

At about the same time, with the increase of colonialism, many species of plants of potential agricultural value were being transported from place to place. Frequently, the plants flourished in their new surroundings, as did some of the insects that were inadvertently transported with them. In some cases, native insects found the introduced plants much to their liking—with disastrous results. Agriculture in the colonies, particularly in the United States, suffered severe losses as a result of damage by insects; consequently, men such as Colonel London Carter (1771) and W. D. Peck (1795) began to study the biology and control of economically important insect pests.

Although much of this applied entomology was conducted in Europe by Du Monceau (1762), William Curtis (1782), and others, American entomology came into being about the beginning of the nineteenth century. One of the first American workers in the field was W. D. Peck, who published a series of articles on insect pests of the New England states. But the title of "Father of American Entomology" was to fall to Thomas Say, the entomologist among a group of naturalists who in 1812 founded the Academy of Natural Sciences of Philadelphia. Say's major contribution consisted of three volumes, entitled *American Entomology, or Descriptions of the Insects of North America.* A little later, T. W. Harris, a student of Peck, published a series of papers on economically important insects that culminated in the printing in

Thomas Say (1787–1834), father of American entomology.

1841 of *Report on Insects Injurious to Vegetation,* which in later editions comprised the first textbook of economic entomology.

The work of Say and Harris, in particular, stimulated an interest in entomology in this country that led to the establishment of the Federal Bureau of Agriculture in 1853 and to the appointment of Townsend Glover as the first

Thaddeus William Harris (1795–1856), father of American economic entomology.

federal entomologist. The United States Department of Agriculture was created in 1878 and was elevated to an executive department ten years later.

The continual widespread agricultural losses caused by insect outbreaks stimulated entomological activities at the state level. In 1854, Asa Fitch was appointed as the nation's first state entomologist in New York, and was provided a budget of $1,000 for entomological research. Illinois and Missouri soon followed with the appointment of Benjamin Walsh and C. V. Riley in 1867 and 1868, respectively, as state entomologists.

During his tenure as state entomologist of Missouri, Riley produced the "Nine Missouri Reports" (1869–1877), which are considered to be the finest nineteenth-century contributions to economic entomology. In 1878, Riley was appointed to the position of federal entomologist and subsequently established the United States Entomology Commission, the forerunner of today's diverse federal entomological programs. It was also Riley who organized the American Association of Economic Entomology in 1889. As prominent as Riley was in American applied entomology, we must not lose sight of the fact that he was guided and greatly influenced by Benjamin Walsh (1808–1870), who was educated at Cambridge University where he knew and worked with Charles Darwin. After coming to the United States at the age of 30, Walsh became a successful lumberman, retired, and devoted the rest of his life to entomology.

Applied entomology in America received additional stimulation with the passage of the Morrill Act (1862) that established land-grant colleges and led to the formation of the entomology departments at Michigan State University, Cornell University, and Kansas State University. The Hatch Act (1887) then established state agricultural stations that also became heavily involved in entomological research and the training of young entomologists.

During this early phase of applied entomology in the United States, biology in general was becoming increasingly sophisticated in Europe. Taxonomists had realized the merits of specializing in more manageable groups and were devoting more attention to improving taxonomic concepts. Darwin and Wallace established the theory of organic evolution and laid the foundation for the development of a much-needed modern concept of the species. Biology thus became less and less descriptive, and theories were developed concerning the physiology of organisms, embryology, and the structure and function of cells. About 1869, Paris green was discovered to be an effective stomach poison for insect pests; that discovery opened the way to the development of chemical insect pest control as a significant field of research. Subsequently, the discovery that insects such as the cottony cushion scale, a serious citrus pest introduced into California (ca. 1868), could not be suppressed with available chemicals led to the first attempt at biological control. As a result of the persistence of W. G. Klee who was the California State Inspector of Fruit Pests, an Australian entomologist, Frazer Crawford, shipped some live parasitic flies of the genus *Crytochaetum* to California in 1888. Later, C. V. Riley sent Albert Koebele to Australia, ostensibly as a representative of the United States State Department to the 1888 International Exposition at Melbourne. While there, Koebele discovered and shipped the lady beetle *Vedalia cardinalis* to Los Angeles. *Crythochaetum* and *Vedalia* saved the citrus industry in California and established biological control as a viable concept.

By the beginning of the twentieth century, numerous insect pests had been

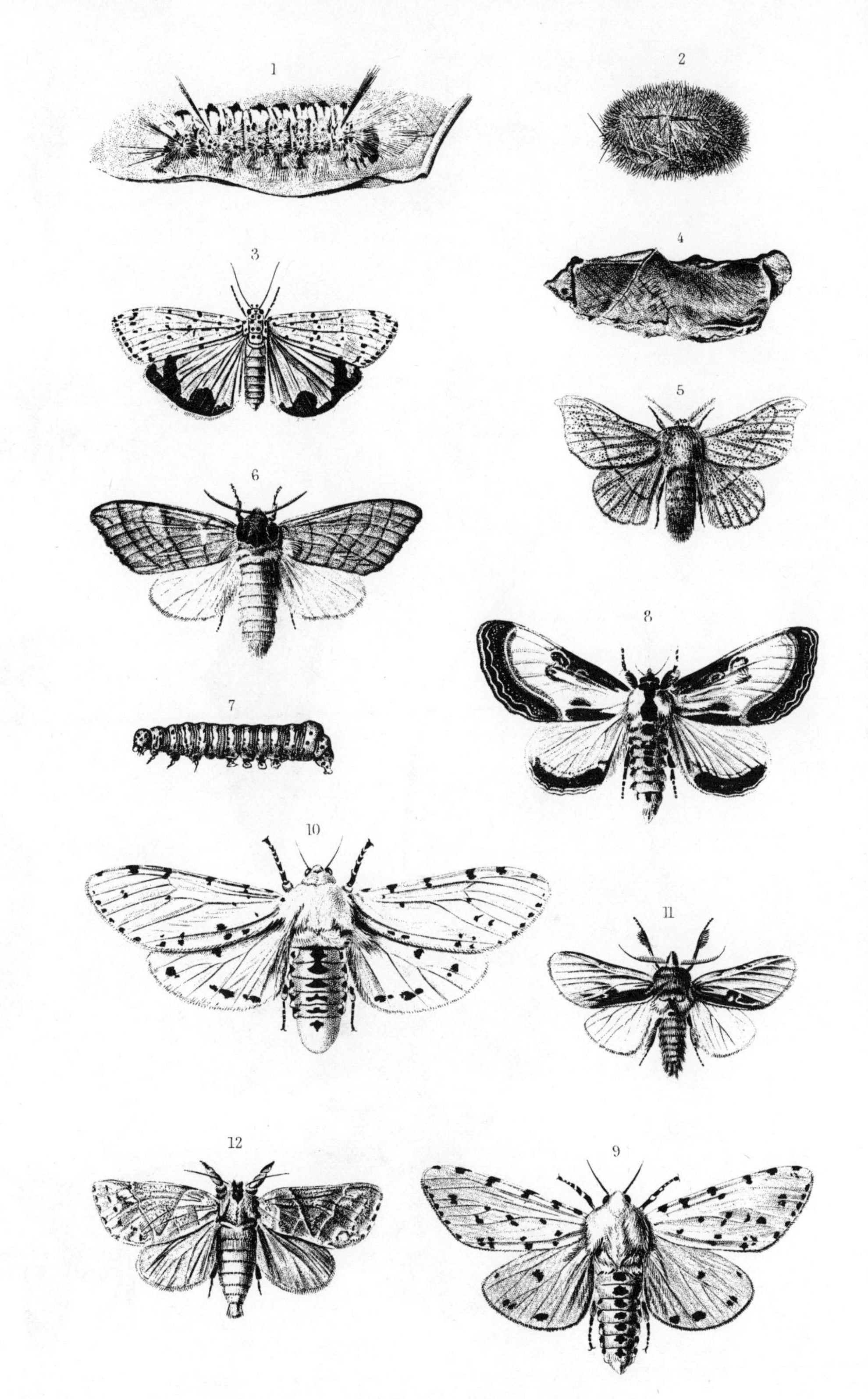


A copy of plate VI from Harris' book, *Insects Injurious to Vegetation.* Drawings by Antoine Sonrel.

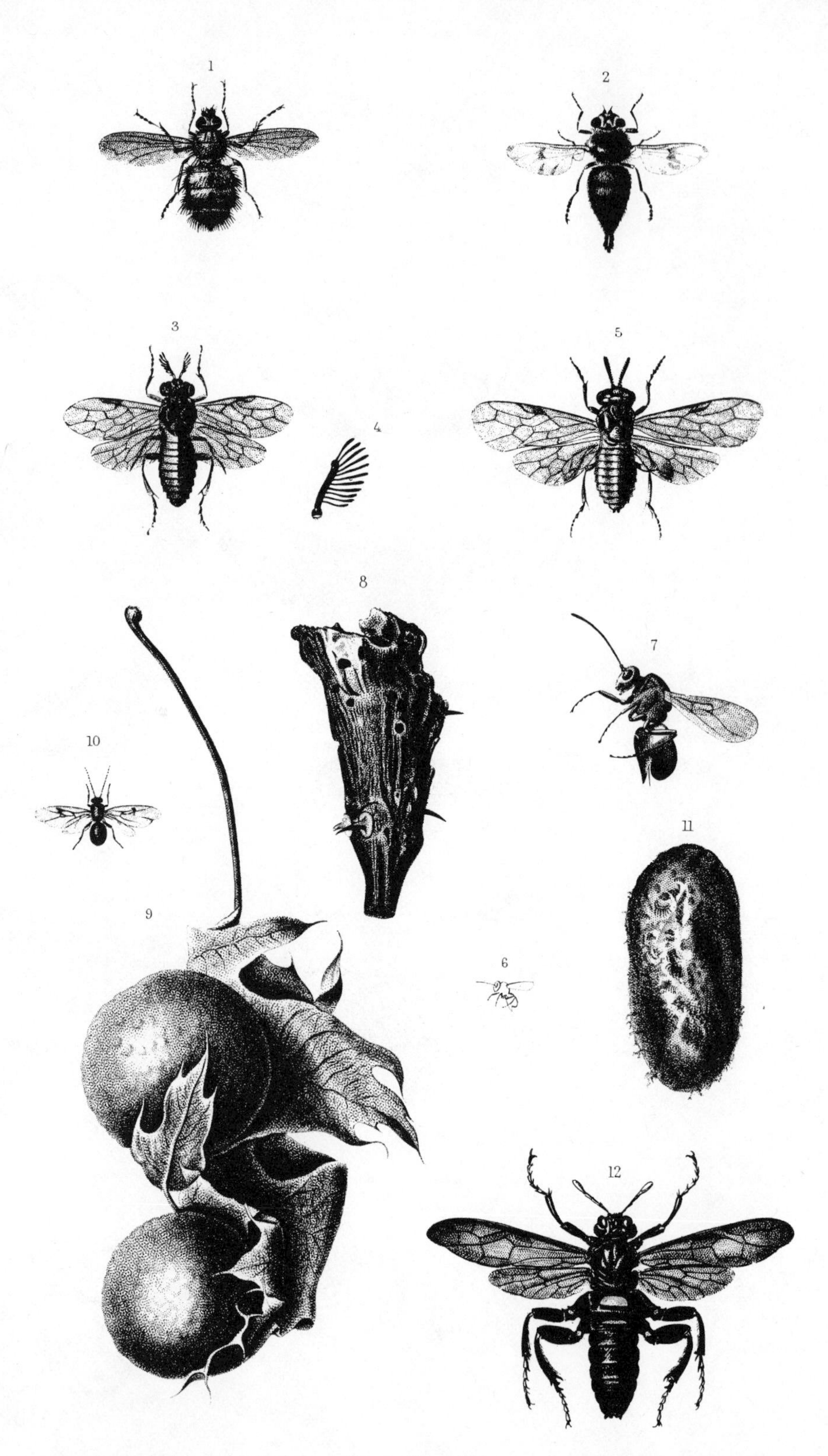


A copy of plate VIII from Harris' book, *Insects Injurious to Vegetation.* Drawings by Antoine Sonrel.

Asa Fitch (1809–1878), the first state entomologist.

Charles Valentine Riley (1834–1895) is considered the greatest nineteenth century American economic entomologist.

introduced and had established themselves in the forests and agricultural areas of the United States. The ongoing need to reduce the damage caused by these pests provided an ongoing stimulus to the growth of applied entomology in this country and in Canada. Thereafter, entomology continued to become an increasingly important and complex field of biology, punctuated with landmarks too numerous to mention here. Entomology has since grown into a field of science of great importance to human welfare that now relies heavily upon the input of scientists from a wide array of disciplines. Chemists and toxologists are concerned with the development of new and safer insecticides, hormone mimics, and attractants for use in pest control programs. Physicists and engineers contribute to the design of more efficient spray equipment and detection devices. Geneticists are searching for ways to induce lethal mutations in pest populations and to contribute to the breeding of new varieties of crop plants that are resistant to insect attack. Ecologists, meteorologists, and horticulturists contribute to the development of over-all management programs for crop and forest ecosystems and are aided greatly by systems engineers and computer scientists. This is just a partial list of the disciplines contributing to the solution of insect problems today, but it serves to illustrate how complex a subject applied entomology has become. This complexity will become increasingly clear as we examine the wide variety of ways in which insects, both harmful and beneficial, impinge upon our daily lives.

## General References

Essig, E. O. 1931. A history of entomology. Macmillan Publishing Co., Inc., New York, N. Y.

Smith, R. F., T. E. Mittler, and C. N. Smith, editors. 1973. History of entomology. Annual Reviews Inc., Palo Alto, Calif.

# Insect or not—that is the question

# 2

The insects comprise but a single class of a highly diverse assemblage of animals that are grouped together as the phylum Arthropoda. Judged on the basis of their large numbers, their species diversity, the wide range of habitats they occupy, and the endless variety of foods they consume, the arthropods are the most successful group of creatures in the animal kingdom. The major classes of this phylum include the Crustacea (crayfish, lobsters, shrimps, crabs, waterfleas, and barnacles), the Chilopoda (centipedes), the Diplopoda (millipedes), the Arachnida (spiders, scorpions, ticks, and mites), and the Insecta.

All arthropods have a more or less well-developed outer covering, or cuticle, grooved to form a series of body segments. Some of these segments bear one or more pairs of jointed appendages (*arthro*, joint; *poda*, foot). The embryos of arthropods have a pair of lateral limb buds on each segment except the first and the last, but many of these disappear during development. In the adults of the different groups, the number of limbs is highly variable, and within groups, they may look different and serve different functions on different parts of the body. These appendages, in spite of their differing appearance, arise in the same way throughout the group and therefore are said to be **homologous.** Most workers agree that this unifying segmented body plan and the arrangement of paired, jointed appendages indicates a common ancestry—probably among a group of segmented worms belonging to the phylum Annelida. Beyond this basic agreement on a common ancestry, the phylogeny of the arthropods has been a subject of considerable controversy, although there is growing acceptance of the concept that there were probably two major lines of evolution, one in the water and one on land. So although we can hypothesize an annelid ancestry, the two lines probably began at different times from annelids with different ways of life. When we propose more than one origin for a related group of animals, we say that it is **polyphyletic.**

The aquatic, or more specifically marine, line of evolution involved a progressive development toward a more motile group of animals that either crawled on the sea floor or swam about in the intertidal zone. This way of life would have required further development of an anterior sensory center, appendages suitable for crawling or swimming, and perhaps some specialized appendages associated with the mouth to assist in the capture and handling of food. A fossil group that displays this level of development, the trilobites, make up one class of the arthropods (Figure 2-1).

### *Class Trilobita*

Trilobites are abundant and widespread as fossils in shallow marine deposits, laid down between 600 and 400 million years ago (Paleozoic era). Members of this group are believed to have had a variety of life styles, ranging from free-swimmers to burrowers and from scavengers to herbivores. What caused the end of their domination of the coastal marine environment remains unknown. Probably the rise of some more efficient organisms, such as the crustaceans, was involved.

The fossil record shows that the trilobite body consisted of a head shield, thorax, and terminal pygidium, all dorsoventrally flattened and divided into

FIGURE 2-1. Ventral view of a trilobite reconstructed from fossil evidence.

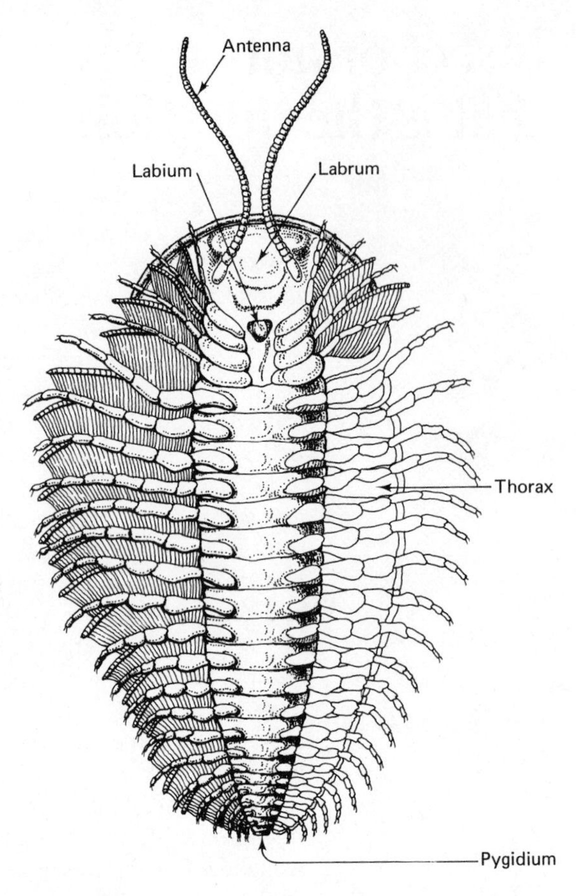


three lobes by two longitudinal grooves. The head consisted of a primitive segment, the **prostomium,** bearing a pair of long, multisegmented feelers (**antennae**), plus the first three body segments, each with a pair of slightly shortened, double-armed (biramous), jointed appendages. Each of the remaining body segments, except the last, bore a pair of similar but longer appendages. The upper featherlike arm of these limbs apparently served in swimming and gas exchange, whereas the lower naked branches were used as walking legs (Figure 2-2).

A recent article (Cisne, 1974), which reveals previously unknown anatomical details, suggests that the trilobites could have been the ancestral stock for two further lines of arthropod evolution which began as marine groups. One line led to the chelicerate (bearing pincerlike feeding organs) arthropods, including the strange horseshoe crab (Class Merostomata) and the modern arachnids which later became primarily terrestrial. The other line led to the jawed crustaceans which are almost all either marine or fresh-water forms.

FIGURE 2-2. Diagram of a typical biramous trilobite appendage based on fossil evidence.

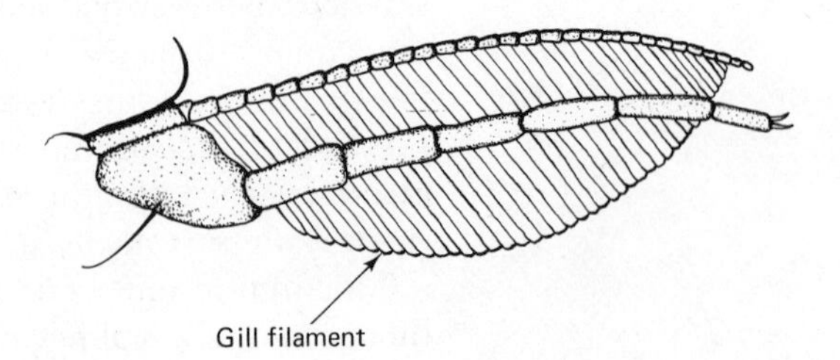

FIGURE 2-3. Dorsal view of a horseshoe crab (Class Merostomata). Note that the large carapace completely covers the head and thorax which are united.

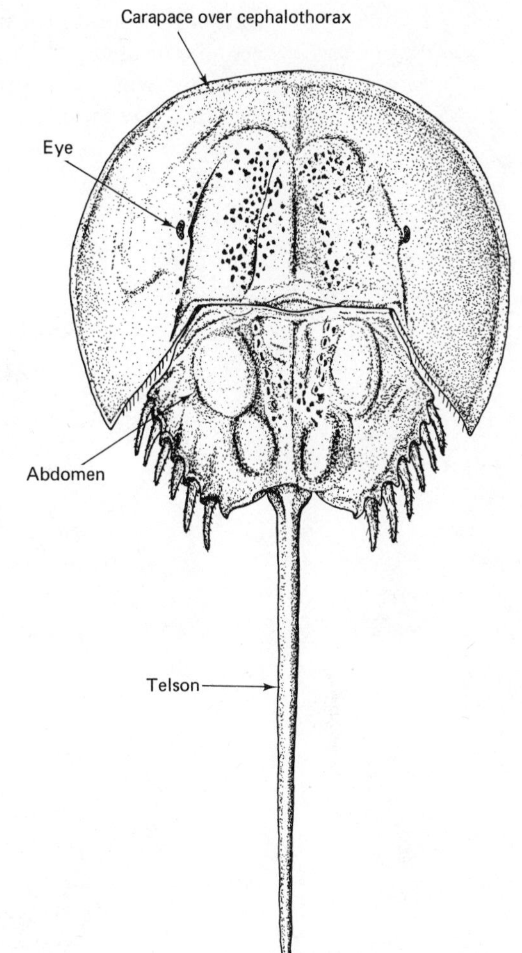


***Class Merstomata*** The horseshoe crabs (Figure 2-3) are primitive marine arthropods, represented today by just a few species of the genus *Limulus.* All live in shallow coastal marine environments where they burrow in the sand and mud in search of the small marine animals upon which they prey. These strange animals are related to the spiders in that they have breathing structures (**gill books**) attached to their flattened abdominal appendages. These gill books are strikingly similar to the respiratory organs called book lungs that are found in the more primitive terrestrial arachnids. Like the arachnids, the body of the horseshoe crab is divided into a completely fused head and thorax (**cephalothorax**) and an abdomen. The cephalothorax has six pair of appendages, the first pair of which are pincerlike **chelicerae.** The abdomen completely lacks legs and outward signs of segmentation.

***Class Arachnida*** The spiders, scorpions, ticks, mites, and several less common animals comprise a closely related group of largely terrestrial arthropods for which no description can be provided to cover all of the forms. In general, the body is divided into a cephalothorax with six pair of jointed appendages and an abdomen with no appendages. The head region lacks antennae and compound eyes, although a variable number of simple eyes may be present. Adjacent to the front of the mouth, the first pair of appendages form a pair of chelicerae. The second pair of appendages sometimes form long sensory structures called the **pedipalps**, which in scorpions are used for grasping prey. In addition, all adult arachnids have four pair of rather long walking legs (Figure 2-4).

This group of arthropods is usually of considerably more interest to entomologists than are the other noninsect classes. Some spiders, ticks, and

**FIGURE 2-4** Representatives of the Class Arachnida. **A.** Scorpion, **B.** Spider, **C.** Harvestman, **D.** Tick, **E.** Pseudoscorpion, **F.** Mite.

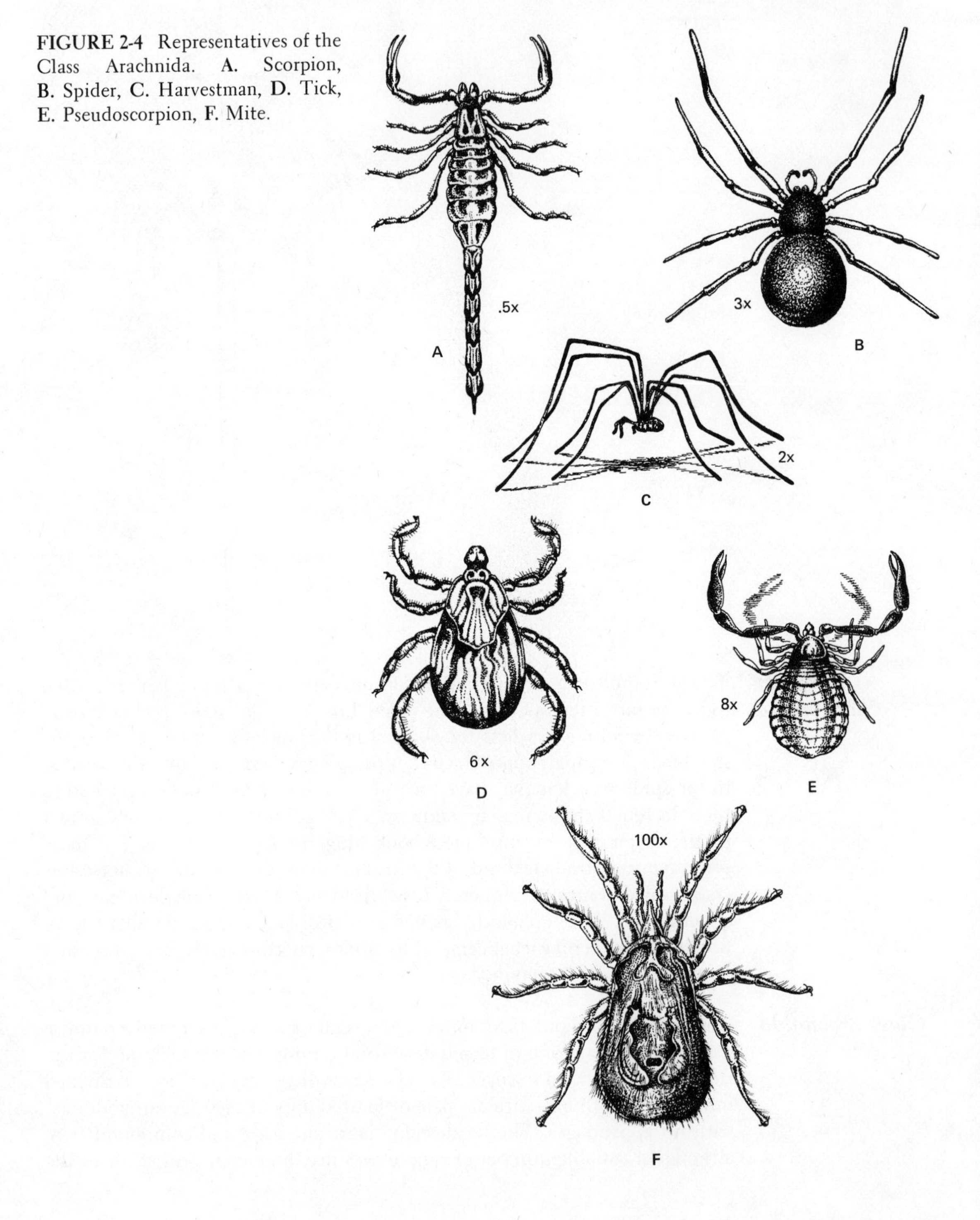

scorpions are venomous and inflict painful, even fatal, bites or transmit diseases of man, and thereby come under the purview of medical entomologists. Mites comprise the largest group of parasitic arthropods; mites and ticks are frequently treated by livestock entomologists along with insect pests. The plant-feeding mites comprise one of the most important groups of pests of trees, agricultural crops, and stored products and must be taken into consideration in the development of pest control programs by entomologists. Spiders and predaceous mites also have some potential in the biological control of arthropod pests.

### *Class Crustacea*

The Class Crustacea contains a wide variety of marine and fresh-water arthropods (Figure 2-5). Once again, no single description is applicable to the entire group, but there are some common characteristics, and the familiar lobster will serve to illustrate the basic form of this group (Figure 2-6). Most crustaceans are characterized by having the body divided into a head, thorax, and abdomen, although the head and thorax are closely joined and sometimes protected from above by a single carapace. Embryological studies show the head to be composed of six segments. The first segment, which bears a pair of stalked eyes, probably represents the ancient premouth segment (*prostomium*). The next five segments bear serially homologous appendages, represented by the uniramous first antenna, the biramous second antennae, the jaws (**mandibles**), and two pair of food-handling structures, the first and second **maxillae.** The thorax may vary from four to twenty segments, each with a pair of limbs. In the lobster there are eight segments which support three pair of food-handling and sensory **maxillipeds,** one pair of large pincer-bearing limbs, and four pair of walking legs. The abdomen has a pair of appendages on every segment except the last, which is flattened to form the middle of the three-lobed tail (**telson**). The abdominal appendages are mainly used for slowly swimming forward and also serve reproductive functions.

The Class Crustacea contains more than 150,000 described species, making it the second largest class in the animal kingdom. This large diversity of kinds has caused some people to call them the insects of the sea, and since they predate the insects, they may have played an important role in preventing the insects from successfully invading the intertidal zone.

Sometime after the evolution of the chelicerates and crustaceans was underway, other annelidlike animals that were perhaps already semiterrestrial began the evolution that would ultimately lead to the rise of the insects. Early on, these ancestral animals evolved a cuticle that provided better protection in a terrestrial environment and provided the opportunity for them to engage in a way of life beyond their aquatic and subterranean habitats. In addition, the cuticle, which had become more rigid, was divided into movable plates with areas for muscle attachment, and the articulation of limbs changed in such a way that the body could be raised off the substrate to permit more rapid locomotion on the surface of the land. The cuticle also became resistant to water loss by evaporation while still allowing for gas exchange from the air rather than from the surrounding water, thus enabling these creatures to better exploit truly terrestrial environments. A major breakthrough occurred with the ingrowth of the cuticle to form a series of branching air tubes that could carry air to all of the body tissues.

Unfortunately, we can only speculate as to the nature of the evolutionary steps that occurred and the appearance of some of the intermediate animals.

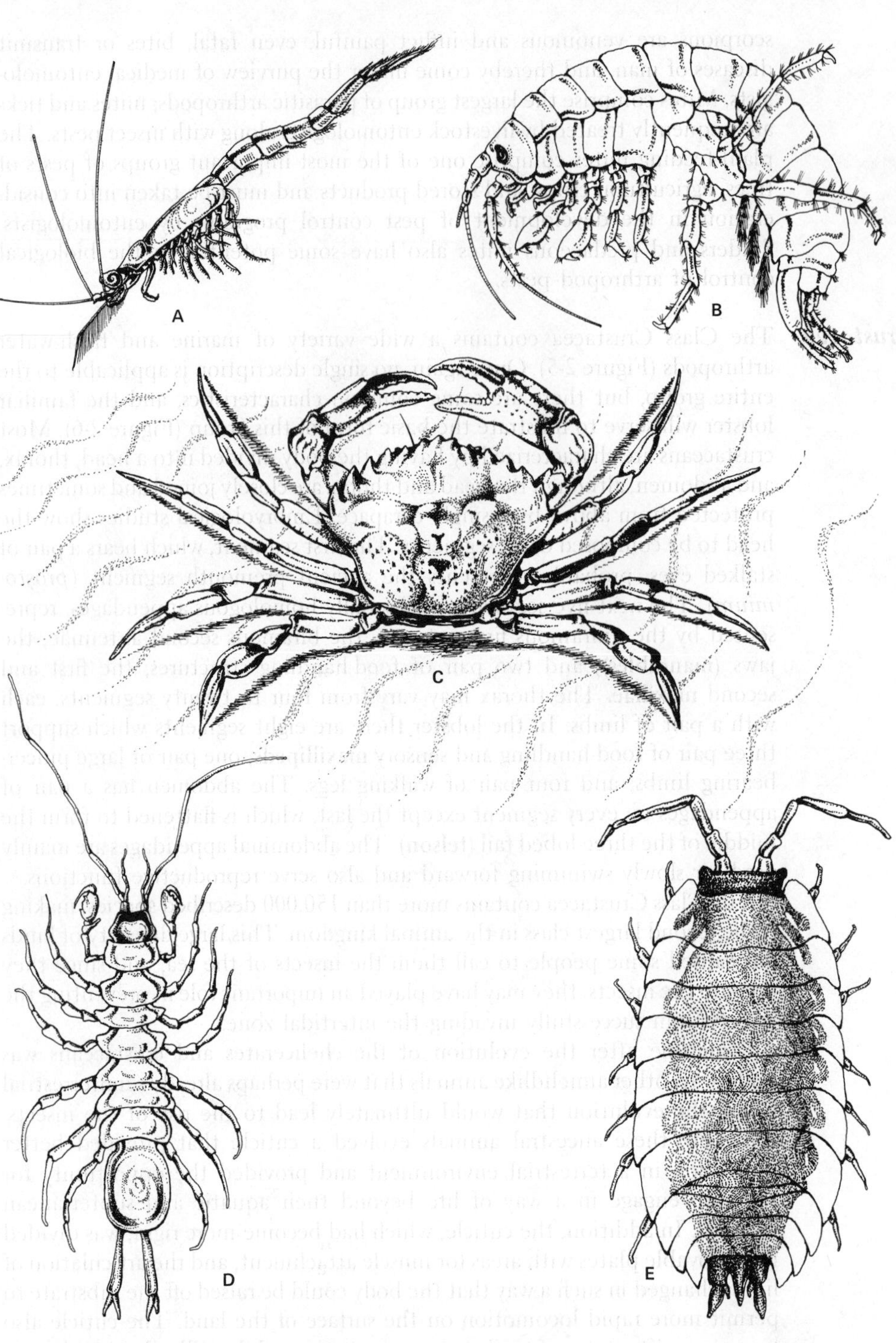


FIGURE 2-5. Representatives of some of the major groups of the Class Crustacea. A. Freshwater mysid shrimp, B. Sand flea (amphipod), C. Shore crab (decapod), D. Freshwater isopod, E. Sow bug (terrestrial isopod).

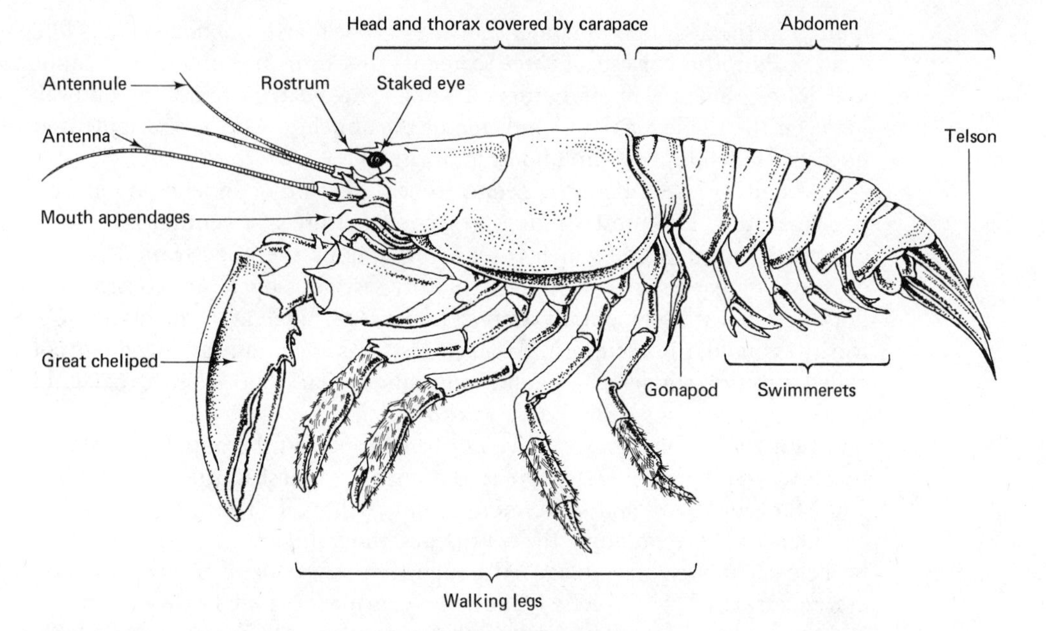


**FIGURE 2-6.** Lateral view of the American lobster *Homaris americana* showing wide variety of appendage modifications.

These creatures apparently did not fossilize well, but we are fortunate in that one strange half-annelid, half-arthropod creature called an onychophoran has survived for us to examine. These rare animals are found in the moist litter that blankets the floor of tropical forests. They are caterpillarlike but have many pairs of short unsegmented legs and thick fleshy antennae (Figure 2-7).

The onychophoran genus *Peripitus* might be considered as intermediate between the annelids and some terrestrial arthropods. The annelid features of *Peripitus* include little outward evidence of segmentation, although there is one pair of legs for each segment represented internally. The legs are not

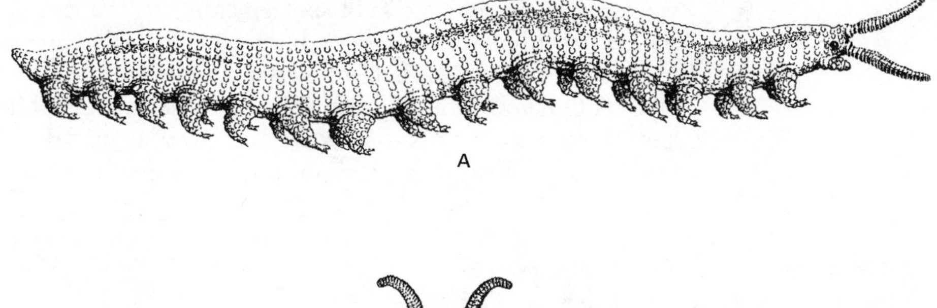


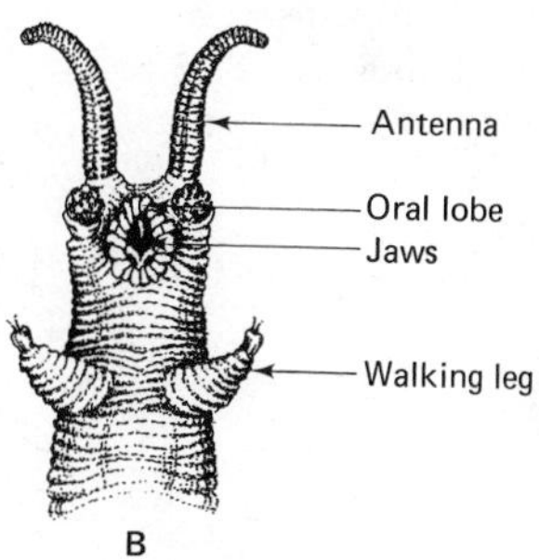


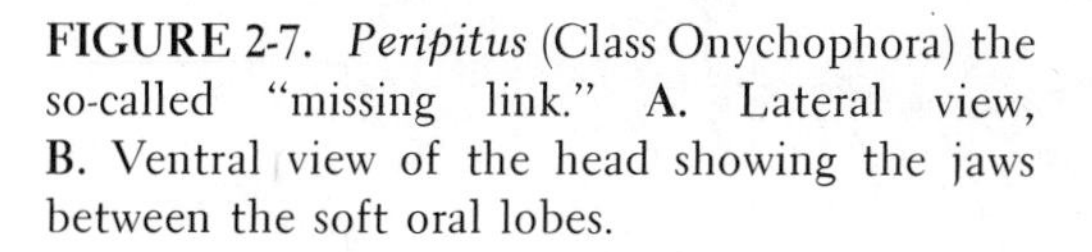
**FIGURE 2-7.** *Peripitus* (Class Onychophora) the so-called "missing link." **A.** Lateral view, **B.** Ventral view of the head showing the jaws between the soft oral lobes.

jointed in the arthropod manner, but they do bear arthropodlike claws. The head of *Peripitus* consists of three segments that bear the antennae, a pair of oral lobes, and a pair of hardened jaws. The head is therefore midway between the single-segment condition of the annelids and the six-segmented head of the chilopods, diplopods, and insects.

Internally, as well, *Peripitus* seems to be a mixture of annelid and arthropod anatomy. The most striking arthropod feature is a ventilatory system comprised of air tubes, which extend throughout the body from a pair of openings on each segment. The circulatory system is open and consists of a dorsal vessel with segmentally arranged openings, much as in the insects. On the other hand, the excretory system consists of segmentally arranged pairs of coiled tubules, similar to annelid nephridiae, that collect and expel fluid waste through pores near the bases of the legs.

Although the Onychopora have not been a successful group, they seem to be on the same evolutionary line as the insects, at least in the view of Tiegs and Manton (1958), and some more recent authors. In fact, Manton (1972) considers the Onychophora, the centipedes, the millipedes, and the insects to be a closely related assemblage. Although they are probably not the immediate ancestors of the insects, Chilopods (centipedes) and Diplopods (millipedes) seem to illustrate another evolutionary step that brings us just a little closer to the insect body plan.

**_Class Chilopoda_**

The centipedes are elongated, flattened, multisegmented animals with one pair of jointed legs on each segment (Figure 2-8). The head is strikingly similar to that of the insects; it is comprised of six segments and bears one pair of antennae, a pair of mandibles, a pair of separate first maxillae, and a pair of united second maxillae. In addition, centipedes have an insectlike excretory system, consisting of a group of long tubules that empty into the hind part of the gut. A tracheal system supplies oxygen directly to the tissues, and the blood is pumped through the body cavity by a pair of dorsal arteries.

**_Class Diplopoda_**

The millipedes (Figure 2-9) are similar to centipedes except that the body segments have fused in pairs, giving the appearance of two pair of legs per segment. The head consists of six segments, but the mouth parts, although similar to centipedes and insects in the embryo, lack the separate first pair of maxillae in the adult. Their internal anatomy resembles that of centipedes. The centipedes and millipedes, although terrestrial, inhabit the moist litter and humus-rich layer of soil where they feed as predators and scavengers, respectively. Their body plan, although adequately suited for this type of

FIGURE 2-8. A typical centipede (Class Chilopoda). (Redrawn from Snodgrass, 1935.)

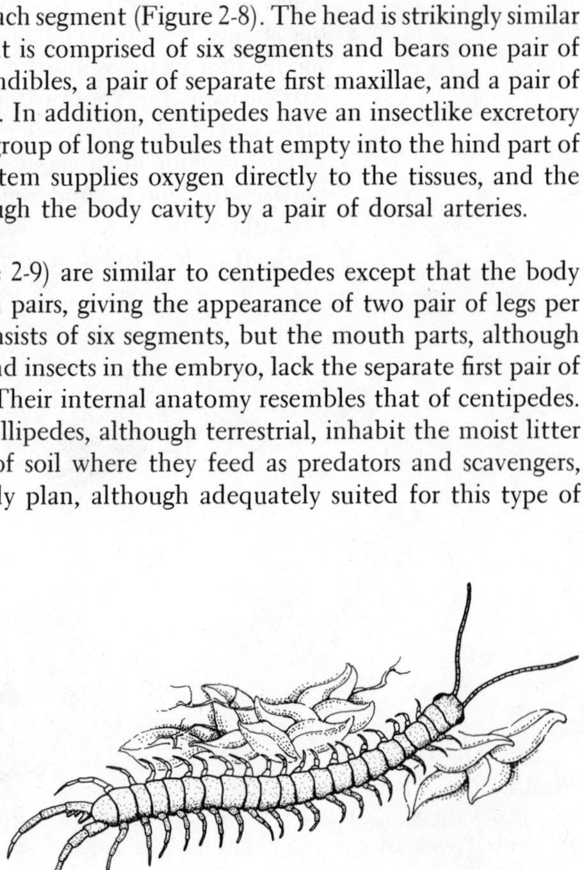

FIGURE 2-9. A typical millipede (Class Diplopoda).

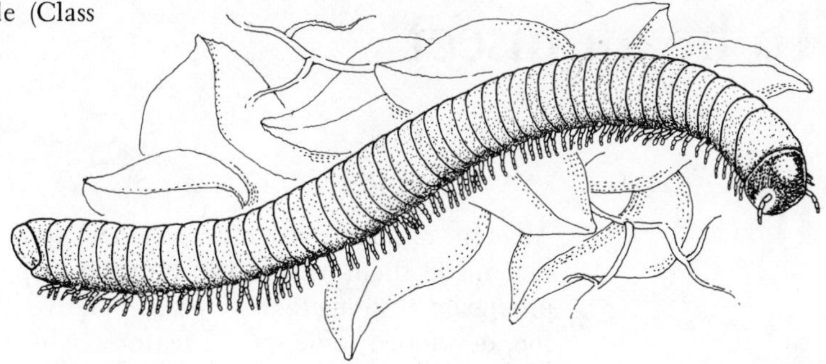

existence, is not as efficient or as well adapted as that of the wingless hexapods, which are shorter and have fewer pairs of legs.

Two other less common groups (the Symphyla and Pauropoda) show a body plan similar to centipedes but with a reduction of the total number of segments to 11 or 12. Unfortunately, there are no fossils or living groups that show an intermediate step between many pairs of legs and the six-legged condition of the insects. Several small hexapods (the Protura, Diplura, and Collembola) are often considered to be insects but probably more accurately represent the end of the sequence of developments that lead to the fascinating group of animals to which the balance of this book is devoted.

# References Cited

CISNE, J. L. 1974. Trilobites and the origin of arthropods. *Science* 186: 13–18.

MANTON, S. M. 1972. The evolution of arthropodan locomotory mechanisms. 10. Locomotory habits, morphology and evolution of the hexapod classes. *J. Linn. Soc. London (Zool.)* 51: 203–400.

TIEGS, O. W., and S. M. MANTON. 1958. The evolution of the Arthropoda. *Biol. Rev.* 33: 255–337.

# General References

BARNES, R. D. 1974. Invertebrate zoology, 3rd ed. W. B. Saunders Company, Philadelphia, Pa.

SNODGRASS, R. E. 1952. A textbook of arthropod anatomy. Comstock (Cornell University) Press, Ithaca, N. Y.

# To be an insect

# 3

All of the animal groups discussed in Chapter 2 possessed some special feature that made them just a little better at coping with their environment and producing offspring than were their relatives that became extinct. The insects, too, developed some special features, which apparently gave them both an initial advantage in the terrestrial environment and the evolutionary flexibility to adapt to a wide range of new situations as they came about. In order to understand why and how the insects were able to diversify in response to changes in their environment, we must be familiar with their basic anatomical plan.

## The External Anatomy

### *The Insect Exoskeleton*

Insects, like all arthropods, have a hardened outside shell, or **exoskeleton.** This covering (the **integument**) consists of a layer of living epithelial cells plus a series of layers of nonliving material (the **cuticle**) that these cells secrete.

When first secreted, the cuticle is pale colored, soft, and pliable. After a short time, it darkens and hardens by a complex chemical process known as tanning. The hardened cuticle provides support and protection for the soft tissues of the body, and invaginations (**apodemes**) provide areas for the attachment of muscles. The surface of the cuticle is covered with wax or oil, which forms an effective barrier against water loss, a major problem for small terrestrial organisms. A number of special hairlike surface structures called **setae** serve as sensory devices. The integument not only covers the entire surface of the insect but also forms the lining of the fore and hind portions of the gut, and it invaginates to form the air tubes of the ventilatory system.

If the entire cuticle became hardened, insects would be rigid, immobile creatures. However, the exoskeleton is broken up into a series of hardened areas (**sclerites**), interconnected by membranous material that provides a reasonable amount of flexibility. Even with these areas of flexible membrane, expansion of the exoskeleton is limited. The exoskeleton cannot stretch sufficiently to accommodate complete growth, so growing insects must periodically cast off, or molt, the skin that has become too small and produce a new one. The new cuticle, which has been laid down beneath the old one prior to the molt, is flexible and can be expanded through the action of special muscles and the intake of air or water. Thus, by the time hardening begins, the new exoskeleton has been stretched to accommodate the next phase of growth and development. In some softer-bodied forms, like caterpillars, the new cuticle is considerably folded, so growth can occur until all the folds have been straightened out.

In the long course of insect evolution, many changes have taken place in the form and the function of different parts of the exoskeleton. Some modifications proved to be advantageous and were maintained; others have been eliminated. Each change opened the way for further adaptations and closed the way to others. If we could retrace all the steps of insect evolution, we would find that changes in external anatomy often were involved. Thus, the entire surface configuration of insects expresses the selective pressures that various species have dealt with during their long history. The resultant features consequently provide the basis for arranging the insects into an orderly classification. The integument will be considered in more detail in the second part of this book.

Insects have no internal skeleton of the type that characterizes the vertebrates. Their entire rigidity comes from the exoskeleton and its invaginations into the body cavity. Consequently, all the muscles are attached to the inner surface of movable plates, rather than on the outer surface of internal supporting bones. Most of the sclerites that comprise the exoskeleton have been named, as have many of the lines and grooves (**sutures**) that occur in and between them. Unfortunately, there is a great lack of consistency in the naming of these parts in different groups, but the terms used in this book should find general acceptance.

## *Organization of the Insect Body*

Insects bear the impress of their segmented annelidlike ancestor in that they exhibit a completely segmented morphology during embryonic development. However, as adults, they exhibit a considerable reduction in the number of apparent segments as a result of the amalgamation of some segments into unified structures. This grouping together of segments into functional subunits of the body is called **tagmosis.**

An insect is thus characterized by a body composed of a total of 20 segments, associated to form three well-defined regions: the **head,** the **thorax,** and the **abdomen** (Figure 3-1). Although it displays little external evidence of segmentation, the head is believed to consist of six segments. The thorax always consists of three segments, which, although joined in a group, are fairly well delineated and bear one pair of legs each. The abdomen is composed of

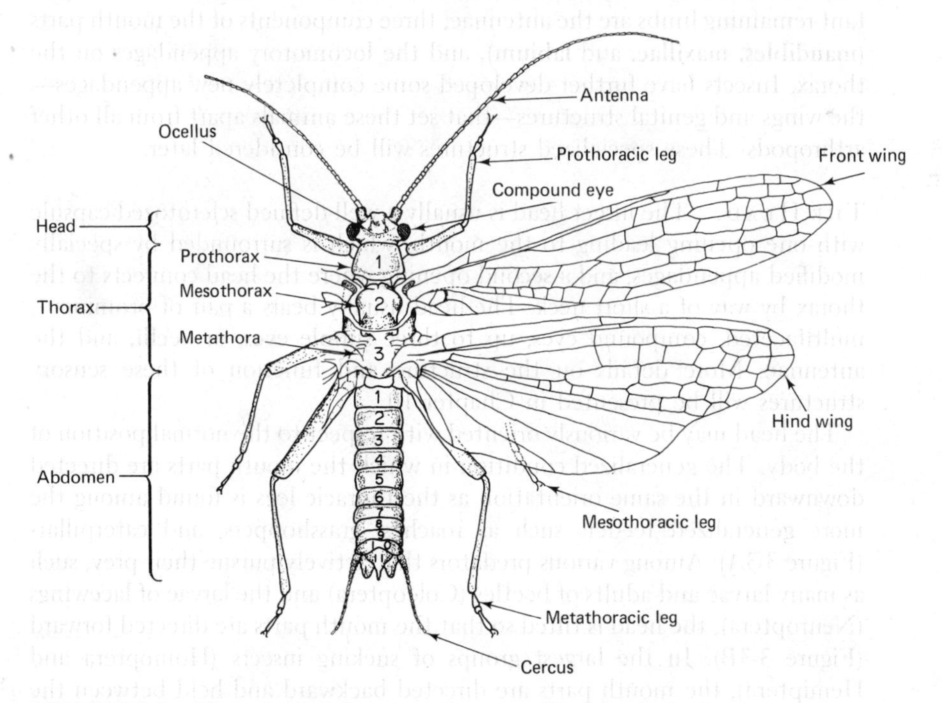


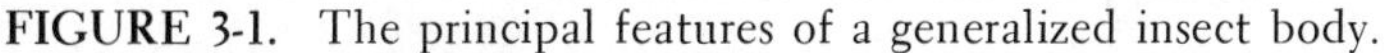
**FIGURE 3-1.** The principal features of a generalized insect body.

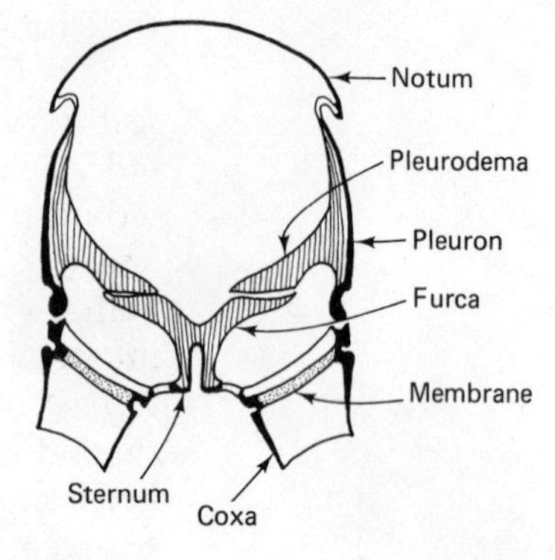


**FIGURE 3-2.** Diagram of a transverse section of a typical thoracic segment showing the invagination of the cuticle to form internal skeletal structures and areas for muscle attachment. (Adapted from Snodgrass, 1935.)

11 segments, although few insects have retained this many; most insects lack abdominal appendages.

Only in the abdomen can we see the primitive condition in which each segment has an upper or dorsal plate (**tergum,** pl. **terga**) and a ventral plate (**sternum,** pl. **sterna**) interconnected by lateral membranes in which we can often find the small segmentally arranged, paired openings (**spiracles**) to the ventilatory system. In the thorax, the membranous side walls have become replaced by a number of closely united **pleural sclerites.** The fusion of these pleural sclerites tends to make the thorax boxlike, with a fairly rigid top, bottom, and sides. Within the thorax and the head, there are large invaginations of the integument that serve as strengthening struts and provide more area for muscle attachment (Figure 3-2).

In insects, the rather primitive segmentally arranged pairs of appendages that characterize the centipedes have been lost from most of the segments. Those that remain, however, are clearly homologous with other arthropod limbs, even though they vary considerably in form and function. The important remaining limbs are the antennae, three components of the mouth parts (**mandibles, maxillae,** and **labium**), and the locomotory appendages on the thorax. Insects have further developed some completely new appendages—the wings and genital structures—that set these animals apart from all other arthropods. These specialized structures will be considered later.

THE HEAD. The insect head is usually a well-defined sclerotozed capsule with one opening leading to the mouth, which is surrounded by specially modified appendages, and a second opening where the head connects to the thorax by way of a short neck. The head usually bears a pair of prominent, multifaceted, compound eyes, up to three simple eyes, or **ocelli,** and the antennae. More details on the structure and function of these sensory structures will be presented in Chapter 10.

The head may be variously oriented with respect to the normal position of the body. The generalized condition in which the mouth parts are directed downward in the same orientation as the thoracic legs is found among the more generalized feeders such as roaches, grasshoppers, and caterpillars (Figure 3-3*A*). Among various predators that actively pursue their prey, such as many larvae and adults of beetles (Coleoptera) and the larvae of lacewings (Neuroptera), the head is tilted so that the mouth parts are directed forward (Figure 3-3*B*). In the largest groups of sucking insects (Homoptera and Hemiptera), the mouth parts are directed backward and held between the

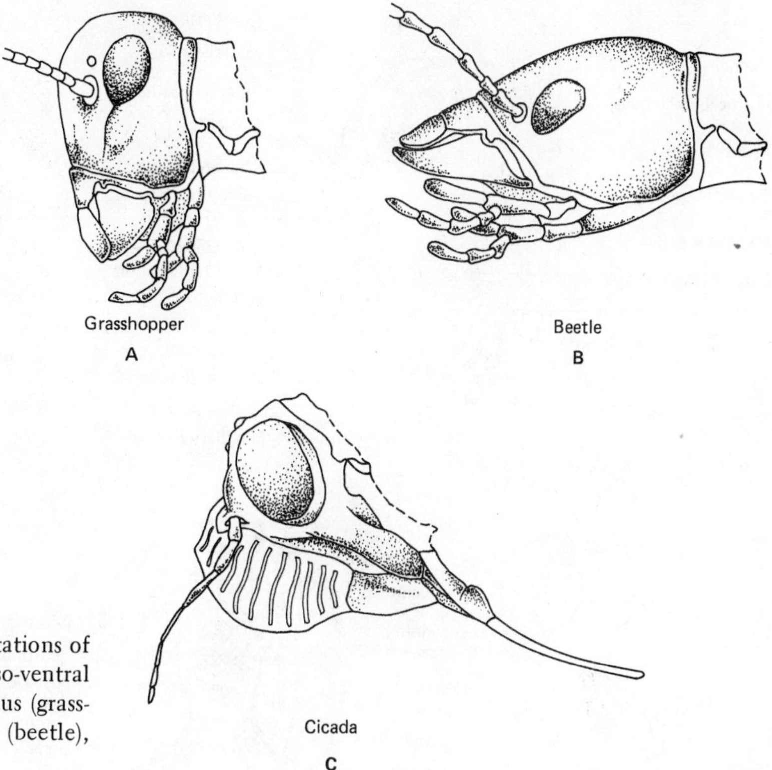


**FIGURE 3-3.** Different orientations of the head relative to the dorso-ventral axis of the body. **A.** Hypognathus (grasshopper), **B.** Prognathus (beetle), **C.** Opisthognathus (cicada).

legs as a result of a downward and backward deflection of the front of the head (Figure 3-3C).

The head capsule, or **cranium,** appears to be composed of a number of sclerites, separated by fine grooves or sutures, but these lines do not represent the boundaries of segments. The sclerotized areas and the sutures have all been named and are of considerable value in insect identification and classification. Details on the important structural components of the generalized insect head are included in the caption of Figure 3-4.

Several structures, including three pair of modified appendages that surround the mouth and form the **pre-oral** cavity in front of it, are collectively referred to as the mouth parts. These sensory and food-handling structures include an upper lip, or **labrum,** the paired mandibles, the paired maxillae, the fused pair of second maxillae, or labium, and a tonguelike **hypopharynx.** The relative position and anatomy of these structures is shown in Figure 3-5.

The labrum is a flaplike structure suspended from the clypeus that forms the front wall of the pre-oral cavity. The anterior surface is usually smooth, whereas the inner (posterior) surface is fleshy and richly endowed with sensory structures. This fleshy surface is sometimes greatly expanded into a lobe called a **labrum-epipharynx.**

The mandibles are unsegmented, broadly based, and heavily sclerotized structures, which move back and forth in front of the mouth immediately behind the labrum. The median edge of each mandible is differentiated into an apical **incisor area,** which bears several teeth for cutting and a proximal ridged **molar area** for grinding.

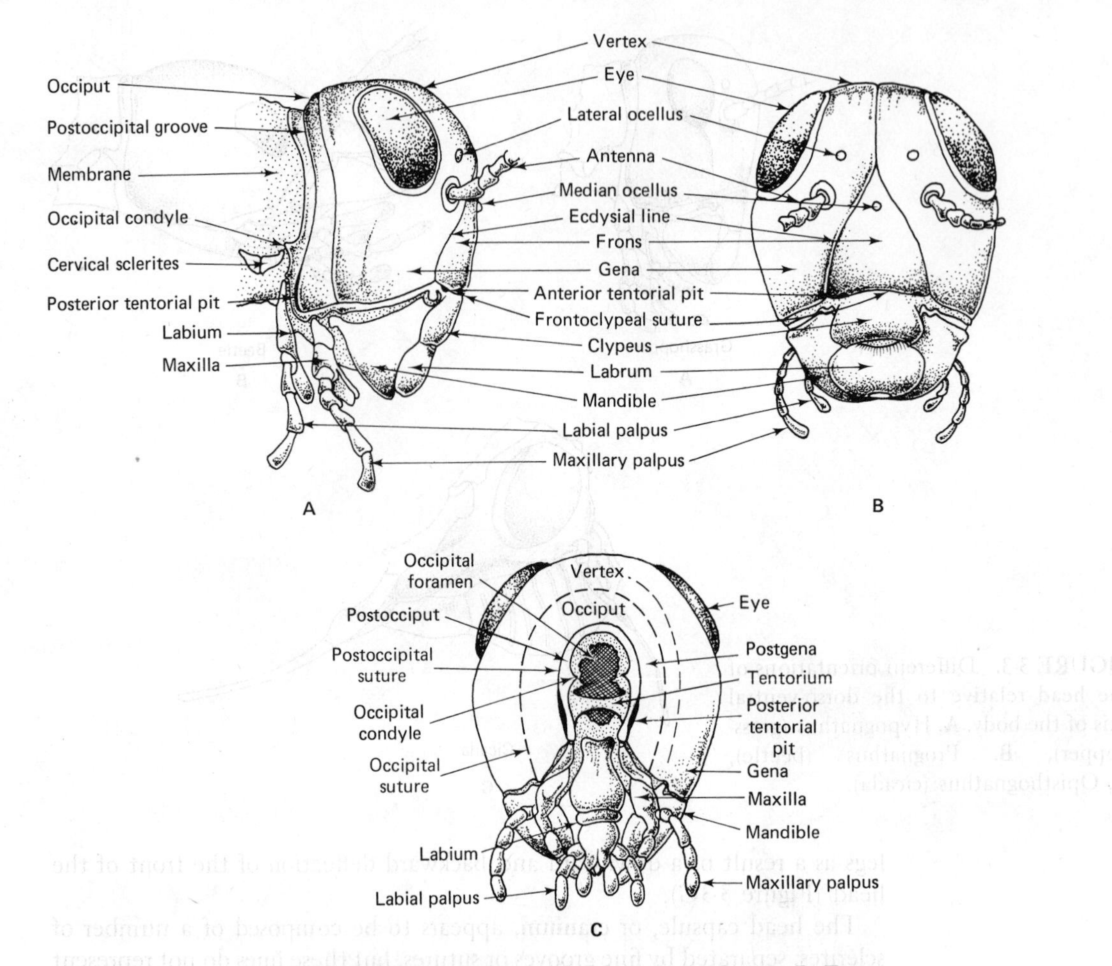


**FIGURE 3-4** Structural details of a typical insect head. **A.** Lateral view, **B.** Front view, **C.** Rear view. Note that although the head is comprised of six body segments there is no outward evidence of segmentation. The inverted V-shaped ecdysial line on the front of the head represents a line of weakness that in many insects splits at the time of a molt. The tentorial pits represent points of cuticular invaginations that form the internal skeleton of the head capsule, the **tentorium.** (Adapted from Snodgrass, 1935.)

The maxillae lie directly behind the mandibles. Each maxilla consists of a basal **cardo** that articulates with the head, a central body called the **stipes** that gives rise to a five-segmented sensory **palp**, a hardened inner toothed **lacinea** (pl. **lacineae**), and an outer fleshy lobe called the **galea** (pl. **galeae**).

The labium, or lower lip, is formed by the fusion of a pair of second maxillae, each with the same basic parts as the first maxillae. The labium articulates with the head by the **submentum,** which is hinged to the **mentum.** The mentum gives rise to a pair of usually three-segmented palps, a pair of lateral lobes, the **paraglossae,** and a pair of median lobes, the **glossae.** The labium is primarily sensory, but also serves to close the pre-oral cavity from behind, and manipulates food into the mouth.

The hypopharynx is an unsegmented fleshy lobe that arises from the membranous floor of the cranium and lies behind the mouth. A salivary duct

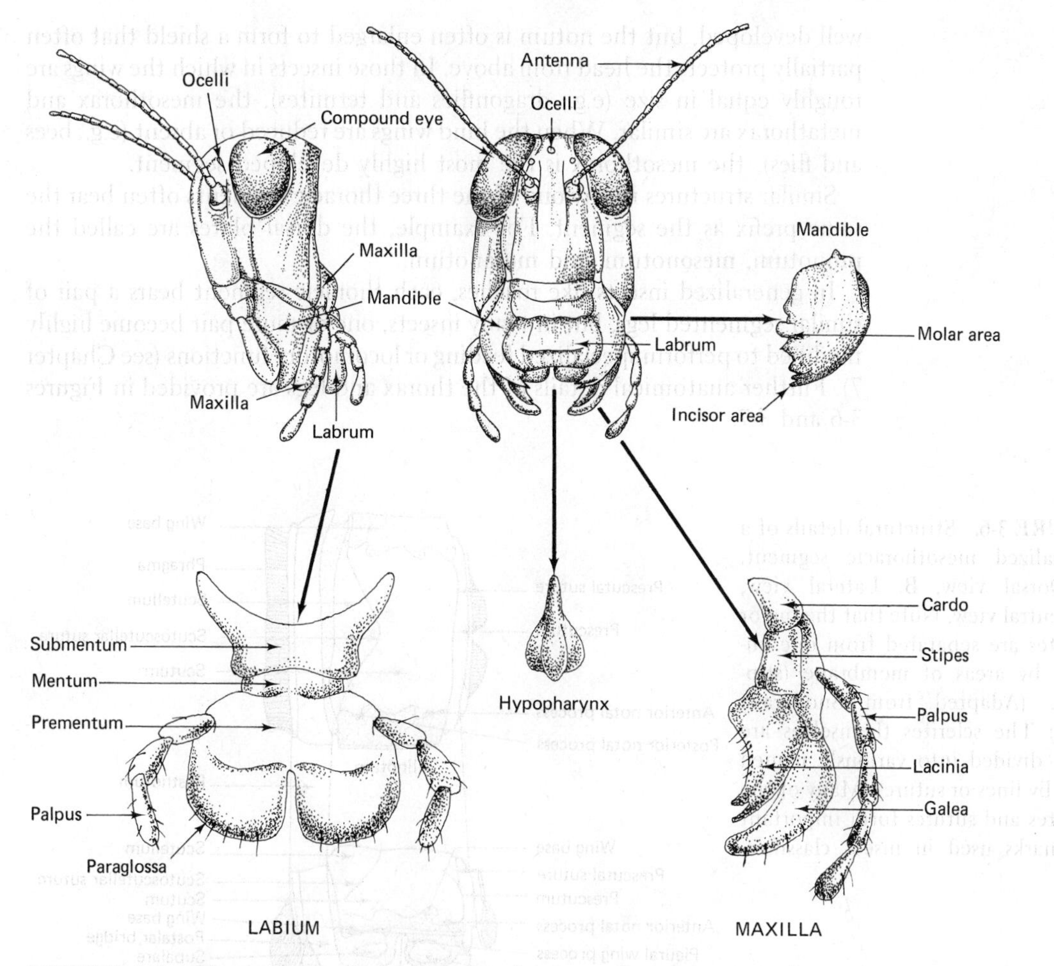


**FIGURE 3-5.** The generalized chewing mouthparts as found in a grasshopper. **A.** Lateral view of the head showing the relative position of the mouthparts, **B.** Front view of the head and structural details of the major components of the mouthparts. (Adapted from Snodgrass, 1935.)

that delivers the secretions of the paired salivary glands generally opens on the back of the hypopharynx.

THE THORAX. The thorax consists of three reasonably easily distinguished leg-bearing segments, named from the head backward as the **prothorax,** the **mesothorax,** and the **metathorax.** In most insects, the mesothoracic and metathoracic segments bear a pair of dorsolateral membranous wings or structures derived from membranous wings. The prothorax never bears wings, and there is no evidence that insects ever had wings on the first thoracic segment.

On the thorax, the dorsal plate is called the **notum** (pl. **nota**), and the ventral plate is called the **sternum** (pl. **sterna**). The side walls of each segment have increased in height and become well sclerotized so that each segment is boxlike in cross section; this change in proportion seems to have been an adaptation associated with the evolution of the thorax as a locomotor center. Since the prothorax does not bear wings, its side, or pleural, walls are not so well developed, but the notum is often enlarged to form a shield that often partially protects the head from above. In those insects in which the wings are roughly equal in size (e.g., dragonflies and termites), the mesothorax and metathorax are similar. When the hind wings are reduced or absent (e.g., bees and flies), the mesothorax is the most highly developed segment.

Similar structures in any one of the three thoracic segments often bear the same prefix as the segment. For example, the dorsal plates are called the **pro**notum, **meso**notum, and **meta**notum.

In generalized insects like roaches, each thoracic segment bears a pair of similar segmented legs, but in many insects, one or more pair become highly modified to perform specialized feeding or locomotory functions (see Chapter 7). Further anatomical details of the thorax and legs are provided in Figures 3-6 and 3-7.

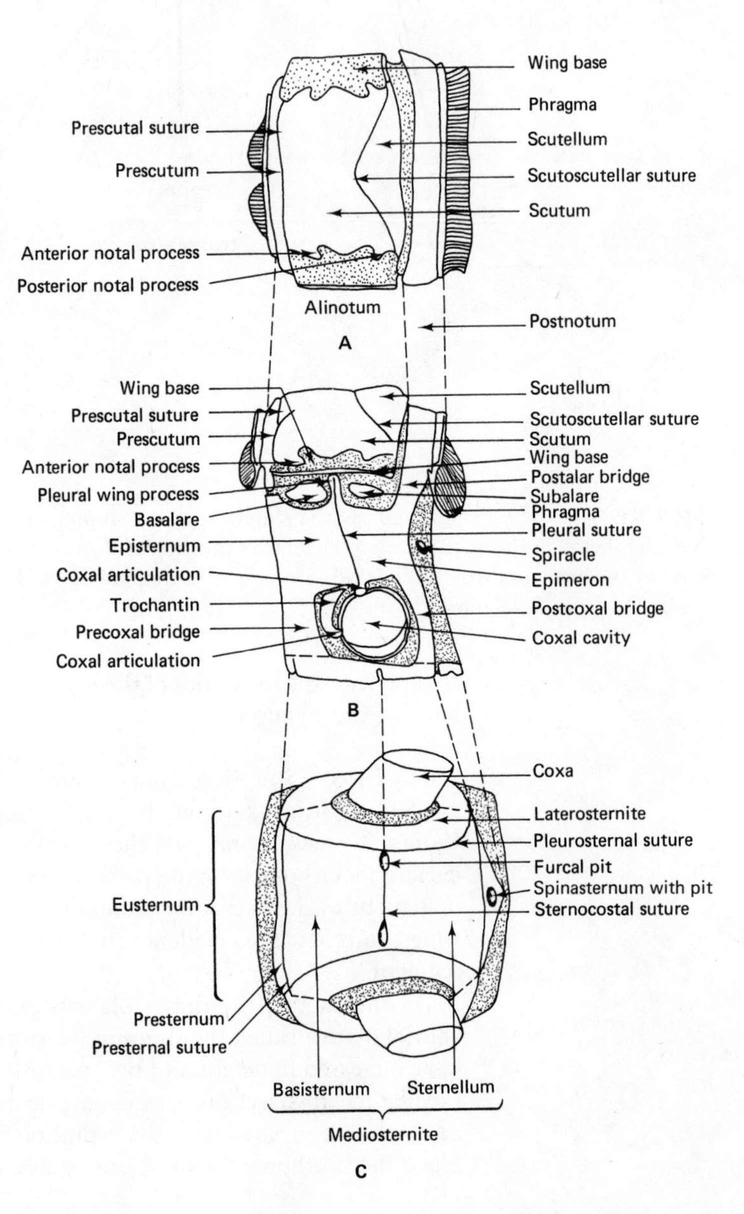


**FIGURE 3-6.** Structural details of a generalized mesothoracic segment. **A.** Dorsal view, **B.** Lateral view, **C.** Ventral view. Note that the major sclerites are separated from one another by areas of membrane (stippled). (Adapted from Snodgrass, 1935.) The sclerites themselves are often divided into variously named areas by lines or sutures. Many of the sclerites and sutures form important landmarks used in insect classification.

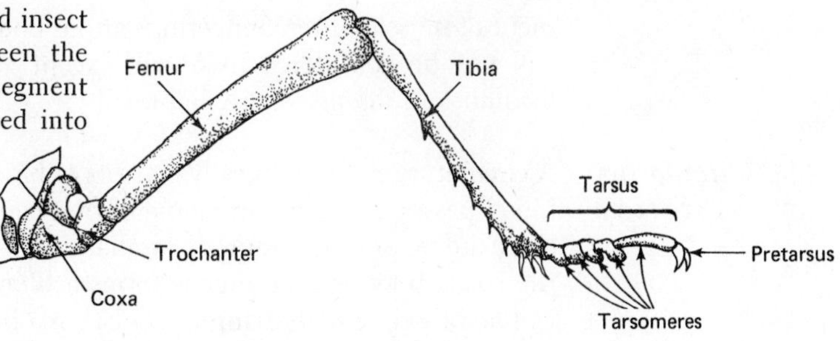


**FIGURE 3-7.** A typical unmodified insect leg showing the relationship between the segments. Note that the tarsus is a segment of the leg that is usually subdivided into tarsomeres.

THE ABDOMEN. The abdomen primitively consists of 11 segments, plus a terminal structure, the **periproct**, through which the anus opens. In most insects, there has been some degree of tagmosis in the abdomen, and we never encounter the full complement of segments among living species.

Each abdominal segment is a relatively generalized unit, consisting of a dorsal tergum, which may wrap downward to provide some lateral protection, and a relatively flat ventral sternum. The terminal segment may bear a pair of sensory appendages called **cerci** (sing. **cercus**), which may represent the only abdominal limbs that have persisted in a number of insect groups. The same segment may be modified into a dorsal lobe, the **epiproct**, above the anus, and a pair of lateral lobes, the **paraprocts** (Figure 3-8).

Most insects do not have abdominal appendages that are clearly homologous with the other jointed limbs. Many insects have paired jointed appendages that arise from the ventral surface of the eighth and ninth segments, but many morphologists are convinced that these reproductive structures (see Chapter 9) are entirely new and distinguish the insects from all other arthropods. In the generalized plan, the female has paired structures arising from both the eighth and ninth segments that form an egg-laying device, or **ovipositor;** the male has one pair arising from the ninth segment that forms a copulatory apparatus.

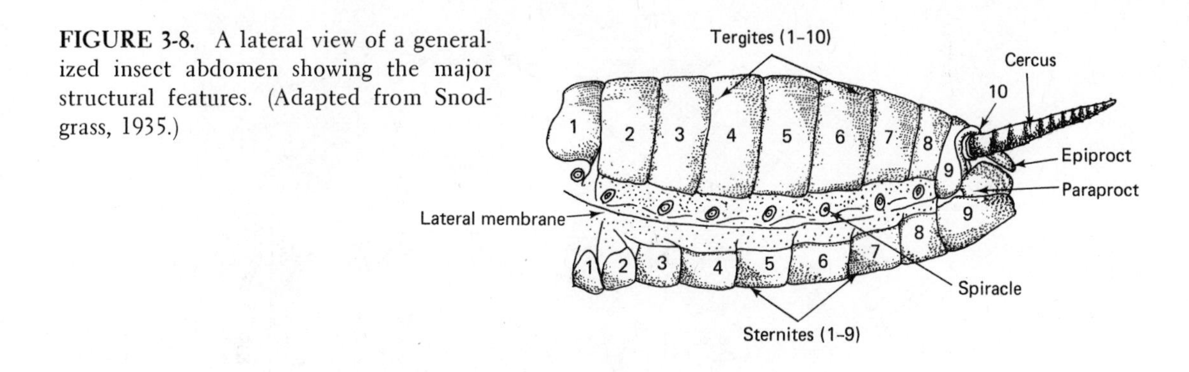


**FIGURE 3-8.** A lateral view of a generalized insect abdomen showing the major structural features. (Adapted from Snodgrass, 1935.)

# The Internal Anatomy

The internal anatomy of insects consists of all of the organ systems concerned with the vital functions of life, nicely protected from without by the exoskeleton. As the skeleton is external, much of the internal bulk, particularly in the thorax and head, consists of bundles of muscle fibers that move the sclerites and appendages in a functionally organized manner. This elaborate arrangement of muscles interconnecting various body parts is extremely complex and will not be considered in detail, except as it applies to specific kinds of locomotion, discussed in Chapter 7.

### *The Circulatory System*

A circulatory system typically consists of blood, the vessels through which the blood passes, and a pumping organ that causes the blood to flow. In insects, there are no arteries, veins, or capillaries arranged as a closed system; instead, the space between the organs forms a blood cavity, or **hemocoel.**

The blood, or **hemolymph,** consists of a liquid portion, the **plasma,** and an assortment of cells called **hemocytes.** Unlike the blood of man, there is not an abundance of red corpuscles that transport oxygen chemically bonded to hemoglobin.

There is a single, dorsal circulatory vessel that extends the full length of the body. This vessel is divided into a posterior **heart** and an anterior **aorta.** Generally, the heart portion is swollen in each of the first nine abdominal segments to form a series of nine chambers, each with a pair of lateral openings called **ostia** (sing. **ostium**). A pair of fanlike **alary muscles** is attached to the ventral side of each chamber (Figure 3-9). The contraction and relaxation of these muscles change the shape of the chambers in a pulsating rhythm that draws blood in through the ostia and forces it forward into the

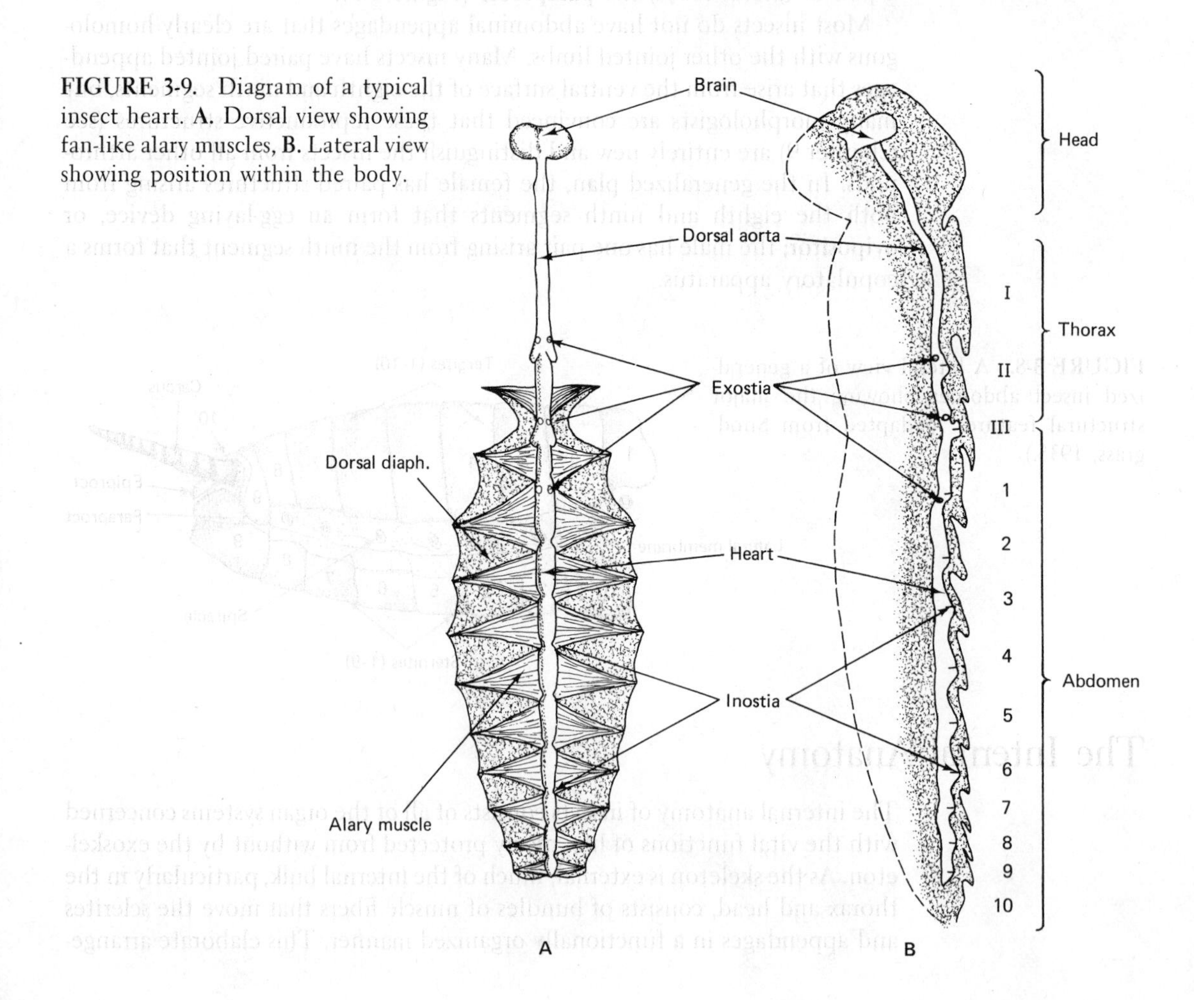


FIGURE 3-9. Diagram of a typical insect heart. A. Dorsal view showing fan-like alary muscles, B. Lateral view showing position within the body.

aorta. The aorta is a simple tube that carries the blood anteriorly to the head, where it empties back into the hemocoel. Thus, the blood flows backward through the body, irrigating the tissues as it goes, and then it is returned forward by the heart and the aorta. The circulation of the blood is further aided by the normal movement of the body and sometimes by special accessory pumping organs.

### *The Digestive System*

The digestive system of insects consists of a conspicuous, tubular alimentary tract and a series of less conspicuous glands. The generalized alimentary tract consists of a compartmentalized tube that extends from the mouth at one end to the anus at the other. This canal is divided into three distinct regions: the anterior **stomodeum** and posterior **proctodeum,** the linings of which are formed by the invagination of ectodermal cells capable of secreting cuticle, and a middle region, the **mesenteron.** In a few insects, the gut is a relatively simple tubular structure, but in most, the three main regions of the digestive tract become subdivided into a series of functionally specialized components (Figure 3-10).

The fore-gut, or stomodeum, is usually divided into an anterior tubular **oesophagus,** an enlarged food storage chamber called the **crop,** and a valvelike **proventriculus.** Where the proventriculus joins with the mesenteron, there is a flaplike **stomodeal valve** to prevent the return flow of partially digested food. In some insects, like the roach, the proventriculus may be muscular and

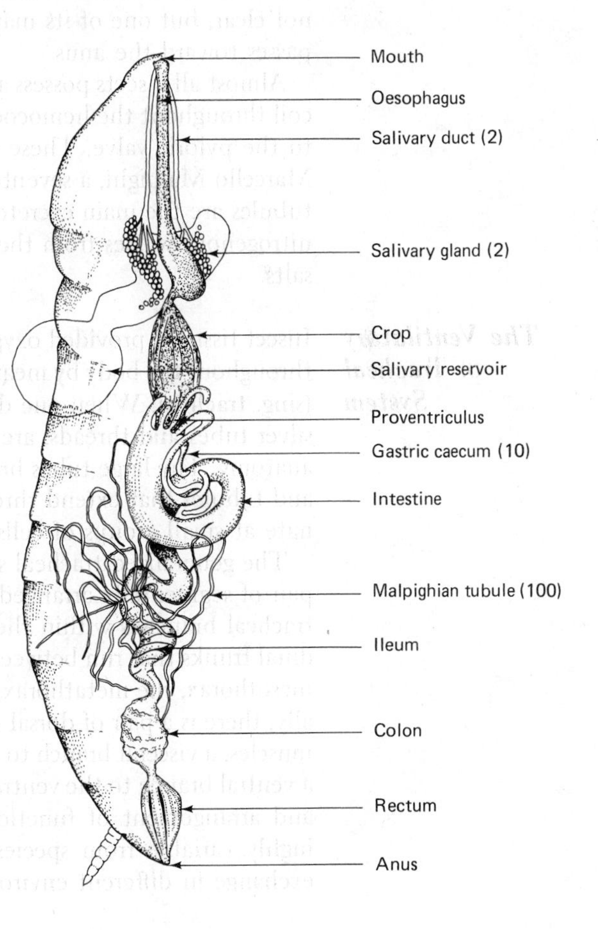


FIGURE 3-10. The digestive tract of the roach *Nauphoeta uneorea.*

**FIGURE 3-11.** Transverse section of the gastric mill of the roach *Nauphoeta uneorea.*

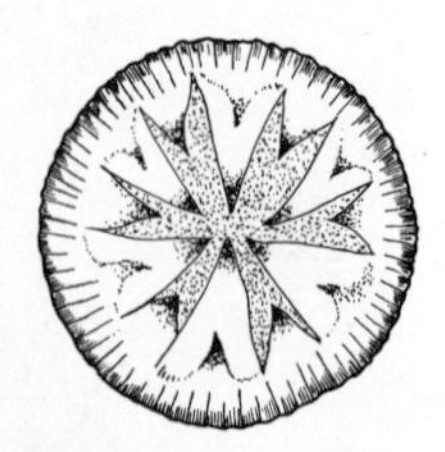


contain a food-grinding **gastric mill,** consisting of a series of sclerotized teeth (Figure 3-11).

The mid-gut, or mesenteron, is often little differentiated and consists of a large stomach, or **ventriculus,** with several anterior fingerlike projections, the **gastric caecae.** The exact function of the caecae is unknown, but they may produce enzymes or harbor micro-organisms important to the digestive process. The ventriculus is the main site of digestion and absorption and may be either a single chamber or divided into several subchambers by superficial constrictions.

The hind-gut, or proctodeum, is highly variable, but it is generally subdivided into a tubular **intestine** and a short expanded **rectum** that is connected to the anus. At the junction of the ventriculus and the intestine is another valve, the **pyloric valve.** Since most of the digestion of food and absorption of nutrients occurs in the ventriculus, the complete function of the hind-gut is not clear, but one of its major roles is to absorb water from the waste as it passes toward the anus.

Almost all insects possess a variable number of long, slender tubules, which coil throughout the hemocoel and join with the digestive tract just posterior to the pyloric valve. These structures are named **Malpighian tubules** after Marcello Malpighi, a seventeenth-century microanatomist. The Malpighian tubules are the main excretory organs, which by their special design remove nitrogenous wastes from the blood while at the same time conserving vital salts.

### *The Ventilatory or Tracheal System*

Insect tissue is provided oxygen by the direct conduction of atmospheric air throughout the body by means of an elaborate system of tubes called **tracheae** (sing. **trachea**). When one dissects an insect, the tracheae, which appear as silver tubes and threads, are one of the prominent features of the internal anatomy. The large tubes branch and rebranch into numerous smaller tubes and tubules that extend throughout the tissues until they ultimately terminate at small groups of cells.

The generalized tracheal system, illustrated in Figure 3-12, consists of ten pair of segmentally arranged openings, or **spiracles,** that lead to a system of tracheal branches within the segment, plus several interconnecting longitudinal trunks that run between the segments. The spiracles are located on the mesothorax, the metathorax, and the first eight abdominal segments. Generally, there is a pair of dorsal tracheal branches to the heart, aorta, and dorsal muscles, a visceral branch to the digestive tract and reproductive system, and a ventral branch to the ventral muscles and nerve cord. However, the number and arrangement of functional spiracles, tracheal trunks, and branches is highly variable from species to species. Some special adaptations for gas exchange in different environments are discussed in Chapter 9.

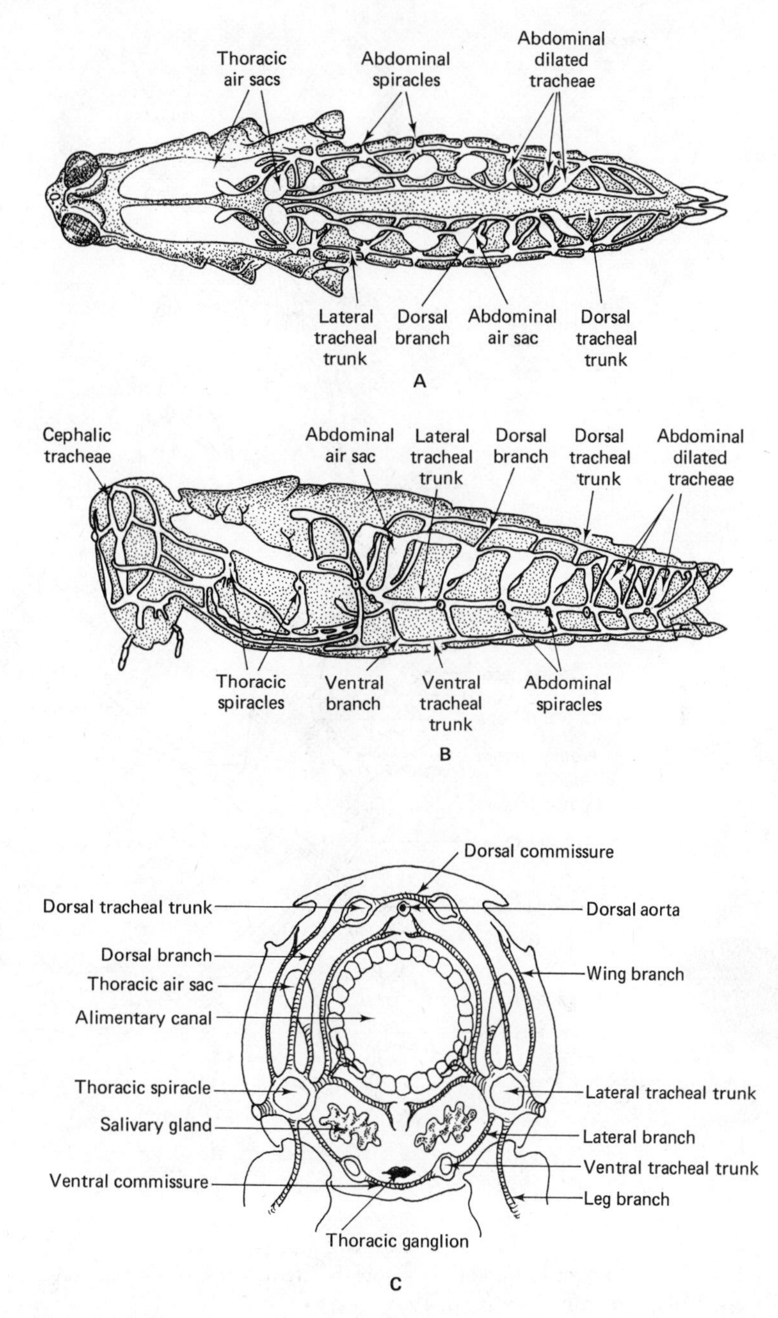


**FIGURE 3-12.** The generalized tracheal system as seen in the grasshopper. **A.** Dorsal view, **B.** Lateral view, **C.** Transverse section through a thoracic segment. (Adapted from Snodgrass, 1935.)

## *The Nervous System*

An insect has an easily seen **central nervous system,** consisting of a ganglionic mass, or **brain,** situated in the head, and a chain of ventrally placed **ganglia** (sing. **ganglion**), interconnected to form the **ventral nerve cord.** These ganglia form the main coordination centers of the body. They receive information directly from the sense organs, located on the associated segments of the body, and generate the motor impulses that flow outward to the appropriate

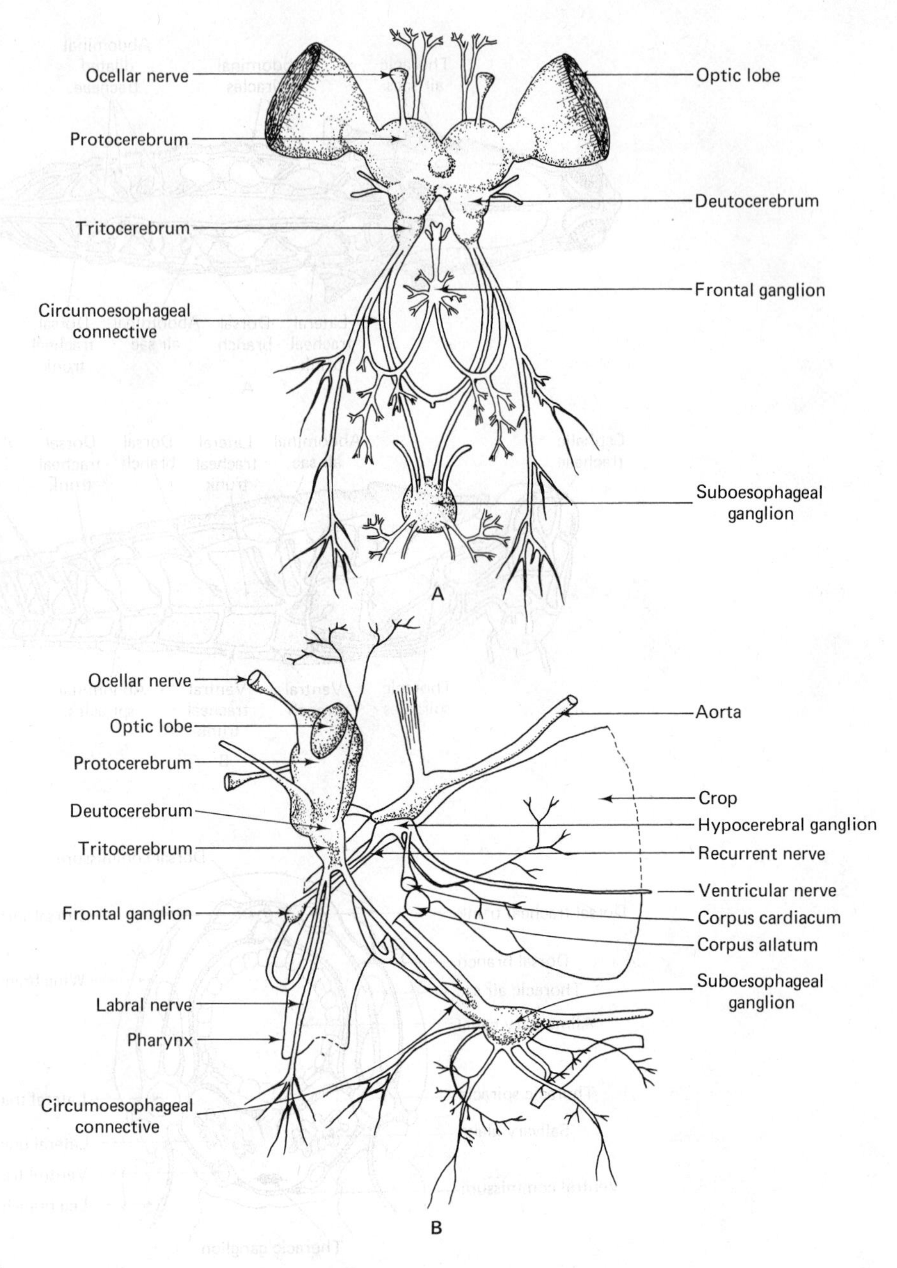


FIGURE 3-13. A typical insect brain and its associated structures. **A.** Dorsal view, **B.** Lateral view. (Adapted from Snodgrass, 1935.)

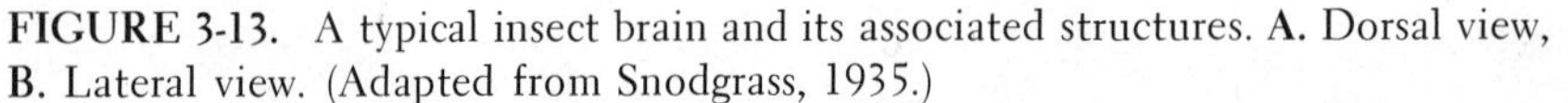

muscles. The brain is divided into three major regions. The front part, or **protocerebrum,** gives rise to two large **optic lobes**; special cells within the protocerebrum secrete a substance that mediates the insect's hormone system (Chapter 9). The mid-portion, or **deutocerebrum,** consists mainly of a pair of lobes that contain the sensory and motor pathways to and from the antennae. The base of the brain, or **tritocerebrum,** consists of a pair of lobes that lead into thick connectives, which extend around the oesophagus to the first ganglion of the ventral nerve cord, Figure 3-13.

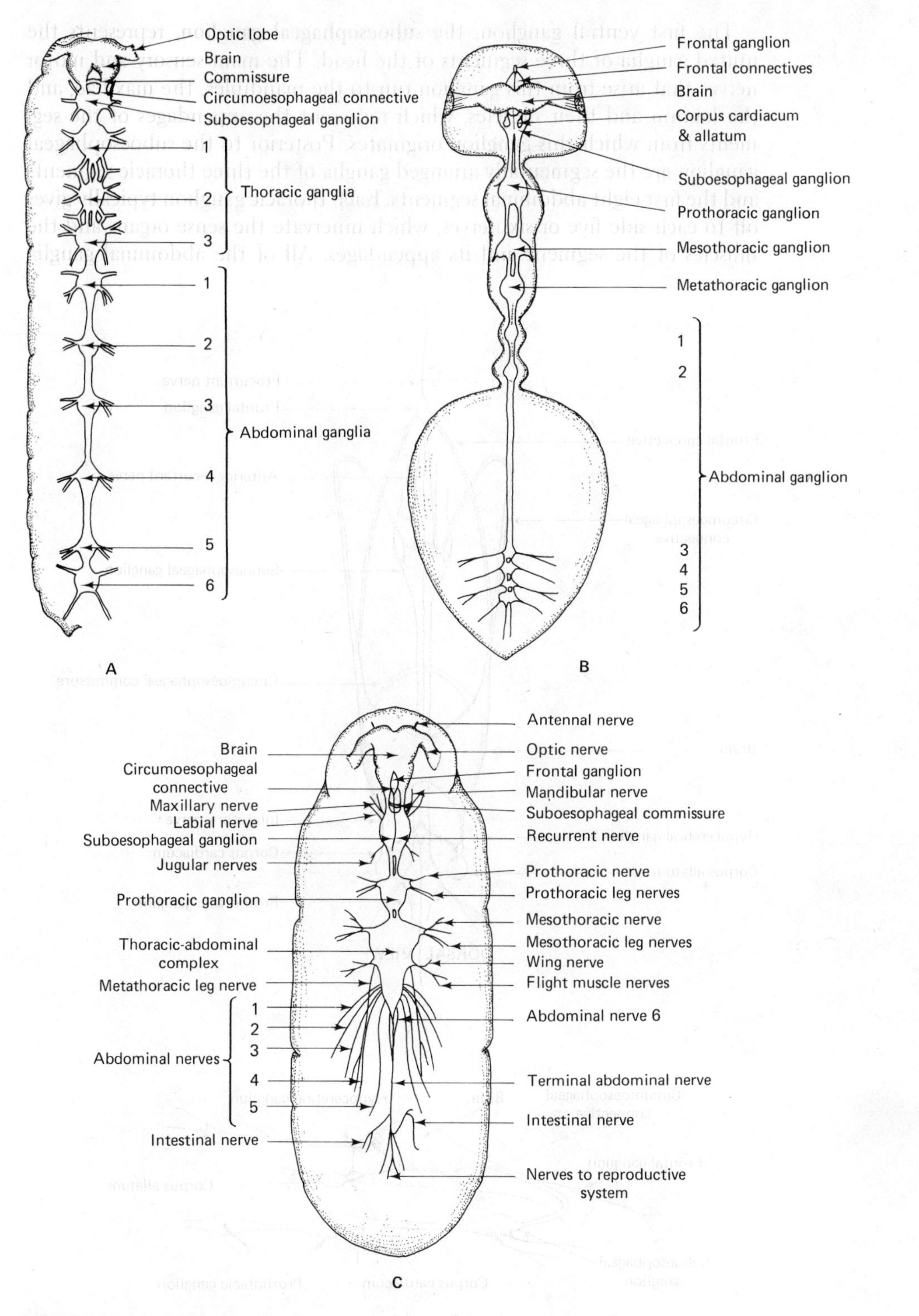


FIGURE 3-14. Variations in the arrangement of the insect central nervous system showing several changes in the generalized segmental arrangement of ganglia. A. Generalized condition as seen in a typical caterpillar, B. Ant, C. Scolytid beetle showing unification of the ganglia.

The first ventral ganglion, the **suboesophageal ganglion,** represents the united ganglia of three segments of the head. The main sensory and motor nerves that arise from this ganglion run to the mandibles, the maxillae, and the labium and their muscles, which represent the appendages of the segments from which this ganglion originates. Posterior to the suboesophageal ganglion are the segmentally arranged ganglia of the three thoracic segments and the first eight abdominal segments. Each thoracic ganglion typically gives off to each side five or six nerves, which innervate the sense organs and the muscles of the segment and its appendages. All of the abdominal ganglia

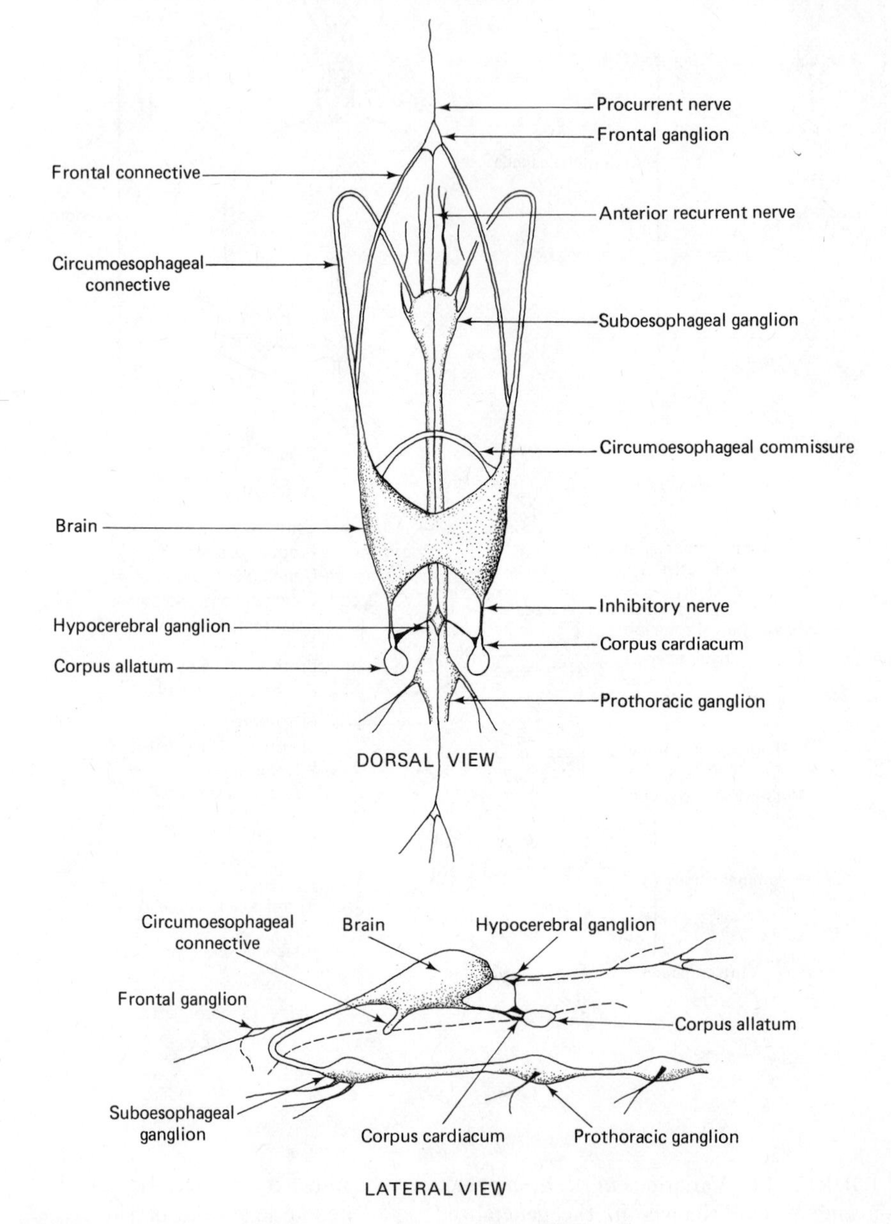


**FIGURE 3-15.** The stomodeal nervous system of the larva of a lucanid beetle. **A.** Dorsal view, **B.** Lateral view.

except the last give off one or two pair of nerves to the relatively simple sensory and muscular system of the abdominal segments. The last ganglion, representing the fused ganglia of the four terminal segments, gives off nerves to the muscles of these segments and to the reproductive organs and accessory structures. In most adult insects, tagmosis has resulted in some reduction in the number of segments and in a fusion of some or all of the ganglia behind the head. The basic plan of the central nervous system and two variations are illustrated in Figure 3-14.

The portion of the peripheral nervous system that is called the **stomodeal nervous system** consists of a number of small ganglia, some relatively slender nerves which innervate mainly the digestive tract, plus several specialized components of the neural-endocrine system. The generalized structure of the stomodeal system is illustrated in Figure 3-15. The function of the neural-endocrine elements is discussed in Chapter 9.

### *The Reproductive Systems*

In almost all insects, the males and females are separate; insects are therefore **dioecious.** The two reproductive systems are basically similar, the major difference being the larger size of the mature ovaries compared to the mature testes. Both the male and female organs are located in the hind portion of the abdomen and open posteriorly.

The female system (Figure 3-16) consists of a pair of **ovaries,** each made up of a group of egg-producing tubules, the **ovarioles.** The ovarioles empty into the **lateral oviducts,** which unite to form a median **common oviduct.** This opens through the **gonopore** into a chamber, the **vagina,** which in turn opens to the exterior directly or into the egg-laying device, the **ovipositor.** Two additional structures open dorsally into the common oviduct or the vagina. One, the **spermatheca,** receives and stores the sperm following copulation; an associated **spermathecal gland** provides nutrients for the sperm. The other consists of a pair of **accessory** or **colleterial glands** which secrete special materials for gluing eggs to the substrate or providing them with a protective covering.

The male system (Figure 3-17) consists of a pair of **testes,** each composed of a group of sperm-producing tubes that unite through small tubules (**vas efferentia**) into a lateral **vas deferens.** Each vas deferens widens into a **sperm**

FIGURE 3-16. Generalized female reproductive system. (Adapted from Snodgrass, 1935.)

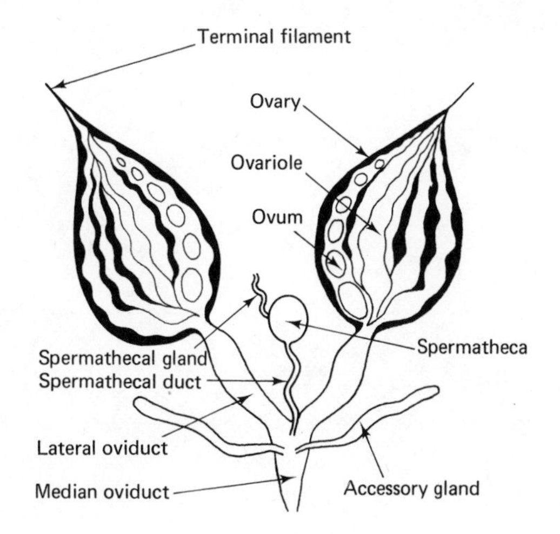

FIGURE 3-17. Generalized male reproductive system. (Adapted from Snodgrass, 1935.)

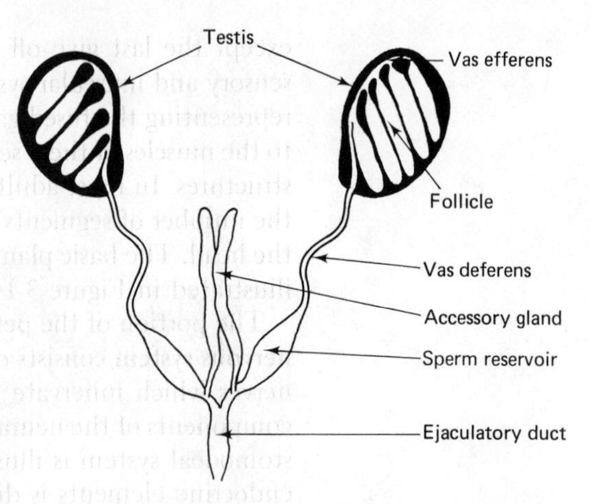


**reservoir,** which empties through a duct into a median **ejaculatory duct** with a muscular tip called the **penis.** A pair of accessory glands that produce the nonsperm components of the **semen** empty their secretions through ducts into the sperm reservoirs.

# General References


CHAPMAN, R. F. 1969. The insects—structure and function. American Elsevier Publishing Co., Inc., New York, N. Y.

MATSUDA, RYUICHI. 1965. Morphology and evolution of the insect head. *Mem. Amer. Entomol. Inst.* 4. 334 pp.

MATSUDA, RYUICHI. 1970. Morphology and evolution of the insect thorax. *Mem. Entomol. Soc. Can.* 76. 431 pp.

SNODGRASS, R. E. 1935. Principles of insect morphology. McGraw-Hill Book Company, New York, N. Y.

# Insects are not all alike

# 4

Having just described the anatomical plan of a more or less hypothetical, generalized insect, it may seem a little contradictory to state that one apparent characteristic of the insects is their great diversity of form. But if we analyze the basis for this contradiction, we discover one of the more fascinating aspects of the study of insects. Insects are actually a rather stereotypic group in terms of the functional arrangement of their body parts and organ systems. In fact, there are few species that could not be correctly placed in the Class Insecta on the basis of the external anatomy described at the beginning of Chapter 3. Yet, evolutionary experimentation with the form and the function of various body components has resulted in the greatest diversification found among the animals. The modification of some major structures, such as the wings and the mouth parts, allows for the rather simple separation of the insects into more than 25 orders. Less pronounced structural modifications permit the subdivision of these orders into more than 500 families, and still more subtle changes have resulted in the description of nearly 1,000,000 species. The actual number of each of these taxonomic divisions varies from book to book, depending on the classification preferences of the author. At the species level, the number is changing daily as entomologists learn more about the biology of insects and obtain specimens from areas where the insect fauna is poorly known. The number of species of insects that have been described and named probably represents only a small fraction of those in existence. About 10,000 new species of insects are described every year. To express the extent of this diversity in a more familiar way, there are more than 28,000 species of beetles (Coleoptera) in the United States, compared with about 6,000 species of mammals in the entire world.

Although such numerical comparisons are interesting, the real fascination of insect diversity is uncovered when we begin to investigate why this particular group of animals became so much more successful in terms of special adaptations than did any other group. The brief account in Chapter 2 of the probable evolution of the insects suggested several important evolutionary breakthroughs that probably provided the initial impetus for this group of arthropods to totally exploit the land masses of the world. Through the development of a specialized cuticle that protects them against water loss and physical harm, the development of an impervious covering for their eggs, the refinement of a tracheal system for gas exchange, and the consolidation of the thorax as a locomotory center, the insects became the first group of animals almost perfectly adapted to a terrestrial way of life.

The early fossil record of insects is extremely fragmentary, as there are only about 13 rich insect fossil beds known in the entire world. The earliest known hexapod is a wingless, insectlike creature from the middle Devonian, and the oldest known true insect is represented by a fossil found in later Devonian deposits. On the basis of this fossil evidence, entomologists have placed the origin of the insects somewhere in the order of 350 to 400 million years ago. These earliest insects were clearly of a primitive form. That is, they displayed the basic generalized characteristics from which we believe all of the more specialized insects were derived in the long course of evolution.

Evolution is an irreversible process. Once an organism has begun to specialize, it must either live with whatever degree of specialization it has achieved or it must specialize still further. The survival of an organism that has specialized depends upon the continuation of the conditions in which the specialization proved to be beneficial or, at least, not detrimental. When the conditions change, the organism must either adapt by specializing further, move to a new location where tolerable conditions have persisted, or become extinct.

When we piece together the fossil record of the insects, as incomplete as it is, we witness the results of this evolutionary process. Usually we can find evidence for the existence of a generalized group with a large number of adaptive possibilities. As environmental conditions changed, we find that new species evolved. Thus, we may encounter in the fossil record a series of adaptive experiments that were brought to bear on the problem of survival under a particular set of circumstances. Ultimately, the group or groups with the best solution to the problem survived, while those with less suitable solutions died out along the way. Some groups specialized to such an extent that they lost virtually all of their adaptive plasticity, so that when the environment changed further, the entire group became extinct. The fossil record indicates that some groups of organisms remained unchanged for long periods. Many fossil insects of great age are closely allied to living groups we call generalists because they can utilize a wide range of requisites for life. There are also fossil groups that survived for only a short time and then became extinct. Between these extremes, there is fossil evidence that suggests many groups blossomed out and enjoyed great success during one geological time but then survived only as relict groups that no longer display any degree of dominance in the fauna.

The insects, more than any other group, might be expected to show these patterns of trial and error and of success and failure. After all, they were the first group to have a truly terrestrial beginning and capability. During the long expanse of time through which the insects have been evolving, the world has undergone a number of severe climatic changes that must have had a more profound impact upon land environments than on marine and fresh-water environments where the impact of some important physical factors is buffered by the water. In fact, some of the early insect groups may well have survived climatic change by becoming aquatic for at least part of their life and thereby less subject to the fluctuations in temperature and moisture that occurred on land.

Unfortunately, the fossil record for insects is sparse. The first rich fossil-bearing strata, the coal measures of the late Carboniferous, contain the remains of an already diverse fauna of winged forms. Consequently, we can only speculate on what transpired in the 50 million to 80 million years between the occurrence of the first true insect and the time when the Carboniferous fossils were preserved. Obviously, the insects as a group enjoyed great success as indicated by the kinds of specialization that led to the wide variety of orders represented in coal measures (Figure 4-1).

Clearly, the insects had experimented successfully with at least two major kinds of adaptation. First of all, the presence of membranous wings is a

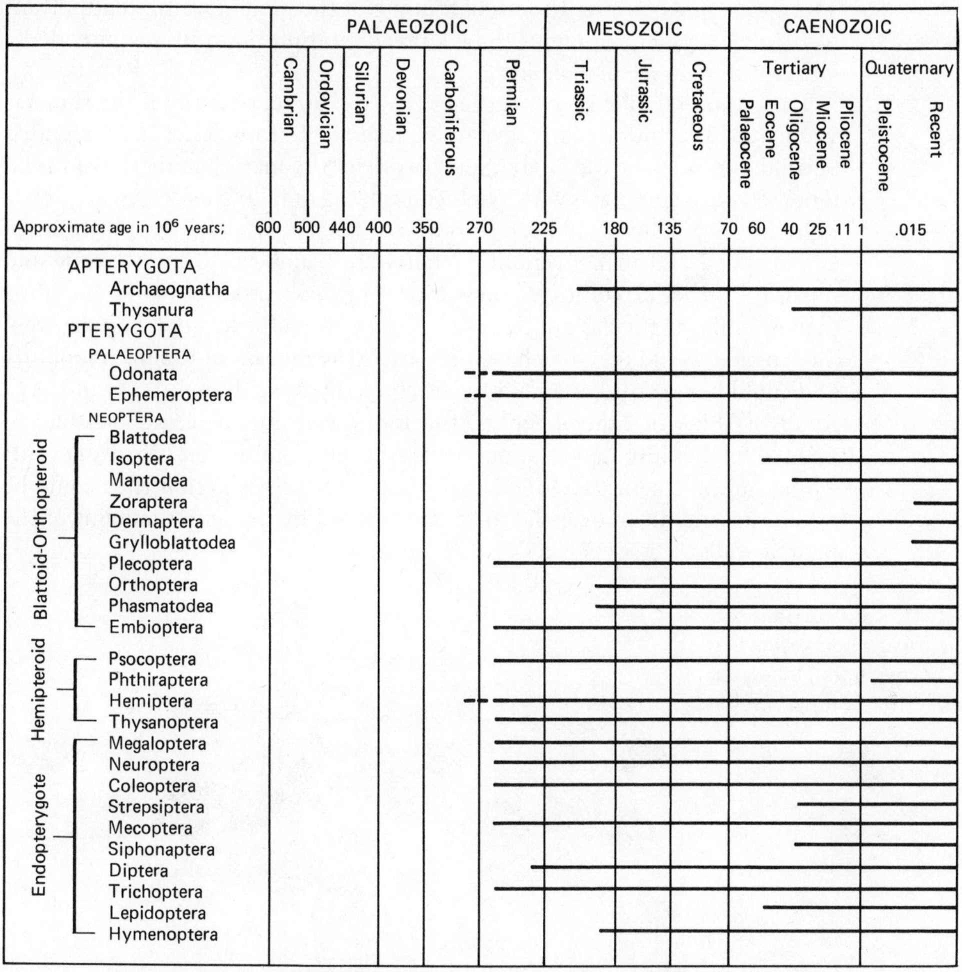

PALAEOZOIC | MESOZOIC | CAENOZOIC

Cambrian, Ordovician, Silurian, Devonian, Carboniferous, Permian, Triassic, Jurassic, Cretaceous, Tertiary (Palaeocene, Eocene, Oligocene, Miocene, Pliocene), Quaternary (Pleistocene, Recent)

Approximate age in $10^6$ years; 600 500 440 400 350 270 225 180 135 70 60 40 25 11 1 .015

APTERYGOTA
Archaeognatha
Thysanura
PTERYGOTA
PALAEOPTERA
Odonata
Ephemeroptera
NEOPTERA
Blattoid-Orthopteroid: Blattodea, Isoptera, Mantodea, Zoraptera, Dermaptera, Grylloblattodea, Plecoptera, Orthoptera, Phasmatodea, Embioptera
Hemipteroid: Psocoptera, Phthiraptera, Hemiptera, Thysanoptera
Endopterygote: Megaloptera, Neuroptera, Coleoptera, Strepsiptera, Mecoptera, Siphonaptera, Diptera, Trichoptera, Lepidoptera, Hymenoptera

Age estimations by Bureau of Mineral Resources, Geology and Geophysics, Canberra, 1960.

**FIGURE 4-1.** The geological history of the Class Insecta. Broken portions of lines indicate that the fossil record is uncertain. About ten other orders that are now extinct were present in the Carboniferous. (Modified from Riek, 1970, *Insects of Australia.*)

common characteristic of late Carboniferous fossil insects. The fossil record for the early Carboniferous is so poor, however, that we know nothing about the actual evolutionary sequence that led to the winged condition. Entomologists have proposed several theories concerning how this major structural modification came about, but the postulation known as the paranotal lobe theory discussed below is the most widely accepted one at the present time.

Prior to the development of functional wings, several fundamental changes in the structure of the thoracic segments must have occurred. These changes, which involved the vertical expansion and consolidation of the sclerites that formed the sockets into which the legs articulate, may have been evolutionary steps related mainly to walking. This unification and development of the walls or pleural regions of the thorax would have enabled the insects to raise themselves off the substrate and to move about more quickly and agilely.

Thus, we could consider the strengthening of the pleural walls, so important to the development of flight, to be a **preadaptation** that initially provided a rather different set of benefits.

The paranotal lobe theory suggests that the dorsal sclerites of the thoracic segments, called notal plates, expanded laterally to form lobes that extended beyond the sides of the body. Some workers theorize that these **paranotal lobes** served as stabilizers which helped jumping or falling insects to alight the right way up. This would have conferred some advantage in the form of added speed and maneuverability that were beneficial to both escape and pursuit. Further expansion of these lobes may have produced a set of gliding planes similar to the flaps of skin stretched between the legs of flying squirrels. Such planes would certainly have been useful as a means of localized mobility and could have served as an aid to widespread passive dispersal by wind. As a result of these or other benefits, the lobes were retained and became expanded further into membranous flaps. At some point in this course of events a hinge arrangement developed at the base of the flaps so that they could be moved up and down over the fulcrum provided by the upper margins of the pleural walls.

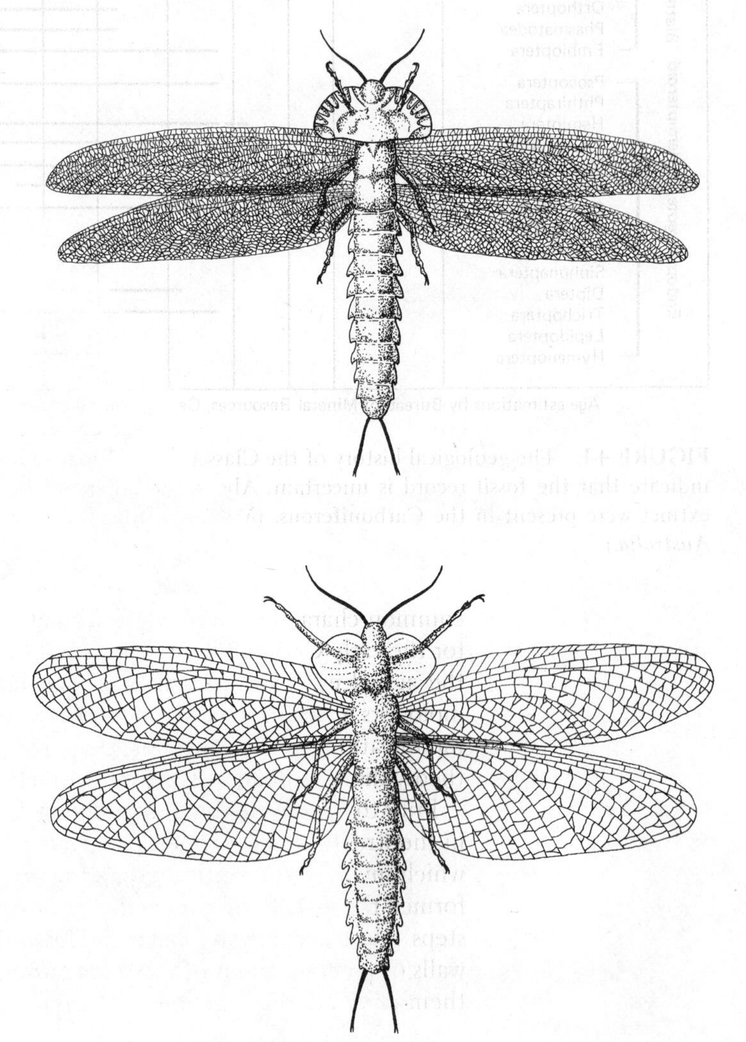

**FIGURE 4-2.** Reconstructions of Carboniferous fossil insects showing the presence of paranotal lobes on the prothorax. (After Jeannel, 1960.)

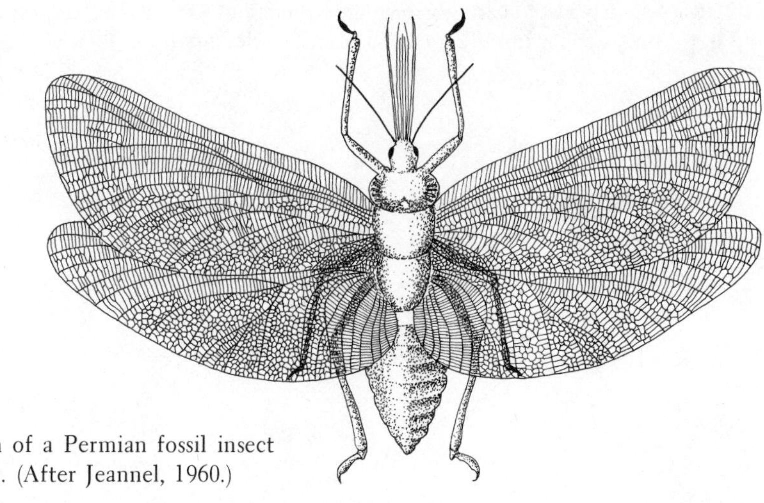

**FIGURE 4-3.** Reconstruction of a Permian fossil insect showing stylet-like mouthparts. (After Jeannel, 1960.)

Perhaps the paranotal lobes were not beneficial mainly as locomotory stabilizers but were used in display behavior that enhanced mate location and selection, as some workers have proposed. But for some unknown reason, wings apparently never occurred on the prothorax. Nevertheless, paranotal lobes occurred on the prothorax of some early insects, and similar structures were probably the forerunners of wings on the mesothorax and the metathorax (Figure 4-2). Perhaps someday someone will find some fossils that will illustrate the intermediate stages in the evolution of insect wings; until then we can only speculate on the origin of flight and wonder at the marvel of it as displayed by the insects around us.

The second major area of experimentation clearly involved the posture of the head and the modifications of the mouth parts that made possible the utilization of a variety ot untapped food resources. The ancestors of the insects probably had mouth parts directed downward (hypognathus). As the thorax and legs changed to allow the animals to raise themselves off the substrate, downward-pointing mouth parts were in a suitable position for feeding on the substrate on which the insect was standing. However, the forward direction of the mouth parts (prognathus) was better suited to tunnelling or to the active pursuit of prey. Some early fossil insects showed these adaptations as well as some modifications of the generalized form of the mouth parts. Quite early in their evolution the mouth parts became modified into elongate stylets that permitted feeding on plant juices by penetration and sucking rather than by chewing (Figure 4-3). In some of the hemipteroids these long stylets have been moved out of the way by a backward deflection of the front of the head (opisthognathus).

These two great developments, the evolution of flight and the modification of the feeding apparatus, which will be considered in more detail later, opened the way to an almost limitless variety of specializations and to the rapid spread of the insects into many new habitats. In the 50 million years after the end of the Carboniferous, the insects diversified greatly. Several new orders came into existence, and many of the fossils preserved during that geological period (Permian) are clearly recognizable as being closely allied to present-day families and genera (Figure 4-4).

Many of the major opportunities for diversification still lay ahead of the

**FIGURE 4-4.** Elytra of beetles belonging to the family Cupedidae. **A.** and **B.** Reconstructions from Permian fossils. **C.** *Cupes moultoni* (modern).

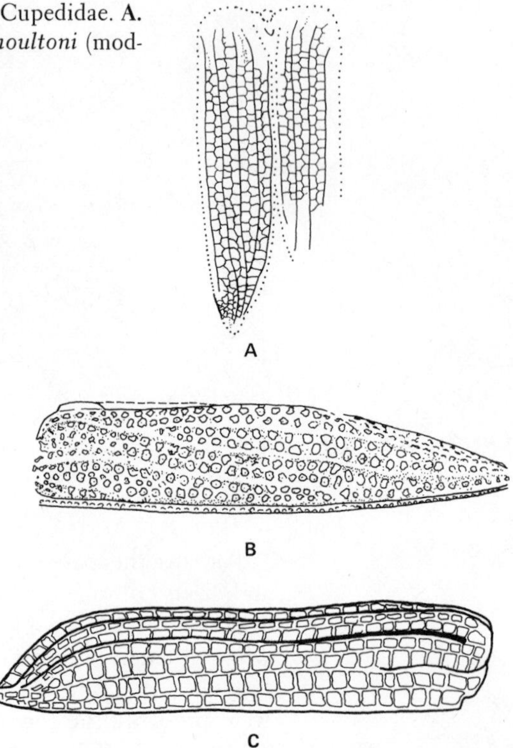


insects. Several modern-day orders do not appear in the fossil record until much later because they evolved and specialized in response to the evolutionary developments that occurred among the plants and other groups of animals. For example, some of the more recent Hymenoptera (bees and wasps) and the Lepidoptera (butterflies and moths) to evolve seem to have diversified in relation to the development, spread, and diversification of the flowering plants. The wide variety of insect-pollinated plants we see all around us did not come about as a matter of chance. Plants expend considerable energy in the production of brightly colored flowers, nectar, and perfume to attract insects and other organisms that assure their reproduction. Apparently the energy is well spent in that it reduces the waste of a large amount of pollen so characteristic of wind-pollinated plants such as pine trees. However, we must recognize that there had to be benefits to both the plants and the insects for such a relationship to have evolved. Colored petals and perfume would not have served a useful purpose had some insects not evolved the sensory apparatus and behavior that made them responsive to new colors, odors, and tastes. Conversely, the new insect behavior patterns would not have persisted if the insects had not benefited from the new sources of nutrition provided by the pollen and nectar. Following the proliferation of the flowering plants came the diversification of the warm-blooded vertebrates that provided still different feeding opportunities and opened the way to the evolution of ectoparasites such as the lice, fleas, and blood-sucking flies.

The foregoing rather general discussion of the diversification of the insects is an example of a common evolutionary phenomenon known as **adaptive radiation.** Simply, this implies that interbreeding groups we call populations display some degree of inherent variability in their structure, physiology, and behavior. This variability serves as a buffer against normal fluctuations in the components of the environment in which the population lives. Fluctuations in the requisites for life (physical factors, food, shelter, and other organisms) may favor certain genetic combinations (**genotypes**) in the population more than others. Consequently, the favored genotype leaves more offspring than do less favored genotypes, some of which may not leave any offspring at all. This selective process of the environment on the capacity of various genotypes to leave offspring is referred to as **natural selection** and results in a gradual change in the genetic composition of the population over a period of time. This gradual change may result in the population eventually taking on quite a different appearance or biological role and may even divide the population into subgroups, some of which can no longer interbreed. The same basic process follows a genetic change called a **mutation.** The mutation may be preserved if it provides the organism (**the mutant**) with a characteristic of benefit in its normal way of life or if it creates an opportunity for a new way of life.

We have been able to study how this process of natural selection operates as a result of several rather severe and short-term changes inflicted upon the environment of insects by man, namely, industrial pollution and pesticide-induced mortality.

The phenomenon called **industrial melanism** has been well documented. The British moth *Biston betularia* was originally described as a predominantly white moth with black speckles on the wings, although some almost black individuals of the same species were known (Figure 4-5). During the latter part of the nineteenth century, the black form became increasingly common around industrial centers until by the early 1900s a large portion of the population in these localities consisted of the dark (melanistic) form. Kettlewell (1961) reported the results of a study in which he released equal numbers of light and dark individuals in areas adjacent to cities and in the wooded countryside away from cities. He found that the survival of the light moths was greater away from the cities whereas the survival of the dark moths was greater near the cities. Kettlewell then demonstrated that the dark moths were better camouflaged when they rested on the lichen-encrusted bark of trees, darkened by soot and other industrial pollutants. The light moths were better camouflaged on trees in the country where the light bark and light grey lichens were not coated with pollutants. These moths are naturally preyed upon by birds, so those that more perfectly blend into the background of their resting places have a distinct survival advantage. In areas where no industrial development has occurred, the moth populations have remained predominantly light colored. In recent years, the reduced output of pollutants as a result of controls imposed upon industrial concerns has resulted in an increased number of light moths adjacent to the cities. These changes in the proportion of the two color types in the population demonstrate how a **selective pressure** (predation) operates upon the **genetic variability** (color pattern) inherent in a population.

A much more intensive and rapid selection has occurred in the response of some insect populations to treatment with various insecticides. In the three years following the widespread use of DDT, the common housefly and several other insect and mite species were found to be able to survive applications of synthetic organic pesticides that had previously killed them. Apparently, the populations of these organisms contained genotypes that were not susceptible

**FIGURE 4-5.** Light and dark forms of the peppered moth *Biston betularia* resting on a soot-covered tree trunk. (Photo by H. B. D. Kettlewell.)

to these poisons. Some species, such as the housefly, had genotypes with the capacity to detoxify the DDT and thereby avoid death. Others, such as certain mosquitoes, had behaviorally different genotypes, which resulted in their not contacting the poison because of a preference for particular resting places. Regardless of whether the reason that some individuals were not killed was biochemical or behavioral, the process of selection that led to total population resistance was basically the same. With repeated pesticide use, the number of individuals of susceptible genotypes was reduced more and more until only those of the resistant genotype remained.

Industrial melanism and pesticide resistance are rather special cases of change through natural selection, because they have occurred rather quickly in response to severe environmental modification. Nevertheless, they have

provided us with the opportunity to witness evolution in progress, as though studied with time-lapse photography. The same process brought about the gradual adaptation of insects in response to changes in the climate and in the flora and fauna that have occurred throughout their long history.

Sometimes significant changes occur in the basic ecological relationships of a species or groups of species that do not require associated structural or physiological changes. Examples of this can be found among insects that have become intimately associated with man, his crops, or his animals. This rather special kind of relationship, referred to as **synanthropy**, implies that man and his activity have become almost a requirement for the survival of some nonparasitic species. If man were suddenly to die out or severely curtail some activity, certain insect species might face a survival threat posed by the depletion of one or more requisites for life. The housefly *Musca domestica* is an outstanding example of a synanthropic insect. This now common fly was originally an old-world tropical or subtropical species that fed primarily on the feces of large, hoofed animals. As these animals were domesticated by man, the flies were given an opportunity to pass the winters of more northern climates in human habitations or animal shelters and since have spread with man throughout the world (Greenberg, 1965). The significance of synanthropic insects to human welfare will become more apparent in later chapters.

To summarize, the insects came into existence with a set of rather special characteristics that enabled them to exploit effectively the terrestrial environment before any other group of animals. Because of their initial plasticity and long history, they have been able to experiment with a wide range of structural, physiological, and behavioral specializations in their endless quest for food and shelter in an ever-changing world. As a result, the insects as a group now exploit virtually every habitat capable of supporting life, with the exception that only a few insects live on or in the ocean. Insects live in and on all forms of plant tissue including fungi and on and within all other terrestrial animals, plus some marine mammals. They live deep in the soil as well as on its surface. They drift as aerial plankton up to 2,000 meters above the earth. They range from the tropics to the absolute limits of life in the arctic and antarctic. They abound in most bodies of fresh, foul, and brackish water as well as in the deserts where water is scarce and the temperature exceeds 55 degrees centigrade. Wherever they live, and regardless of how rigorous physical conditions seem to be, insects seem to have become well adapted for survival. Through the development of special physiological and behavioral mechanisms, they can survive wide fluctuations in temperature and long periods without food or water. They can even tolerate noxious vapors and high concentrations of chemicals harmful to many living organisms.

All of these specializations have led to the development of an enormous group of animals united by a number of common characters, yet different in physiology, appearance, behavior, and size. They range from drab to exotically colored, from solitary to complexly social, and from the size of protozoans to the size of small birds and mammals (Figure 4-6).

## Keeping Things Straight

In dealing with this enormous diversity, man requires some means of organizing or systematizing the information he accumulates about other living

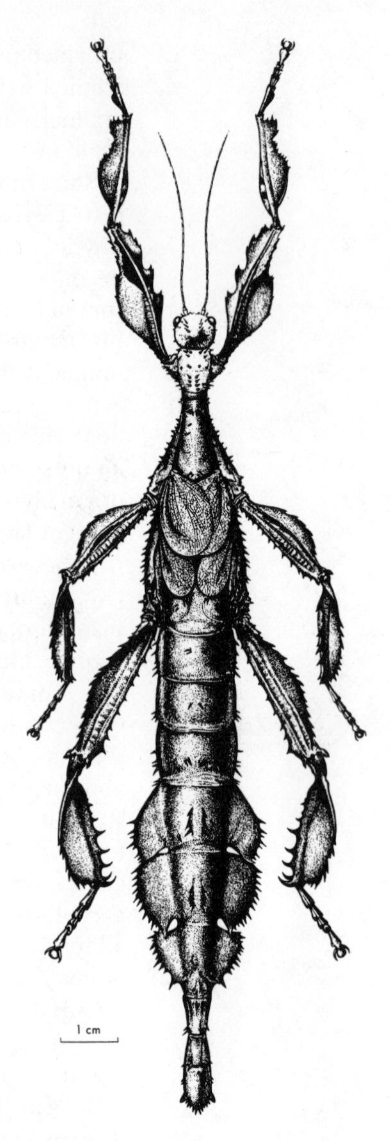


**FIGURE 4-6.** The large and bizarre Australian phasmid *Extatosoma tiaratum.* (Drawing by F. Nanninga reproduced by permission of Melbourne University Press.)

things. First, we need to facilitate the exchange of information by being able to refer to different organisms by a name that other workers recognize. In order to assign the correct name to an organism, the name already given to it must be accompanied by a basic description. For a group containing just a few different kinds, the descriptions and names might provide the basis for an adequate filing system, but for large groups, we need a system for grouping related kinds. The task of describing, naming, identifying, and initially grouping organisms is called **taxonomy.** But ultimately we must place organisms in a classification arrangement that is based upon natural relationships developed from a thorough understanding of evolution and speciation. The studies that provide this understanding are part of what we refer to as **systematics.**

The **species** is the taxonomic unit that is fundamental to our system of organizing a large number of different types of organisms into a manageable scheme. A species is a genetically distinctive group of individuals within which group there is or can be an effective flow of genetic material. Although some workers divide species populations into subspecies and varieties, the species is the basic objective entity of classification for which there is an acceptable definition based on sound evolutionary theory and biological evidence. Each species is assigned along with other closely related species into a group called a **genus.** Genera are then grouped into a larger assemblage and so on through an ascending hierarchy of subjective categories, each with a series of unifying characteristics. The classificational hierarchy that we use today was brought into prominence by Linnaeus. It consists of a series of obligate (capitals) and optional (lower case) categories, as follows:

KINGDOM
PHYLUM
subphylum
CLASS
subclass
superorder
ORDER
suborder
FAMILY
subfamily
tribe
subtribe
GENUS
subgenus
SPECIES
subspecies

As an illustration of how this system is used, the full classification of the common honeybee, using just the obligate categories, would be as follows: KINGDOM-Animal, PHYLUM-Arthropoda, CLASS-Insecta, ORDER-Hymenoptera, FAMILY-Apidae, GENUS-*Apis,* SPECIES-*mellifera.*

The higher the category, the easier it is to make the correct determination. It would be unusual to encounter an adult insect that could not be placed in the correct phylum or class on the basis of general appearance. With relatively little study, most insects can be placed quite easily into the right order, as you will soon realize. Below the ordinal level, identification becomes increasingly more difficult, until at the species level only a few specialists are really competent to make accurate determinations.

The aspect of taxonomy that is really of most importance to applied or field entomologists is the correct identification of previously described and named species so that workers can communicate about the insect with which they are concerned. In order to minimize confusion, each species has been given a distinctive latinized name in accordance with The International Rules of Zoological Nomenclature. The system of scientific names is a binomial one that involves the use of both the genus name and the species name. Thus, in the scientific literature, a honeybee would be referred to as *Apis mellifera,* in which *Apis* is the genus and *mellifera* is the species. According to convention, all scientific names are printed in italics, and the first time one is used in a publication, it is printed in full and is followed by the name, or an abbreviation of the name, of the taxonomist who first described and named the species. Thus, the first time an author of a paper refers to the honeybee he would designate it as *Apis mellifera* L. The capital letter *L* indicates that the honeybee was first described by Linnaeus. If the author's name appears in parenthesis, it means that the species was originally placed in a different genus and was later assigned to the current one. When the scientific name is used later in the same publication, an abbreviated form such as *A. mellifera* can be used.

In addition to scientific names, many frequently encountered insects are given common names. Although some common names are used uniformly within a single language, many vary from place to place within a country. For example, small biting flies of the family Ceratopogonidae are called "midges," "punkies," "no-see-ums," or "sandflies," depending on the local vernacular. Nevertheless, some common names are useful in applied entomology in that they better facilitate communication between professional entomologists and members of the public who share their concern about a local insect problem. This need has been recognized more in recent years, and an attempt has been made to standardize the common names of economically and medically important pests in several countries. Both the Entomological Society of America and the Entomological Society of Canada have Common Names Committees that screen proposed common names and publish lists of those of which they have approved.

Before proceeding to the subsequent chapters, the reader should become familiar with the common orders of insects, as treated in the Appendix. Throughout the remainder of this book, it will be necessary to use examples in order to clarify some of the facts and theories to be discussed. The Appendix was developed to provide the basis for an understanding of the way in which the insects are classified at the ordinal level so that the reader can associate most insects with the name of the group to which they belong. No attempt was made to make the Appendix comprehensive. Nevertheless, the descriptions and illustrations should provide a means of developing a mental picture of the various groups. It is hoped this will make the discussions that follow more clear and exciting. With relatively little practice one should soon be able to assign most insects encountered on a day to day basis to the appropriate order. In addition several excellent books are available for those wishing to delve more deeply into insect identification. For classifying the insects of North America to the family level, Borrer and DeLong (1976) is particularly useful.

## References Cited


Borror, D. J., and D. M. DeLong. 1976. An introduction to the study of insects, 4th ed. Holt, Rinehart and Winston, Publishers, New York, N. Y.

Greenberg, B. 1965. Flies and disease. *Sci. Amer.* 213: 92–99.

Jeannel, R. 1960. Introduction to entomology. Trans. by H. Oldroyd. Hutchinson & Co., Ltd., London.

Kettlewell, H. B. D. 1961. The phenomenon of industrial melanism in Lepidoptera. *Ann. Rev. Entomol.* 6: 245–262.

# II

# Success Through Specialization

*Relatively few species of insects are generalists. A few such as some species of the roaches or flies could be classed as generalists, but during the struggle for survival most insects have had to avoid competition from other species by becoming specialized to some degree. There is a recognized law which states that only one species can occupy a niche, and this law has applied to insects as it has to other organisms. Thus, emerging species had to become more and more specialized to avoid extinction, and some of the most recently evolved groups like the fleas (Siphonaptera) and lice (Mallophaga and Anoplura) are among the most specialized insects of all. Interestingly enough, most of the insects that we consider to be either pests or beneficial in more than a superficial way are specialists. No one will deny that some native generalists are beneficial or that a few other generalists cause problems, but these are usually of lesser concern. We also have a few problems with imported insects, such as the gypsy moth, which have a wide range of plant hosts. But the insects that seem to become the most serious pests are specialists. Those insects that attack specific kinds of trees or crop plants or vector specific diseases are the ones that come into conflict with man the most. Similarly, those insects that man manages for his own benefit, such as the honeybee or insect parasites, are usually specialists. Consequently, we have no alternative other than to develop an understanding of the specialized adaptations of insects so that we can both exploit them and cope with them more effectively.*

*In the first few chapters we reviewed some of the background and general considerations useful as a foundation for a more detailed discussion of the ways in which insects have become structurally, physiologically, and behaviorally specialized. We must now examine some of the ways that insects have been able to adapt to a number of ecological situations so that we can later examine how insects impinge upon our own activities. The success story of the insects really lies in what they have been able to do through evolutionary changes in the generalized plan of their ancestors.*

*In some respects, the organ systems of insects function much as they do in other invertebrates; therefore no attempt will be made to review some of the more basic aspects of animal structure or physiology unless they have a specific bearing on the discussion of a related topic. It is, for example, beyond the scope of this book to examine how nerve impulses are transmitted or how muscles contract and relax. Although not a complete list, the subjects chosen for inclusion in this section are those which I believe to be of major importance to an understanding of the success of the insects as a group or those which provide an insight into the biology of insects that will enable us to live in greater harmony with them.*

*We will examine the insect integument because it is one feature of the anatomy of insects that has played a major role in their success as terrestrial animals. Furthermore, the integument provides an effective barrier to invasion by some pathogens and to the entry of certain pesticides used for insect control. Related to the integument is the fact that periodically shedding the exoskeleton to permit growth provides an opportunity for structural changes to occur between developmental stages. These changes, called metamorphosis, allow for ecological diversification within the life history. Obviously, this has profound basic and applied implications.*

*Modifications of the generalized mouth parts, which in their original form restricted feeding to solid foods, into completely different looking structures that function in a wide variety of ways not only contributed to the success of the insects but are important to us for several practical reasons. The same is true of insect flight and other locomotory adaptations. There is no need to justify this section further, as I believe its importance will be apparent. However, we should remember that a detailed discussion of how the insects live, compete, survive in the face of adversity, escape their adversaries, and, finally, procreate has much more significance today than the mere stimulation of interest in the "wonders of nature." We ourselves are facing some basic problems of survival, some of which involve our interaction with insects. The insects are highly complex and adaptive organisms which we have learned cannot be taken lightly. We must get to know them well in order to avoid some of the errors in judgment we have made about them in the past.*

# The integument—life in a suit of armor

# 5

As mentioned earlier, the insectan integument is a complex structure, consisting of a layer of living cells and a series of nonliving layers, collectively referred to as the cuticle (Figure 5-1). The living tissue is comprised of a single layer of **epithelial cells** that rest on a thin sheet of connective tissue, the **basement membrane,** that is continuous with the sheaths of the muscles. Many of the epithelial cells are specialized gland cells that secrete the chemicals from which the cuticle is formed. Other cells of the epithelium are responsible for the formation of sensory structures, such as **setae** and their membranous sockets or the ducts that deliver special secretions to the surface of the cuticle.

The nonliving portion of integument is divided into two main layers, a relatively thick inner **procuticle** and a thin outer **epicuticle.** The procuticle is, in turn, often divisible into a darkened outer layer, the **exocuticle,** and a pale inner **endocuticle** that often appears to be composed of a series of fine lamellae. The epicuticle is also composed of several layers, usually three. Immediately adjacent to the exocuticle is a thin layer of lipo-protein called the **cuticulin.** On top of the cuticulin is a **wax layer,** which in turn is covered by a coating of material similar to resin or varnish that is called the **cement layer.** Each of these cuticular layers is chemically distinct, and each fulfills a particular function.

We have already noted that terrestrial arthropods require an exoskeleton that provides some flexibility and yet is rigid enough for muscle attachment, hard enough to offer protection against physical injury, and impervious enough to provide protection against water loss. The various layers of the integument are beautifully adapted to fulfill all of these requirements.

The conservation of body water is clearly one of the major problems confronting terrestrial organisms. Any animal or plant that is exposed to the drying properties of the air will lose water by evaporation from its surface. The smaller the organism, the greater is the surface area from which water can be lost relative to the volume of the animal, which in turn limits the water it can hold. As most insect species are less than a centimeter long, they contain only a tiny quantity of water that would be lost very quickly if it were not separated from the air by an effective barrier. Semiterrestrial animals usually live in the moist, organically rich upper layer of the soil that provides them with some protection against desiccation. Such animals, among which the earthworm is a good example, are often protected further by a thin layer of a sugar-protein material called mucus. However, mucus is subject to drying, as illustrated by the partially desiccated earthworms we find caught in the sun between rain showers. In insects, the sugar component of the mucus has been converted into a tough polymer called **chitin,** which combines with cuticular proteins to form an important part of the cuticle. Not all insects have chitin in their cuticle, but when present, it is a component of the procuticle, particularly the endocuticle, where it may comprise more than half of the total constituency of that layer. Although very tough and resistant to attack by many chemicals, chitin is not an exceptionally good barrier against tran-

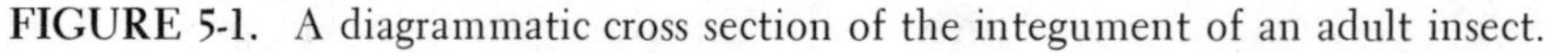

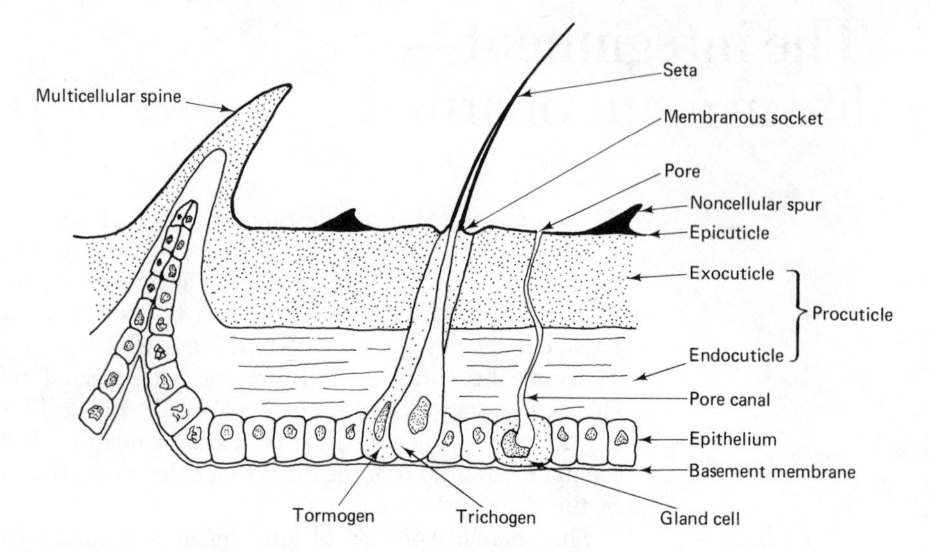


**FIGURE 5-1.** A diagrammatic cross section of the integument of an adult insect.

spiration, and it is too flexible to serve as the main component of body parts that need to remain rigid.

The strong bundles of muscle that function as powerful movers of the wings, legs, mandibles, and other structures are, of course, attached to the exoskeleton. Some additional substance is needed to make the integument more rigid so that it is not pulled out of shape when these muscles contract. Additional hardening also provides protection against mechanical injury. This hardening is provided by a substance called **sclerotin;** we thus refer to the hardened areas of the cuticle as being **sclerotized.** Sclerotin is formed and incorporated into the cuticle by a biochemical process far too complex to present in detail here. Briefly, a constituent of insect blood, **tyrosine,** moves into the cuticle at the appropriate time and is converted to quinones by oxidation. The quinones, so formed, then react with free amino groups of long-chain proteins, resulting in their becoming cross-bonded through the benzene rings of the quinones to form a hard plasticlike material that is somewhat similar to the keratin of our fingernails (Figure 5-2). When sclerotin becomes fully hardened, or tanned, it forms one of the hardest animal-made substances.

Other proteins in the cuticle have a different arrangement of amino groups that results in an irregular pattern of protein-chain bonding. The amino acid sequence in these proteins seems to prevent the establishment of additional cross links and results in the formation of an elasticlike material called **resilin** (Figure 5-2). Resilin can be stretched under tension and stores energy until the tension is released, at which time it returns to its original condition. This material is ideally suited for the formation of springlike ligaments and hinges such as those found at the base of the wings.

The layer of the cuticle that is primarily responsible for water retention is the epicuticle, particularly the wax layer, although some heavily sclerotized insects seem to be adequately protected against desiccation without wax. The wax layer is deposited on the surface of the cuticulin by way of a series of fine canals that extend from the epithelium through the procuticle and empty through pores in the cuticulin. Although the wax layer is usually several

FIGURE 5-2. Diagrams of the chemical structure of A. Sclerotin, B. Keratin, C. Resilin.

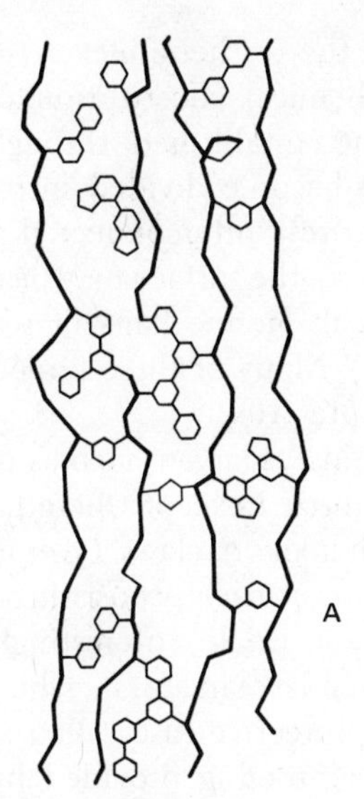


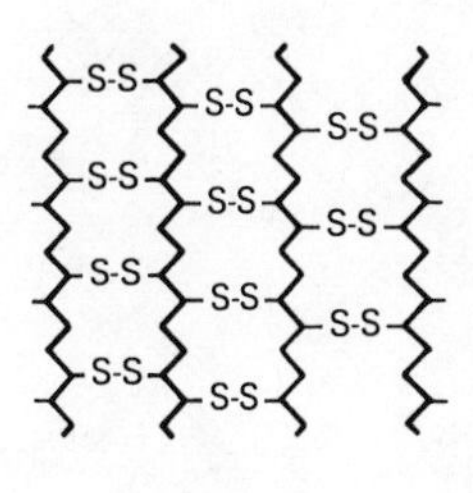


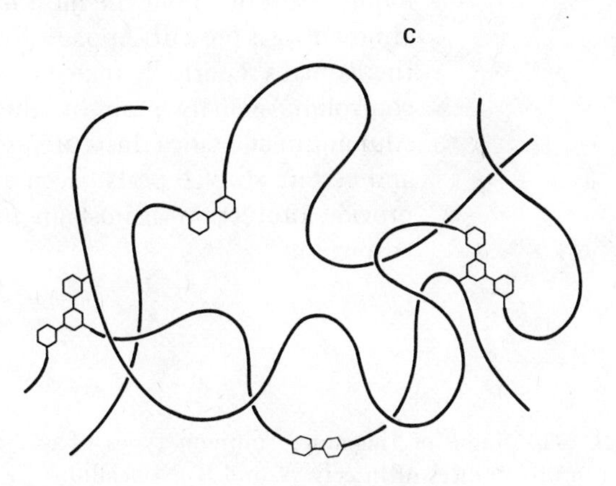


molecules thick, most of its water retention properties are the result of the formation of a single layer of wax molecules so densely packed that water molecules cannot pass between them. The wax molecules are bipolar; that is, they have a chemical group at one end that has an affinity for water (hydrophile) and a different group at the other end that rejects water (hydrofuge). The hydrated protein of newly formed cuticulin apparently has a high affinity for the hydrophile pole of the wax and thereby maintains the orientation of the monomolecular layer (Beament, 1960). Such a fine layer of wax is very fragile and can be abraded away easily by many gritty materials that are common in the environments of insects. However, the wax is usually protected by the varnishlike properties of the cement layer secreted over it.

In a typical adult insect then, we find an integument with inherent properties of toughness (chitin), flexibility (chitin and resilin), hardness (sclerotin), water conservation (wax), and resistance to mechanical abrasion (cement). However, not all of these properties are required either by all developmental stages or over the entire body. For example, if the entire exoskeleton were hardened by sclerotization, insects would be unable to move. The cells of the epithelium have the capacity to grow and to secrete the various chemicals incorporated in the cuticle immediately overlying them.

In this way the epithelium exercises control over the particular configuration and the chemical constitution of each area of cuticle according to the requirements established through evolutionary specialization. Thus, the cuticle has become divided into various arrangements of hard protective plates (sclerites) interconnected by areas of flexible membrane or hinge material. Also, the surface may bear stout spurs and spines, long sensory setae, a dense cloak of fine hairs or scales, and other specialized protuberances (Figure 5-3). Many of these projections serve as sensory receptors, as we will see in Chapter 10.

In many insect larvae, such as maggots and caterpillars, the cuticle is thin and unsclerotized except where hardness is really necessary as in the mandibles, mouth hooks or claws. Over most of the body the cuticle consists mainly of colorless procuticle, consisting of chitin and proteins, covered by a thin pigmented epicuticle. Interestingly, when some fly maggots pupate, they do so inside their last larval skin, which becomes sclerotized after the pupal molt to form a protective case called a **puparium** (Figure 5-4).

The waterproofing provided by the wax layer, however, seems to be a requirement of all but the most heavily sclerotized insects, so the wax layer is almost always present. Apparently, this has been recognized for a long time, as the Romans reportedly mixed road dust with their stored grain as a means of controlling granary pests by abrading away their cement and wax layers. Aluminum and silica dusts are used today as a means of controlling termites and certain storage pests. Even aquatic insects need some waterproofing to provide protection against an *inflow* of water that would result in their "drowning."

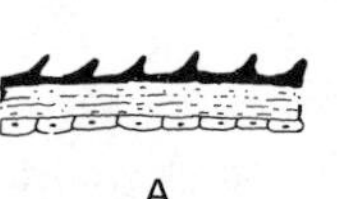


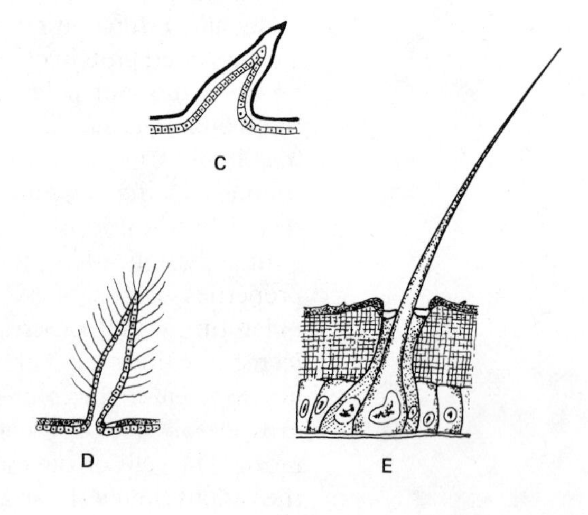


**FIGURE 5-3.** Some of the more common types of cuticular protuberances of insects. **A.** and **B.** Noncellular surface configurations such as ridges and spurs, **C.**, **D.** and **E.** Cellular processes such as setae and multicellular spines. (Adapted from Snodgrass, 1935.)

**FIGURE 5-4.** Fly puparium formed from the last larval skin. **A.** External appearance, **B.** Pupa within a cut-away puparium.

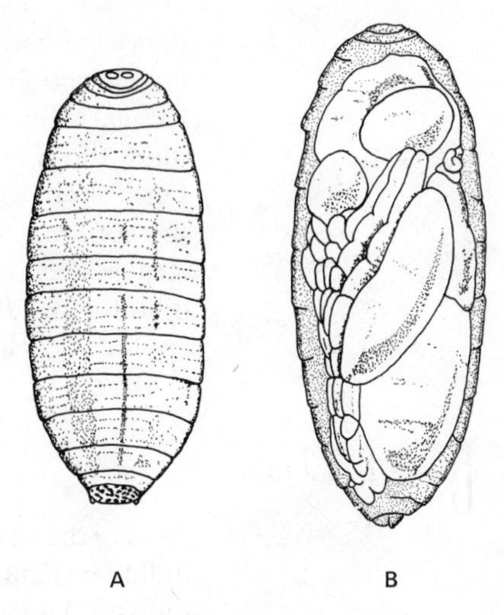


## Molting

Although the membranous areas between sclerites permit some elasticity, it is only enough to accommodate the degree of swelling associated with a large intake of food or the increase in volume of ripening ovaries, but not growth. Growth must be accommodated by a periodic shedding of the cuticle known as **molting** or **ecdysis.** Thus, insects increase in size in a stepwise fashion, each growth stage beginning with the expansion of a new untanned cuticle and ending with its eventual discard once it has become too small. Insects do not molt as adults, so all of their growth must occur throughout a series of immature stages called **instars.**

Each molt involves a sequence of biochemical steps that begins when the brain mediates information from the body and environment and stimulates the release of the molting hormone, **ecdysone,** into the hemolymph. The ecdysone stimulates the epidermal cells to divide mitotically and to greatly increase their production of chemical secretions. This cell activity results in a separation of the cuticle from the surface of the epidermis. Following separation, secretory cells produce a thin layer of lipo-protein, which becomes the new cuticulin layer on top of the epidermis. Then another secretion called **molting fluid** that is rich in proteinase and chitinase enzymes is released between the new cuticulin and the old endocuticle. The enzymes break down the endocuticle into amino acids and saccharides that can be resorbed by cells of the epidermis and used in formation of the new cuticle. While the inner layer of the old skin is being dissolved away, new proteins and chitin are being laid down beneath the new cuticulin layer. When all but a thin layer of endocuticle, the **ecdysial membrane,** has been dissolved, the insect is ready to discard its old skin, consisting mainly of the exocuticle and the epicuticle. The insect becomes quiescent and distends itself by swallowing air or water or increases the blood pressure in the anterior part of the body by contracting the abdomen. The increased pressure from within causes the old cuticle to split along lines of weakness; the most common such line forms an inverted Y on the front of the head capsule. Some fly pupae have a special inflatable sac, the **ptilinum,** at the front of the head, which can be expanded by increased blood pressure to split a cap off the anterior end of the puparium. After the insect has withdrawn itself completely from its old skin, it expands itself by muscle action and increases its volume with an intake of air or water before the newly formed cuticle loses its elasticity. Almost immediately following the molt, the chemical processes that harden and darken the new cuticle begin. For a short time, newly molted insects are highly vulnerable to attack and susceptible to injury. Also, prior to the completion of tanning, the monomolecular layer of wax is imperfectly oriented, allowing a somewhat higher level of water loss. But within a short time the cuticle provides full protection, and another developmental step has been completed.

## Insect Colors

Among the striking characteristics of insects are their colors which range from dull to brilliant and the effectiveness of their color patterns in protection and display. The importance of insect coloration will be discussed in a later chapter, but the basis of their colors will be considered here because it is almost entirely a feature of the integument. The color patterns of insects result almost entirely from either structural properties of the cuticle or pigments distributed in the cuticle, the epidermal cells, and, in a few cases, the blood.

Most blue, either metallic or iridescent, and some white coloration results from surface characteristics of the integument that cause scattering, interference, or diffraction of the light that strikes it.

If the integumentary surface is irregular, the light striking it may be reflected in all directions or, more properly, scattered. This results in a dull white appearance, typical of the wings of cabbage butterflies, which are clothed with light-scattering, striated scales.

If the surface consists of a series of superimposed layers, it may produce a phenomenon known as interference. Depending upon the distance between the layers, certain wavelengths of light may be reflected in phase with each other so that they are reinforced, while other wavelengths that are out of phase cancel each other. The surface then appears as the color of the reflected wavelengths. When viewed obliquely instead of from directly above, the effect is to reduce the distance between the superimposed layers, resulting in reception of a different wavelength. This change in color associated with a change in the viewing angle is called **iridescence.** A very simple example of this occurs when light strikes the superimposed surfaces of membranous wings. However, the brilliant, iridescent blue wings of the *Morpho* butterflies result from the fact that their wing scales have a flat base that supports a series of vertical vanes, each with 12 evenly spaced thickenings that decrease in size from the base to the tip. These thickenings produce a series of superimposed reflecting surfaces (Figure 5-5) that cause interference.

Although not a common color phenomenon among insects, light diffraction also occurs. Diffraction results when white light is split into spectral colors to produce a rainbowlike pattern. Some beetles have wing covers bearing appropriately spaced striations to produce such an effect.

In addition to the **structural colors** just mentioned, insects also display a wide range of colors that result from the presence of one or more pigments deposited in their integument. The chemical nature and origin of all of these substances is not known, but most are metabolic by-products of food sub-

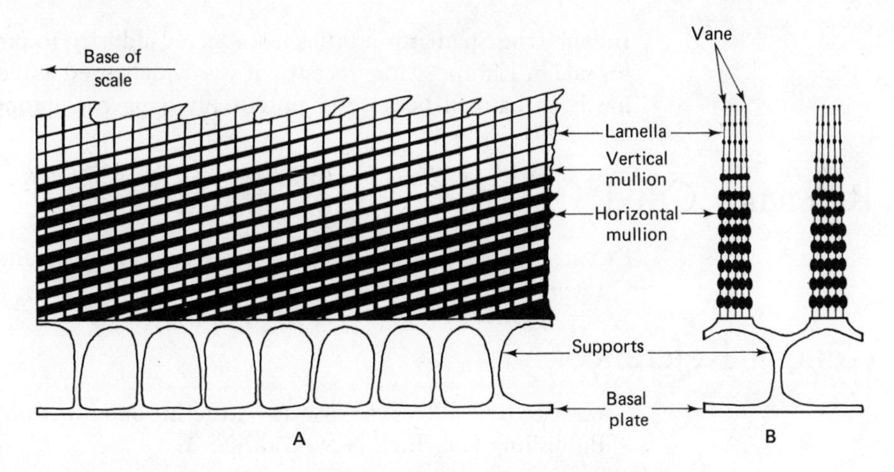


**FIGURE 5-5.** Diagrammatic representation of the multi-layered structures of a scale of the *Morpho* butterfly that produces iridescence. **A.** Lateral view, **B.** Transverse section of part of a scale. (Modified from Chapman, 1969, after Anderson and Richards, 1942.)

stances accumulated in the integument as a form of storage excretion. That is, they are bound up and stored in a safe manner to be disposed of each time the insect molts.

The pigments of insects are far too numerous to include here, but the following include the more common and important groups:

MELANIN. The brown and black color so characteristic of many insects is the result of the pigment **melanin,** which is the same or similar to the dark pigment in our own skin and hair. In most animals, melanin occurs in the cells in a granular form, but in most insects, it is incorporated directly into the endocuticle through a chemical reaction with cuticular proteins. Hardening and darkening of insect cuticle often occur together, suggesting that melanin may sometimes be involved in the tanning process. However, hardening and darkening may occur independently, and not all hardened cuticle becomes dark.

PTERINES. The **pterines** are the most common pigments contributing bright colors to insect cuticle and are also important as the pigment that separates the individual receptor elements that make up the compound eyes. **Leucopterin** is white, **xanthopterin** is yellow, and **erythropterin** is red.

CAROTENOIDS. The **carotenoid** pigments are acquired directly from plants and are not synthesized by insects. The most common is **β-carotene,** which is an orange-yellow pigment that often combines with blue pigments to produce a group, the **insectoverdins,** which make insects many shades of green. Other carotenes are responsible for some shades of red displayed by insects.

ANTHRAQUINONES. The **anthraquinone** pigments are mentioned because they include a few materials of commercial value. Cochineal, produced by scale insects which feed on prickly-pear cactus, was so highly prized as a red dye for cloth that following the discovery of its use among the Mexican

Indians, the Spaniards established a secret industry to produce the pigment for sale in Europe. More recently it was widely used as red food coloring but has been banned because of potentially dangerous properties.

## Reference Cited

BEAMENT, J. W. L. 1960. Wetting properties of insect cuticle. *Nature, Lond.* 186:408–409.

## General Reference

CHAPMAN, R. F. 1969. The insects—structure and function. American Elsevier Publishing Co., Inc., New York, N. Y.

# Modes of feeding and related adaptations

In the embryonic stage of all insects, several head segments bear a similar set of appendages that become the structures we collectively call the mouth parts. However, in the larval and adult stages, these structures are highly variable in appearance and reveal the degree to which the insects have adapted to the new feeding opportunities that arose during their long evolutionary history. Morphologists consider the chewing mouth parts of the primitive groups such as dragonflies, roaches, and grasshoppers to represent the basic condition from which all the other structural feeding specializations evolved. Regardless of how different the mouth parts are in appearance, they usually can be homologized with the chewing condition of the primitive groups.

Although the primitive condition of the mouth parts was presented in Chapter 3, it will be reviewed again here as a basis for a consideration of some of the derived types that are so important. The **labrum** is a flaplike upper lip suspended from the lower margin of the clypeus. The exterior labral surface is sclerotized, but the inner surface is fleshy and highly endowed with sensory structures; this fleshy area may form a lobe called the **labrum-epipharynx.** Behind the labrum are a pair of triangular jaws, the **mandibles,** which move back and forth in front of the mouth. Posterior to the mandibles are the paired **maxillae,** of which the **palps, lacinea,** and **galea** frequently become greatly modified. Closing the pre-oral cavity from behind is the **labium,** or lower lip, which is a structure that has resulted from the fusion of a second pair of maxillae. Like the maxillae, the labium has a series of components, the **palps, glossae,** and **paraglossae** that may be modified independently. The fleshy tongue, or **hypopharynx,** lies between the mouth and the labium. For a more complete review of the structure and arrangement of the generalized chewing mouth parts, refer to Figure 3-5.

The chewing mouth parts developed very early in the evolution of the insectan line and are only slightly changed from the condition found among the centipedes. Apparently this arrangement was and is particularly effective for feeding on particulate food, such as the leaves of plants, which is well illustrated by a grasshopper devouring a piece of lettuce. Not only does such a feeding apparatus occur among the groups that evolved early but it has persisted in several groups that evolved much later, including the beetles, the caterpillars (Lepidoptera), and the larvae of less advanced Hymenoptera we call sawflies. But there were other kinds of food available that could be utilized effectively when mouth part modifications occurred. For example, many of the earliest insects, like their modern descendants, fed mainly on the tissue of plants. Much of the bulk of plants, however, is made up of cellulose and a stiffening material called lignin, both of which are poor-quality foods. Some insects manage to derive nutrients from the breakdown of cellulose with special enzymes or by harboring micro-organisms that break it down for them. But most plant feeders obtain their nutrition from the chemical components of the plant sap, and most of the fibrous cellulose is eliminated from the gut as fecal pellets. One only has to watch a feeding caterpillar to gain an appreciation of the volume of plant material that passes in one end and out the other. These insects handle an enormous amount of unusable material in order to obtain the nutrients needed to sustain life and to grow. Clearly, it would be more efficient to extract the plant juice and leave the fiber on the plant, but this would require some modification of the mouth parts to permit piercing and sucking instead of munching. The line of evolution leading to the orders Hemiptera and Homoptera displays a sequence of mouth part specializations, culminating in the piercing-sucking stylets seen in all of the true bugs.

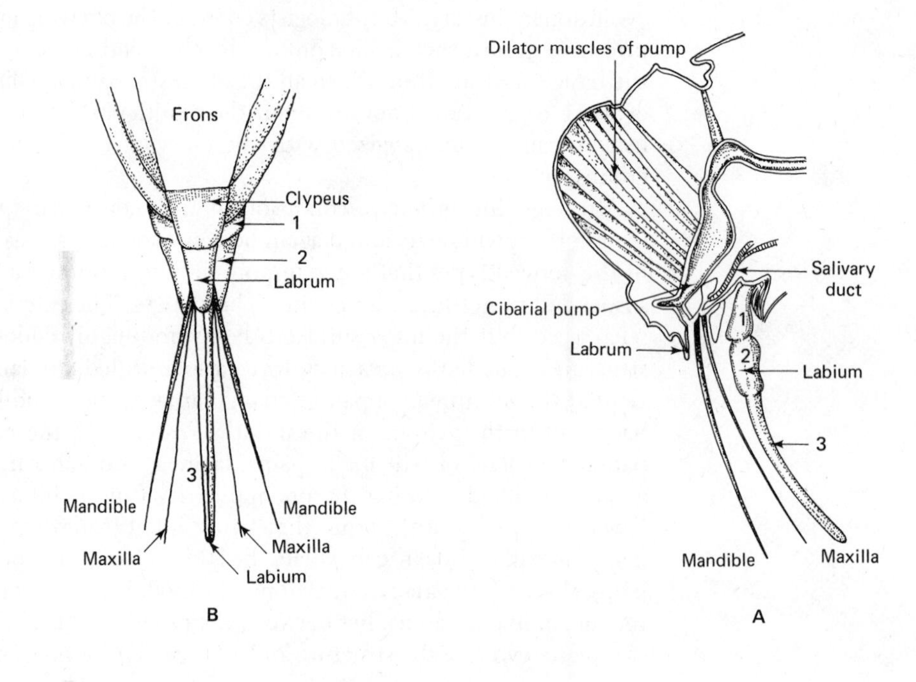


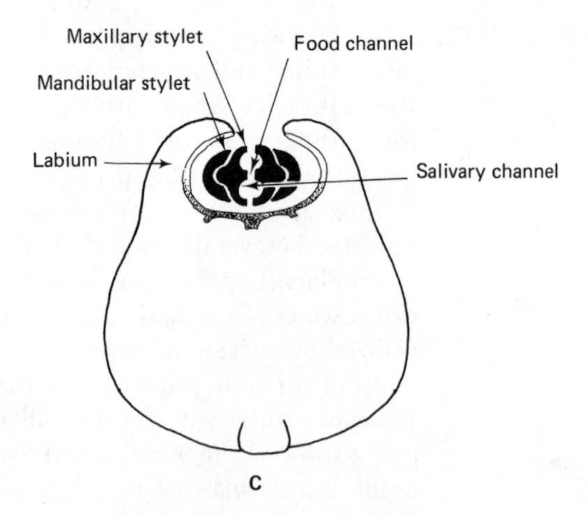


**FIGURE 6-1.** The piercing-sucking mouthparts of the cicada (Homoptera). **A.** Lateral view showing the position of the cibarial pump, **B.** Frontal view, **C.** Transverse section diagram showing the relative position of the stylets in the resting position. (Adapted from Snodgrass, 1935.)

In the bugs, the head has assumed the opisthognathus position, and the mouth parts consist of stylets ensheathed by a segmented beak that usually lies back between the front legs when the insect is not feeding. The labrum is little changed, forming a roughly triangular flap that lends support to the front base of the beak. The segmented beak is formed by the labium, which is medially grooved the length of its anterior face. The sheath so formed houses two pair of stylets formed from the mandibles and the maxillae. The maxillary stylets have a pair of grooves on their medial faces such that when the two are held together a food canal and salivary canal are formed. The palps are reduced to lateral supporting structures at the base of the beak (Figure 6-1).

During feeding, the mandibular stylets work up and down to penetrate the host and to allow the united maxillary stylets to reach the liquid food in and between the plant cells. As the two pair of stylets enter, the segmented labium bends backward out of the way, its tip serving as a guide for the stylets (Figure 6-2). A salivary syringe in the head near the base of the beak forces saliva that contains enzymes into the host. A mixture of plant juice and saliva is drawn upward into the mouth by the action of the **cibarial pump,** which consists of a large chamber with well-sclerotized walls and posterior surface, and a flexible anterior surface. In the cicada (Homoptera), for example, the front of the head (clypeus) bulges forward to accommodate a large bundle of muscles that run from its inner surface to the flexible anterior surface of the pump chamber (Figure 6-1*A*). These muscles are the dilators of the pump. When they contract, they raise the anterior wall of the chamber, thereby increasing its volume and creating an inward suction that draws the food up the food canal. When the muscles relax, the chamber wall returns to its previous position by its own elasticity and the food in the pump chamber is expelled backward into the adjacent portion of the digestive tract. In very small sucking insects such as aphids, however, the plant juice rises up the food canal as a result of either pressure in the plant or capillary action, so no special sucking pump is necessary.

Although the evolution of the hemipteran mouth parts probably first served as an adaptation that permitted plant juice feeding, the same struc-

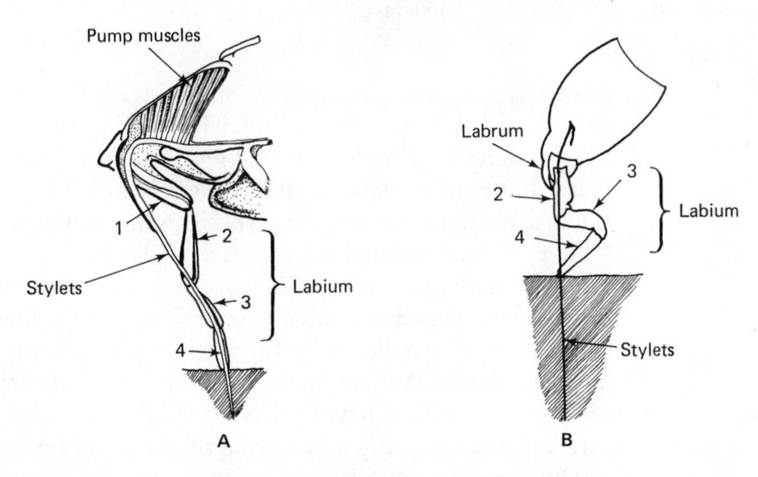


**FIGURE 6-2.** Diagram of stylet insertion by a hemipteran showing the manner in which the jointed labium folds backward while its tip serves as a guide for the stylets during penetration, **A.** Stinkbug, **B.** Bedbug (Adapted from Snodgrass, 1935.)

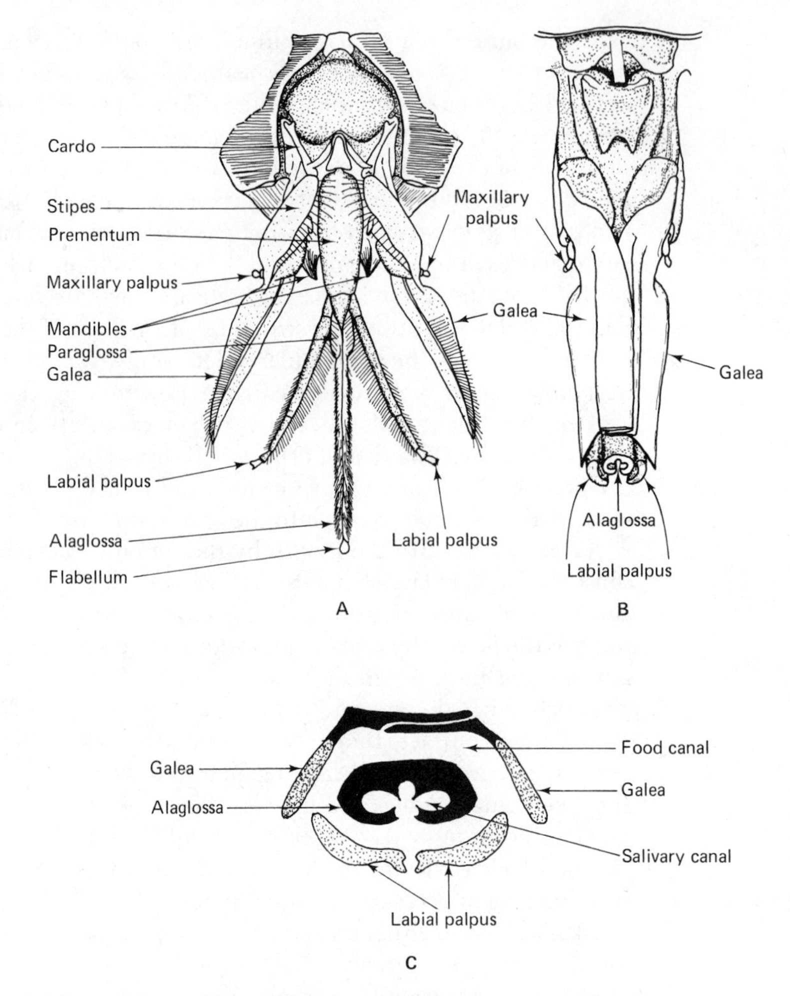


**FIGURE 6-3.** The multifunctional mouthparts of the honey bee. **A.** Posterior view of the mouthparts spread apart, **B.** Frontal view of the mouthparts folded together, **C.** Transverse section diagram showing the relative position of the components during liquid feeding. (Adapted from Snodgrass, 1935.)

tures were equally well suited for penetrating the integument of other animals and extracting their juices. Thus we find families of bugs with species closely related to plant feeders that feed as piercing-sucking predators of other invertebrates or as blood feeders on vertebrates. The stinkbug family Pentatomidae contains similar looking species, some of which are serious plant pests and others that are highly beneficial predators.

In the early Mesozoic, 150 million years or so after the origin of the primitive insects with chewing mouth parts, the first flowering plants evolved and were soon paralleled by the evolution of a wide variety of flower-frequenting insects. Among the best known flower-visiting insects are the bees, some species of which have evolved complex social patterns, based on parental care and the gathering and storing of nectar and pollen.

The mouth parts of the honeybee provide a superb illustration of a multipurpose adaptation of generalized structures. The labrum and mandibles are reduced in size. The latter have blunt, rather than toothed, tips, which facilitate their use in molding wax into the six-sided cells of the honeycomb. However, the maxillae and labium are highly modified. The paraglossae are reduced, whereas the glossae are elongated and fused into a long flexible tongue, the **alaglossa,** used for collecting nectar from flowers and for manipulating it within the hive. The lateral margins of the alaglossa are curved downward and inward to form a tube with a slitlike opening ventrally. The laciniae of the maxillae are lost, but the galeae form enlarged, flattened scoops that can be fitted closely over the alaglossa and, in combination with the flattened labial palps, form a second tubular structure probably used in lapping up water or other liquids that are more or less available on the surface of the substrate (Figure 6-3).

The Lepidoptera, many of which feed on nectar as adults, evolved one of the simplest, yet most effective feeding mechanisms to be found among the insects. In almost all butterflies and moths, the mandibles have been lost. Feeding is accomplished by a long siphoning tube comprised of the galeae (Figure 6-4C). This tube coils as a watch spring beneath the head (Figure 6-4*A*,*B*). The labium is greatly reduced to a small flap from which the large labial palps arise to stand erect in front of the face; the maxillary palps are usually small or absent. In its relaxed state, the siphon is coiled because of a ridge of elastic cuticle along the upper surface of the paired galeae. The siphon is extended by special muscles that arch the ridge of cuticle, causing the siphon to uncoil (Figure 6-4C). At a point about one third the way along the siphon, there is another small group of muscles that oppose those that arch the cuticular ridge. This produces an elbowlike bend in the siphon that

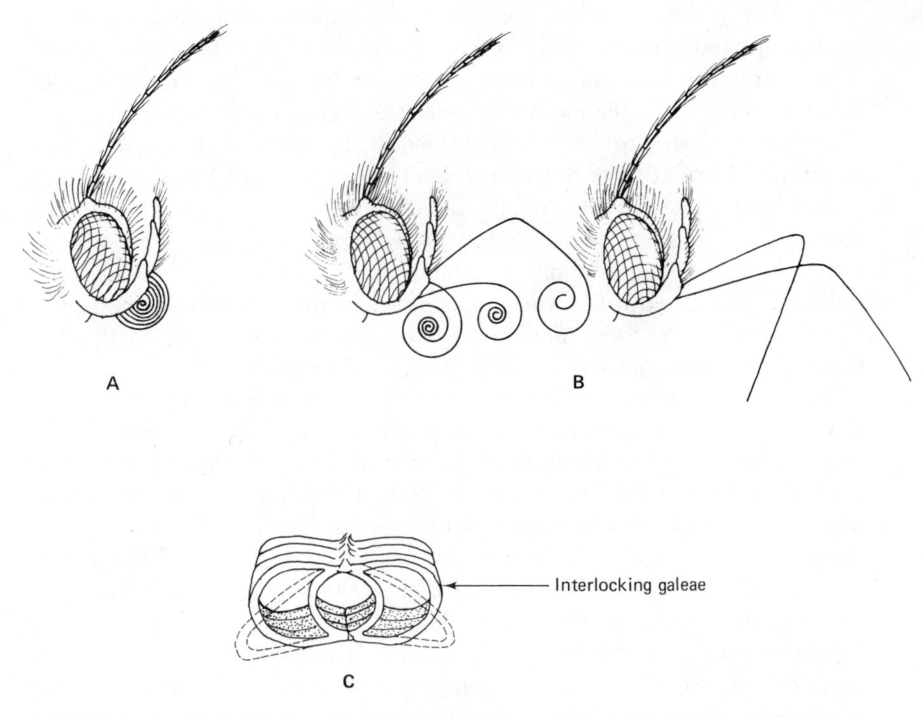


**FIGURE 6-4.** The siphoning tube of adult lepidopterans. **A.** The siphon in its coiled resting position, **B.** Diagrams of the uncoiling and bending of the siphon, **C.** Diagram of a transverse section of the galeae. The broken line shows how the cross section shape of the siphon is changed to produce its uncoiling.

facilitates the positioning of its tip in the flower nectaries (Wigglesworth, 1964).

About the same time that the flowering plants were radiating out over the earth, the birds and mammals began the ascent that led to their becoming the predominant large land animals by the beginning of the Tertiary (about 70 million years ago). These warm-blooded animals provided the insects with an abundance of somewhat new feeding opportunities in the form of their blood, flesh, and excrement. For some insects, these new animals and their waste products posed no new feeding challenge, especially if they had previously been associated with large terrestrial reptiles. However, there were at the very least some behavioral adjustments that probably developed following relatively little evolutionary experimentation. The Hemiptera, for example, could employ the piercing-sucking mouth parts, which had already proved to be effective instruments for extracting plant sap or the blood of invertebrates, to extract vertebrate blood. Likewise, various insects with chewing mouth parts, particularly the beetles, became efficient carrion and excrement feeders. Nevertheless, there was clearly room for some additional specialists that soon evolved largely as a result of changes in their mouth parts. The order Diptera displays a beautiful array of specialized adaptations for utilizing flesh, blood, and excrement as food.

Among the more primitive flies (suborder Nematocera), we find one of the more complex adaptations of the mouth parts for the extraction of blood. This is well illustrated by the female mosquito, which normally requires a blood meal to obtain certain nutrients required for the development of ripe eggs. As in the hemipterans, the mouth parts have become elongated into stylets, but in the mosquito all of the basic components are involved in the feeding apparatus (Figure 6-5). What we often refer to as the proboscis of a resting mosquito is a labial sheath, open down the anterior surface, with six stylets inside. When the mosquito feeds, the sheath bends to one side as its tip guides the stylets into the flesh of the host. The needlelike mandibles and saw-tipped maxillae work together to achieve penetration. The hypopharynx is also elongated and is traversed by the salivary duct that opens at its tip; the salivary glands produce enzymes and an anticoagulant that prevents the blood from clotting during feeding. The labrum-epipharynx, which is long and hollow, is positioned within the groove of the labium and forms a food canal when appressed to the hypopharynx. The maxillary palps are retained as sensory structures, but the labial palps are wanting.

Among the higher Diptera (suborder Brachycera), we find some other rather interesting mouth part modifications. In the horse flies such as *Tabanus* (Figure 6-6), the mandibles have become flattened blades that work as a pair of scissors to cut away a piece of flesh. These flies do not eat the flesh but feed on the blood that seeps from the wound. The maxillae form a pair of sharp probes that work up and down to keep the wound open. When at rest, the maxillae and mandibles are partially enclosed in a groove down the anterior face of the stout labium. The apical portion of the labium is expanded into a pair of fleshy lobes known as the **labellum,** which have a series of transverse channels that converge upon a median cleft. When the labellum is pressed into the seeping wound, the fluid flows through the transverse channels to the cleft. The hypopharynx forms a slender tongue, which in combination with the elongate labrum forms a food canal. The tip

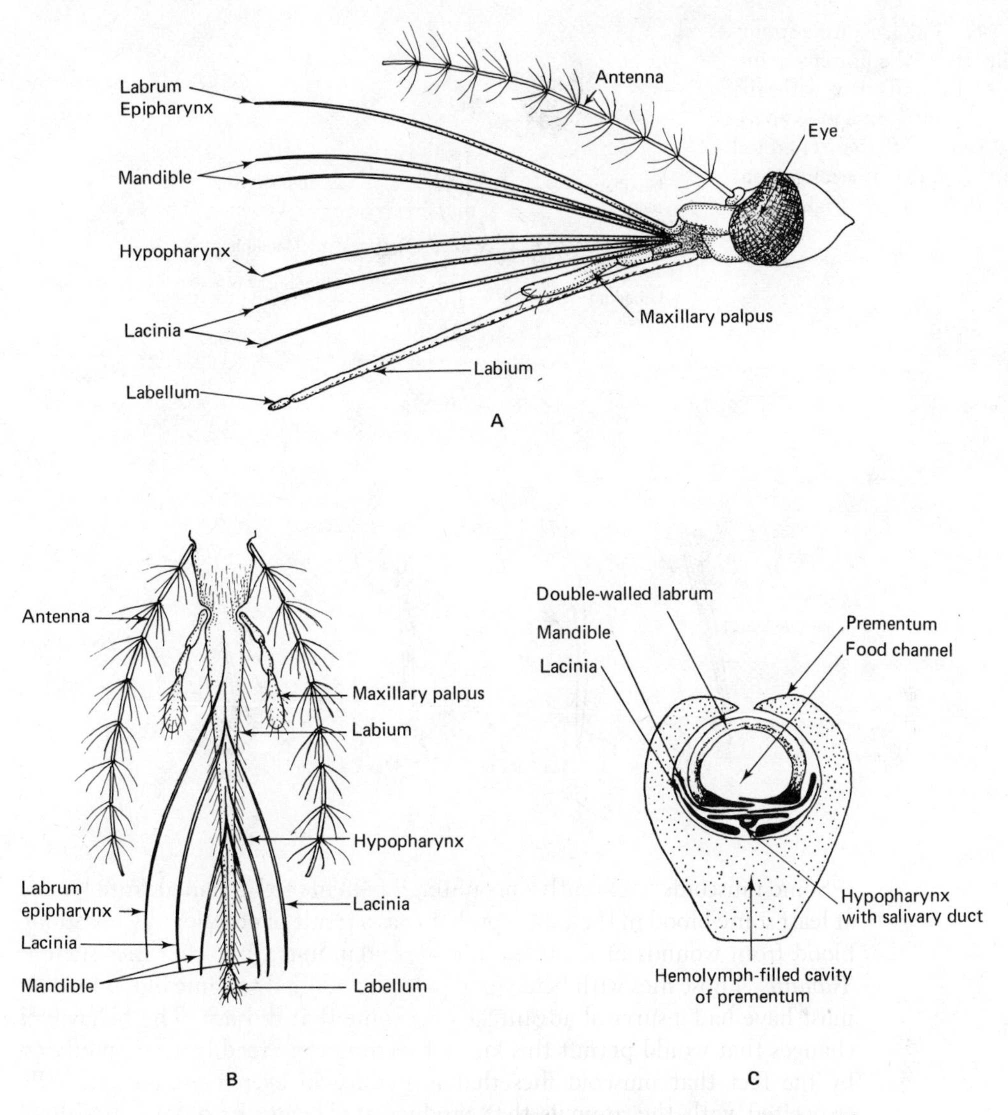


**FIGURE 6-5.** The piercing-sucking mouthparts of a mosquito. **A.** Lateral view of the mouthparts artificially spread apart, **B.** Frontal view of the spread components, **C.** Transverse section diagram showing relative position of the components in the resting position.

of the canal receives the liquid food from the upper margin of the median cleft and carries it to the mouth.

In the more advanced muscoid flies, the cutting mandibles and probing maxillae have been lost and the spongelike labella have become larger. Such mouth parts are well suited to feeding on the exudate from fermenting excrement and putrifying flesh or on substances such as sugar that are readily dissolved in saliva. Thus, in the house fly, *Musca domestica,* and its relatives, the mouth parts consist mainly of the well-developed labium, a pair of sensory palps, and the sucking tube composed of the hypopharynx and labrum (Figure 6-7*A*,*B*,*C*).

FIGURE 6-6. The lacerating-sponging mouthparts of the horsefly genus *Tabanus*. **A.** Frontal view with the components not spread apart, **B.** Lateral view of the separated components in order of arrangement from front (left) to rear (right).

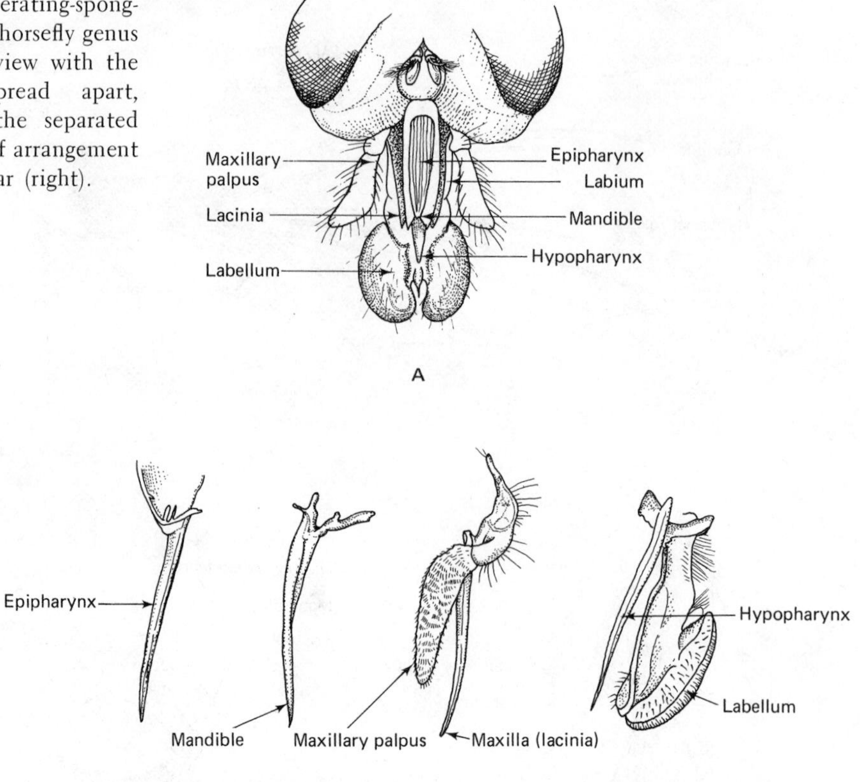


Some muscoids apparently encountered benefits to be gained from having at least some blood in their diet, probably as a result of occasionally obtaining blood from wounds or from feeding sites adandoned by other flies such as *Tabanus*. Those flies with behavioral patterns that led to some blood feeding must have had a survival advantage over some that did not. The behavioral changes that would permit this kind of feeding were readily accommodated by the fact that muscoid flies that reproduce in excrement are naturally associated with the animals that produce it. Having to obtain the blood opportunistically would not have been as efficient as being able to create a wound and obtain the blood as required. However, according to *Dollo's Law*, there is a very low probability that structures lost in the course of evolution can be recreated by a subsequent mutation. If a function previously performed by a structure that has been lost becomes advantageous once again, it must then be performed by another structure still in existence. The muscoid flies were able to return to blood feeding by penetration, because new structures evolved to perform the function that was once performed by the mandibles and the maxillae. Thus, several species have developed a series of teeth on the tip of a reduced labellum. In the stable fly, *Stomoxys*, the labium has become stiffened and bears two rows of prominent teeth on the labellum that is rapidly everted and retracted several times until the skin of the host has been punctured (Figure 6-7*D,E*). In the infamous tsetse fly, *Glossina*, the labellar teeth are more numerous and reduced in size (Figure 6-7*F*). This apparently permits the fly to penetrate the skin quickly and painlessly, thereby avoiding a disturbance reaction by the host.

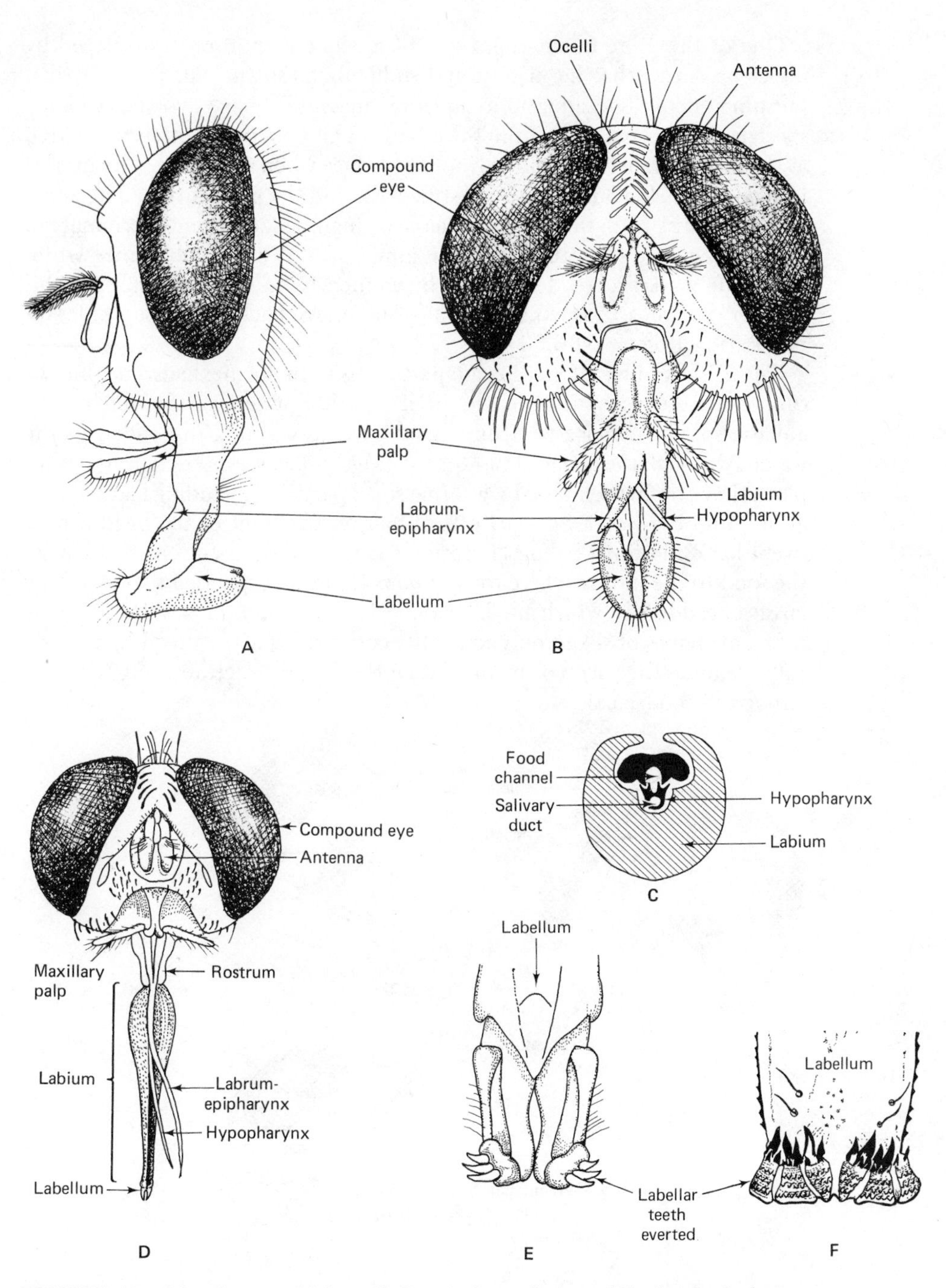


FIGURE 6-7. Mouthparts of Muscoid flies. **A.** Lateral view of the head of a house fly, **B.** Frontal view of the head of a house fly, **C.** Diagram of a transverse section through rostrum above the labella showing the normal position of the components, **D.** Frontal view of the head of the stable fly, *Stomoxys*, **E.** Enlargement of the labella of the stable fly showing the eversible teeth, **F.** Enlargement of the labella of the tsetse fly *Glossina* showing the more numerous fine eversible teeth.

One of the more fascinating types of feeding adaptation is displayed by insect predators that lie in wait and suddenly grasp their prey rather than pursuing them. The aquatic nymphs of dragonflies have evolved a highly specialized raptorial organ from the labium. This structure is often referred to as the **mask** because it is held in front of the face when not in use. The mask is formed by an elongation of the postmentum and the prementum. The labial palps that arise from the tip of the mask are highly modified and terminate in a pair of incurved sclerotized hooks (Figure 6-8*A*). When a prey comes within reach, the mask is extended rapidly by an increase in blood pressure. In this manner the prey is grasped by the palps and drawn back to the mandibles for maceration (Figure 6-8*B*,*C*).

The subject of insect feeding adaptations is virtually inexhaustible, but we cannot leave it without at least a brief mention of the way in which some aquatic species manage to utilize small particles of food suspended in the water. Although the larvae of mosquitoes have more or less generalized mouth parts, they obtain their food by filtering small particles, including bacteria and algae, from the water. Several tufts of hair on the front of the head can be swept back and forth by muscle action to create a water current that carries the food to the mouth where rows of hairs trap the suspended particles. The larvae of caddisflies, which also have typical chewing mouth parts, build many different shapes of silken nets across the current. They then feed by periodically cleaning their web or by devouring the entire structure along with the entrapped food particles.

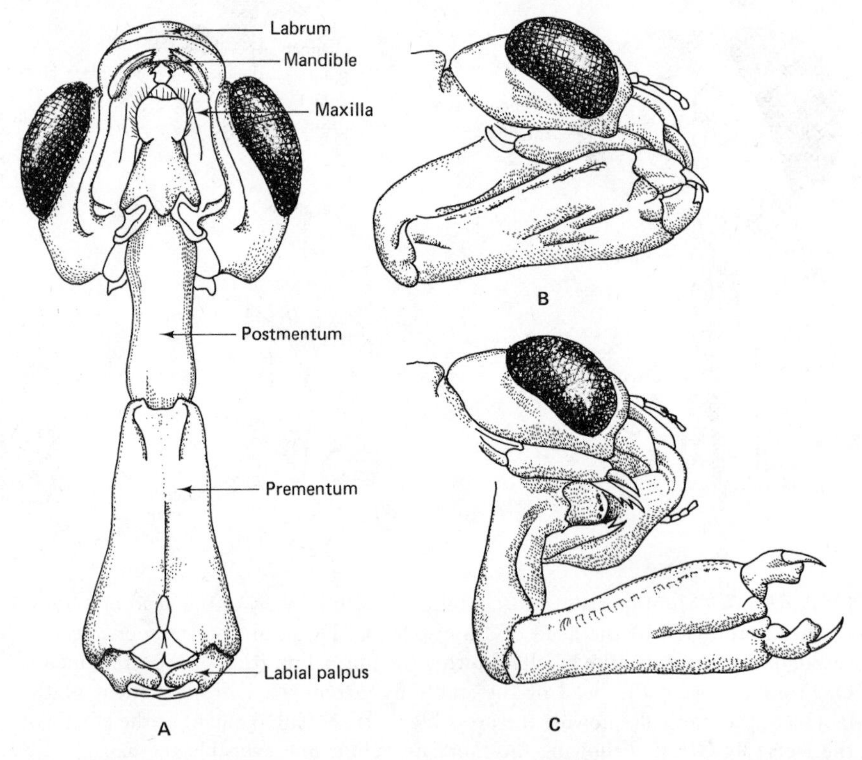


**FIGURE 6-8.** Mouthparts of a dragonfly larva. **A.** Ventral view of the head with the labium extended backward to show structural details, **B.** Head with the labial mask in the normal resting position, **C.** Head with the labial mask in partially extended capture position.

The foregoing examples provide an overview of some of the ways that insects have exploited the adaptability of the generalized arrangement of mouth parts to facilitate the use of a variety of types of food. There are many more that could be presented, but these should serve to illustrate that mouth part specialization is one of the reasons why the insects have become such a diverse and successful group. It also shows why they are both in conflict with us and of benefit to us in our attempts to manipulate various kinds of ecosystems. The key to the wide variety of feeding structures that the insects display lies in the fact that there are so many anatomical elements that can be modified, even to the point where functions lost when elements were lost can be regained through the further specialization of other elements. If the insects had begun with nothing more than a single pair of jaws, like a bird, for instance, their feeding opportunities would have been much more limited.

The feeding opportunities provided by the specialization of the various mouth part components simply enhanced the variety of foods that insects have been able to introduce into their alimentary tract. Without some associated modifications of the digestive organs, some of these foods could not have been utilized effectively. Although details of digestive physiology, excretion, and solid waste elimination are beyond the scope of this book, this is an appropriate place to consider a few of the more fascinating internal adaptations.

THE GASTRIC MILL. Even among some of the insects with generalized chewing mouth parts, there is a special structure that grinds the food before it passes into the stomach. This structure that lies between the crop and the stomach, or ventriculus, is called the **gastric mill.** This grinding organ consists of an arrangement of sclerotized teeth that project inward from the cuticular lining of the proventriculus. These teeth, which move back and forth because of contractions of the muscles in the wall of the proventriculus, can be seen readily by cutting open the fore-gut of a roach (see Figure 3-11).

THE FILTER CHAMBER. Insects such as the homopterans and some hemipterans have a problem with the highly dilute nature of the plant juice they feed upon. A large volume of food must be ingested so that enough nutrients for growth, development, and reproduction can be extracted from the sap. Modifications of the gut are needed to provide a means of eliminating the excess water taken in. If the entire volume of plant juice ingested passed through the part of the gut where digestion and absorption occurs, there would be a tremendous dilution of digestive enzymes, potential problems of absorption, and dilution of the hemolymph. These problems have been averted in some homopterans by a looping of the alimentary tract so that the anterior part of the hind-gut and the Malpighian tubules come in contact with the anterior portion of the mid-gut in a bladderlike structure called a **filter chamber** (Figure 6-9). Much of the water content of the plant juice passes directly from the anterior mid-gut to folds of the hind-gut by osmosis, thereby reducing the flow of water through the regions of the mid-gut where most of the digestion and absorption occurs. The water crosses over to the hind-gut and becomes mixed with the substances leaving the mid-gut. The unused nutrients (particularly sugars), waste, and excess water then pass to the exterior through the anus as a rather sweet liquid called **honey dew,** which is favored as food by other insects such as ants and bees.

FIGURE 6-9. Diagram of the filter chamber of a homopteran. (*Left.*) Gross anatomy of the digestive tract. (*Right.*) Transverse section of the filter chamber. (Modified from Chapman, 1969, after Snodgrass and Imms.)

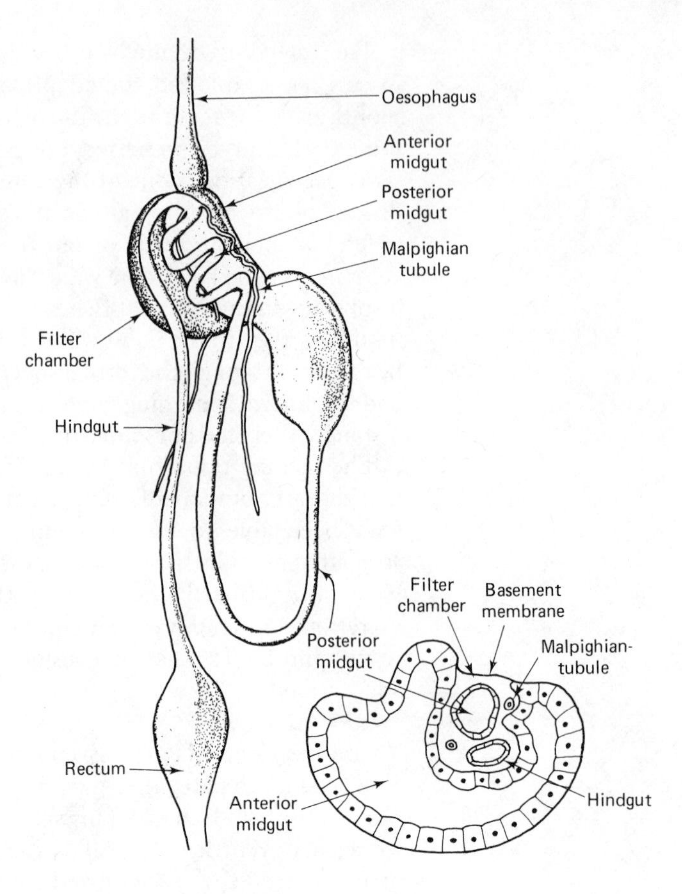


THE MALPIGHIAN TUBULES. Depending on the species, two to more than 250 slender, blind Malpighian tubules of complex histological structure arise from the digestive tract at the junction of the mid- and hind-gut. These tubules, which have a total surface area often in excess of 100,000 square millimeters, wind throughout much of the body cavity where they are bathed in hemolymph. In most insects, the distal ends of the tubules are free, but in many Lepidoptera and Coleoptera, the ends are imbedded in tissue surrounding the rectum. The Malpighian tubules are involved in the regulation of body water, salt balance, and the excretion of soluble wastes.

Insects that have a diet containing a large percentage of water or that live as aquatics must conserve certain inorganic ions and rid themselves of excess water that is either ingested or absorbed through the integument. Various molecules enter the tubules either by diffusion or active transport, and those substances that must be conserved are then resorbed by the rectum or certain regions of the tubules.

As you are now aware, however, most terrestrial insects do not have a problem with too much water, but instead must conserve it. Ammonia is the primary end product of amino acid metabolism and although highly toxic, ammonia can be excreted in large quantities by aquatic insects and some fly larvae that live in moist environments. But terrestrial insects cannot afford to use water to dilute ammonia, so much of their nitrogenous waste is converted to a harmless insoluble form, **uric acid**, which can be retained as a solid and passed out with the feces. The synthesis of uric acid occurs in several ways but frequently involves an input of adenosine-triphosphate (ATP) and therefore has an energy cost. This appears to be a satisfactory trade-off for organisms that can afford to expend energy better than water. Uric acid may be found in many insect tissues, particularly in the fat body where it occurs in **urate cells** (Wigglesworth, 1965). Most of the uric acid passes into the Malpighian tubules along with other constituents of the hemolymph. The mechanism is not fully understood, but it may be linked to the movement of potassium. A continuous flow of water down the Malpighian tubules carries the uric acid into the rectum where the excess water is resorbed. The uric acid then crystallizes out of solution and can be retained as a nontoxic solid to be discharged with the fecal matter.

Interestingly, the larvae of parasitic and colonial hymenoptera have incomplete digestive tracts and store their uric acid waste until just prior to pupation. This adaptation prevents the discharge of toxic wastes into the body of the host or into the cells of the nest where the larvae are confined during development.

SYMBIOTIC MICRO-ORGANISMS. The insects as a group possess an amazing array of enzymes for breaking down their food into usable nutrients, but some still have to employ outside help from micro-organisms. In some insects, these **symbionts** are free in the lumen of the gut but often occur in cells called **mycetocytes** that are frequently aggregated to form organlike **mycetomes.** The gastric caecae also harbour micro-organisms, particularly bacteria, and some fungal spores are stored in special integumentary pouches.

Although many insects feed on plant tissue, only a limited number of wood-boring beetles and wasps produce **cellulase** for the breakdown of cellulose. The rest either derive their nutrients from the content of the plant cells or rely on symbionts to hydrolyse the cellulose. In a few beetles, cellulose-fermenting bacteria provide the breakdown, whereas in wood-eating cockroaches and termites, intestinal flagellates provide the assistance. The flagellate *Trychonympha* occurs in large numbers in the hind-gut of the common termite *Zootermopsis.* These protozoans, which can be observed readily by making a slide from fecal matter squeezed from the termite's anus, phagocytize particles of wood that they then digest, releasing certain usable nutrients into the termite's gut. Since the flagellates inhabit the hind-gut, which is lined with cuticle, they are lost each time the termite molts. However, the newly molted individuals are soon recolonized with flagellates as a result of feeding on feces that contain them.

In addition to being involved in the digestion of cellose, micro-organisms have also been shown to provide essential nutrients. Apparently the only vitamins required by insects are the water-soluble B vitamins, but these are not always available in sufficient quantity directly from the diet. For example, blood contains B vitamins in amounts too small to meet the requirements of insects such as sucking lice and blood-sucking hemipterans that subsist entirely on blood throughout their life. Experiments have demonstrated that the vitamin B deficiency for these insects is made up by symbionts. In the blood-feeding bug *Rhodnius* a micro-organism called *Actinomyces rhodnii* that inhabits the alimentary tract can be eliminated by surface sterilization of the eggs. If the young *Rhodnius* from these eggs are then reared free of their symbionts, their growth will stop unless they are fed on blood to which B vitamins have been added (Wigglesworth, 1965).

The ambrosia beetles (family Scolytidae) feed on fungi that they culture in tunnels constructed in woody plants. Since the adults disperse between generations, they must carry the fungus spores with them to their new host plants. The spores are carried in special invaginations of the integument called **mycangia** (Figure 6-10). Bark beetles belonging to the same family similarly transport the spores of a fungus that infects the conductive tissue of their host trees. As the fungus grows, it reduces the flow of resin that is a main part of the plant's defensive chemistry, and the survival of the beetles is enhanced.

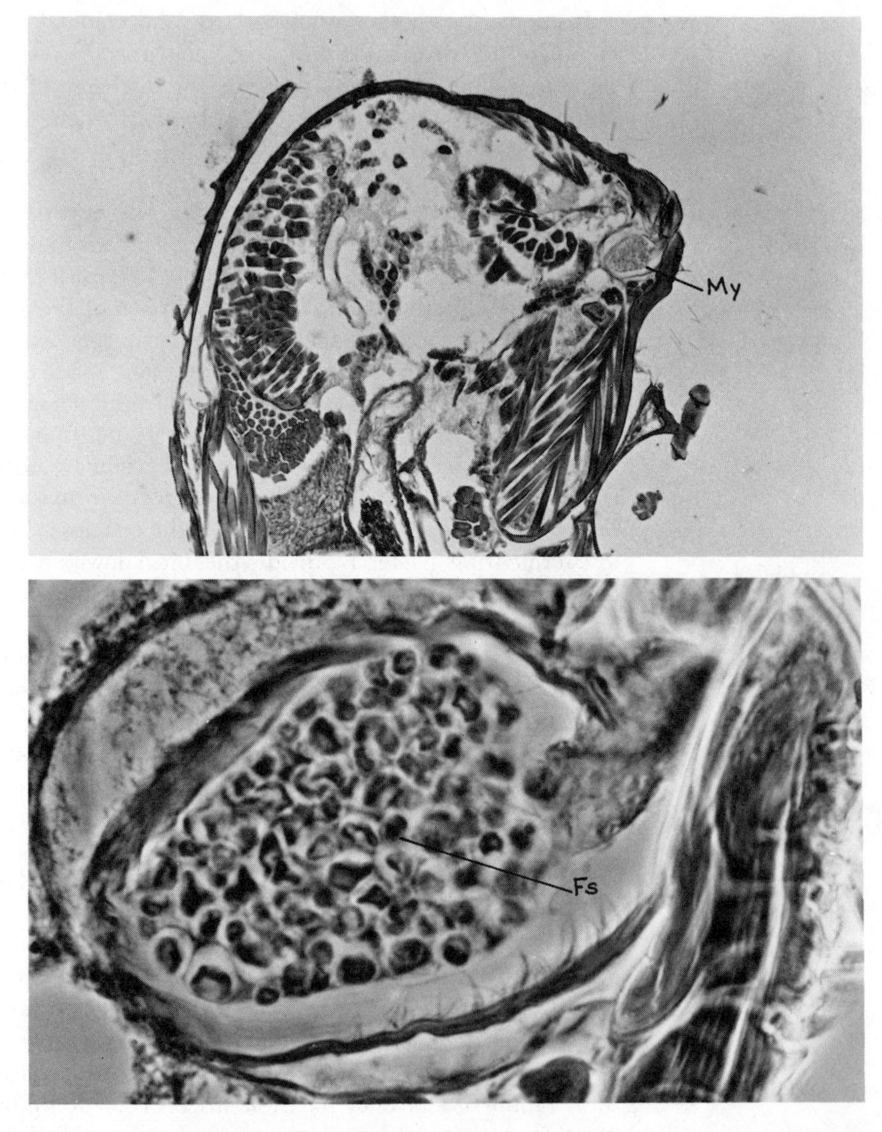


**FIGURE 6-10.** Mycangium of a scolytid beetle. (*Top.*) Section through the head showing a mycangium (My) near the mouth. (*Bottom.*) Enlargement of the mycangium showing fungus spores (Fs) within. (Photomicrographs courtesy of Canadian Forestry Service, Victoria, B. C.)

## References Cited

WIGGLESWORTH, V. B. 1964. The life of insects. The New American Library, Inc., New York, N. Y.

WIGGLESWORTH, V. B. 1965. The principles of insect physiology, 6th ed. Methuen & Co., Ltd., London

## General References

MATSUDA, Ryuichi. 1965. Morphology and evolution of the insect head. *Mem. Amer. Entomol. Inst.* 4. 334 pp.

SNODGRASS, R. E. 1935. Principles of insect morphology. McGraw-Hill Book Company, New York, N. Y.

# Legs, wings, and locomotion

# 7

Although some animals are much more motile than others and many appear to be sessile, locomotion is necessary in at least one stage of an animal's life history in order that it can disperse to colonized new areas and assure a reasonable mixing of genetic material. As already noted, the insects as a group of terrestrial animals took a giant step forward when they evolved an efficient locomotory center in the form of the thorax. This development combined with a modified mode of leg articulation permitted them to raise their bodies off the ground and to move with much greater agility on land and then later, in the water, when they invaded it. Subsequently, the boxlike structure of the thorax led to the evolution of the flight mechanism and a completely new means of locomotion. As a result the insects were better able to move about in response to changing conditions and to the opening up of new habitats than was any other group of animals. In addition, they were able to spread beneficial genetic changes quickly and efficiently. Consequently, the insects spread rapidly from their centers of origin and managed to keep pace with climatic and related biotic changes that occurred over the earth during their long evolutionary history. Often they became established in new situations before other groups had the opportunity. The insects today are no less capable of both local and widespread movement. Frequently they are the first animal invaders of new habitats, and their locomotor ability is often a source of frustration in man's attempts to manage both beneficial and harmful species.

Animal locomotion involves a complex interplay of the organs that perceive various kinds of external and internal stimuli. These organs are the afferent (sensory) nerves that transmit information by way of the central nervous system to the efferent (motor) nerves and the muscles that contract when stimulated to affect the movement of the structures to which they are attached. Details of nervous and muscle physiology will not be discussed because they are treated adequately in most general biology texts. Instead, we will consider the basic movement and adaptations of the legs and wings and briefly review the function of a few accessory locomotor structures.

## *Terrestrial Locomotion*

Even though the structure of the thorax permitted the early evolution of the flight mechanism, the legs are still of great basic importance, and their use in walking or jumping is a characteristic of almost all adult insects as well as of many immature forms. Among apterygotes and those species that have secondarily lost their wings, the legs are the principal locomotor organs; even among some winged groups like the earwigs, walking seems to be preferred to flight.

As noted in Chapter 3, the generalized insect leg consists of a series of segments with joints between them. The proximal segment, the coxa, is attached by an area of membrane in the coxal cavity. The freedom and the plane of movement of the coxa relative to the body are dependent upon the number and position of the points of articulation (Figure 7-1). In most insects, the trochanter and femur are fused, and the trochanter articulates with the coxa by either one (**monochondylic**) or two (**dichondylic**) projections.

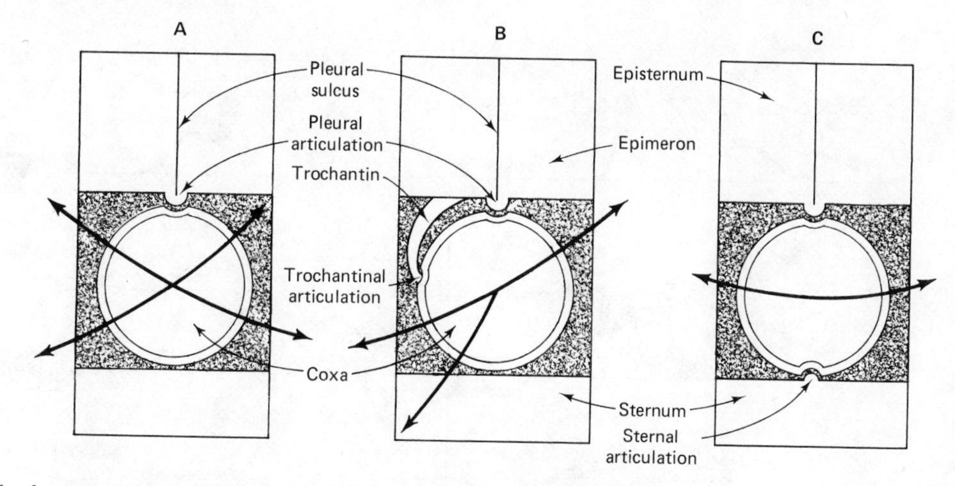


FIGURE 7-1. The basic patterns of articulation between the body and the coxa. Arrows indicate the main planes of movement permitted by each type of articulation. (Adapted from Chapman, 1969.)

The femur articulates with the tibia by way of a dichondylic joint similar to a human knee joint (Figure 7-2). The tibia is then jointed to the tarsus, which usually consists of several tarsomeres, the last of which often terminates in a pair of claws (see Figure 3-7). This arrangement of joints enables the combination of leg segments to move in all planes relative to the body.

The muscles that produce the leg movements lie both outside and within the legs. The coxa is moved by a complex arrangement of muscles that arise in the thorax and attach to the proximal rim of the coxa. The other leg segments are moved by antagonistic muscle pairs that arise in one segment and attach by tendons to a distal segment. Although muscles connect the tibia and the first tarsomere, there are no muscles that run between the tarsomeres or to the pretarsus. The pretarsus is depressed by the contraction of a small muscle in the tibia that pulls on a long tendon; pretarsal levation occurs by elasticity.

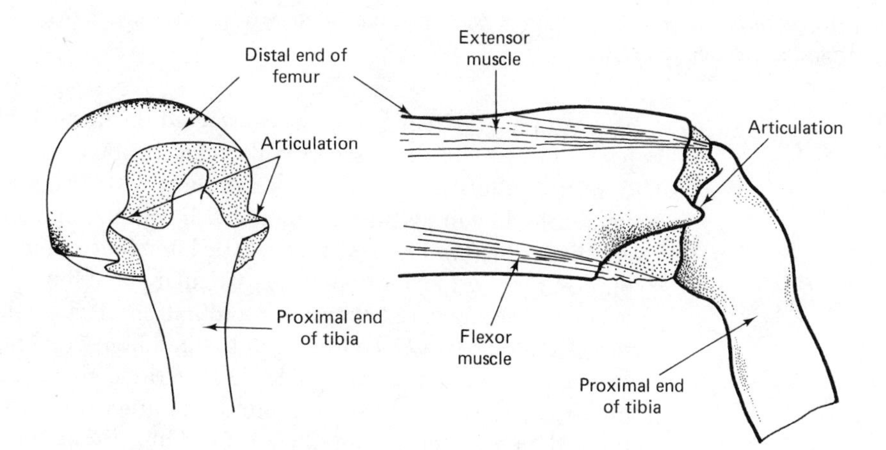


FIGURE 7-2. Diagram of the knee-like joint between the femur and tibia. (Adapted from Snodgrass, 1935.)

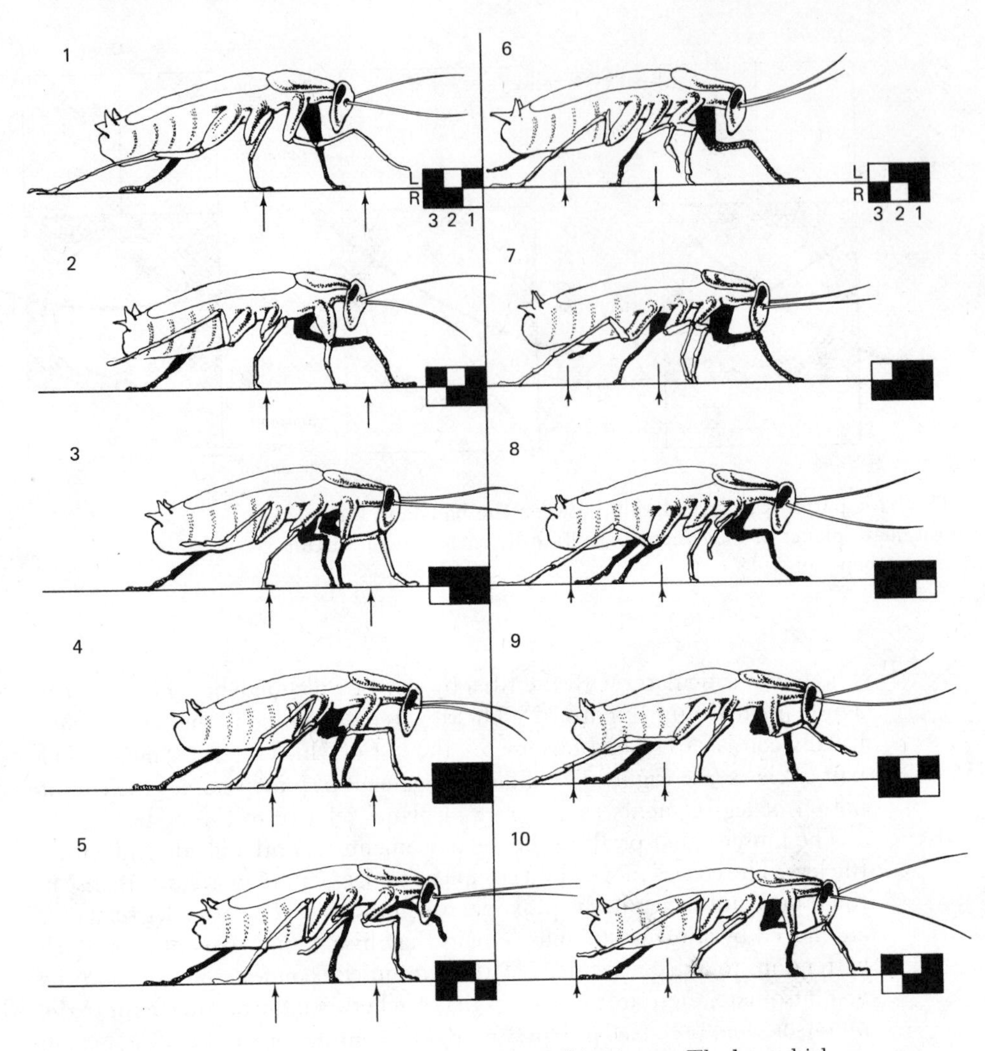


FIGURE 7-3. Basic pattern of leg movement of a walking insect. The legs which are off the ground during each phase are indicated by white squares in the diagram in the corner of each drawing. The shaded legs are on the far side of the body. (From Hughs, 1952, *Terrestrial Locomotion*. Modified and used with permission of the author and Academic Press Inc.)

WALKING. Walking could be considered the normal means of locomotion for all insects that have structurally typical thoracic legs. These include all terrestrial adults and the immatures of those with incomplete metamorphosis (nymphs). The immatures of insects with complete metamorphosis move by crawling, which will be described a little later. Since there are typically six legs on the thorax, locomotion in a particular direction at a particular speed becomes a complicated matter of coordination. In general, walking insects are supported by a tripod, consisting of the middle leg on one side and the front and hind legs on the other side. While these three legs are planted and pushing backward against the ground, the other three are raised and moved forward in preparation for their turn. This alternation of moving tripods results in a slightly zig-zagged path about a line describing the direction of travel (Figure 7-3). The alternating tripod gait allows an insect to stop at any time because its center of gravity is always within the area of the triangle of support on the substrate. When one leg is amputated, there is little change in the gait. If two legs are removed, the unstable walk that one would expect does not occur. Instead, the walking pattern is typical of a tetrapod. If the mesothoracic legs are removed, for example, the pattern becomes one of diagonal pairs. Apparently the removal of the legs results in the loss of sensory signals that are normally provided by the receptors they bear. The new gait results from the sensory signals provided by the remaining legs.

Not all adults and nymphs use six legs for walking. The adults of the family Nymphalidae are often referred to as the four-footed butterflies because they typically walk on their mesothoracic and metathoracic legs; the forelegs are reduced in size and are usually folded against the thorax. In several other groups including the mantids and ambush bugs, the forelegs are modified for capturing prey, with the middle and hind legs serving for locomotion.

Although walking or running insects often appear to be moving at great speed, that is largely an illusion because of their small size. This is clearly illustrated by the fact that you can walk alongside an apparently fast-moving insect and keep pace with it quite easily. The maximum walking speed of insects is about three miles per hour. When they walk slowly, some insects depart from the pattern of alternating tripods and move the legs sequentially as follows: right front, left middle, right rear, left front, right middle, left rear, and so on. For additional descriptions, the interested student should refer to Hughs (1965) or Wilson (1966).

For an animal to walk, it is necessary that pressure be applied to the substrate such that there is some force directed in opposition to the direction of travel. The better the grip on the substrate, the more efficient the locomotion, as less force is dispensed by slipping. If you have ever attempted to walk on ice or up a slippery incline, you can attest to this. Insects seem particularly adept at walking on a variety of surfaces, even steep inclines. When the walking surface is rough, the pretarsal claws provide an ample grip, but gripping by the claws alone will not permit an insect to traverse waxy plant leaves or smooth window panes unless either is coated with a film of dirt. To enable them to walk on such slick surfaces, the insects have evolved accessory adhesive structures, the tarsal pads called the **arolium** or **pullvillus** (Figure 7-4). These pads are either coated with a sticky secretion produced by

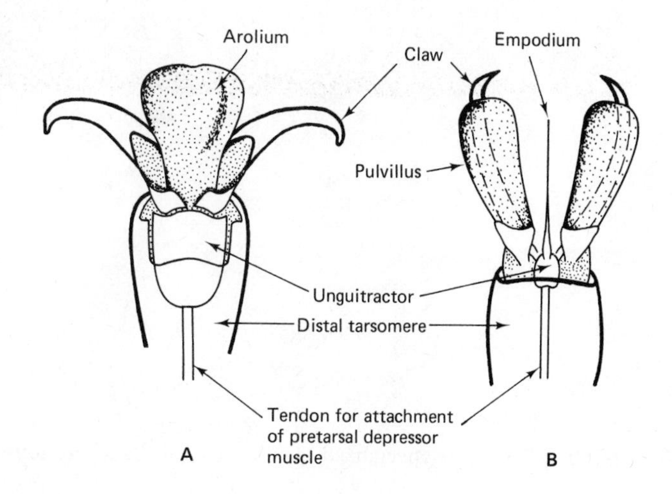


**FIGURE 7-4.** Structural details of the insect pretarsus showing two common variations. **A.** Cockroach, **B.** fly. (Adapted from Snodgrass, 1935.)

gland cells in the pad or are covered with a dense mat of special adhesive hairs. The bug *Rhodnius* has at the tip of each tibia a special climbing organ consisting of a hairy pad that functions like the pulvilli. In the lice, the tarsae and the claws are modified for gripping their host. The sucking lice have an effective adaptation comprised of single-segmented tarsi, each terminating in a powerful claw that combines with a tibial thumblike process to grip around the shafts of the host's hair (Figure 7-5).

CRAWLING. The larvae of many insects that display complete metamorphosis are soft-bodied and therefore do not have a skeleton that can provide for the attachment of the powerful muscles usually associated with the movement of the thoracic legs. In some, notably the active beetle larvae, the thorax is relatively well sclerotized and permits a fairly rapid movement similar to the walking of adults but with the body more or less in contact with the substrate. In sawfly larvae and caterpillars, the thoracic legs are small and are aided by several pair of fleshy prolegs on the abdomen. Thin bands of muscle underlie the flexible exoskeleton and keep the body turgid so that the pressure of the hemolymph within provides a hydrostatic skeleton similar to that found in some burrowing (polychaet) worms. Since the fluid hemolymph cannot be compressed, a change in the shape of the body caused by muscle contraction in one part of the body must result in a compensatory change in shape somewhere else. For example, if the muscles in one segment are contracted to reduce its size, the hemolymph is forced to an adjacent segment that can increase in size while the muscles are relaxed. These changes in the

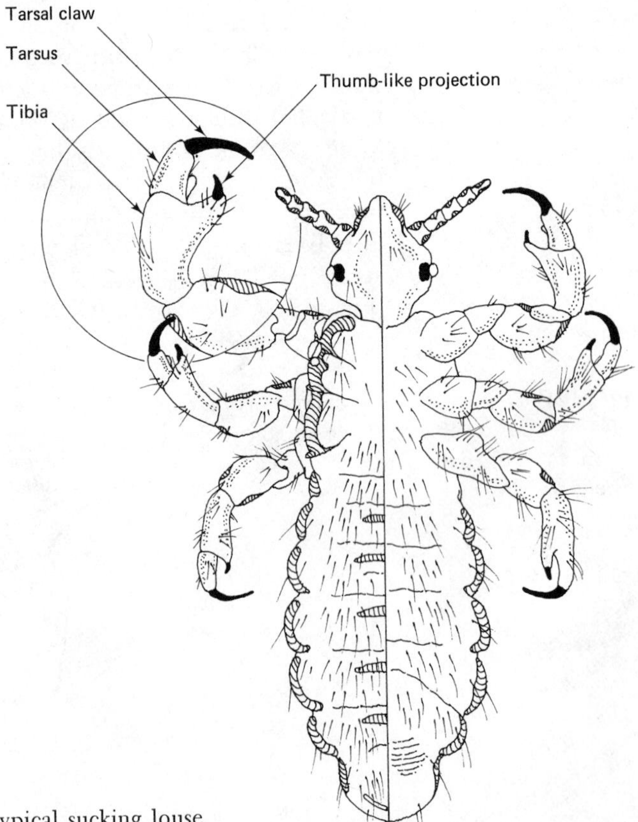


**FIGURE 7-5.** The specialized claw of a typical sucking louse.

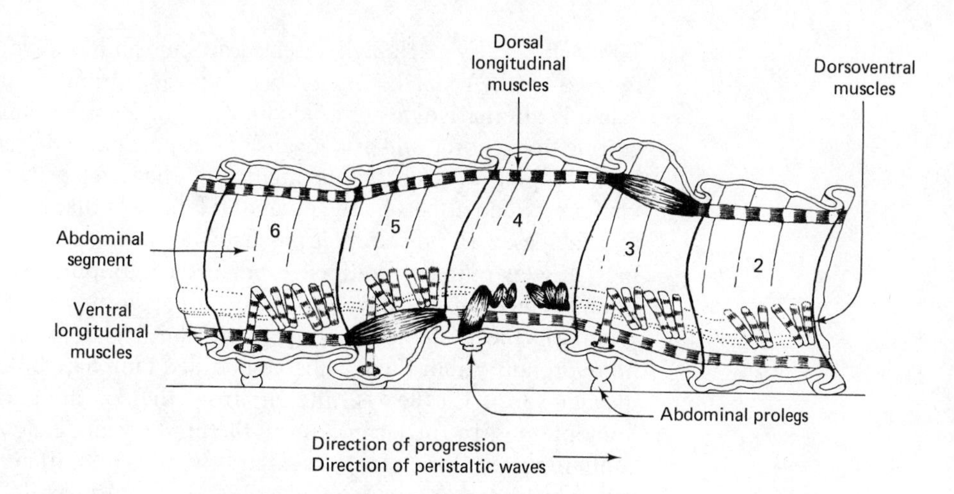


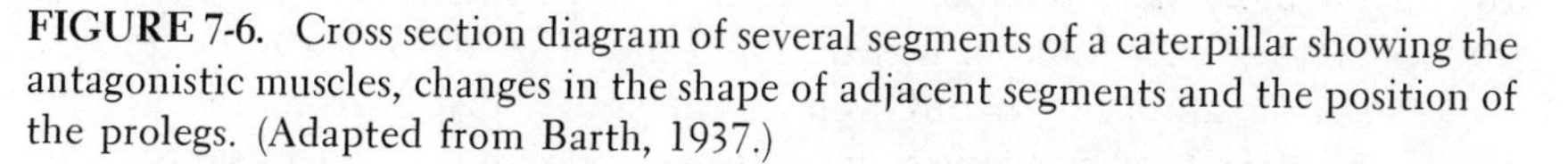

**FIGURE 7-6.** Cross section diagram of several segments of a caterpillar showing the antagonistic muscles, changes in the shape of adjacent segments and the position of the prolegs. (Adapted from Barth, 1937.)

shape of segments can be coordinated to produce undulations of the body as well as the withdrawal or extension of the prolegs.

When a caterpillar crawls, the longitudinal and dorsoventral muscles oppose each other to produce a wave of contraction that passes anteriorly from segment to segment. Since the prolegs can be anchored to the substrate by small hooks on their undersurface, the wave of contraction slowly moves the body forward. As shown in Figure 7-6, three adjacent proleg-bearing segments are in different stages of shape change, resulting in the prolegs being alternately planted, raised, and replanted in a forward progression.

In the loopers (larvae of geometrid moths), the number of segments bearing prolegs is reduced. When crawling, the body is anchored anteriorly by the short thoracic legs, and the abdomen is drawn up in a loop. After the posteriorly located prolegs are anchored, the front of the body is extended forward. With each "step," the larva moves forward a distance equivalent to the distance between the hind thoracic legs and the front pair of prolegs (Figure 7-7).

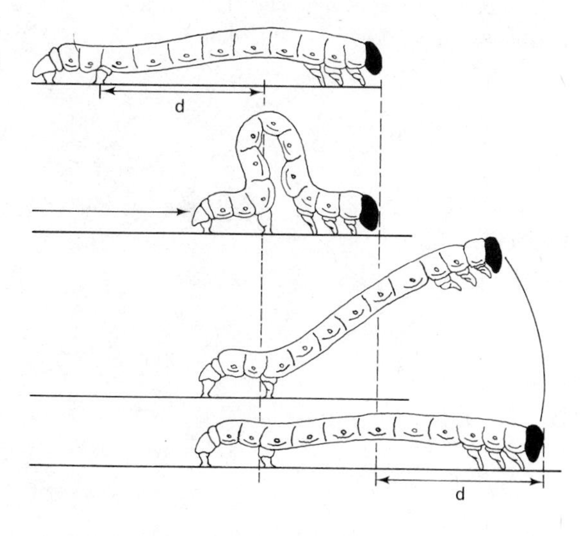


**FIGURE 7-7.** The method of crawling used by caterpillars known as inch worms or loopers; *d* equals distance travelled with each step.

JUMPING. Even though insects can run and change direction quickly relative to their small size, for many species, jumping has proved to be a valuable alternative means of locomotion for regular changes in location, for getting on to hosts, and as a means of escape. Some workers have postulated that jumping evolved before flight and that one of the early advantages conferred upon insects by the paranotal lobes was that they provided stability and increased the distance jumped by providing gliding planes. Although most insects employ modified hind legs in jumping, some rather unusual jumping mechanisms have evolved in a few groups.

Among the best-known jumpers are the grasshoppers, fleas, and certain homopterans. All of these utilized modified hind legs, but, as we will see, the mechanism of the flea is different from that of the grasshoppers and the homopterans. In the latter groups, the hind femora are greatly enlarged and contain powerful muscles that extend the tibia and others that flex them. In the resting position, the femora are held backward parallel to the sides of the abdomen, the "knee joint" is flexed so that the tibiae lie beneath the femora, and the tarsae are firmly planted on the substrate. Just prior to a jump, the tibiae are flexed tight against the femora. This draws the tarsae off the substrate. The flexor and extensor muscles are actually activated simultaneously. Since the flexor muscles have a mechanical advantage over the extensor muscles, the latter develop maximal tension. When the flexor muscles relax, the contracted extensors drive the tibiae against the substrate, and the grasshopper leaps upward (Figure 7-8). If it were not for the advantage of the flexor muscles, the insect would be off the substrate before the extensors developed their maximum tension, and the length of the jump would be reduced considerably (Heitler, 1974). Instead, grasshoppers can jump distances in excess of 20 times their body length.

A study by Rothschild and others (1973) provided an insight into the reasons for the astounding ability of some fleas to jump distances equal to more than 100 times their body length. A pad of resilin (the elastic material normally found in the hinge mechanism at the base of the wings) is present at the top of each metapleural ridge. This pad is compressed when the hind

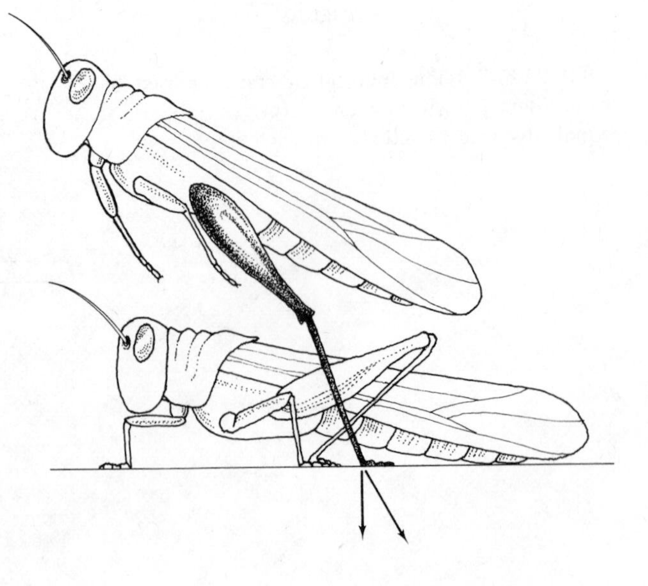

**FIGURE 7-8.** The jumping mechanism of a grasshopper. (From Chapman, 1969, with permission of Elsevier, North-Holland, Inc.)

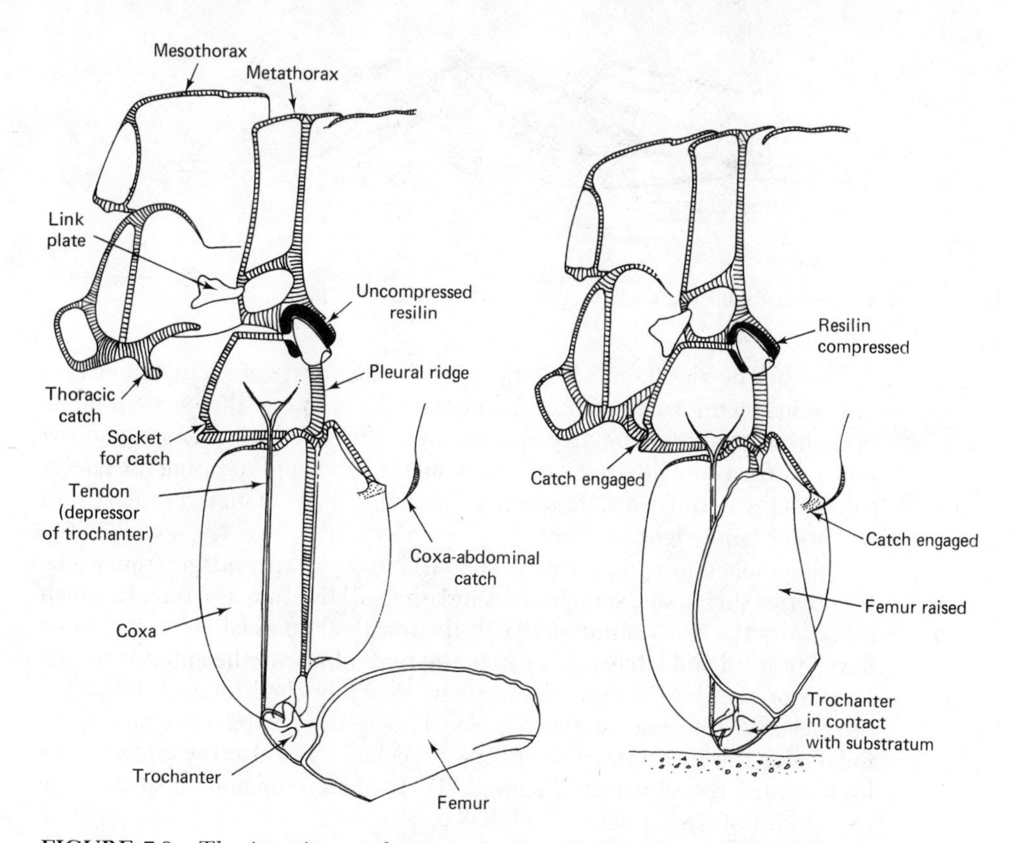


**FIGURE 7-9.** The jumping mechanism of a flea. (*Left.*) The mechanism in the "uncocked" position. (*Right.*) The mechanism "cocked" in preparation for a "flying leap." (Adapted from Rothschild and others, 1973.)

femora are raised to a vertical position alongside the large coxae. When the hind legs are in this position, the trochanters are pressed against the substrate (Figure 7-9). When the resilin pad is compressed, two pair of catches lock the jumping assembly in position. When the catches are released by muscle movement, the energy stored in the compressed resilin drives the trochanters against the substrate, thereby providing the initial acceleration for the jump. As the flea leaves the surface, a thrust of the tibiae against the surface increases the acceleration.

Not all fleas have maintained the ability to jump, and in considering this matter Rothschild and her associates state:

> Certain species of fleas have secondarily lost the resilin in the pleural arch and hence their flying leap, but some of them are still capable of feeble jumping; others are reduced to walking and crawling. One group of mole fleas, which live in underground nests, have lost the pleural ridge as well as the arch. Various bat fleas, which hatch in guano on the floors of caves and crawl up the walls to reach their roosting hosts, often have reduced arches or lack them altogether. Fleas parasitizing arboreal rodents or birds such as swallows, which build high aerial nests, tend to lose the resilin in their pleural arch. It will be readily appreciated that speculative jumpers in those situations would be rapidly eliminated from the breeding population. On the other hand, fleas that are associated with large hosts, particularly those without well-circumscribed nests or lairs, such as deer or sheep, the ubiquitous cat or man himself, have the largest pleural arch. Broadly speaking, the larger the host, the higher the jump and the larger the cap of resilin—a relation that serves to emphasize the importance of the flea's initial jump onto a passing animal.

FIGURE 7-10. A click beetle on its back about to right itself with a body flip.

The click beetles (family Elataridae) have an elongate prosternal spine that can be inserted into a notch in the mesosternum. When the beetle is on its back, it arches itself, compresses a pad of resilin, and locks the spine in the notch. It then straightens itself, unlocking the catch and releasing the energy stored in the resilin, which results in a flip (Figure 7-10). This is repeated until the insect lands right side up.

Collembolans jump by means of special abdominal appendages from which they derive their name **springtails.** A fork-shaped structure, the **furcula,** which arises from the rear ventral surface of the fourth abdominal segment, can be flexed forward and latched in a catch, the **tenaculum,** on the anterior ventral surface of the third abdominal segment. When latched forward, tension is produced at the base of the furcula. This tension is released rapidly by unlatching the apparatus. The insect is propelled forward as the furcula snaps back against the substrate (Figure 7-11). In some semiaquatic species, the forked end of the furcula is paddlelike.

DIGGING. Insects in several different orders spend part of the life history in the soil where they feed on organic debris. To aid in digging underground, the legs are often shortened and modified to form hard toothed surfaces used for scraping away the dirt. The limb modifications of two digging (fossorial) insects are shown in Figure 7-12.

### *Aquatic Locomotion*

Many insects, though primitively terrestrial, have returned to the water for part of their life history. Some that live along the water's edge make brief excursions into the water or across its surface using basically terrestrial means of locomotion. But many are clearly adapted for spending some significant part of their lives in different kinds of aquatic environments where rather different kinds of locomotion are a necessity.

LOCOMOTION ON THE SURFACE. At the interface of water and air, the water molecules become tightly packed because of a phenomenon known as cohesion and produce **surface tension.** This results in a thin flexible film capable of supporting a limited mass under certain conditions. The basic requirements of support of an object on the surface film are that its weight be spread over an area of the film and that the surface of the object does not have an affinity for water. If the surface of the object attracts water, the film is disrupted and the object slips through the opening and sinks. Since insect cuticle is covered with a layer of wax that expels water, an insect normally can rest on the surface of the film. Removal or coating of the wax with a wetting agent such as soap will disrupt the surface film and cause the insect to sink. The effect of weight distribution can be demonstrated by carefully placing a

**FIGURE 7-11.** The jumping mechanism of a springtail. **A.** Furcula cocked in preparation for jump, **B.** Released furcula thrusting down against the substrate, **C.** Furcula fully extended after the jump.

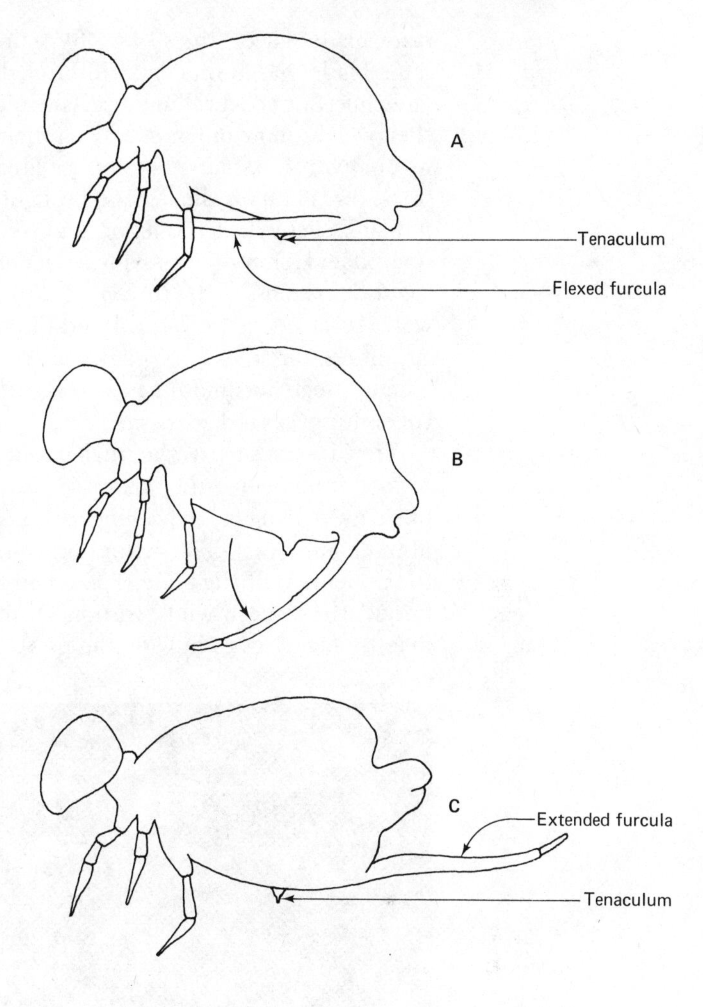


**FIGURE 7-12.** Prothoracic legs adapted for digging. **A.** mole cricket, **B.** cicada larva (nymph).

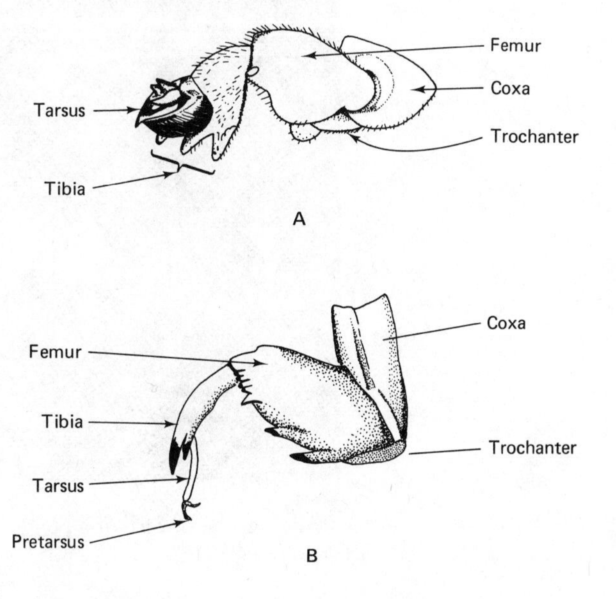

razor blade flat on the surface film, in which case it will be supported; the same blade placed on edge will immediately break through the surface film and sink. Surface-dwelling insects exploit these phenomena by distributing their weight more or less evenly on their tarsae, which are covered with either wax or hydrofuge hairs. The surface film can thus support the weight over any single leg, so unless the legs are brought together in one spot, the insect does not sink even when the legs are moved about. The common water striders (Gerridae: Hemiptera) provide a good example of this. If one looks closely, a slight depression of the surface film can be seen where each leg contacts the water surface (Figure 7-13). In addition, the pretarsal claws arise before the tips of the tarsae and are bent backward to prevent penetration of the film. These insects locomote across the surface as though on ice, synchronously thrusting backward with their legs in a striding motion.

A few insects such as the small elmid beetles are able to exploit the surface tension from below and walk upside down on the underside of the film, much like a fly walking on the ceiling. But not all surface dwellers require that the film remain intact. The gyrinid or whirligig beetles swim at the surface with the upper part of the body exposed and their lower surface submerged. They frequently swim in wild gyrations, using highly modified mesothoracic and metathoracic legs to provide the power. The coxae of these legs are immova-

**FIGURE 7-13.** A water strider on the surface of a pond showing the depression of the surface film beneath the tarsae.

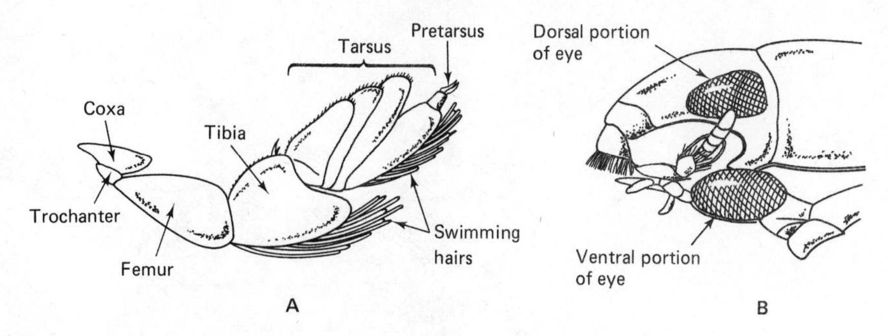


**FIGURE 7-14.** Adaptations for life at the surface of the water as illustrated by the whirligig beetles (Gyrinidae). **A.** The highly modified hind leg used in swimming. **B.** The division of the compound eyes into distinct dorsal and ventral groups of lenses which permit the scanning of both the aerial and aquatic environments simultaneously. (Redrawn from Leech and Chandler, 1968.)

bly fused to the thorax, and the tibiae and tarsae are flattened to form paddles (Figure 7-14*A*). The compound eyes are split into two pair so that one pair can scan the aerial environment above and one pair can scan the aquatic environment below (Figure 7-14*B*).

LOCOMOTION BENEATH THE SURFACE. Subsurface locomotion by insects is highly variable, ranging from bottom dwellers that walk about much like their terrestrial relatives to those that swim and dive with great proficiency. Both walking and swimming underwater are frequently accommodated by modifications of the legs or the employment of accessory locomotor structures.

The immatures of dragonflies spend much of their time walking across the substrate but periodically make sudden movements for escape or pursuit by means of "jet propulsion." This is accomplished by means of an **anal gill**, consisting of a convoluted pouch in the tip of the abdomen, surrounded by a tracheal basket. Water drawn into the pouch for gas exchange is compressed by a contraction of the abdominal muscles while the anus is closed. The anus is then opened slightly, resulting in the release of a fine jet of water that propels the animal forward at 30 to 50 centimeters per second (Hughs, 1958).

The form of the body and the legs of the bottom-dwelling insects are variable depending upon the nature of the environment. Those that occupy trashy environments are often slender and have long stiltlike legs that aid their movement through entangled debris. On the other hand, those that live in running water are often short-legged, have flattened bodies that are held against the substrate by water pressure, and are thus protected against being swept away by the current. Others anchor themselves by suckers or on a silken life line. Still others, such as the larvae of mayflies, apparently supplement their walking on the bottom by moving their abdominal gills (Figure 7-15) as paddles.

Numerous aquatic insects, particularly members of the orders Hemiptera and Coleoptera, are proficient swimmers. This ecological group provides a beautiful illustration of convergent evolution in that they are almost all streamlined, dorsoventrally flattened or keel-shaped, and have one or more pair of legs modified to form swimming appendages. Many of these insects combine their swimming ability with bouyancy to produce a characteristic

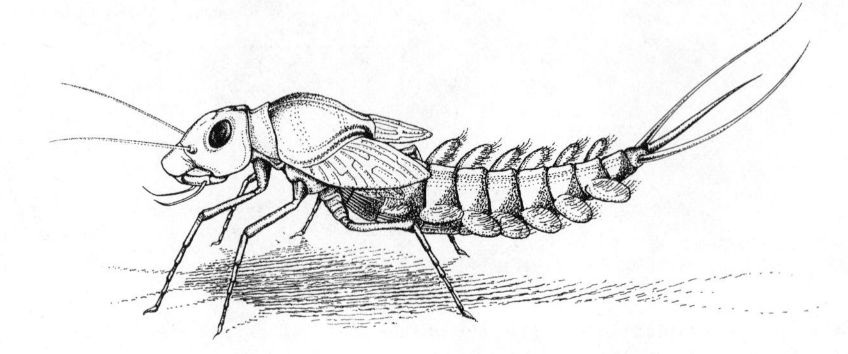

**FIGURE 7-15.** A mayfly immature showing the lateral abdominal gills that may serve as paddles used as a locomotor aid.

pattern of diving downward quite rapidly and then drifting to the surface by floatation. This can be observed readily by placing backswimmers (Notonectidae: Hemiptera) in an aquarium. As their name suggests, these aquatic bugs swim upside down, their folded wings forming a shape like the hull of a boat. The hind legs, which are long and fringed with hairs (Figure 7-16), propel the insect downward by simultaneous thrusts. If the insect does not attach to a submerged object, it floats back to the surface.

The predaceous diving beetles (Dytiscidae) are among the strongest swimmers. Most of their power comes from the modified hind legs that are fringed with dense rows of movable hairs. As the legs are thrust backward, the hairs fan out, increasing the surface pushing against the water. On the return stroke, the tibiae rotate so that the fringed margins become anterior and posterior instead of dorsal and ventral, and the hairs fold back, reducing the resistance (Figure 7-17). In some genera, the first three tarsomeres of the

**FIGURE 7-16.** Dorsal view of a backswimmer (Notonectidae) showing keel-like shape provided by the folded wings, and the fringed hind legs used as oars.

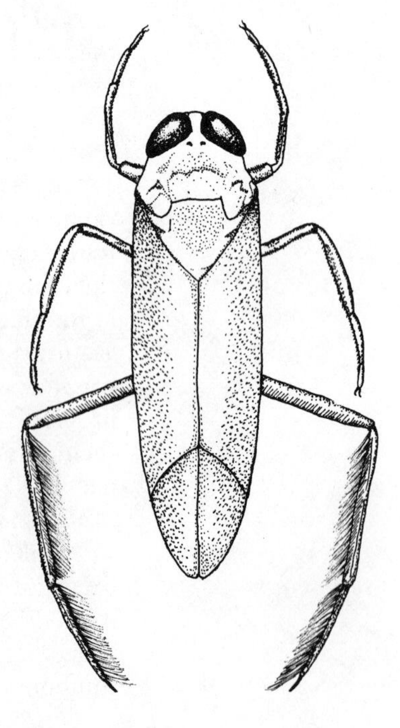

FIGURE 7-17. Ventral view of a diving water beetle (Dytiscidae) showing the progressive positions of the oar-like hind legs on the backstroke (1–4) and the forward-stroke (5–8). Front and middle legs removed. The small diagrams on each side show the position of the fringe on hairs that results from rotation of the tibiae. (Modified from Nachtigall, 1965, Swimming of aquatic insects. Used with permission of the author and Academic Press Inc.

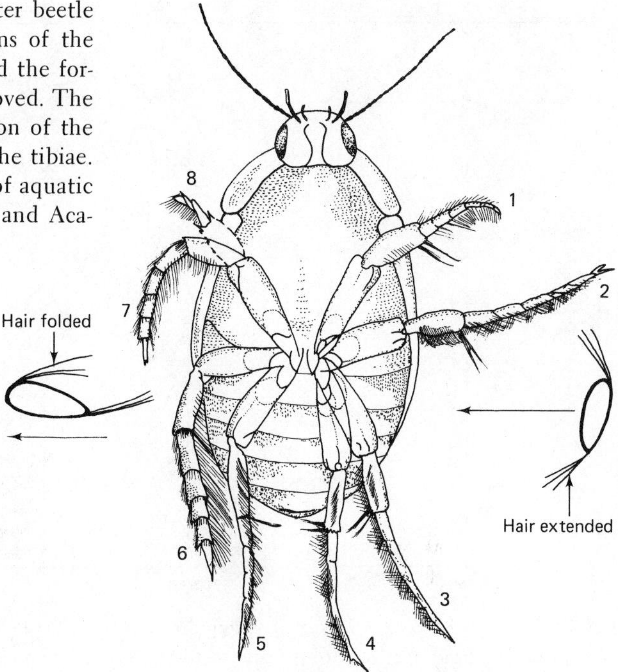


forelegs of males are enlarged and equipped with suckers used for grasping the female during mating (Figure 7-18).

Probably the most bizarre method of swimming occurs among a few Hymenoptera that are parasitic on aquatic insects and among some aquatic Lepidoptera. These insects move through the water by flapping their wings similar to the manner in which penguins swim.

There are numerous other variations on the locomotory themes mentioned for movement in the aquatic environment, but those presented include the basic modes, with the exception of the wiggling movement displayed by soft-bodied larvae. These animals propel themselves by a twitching and undulating movement of their bodies, which they frequently combine with buoyancy such that they wiggle downward and float up. This mode of

FIGURE 7-18. Foreleg of a male diving water beetle (Dytiscidae) showing the modification of the tarsae to form a sucker like structure used to grip the female while mating during swimming.

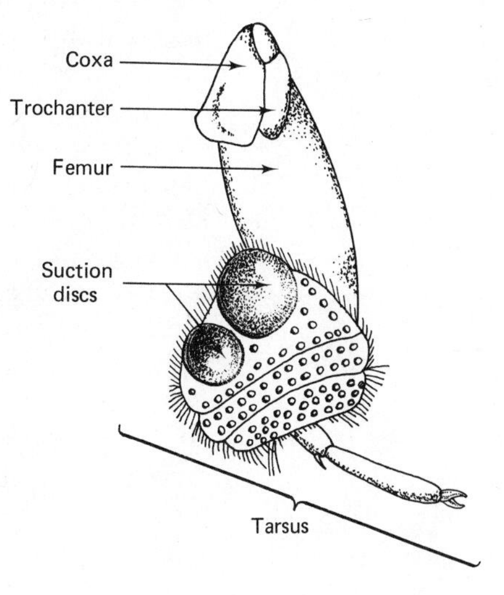

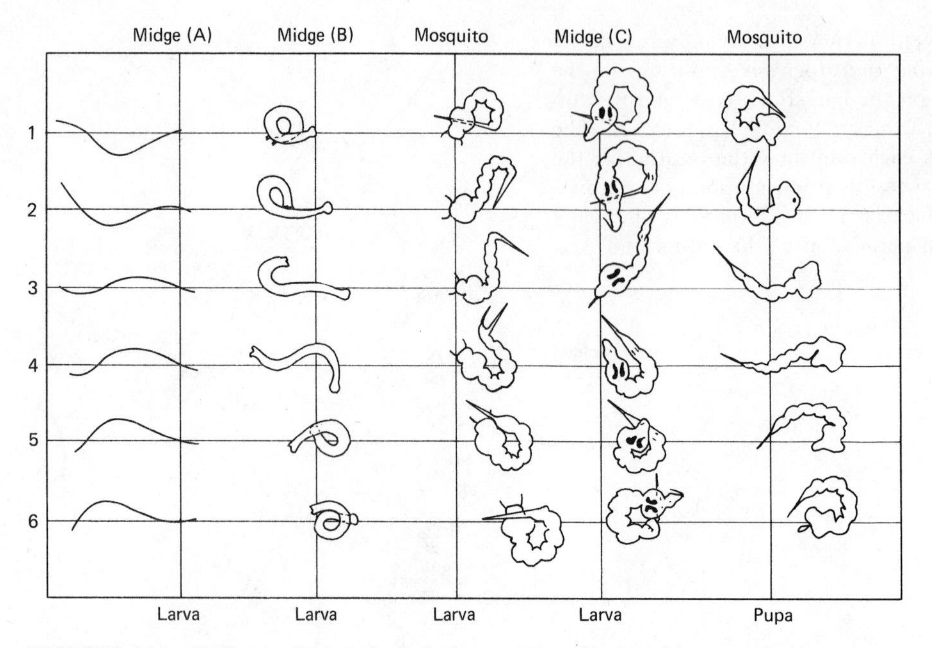


FIGURE 7-19. Different forms of wiggle locomotion displayed by aquatic insects. Note that the final position (6) is approximately equal to the initial position (1) but a mirror image of it. (Reproduced from Nachtigall, 1965, Swimming of aquatic insects. Used with permission of the author and Academic Press Inc.

locomotion is illustrated by the larvae and pupae of mosquitoes and by the larvae of other aquatic diptera (Figure 7-19).

### *Aerial Locomotion*

The insects are the only invertebrates that can fly, and this capability has contributed immeasurably to their success as terrestrial organisms. Flight is achieved by beating the wings up and down almost constantly. Only a few species are able to glide between strokes to any extent. Once airborne, many small insects are carried for long distances on the wind during which the wings probably serve more as sails than aerodynamic structures. We will be concerned here mainly with the major modifications of the flight mechanism and flapping flight. The biological and ecological significance of flight will be discussed in Chapter 12.

Fully developed functional wings occur only in the adult stage. There are a few surviving apterygotes, plus some groups such as the lice and fleas for which wings must have proved disadvantageous and were lost. Although wings of a much reduced size occur in the late immature stages of hemimetabolous insects (see Chapter 9), they are not capable of producing flight. The Ephemeroptera are the one exception in that they have a transitional winged stage, the **subimago,** between the last aquatic larval stage and the full-fledged adult stage. However, mayfly subimagos are only capable of short, weak flights.

The movement of the wings of insects is complex and involves muscles attached to the wing bases (**direct flight muscles**), muscles extending between different parts of the thorax (**indirect flight muscles**), the elastic properties of the resilin of the wing hinges, and the elastic properties of the thorax itself. During flight the wings are not only moved up and down but are also twisted to provide the aerodynamic forces that propel the insect through the air. Since the mechanism varies among the insects and is rather complex, the following description of some of the major features will provide only a fundamental understanding.

In the dragonflies and damselflies (Odonata), the muscles that elevate and depress the wings are attached directly to the wing bases by structures called **cap tendons** and to the lateral margins of the notal plates. The **pleural wing processes** provide fulcrums over which the wings are moved to produce a flapping stroke. The large **dorsoventral muscles** are inserted proximally on the notal sclerites so that when they contract and the **basalar muscles** relax, the wing bases are depressed, causing the wing tips to be elevated (Figure 7-20*A*). Since the cap tendons to which the basalar muscles are attached extend to the wing bases distal to the pleural processes, a contraction of these muscles while the dorsoventral muscles are relaxed pulls the wings downward (Figure 7-20*B*).

The direct flight muscles are also of major importance in the up-and-down movement of the wings in beetles, cockroaches, and some orthopterans. In

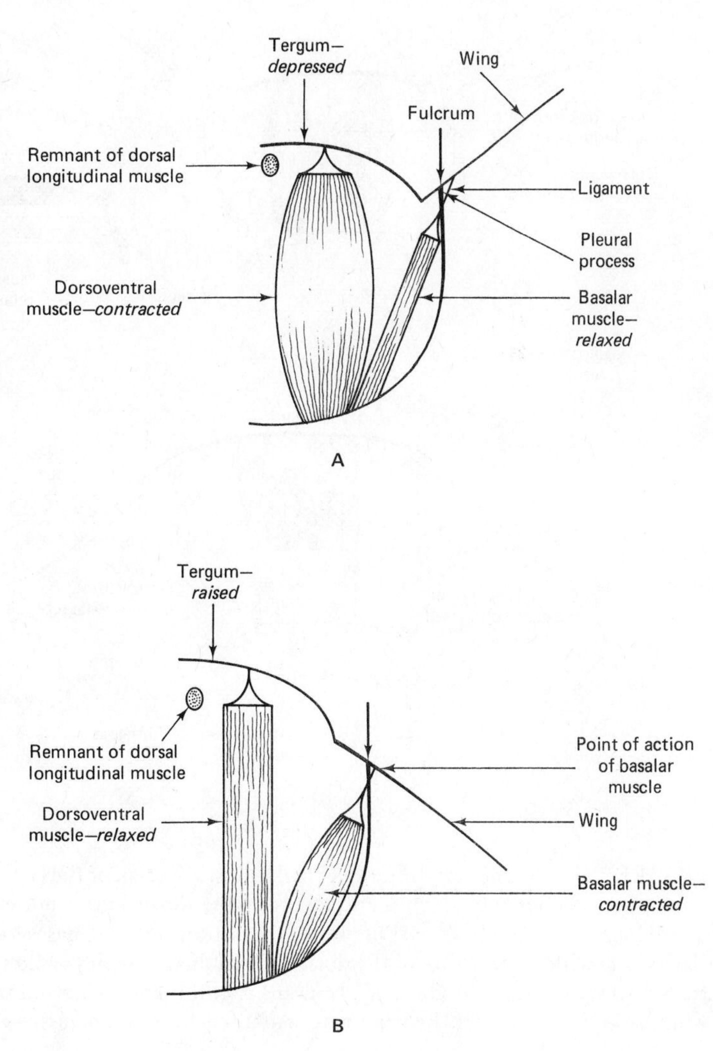


**FIGURE 7-20.** Diagram of the direct flight muscles system of the type found in the dragonflies (Odonata). Contraction of the dorsoventral muscles while the basalar muscles relax, distorts the notum downward and depresses the wing bases below the pleural wing process thereby producing the up-stroke. Relaxation of the dorsoventral muscles allows a contraction of the basalar muscles to pull the wing downward for the power stroke. (Adapted from Snodgrass, 1935.)

most of the more advanced insects, however the up and down movements of the wings are produced by the indirect flight muscles. In the honeybee, for example, the up-and-down stroke is produced as the indirect flight muscles distort the shape of the thorax such that the position of wing bases changes relative to the pivot points provided by the pleural processes. When the dorsal longitudinal muscles relax and the dorsoventral muscles contract, the notum is depressed, thereby pulling the wing bases below the dorsal pleural processes and raising the wing tips (Figure 7-21*A*). When the dorsoventral muscles relax and the dorsal longitudinal muscles contract, the notum is arched upward, thereby raising the wing bases and driving the wings downward (Figure 7-21*B*).

The mechanisms described so far produce only an up-and-down beating of the wings. Clearly, this is not sufficient to produce flight in that the forces

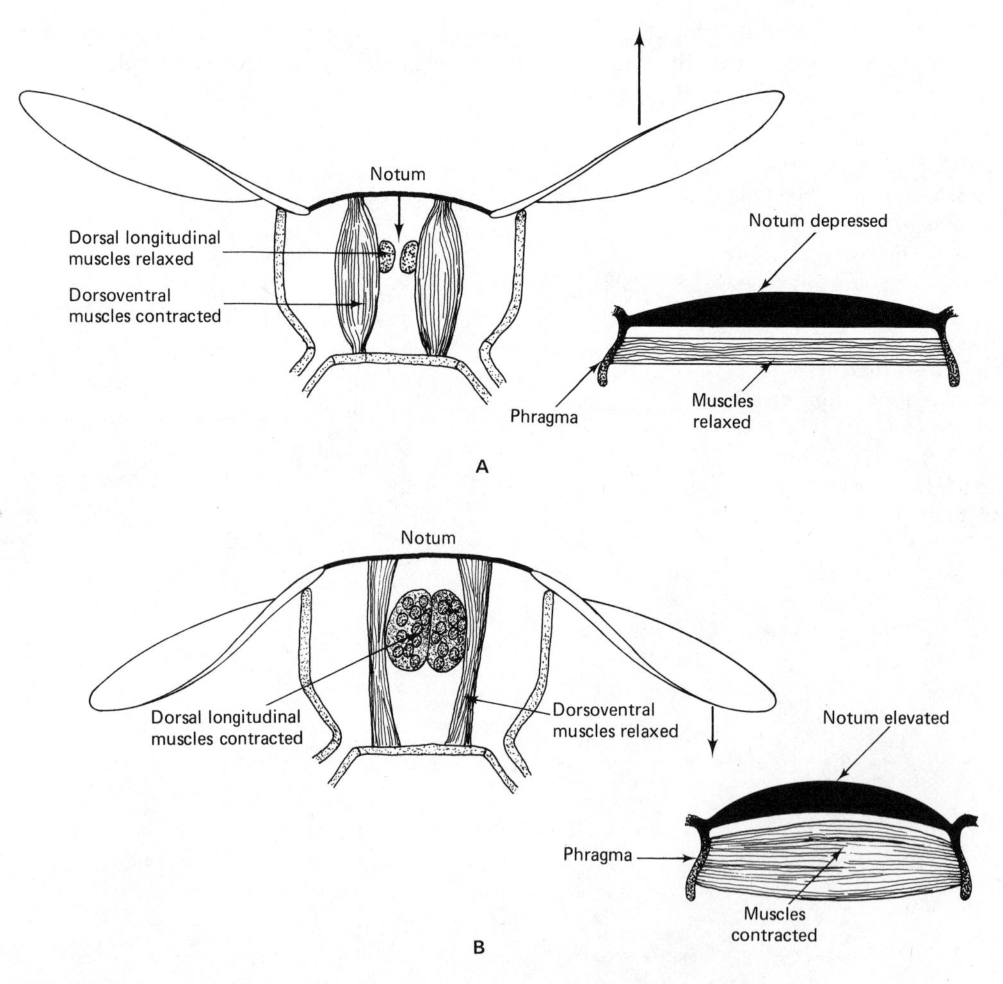


**FIGURE 7-21.** Diagram of the indirect flight muscle system of the type found in the honey bee (Hymenoptera). **A.** Contraction of the dorsoventral muscles when the dorsal longitudinal muscles relax depresses the notum and wing bases thereby raising the wing tips. **B.** Contraction of the dorsal longitudinal muscles while the dorsoventral muscles are relaxed produces a downward distortion of the notum that raises the wing bases and depresses the wing tips. (Adapted from Snodgrass, 1935.)

FIGURE 7-22. Diagram of the manner in which the wings of the honey bee twist during a wing beat cycle. (Adapted from Pringle, 1957.)

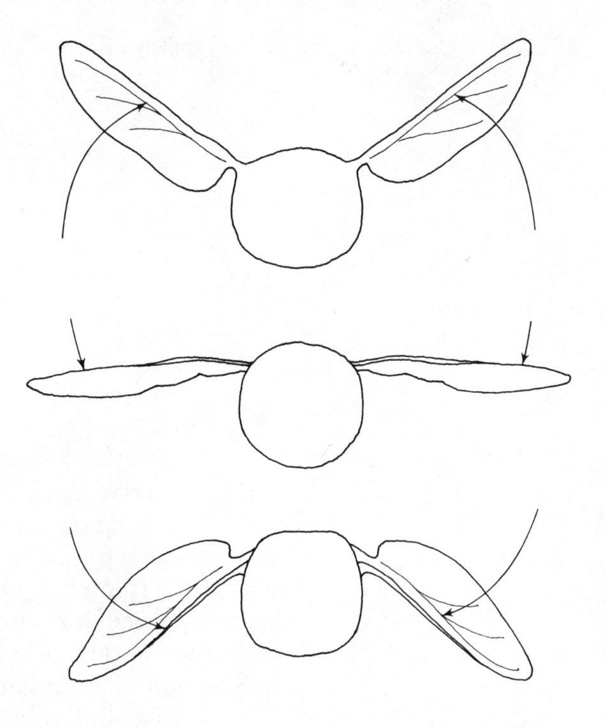

produced by the downstroke would be more or less cancelled out by the opposing forces produced in the subsequent upstroke. In order that more force is exerted on the downstroke than on the upstroke, it is necessary that the wings rotate so that more surface area pushes on the air as the wings are depressed than when the wings are elevated. It is also necessary on the downstroke that the wings push the air backward to propel the insect forward. These secondary movements are produced by a distortion of the nota and by additional muscles that alternately exert pressure on the front and back portions of the wing bases, causing a deflection of the wing membrane. As a result, the wings pronate at the top of the stroke so that a maximum surface area attacks the air as they move downward. At the bottom of the stroke, the wings supernate for their return (Figure 7-22).

The amplitude of the wing stroke is variable between species and also becomes smaller with fatigue. In some Odonata each wing subscribes an arc of only 70 degrees, whereas in some Diptera and Hymenoptera they each pass through about 160 degrees. In insects such as orthoperans, heteropterans, earwigs, and beetles, in which the forewings are modified to form protective covers for the membranous hind wings, the two pair have a stroke of different amplitude. In the grasshopper *Schistocerca,* the fore wings, or **tegmina,** have an amplitude of 60 to 70 degrees while that of the hind wings is 110 degrees. In the beetle *Priacma,* the fore wings, or **elytra,** have an amplitude of only 50 degrees, compared with 150 degrees for the hind wings (Figure 7-23).

The wing beat frequency is also highly variable. Some grasshoppers and lepidoptera have wing beat frequencies below 20 cycles per second. In many Hymenoptera and larger flies, the frequency exceeds 200 cycles per second, and some small flies have frequencies reported to be in excess of 1,000 cycles per second. However, the wing beat frequency varies not only from species to species but also with sex, age, changes in temperature, relative humidity, and

FIGURE 7-23. Diagram of the wing and elytral amplitude of a primitive beetle *Priacma serrata* (Cupedidae).

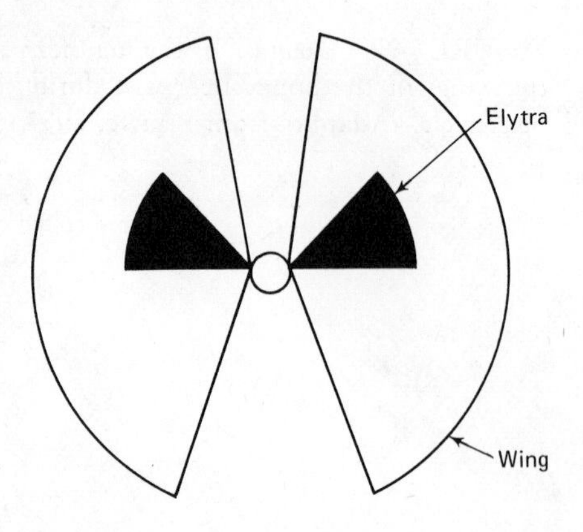


with other characteristics of the air in which the insect is flying. The relationship between temperature and wingbeat frequency for the Douglas fir beetle is shown in Figure 7-24.

Most muscles contract only once for each nerve impulse received. The maximum number of times that such muscles can be stimulated to contract is limited by the basic physiology of nerves to about 40 times per second. Insects such as dragonflies, locusts, and cockroaches that mainly employ the direct flight muscles have wing beat frequencies that are limited by this basic physiological fact. The high wing beat frequencies displayed by bees and flies is possible because of the special properties of their indirect flight muscles. These special striated (fibrillar) muscles may contract several times after each nerve impulse is received. Such **myogenic** contractions are a response to a quick mechanical stretch that is provided by the structure and elasticity of the thorax. The thoracic segment and its indirect flight muscles produce a resonating system such that very high wing beat frequencies are possible.

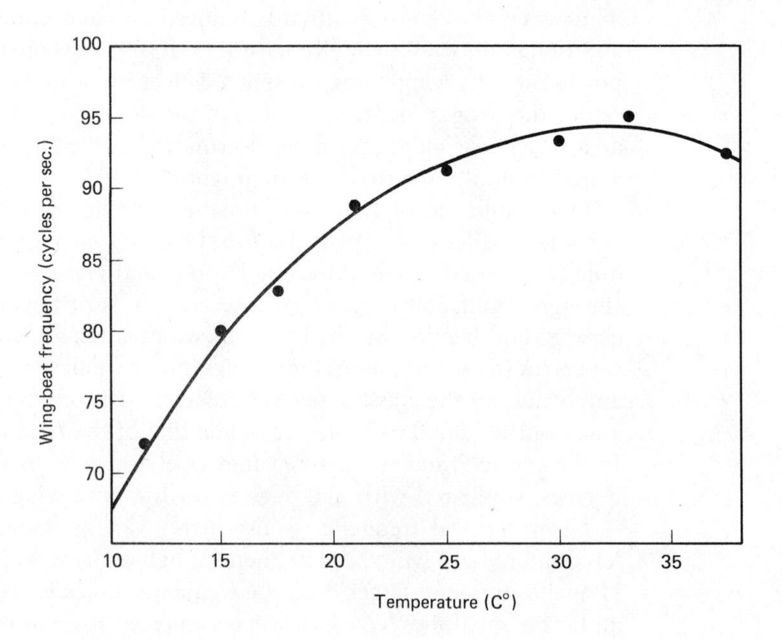


FIGURE 7-24. The relationship between temperature and wing beat frequency for the Douglas-fir beetle (Scolytidae).

TABLE 7-1 **Representative Measurements of the Velocity of Insect Flight**

| Insect | Velocity of flight (meters per second)[1] |
| --- | --- |
| Odonata | |
| *Anax* (dragonfly) | 8 |
| *Libellula* (dragonfly) | 4–10 |
| Neuroptera | |
| *Chrysopa* (lacewing) | 0.6 |
| Coleoptera | |
| *Dendroctonus* (bark beetle) | 1–1.2 |
| *Melolontha* (scarab) | 2.2–3, 2.5 |
| Lepidoptera | |
| *Pieris* (white butterfly) | 1.8–2.3, 2.5 |
| *Sphingids* (hawk moth) | 15 |
| Diptera | |
| *Tabanus* (horse fly) | 4, 14 |
| Hymenoptera | |
| *Bombus* (bumblebee) | 3, 3–5 |
| *Apis* (honeybee) | 2.5, 2.5–3.7 |

[1]Two or more sets of numbers indicate data obtained by different investigators.

Obviously, the speed with which insects fly depends upon the structural characteristics of the insect and the physical properties of the air. Some outlandish claims of high flight speeds have been made, but these have not been substantiated by observations of the time taken by various species to traverse a measured distance or from approximations obtained from flight mill studies. Table 7-1 presents some more reliable estimates of the flight velocity of several insects that have been obtained experimentally.

To this point nothing has been said about orientation in flight. The behavioral aspects of orientation will be discussed in Chapter 11, but how insects maneuver in flight can best be considered here. There are three planes of orientation during flight, the plane of climbing or diving known as **pitch,** the plane of side-to-side tilting known as **roll,** and the plane of right or left turning known as **yaw** (Figure 7-25). An insect must be able to control its movement in all of these planes in order to climb, land, and steer a course. The mechanisms used to maintain stability in flight are too complex for presentation here. In simple terms, they involve changes in the plane in which the wings beat, variations in wing twisting, and the wing stroke amplitude on one side of the body relative to the other side, combined with movements of the legs and the abdomen. Students who wish more information are referred to the review by Pringle (1965).

Many more primitive insects have wings of a generalized design that are not particularly efficient aerodynamically. They tend to be rather fluted and thickened by numerous heavy longitudinal and cross veins. Frequently both pair of wings are large and have no coupling mechanism. These features would characterize the wings of the Plecoptera (Figure 7-26), which, like many primitive insects, are not particularly good fliers. In the course of evolution there has been a trend toward the elimination of fluting, a reduction in the venation, a reduction in the size of the hind wings, and the development of various kinds of wing-coupling mechanisms. In the strong-flying bees and wasps, the hind wings are small and coupled to the fore wings

FIGURE 7-25. Diagram of the planes of flight orientation.

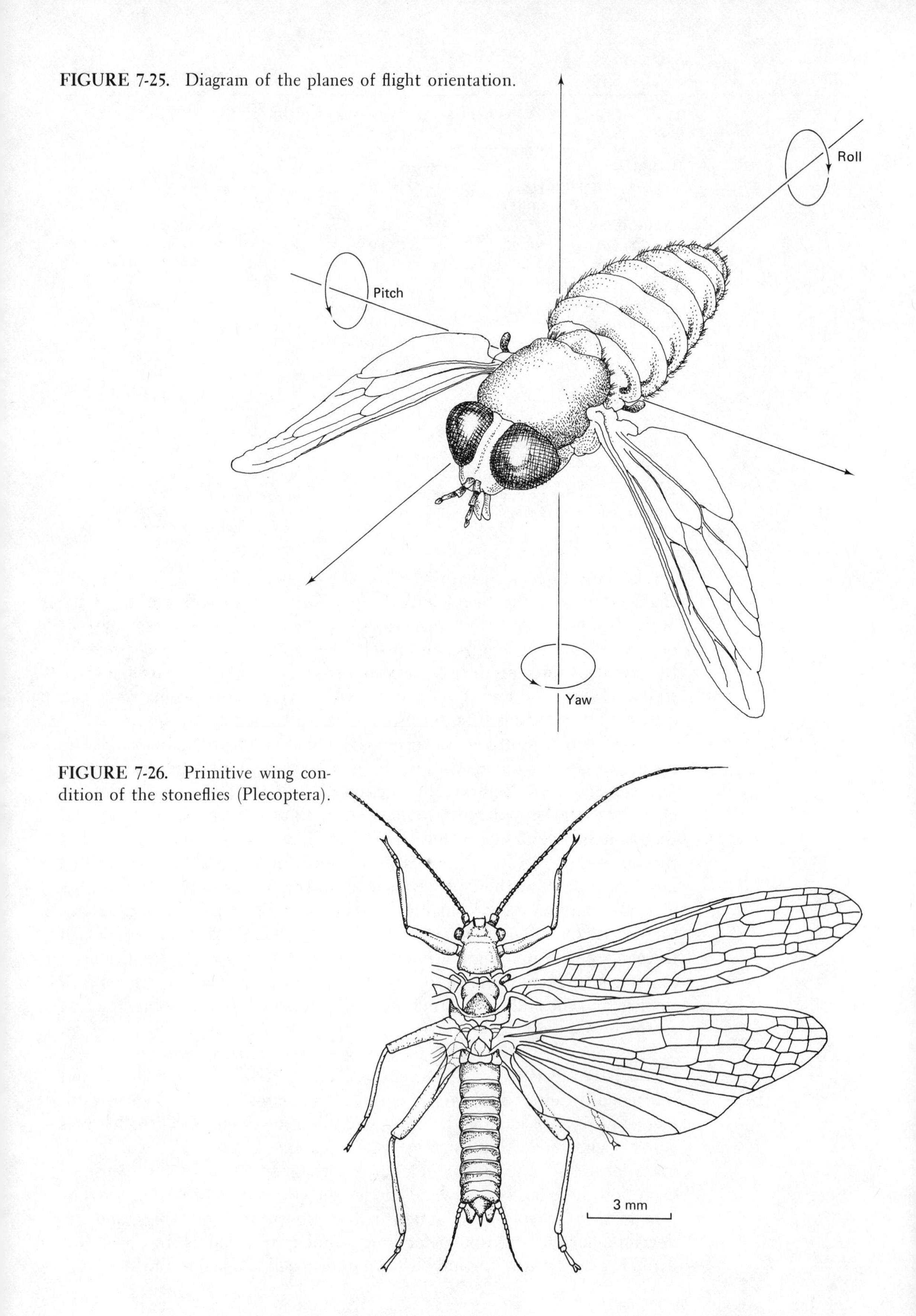


FIGURE 7-26. Primitive wing condition of the stoneflies (Plecoptera).

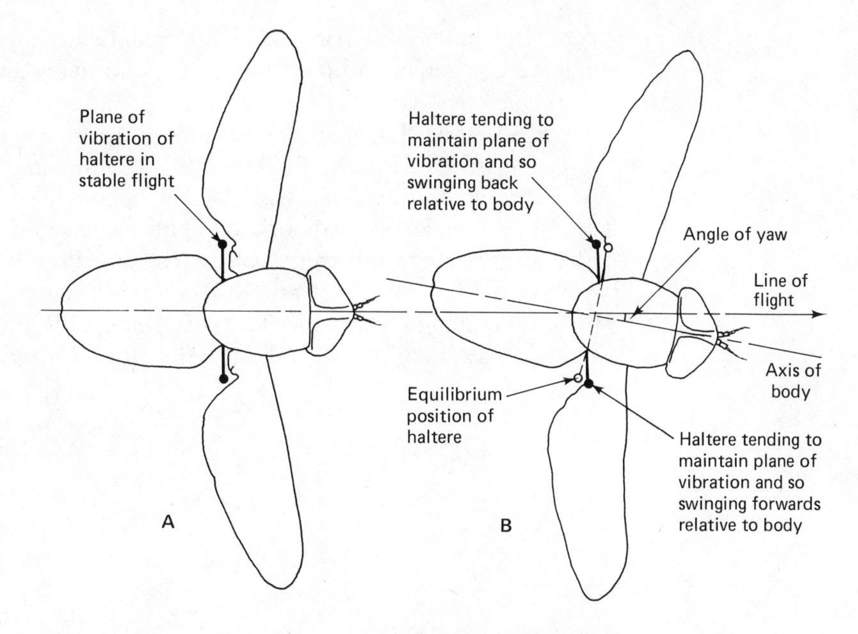


**FIGURE 7-27.** Diagrams illustrating the functioning of the halteres during flight. **A.** During stable flight the halteres vibrate in a plane with their long axes at right angles to the long axis of the fly's body, **B.** During a yawing movement the halteres vibrate in the original plane until yaw is corrected; if it is not corrected the halteres return to their normal plane of vibration. (From Chapman, 1969, used with permission of Elsevier, North-Holland, Inc.)

by rows of hooks so as to produce a very flexible trailing edge. In the Diptera, which are probably the best fliers of all, the hind wings have been reduced to a pair of special balancing organs called **halteres** (Figure 7-27). The halteres vibrate at the same frequency as, but out of phase with, the wings. Their inertia resists any change in their plane of vibration. As changes occur in the various planes of flight, the tendency of the halteres to keep vibrating in their original plane results in a build-up of pressure near their bases. Sensory receptors in the basal plates of the halteres detect these pressures and send impulses to the central nervous system that lead to reflex changes in the twisting of the wings that correct the deviation in the orientation of the body. In a sense then, the halteres act as a pair of gyroscopes, and their importance to flight stability can be demonstrated by carefully clipping one from a house fly and observing the fly in postoperative flight.

# References Cited


Barth, R. 1937. Muskulatur und Bewegungsart der Raupen. Zool. Jahrb. 62: 507–566.

Heitler, W. J. 1974. The locust jump—specializations of the metathoracic femoral-tibial joint. *J. Comp. Physiol.* 89: 93–104.

Hughs, G. M. 1958. The coordination of insect movements. III. Swimming in Dytiscus, Hydrophilus, and a dragonfly nymph. *J. Exp. Biol.* 35: 567–583.

Hughs, G. M. 1965. Locomotion: Terrestrial. Pages 227–254 *in* M. Rockstein, ed. The physiology of Insecta, Vol. 2. Academic Press Inc., New York, N. Y.

Leech, H. B. and H. P. Chandler. 1968. Aquatic Coleoptera. Pages 293–371 *in* R. L. Usinger, ed. Aquatic insects of California. University of California Press, Berkeley, Ca.

Nachtigall. 1965. Locomotion: Swimming (hydrodynamics) of aquatic insects. Pages 255–282 *in* M. Rockstein, ed. The physiology of Insecta, Vol. 2. Academic Press Inc., New York, N. Y.

Pringle, J. W. S. 1965. Locomotion: Flight. Pages 283–329 *in* M. Rockstein, ed. The physiology of Insecta, Vol. 2. Academic Press Inc., New York, N. Y.

Rothschild, Miriam, Y. Schlein, K. Parker, C. Neville, and S. Sternberg. 1973. The flying leap of the flea. *Sci. Amer.* 229(5): 92–100.

Wilson, D. M. 1966. Insect walking. *Ann. Rev. Entomol.* 11: 103–122.

# Gas exchange—a matter of life and breath

All living cells require oxygen for the oxidative breakdown of carbohydrates. This basic metabolic process results in the release of energy and the production of carbon dioxide and water. Because carbon dioxide is toxic to living tissue, it cannot be allowed to accumulate in the body. Consequently, tissues are constantly engaged in acquiring oxygen and giving up carbon dioxide, or more simply, in gas exchange. In minute organisms, the ratio of surface area to volume is great enough that the required gas exchange can be achieved by diffusion of the gases directly across the body wall. As organisms increase in size, the surface area increases by an exponent of two, whereas the volume increases by an exponent of three. This means that in larger organisms gas cannot diffuse through the body wall and internal tissues quickly enough to meet the needs of the increased volume of living tissue. Thus, in all larger animals we find some system for increasing the surface area over which an exchange of oxygen and carbon dioxide can take place. The more active the organism, the higher its metabolism and the more efficient the system of gas exchange must be. As was briefly noted in Chapter 3, the area for gas exchange in the generalized insect consists of a series of invaginated tubes called tracheae that ramify the body and carry air directly to the tissues.

In some small terrestrial insects and among some that live in aqueous environments, all of the gas exchange occurs by diffusion across the surface area of the exoskeleton. However, in the majority of insects, gas exchange is aided by the elaborate system of branched tracheae and tracheoles. The lining of the tracheae is continuous with the integument and therefore must be molted periodically along with the exoskeleton. Histologically, the tracheae consist of a basement membrane that supports a layer of pavementlike epithelial cells, which secrete the cuticular lining of the tracheal lumen. The cuticular lining, or **intima,** is laid down with spiral thickenings called **taenidia,** which allow the tracheae to stretch longitudinally but resist collapse (Figure 8-1). The waxy coating that is characteristic of insect cuticle repels water, thereby reducing the danger of "drowning" as a result of water entering the tracheae by capillary action.

Each time a trachea branches, the diameter of the tubules becomes smaller until they are less than one micron in diameter. The fine tubules are called **tracheoles** and are not lined with cuticle, so are not molted. They carry gases directly to and from small groups of cells.

Many insects have dilations of the tracheae called **air sacs** that lack the supporting taenidia (Figure 8-2). Because they are collapsible, they can be used as bellows for air ventilation. Among both flying and aquatic insects, these air-filled sacs decrease the organism's specific gravity and are thought to improve their flight characteristics and buoyancy. As the efficiency of the flight musculature is directly related to temperature, within certain limits, air sacs surrounding the thoracic muscles may also provide insulation against heat loss. Air sacs may also serve to reserve space needed later for the growth of various organs, particularly those of the reproductive system.

The tracheal system opens to the exterior through apertures in the exo-

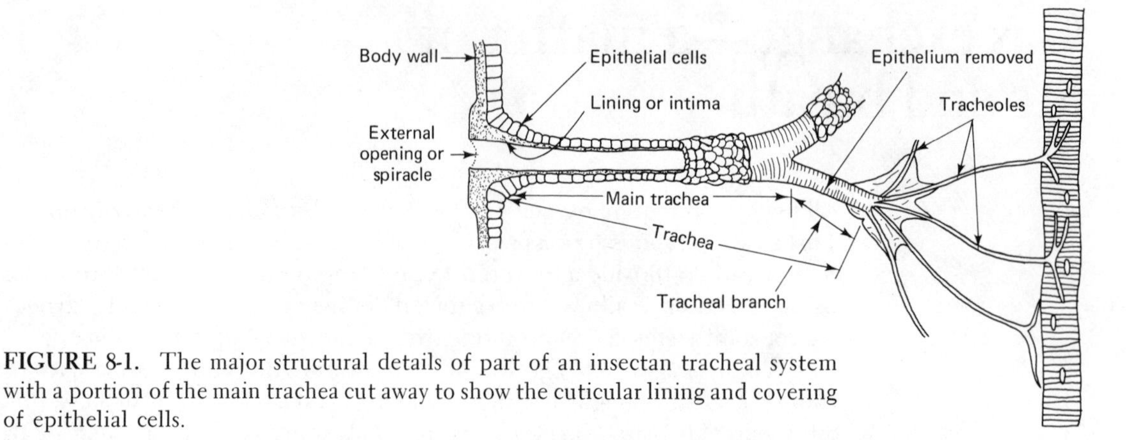


FIGURE 8-1. The major structural details of part of an insectan tracheal system with a portion of the main trachea cut away to show the cuticular lining and covering of epithelial cells.

skeleton called **spiracles.** Among the true insects the maximum number of spiracles is ten pairs. These occur as one pair per segment, laterally placed on the mesothorax, the metathorax, and eight abdominal segments. However, the number and position of **functional spiracles** vary greatly from group to group. In their simplest form the spiracles open directly into a tracheal trunk; such an arrangement has a number of drawbacks that have been overcome by various closure modifications during the course of evolution.

In addition to increasing the surface area for the exchange of oxygen and carbon dioxide, the tracheal system also increases the area over which water loss can occur and provides a route for the inflow of noxious vapors. Two basic types of spiracular closing mechanisms have evolved that reduce these hazards. Both involve an additional invagination of the cuticle to form an **atrium** into which the trachea opens. In one type, the outer lips of the atrium can be drawn together to effect the closure. In the second type, the atrium remains open but spiracular muscles open and close some form of valve at the junction of the atrium and the trachea.

Most terrestrial insects have what is called an open tracheal system in which one or more pair of spiracles can provide a direct exchange of air between the atmosphere and the intertracheal space. However, the spiracles may not be open all of the time; the frequency of opening is regulated according to the

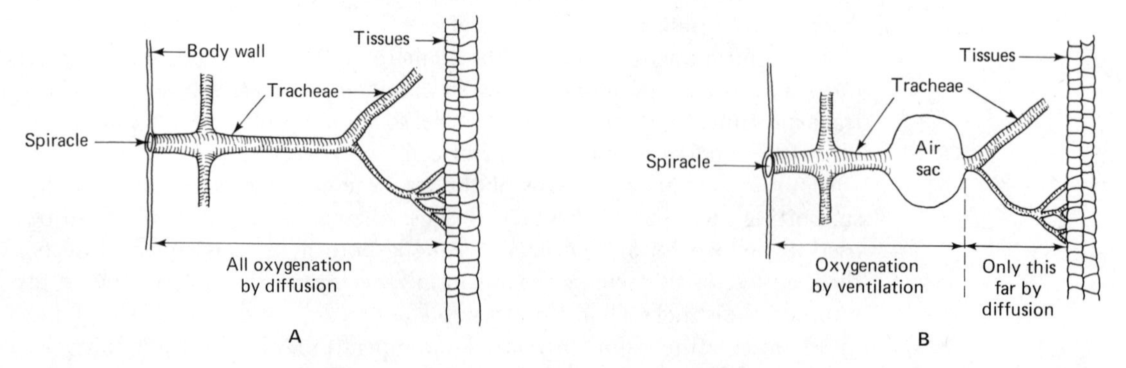


FIGURE 8-2. Two variations of an insectan tracheal system. **A.** Unmodified form in which the air is delivered to the tissue entirely by diffusion, **B.** A modified form in which the expansion of main tracheae to form air sacs provides a mechanism for air movement by ventilation. (Adapted from Snodgrass, 1935.)

animal's oxygen requirements. In small insects, or during periods of low metabolic activity, the gas exchange requirements can be met exclusively by diffusion or by the partial or periodic opening of functional spiracles. The closing mechanisms are apparently under both nervous and chemical control as they have been shown to remain open for longer periods when the oxygen content of the air is decreased but other environmental parameters remain constant. The more the spiracles can be kept closed, the lower the loss of water vapor caused by transpiration. In large insects, or during periods of high activity, diffusion cannot deliver enough oxygen to meet the animal's metabolic needs. As a result, air may be pumped in and out of the system much as the human diaphragm pumps air in and out of the lungs. This is achieved by rhythmic contractions of the abdominal muscles that result in the alternate expansion and compression of the abdomen. These movements, plus the normal movements of the internal organs, such as gut peristalsis, produce changes in the shape of the tracheae and air sacs, thereby inducing active ventilation. In some insects these movements are coordinated with a sequential opening and closing of the functional spiracles so as to create a directional flow of air through the system.

Insect blood does not usually contain hemoglobin or hemocyanin for the chemical transport of oxygen as in vertebrates and many other invertebrates. For the most part, gaseous oxygen is delivered directly to the cells by the tracheoles. This has the effect of speeding up the availability of the oxygen and permits the explosive metabolism that is characteristic of insectan indirect flight muscles.

The carbon dioxide produced by cellular respiration diffuses through the tissues and across the integument more readily than oxygen, but most of it is evacuated through the spiracles. During periods of high metabolic activity, the body must be able to rid itself of carbon dioxide more or less continuously, but during periods of low activity, it can be allowed to build up in the tracheal system for periodic release. The so-called **burst phenomenon** permits the carbon dioxide to build up until it reaches a concentration high enough to trigger the relaxation of the spiracular muscles. This results in a sudden opening of the valve and thus the escape of the carbon dioxide and its subsequent replacement by fresh air. Since the spiracles are open briefly and only at infrequent intervals, the loss of water from the body is greatly reduced. Insects that are in a state of quiescence over a long period of unfavorable conditions, such as occurs during the winter, are unable to obtain water, so must conserve what they have in their bodies. Pupae of some lepidopterans keep their spiracles closed almost entirely during winter quiescence. Because their metabolic rate is extremely low under such conditions, they can get by for long periods on the oxygen trapped in their tracheal system, and the small amount of carbon dioxide produced is dissolved in their body fluid. This removal of oxygen and carbon dioxide from the tracheae creates a partial vacuum that is replaced by air when the spiracles are opened, perhaps as infrequently as once each day. When the body temperature rises, the carbon dioxide in the body fluid is driven out of solution and escapes into the atmosphere through the spiracles during normal breathing.

## *Gas Exchange in Aquatic Insects*

As the development of the tracheal type of gas exchange system was one of the advances that led to the success of the insects as terrestrial animals, one might expect that some major modifications would have been needed before insects could become aquatic for one or more stages of their life history. Aquatic insects exploit both open and closed tracheal systems. Those with an open system, that is, with some functional spiracles, can utilize atmospheric oxygen by regularly coming to the surface to breathe, by carrying a store of air underwater with them, or by extracting the air trapped in the tissue of aquatic plants. Those that have a closed system in which all the spiracles are nonfunctional depend upon the inward diffusion of oxygen dissolved in the water.

The atmosphere is a reliable source of oxygen for aquatic insects because it is always available under natural conditions, except when the surface of the water is frozen. However, acquiring oxygen from the atmosphere is not as simple for aquatic insects as it might seem. The physical properties of the surface film create several problems. Furthermore, a total reliance on the atmosphere for oxygen limits the amount of time that many species can remain submerged, and the more often an individual must come to the surface to breathe, the more likely it is to be exposed to and attacked by natural enemies. It is not surprising then that aquatic insects with open tracheal systems have evolved a variety of means to reduce the dangers of breathing at the surface.

An obvious modification would be to reduce the number of functional spiracles, as it would be impossible for an aquatic insect to expose a full complement of spiracles to the atmosphere at the same time. Most commonly, only one pair of spiracles near the tip of the abdomen is functional, and often the spiracle-bearing segment is extended into a breathing tube or siphon (Figure 8-3). Clearly, these breathing structures must be designed so that they can penetrate the surface film to open into the air and yet not be subject to an inflow of water when the insect submerges. This is usually achieved by a coating of waxy material secreted by special integumentary glands or by a tuft or circle of hydrofuge hairs around the spiracle. These surface characteristics cause the water to flow away from the spiracular opening and reduce the danger of drowning. But if the hairs surrounding the spiracle are numerous or long, they may become so firmly held by the surface tension that the insect has to go through a rapid jerking or twisting movement to free itself (Figure 8-4*A*).

When an insect is actively swimming, the volume of air in the tracheal system is rapidly depleted, making frequent trips to the surface necessary. Such behavior is characteristic of the larvae of some mosquitoes or feeding back swimmers. To reduce the time requirements and the dangers of frequent surfacing, many aquatic species carry an extratracheal air store in the form of a bubble into which one or more pair of spiracles open. The bubble is held in place by a dense mat of hairs as in back-swimmers (Notonectidae) or beneath the elytra as in the diving beetles (Dytiscidae). These air stores are replenished as needed in a variety of ways. Breathing usually involves a characteristic pattern of behavior that brings a group of semihydrofuge hairs surrounding the opening to the air storage site into contact with the atmosphere. For example, the hydrophilid beetles come to the surface headfirst, using their antennae to replenish their bubble (Figure 8-4*B*), whereas the dytiscids come up tailfirst, using the tips of their elytra and abdomen (Figure 8-4*C*).

The bubble carried by aquatic insects serves as more than just an extra supply of oxygen because it forms a **physical gill.** A newly formed bubble contains about 21% oxygen and 79% nitrogen, whereas water in equilibrium with the air contains 33% oxygen, 64% nitrogen, and 3% carbon dioxide. As

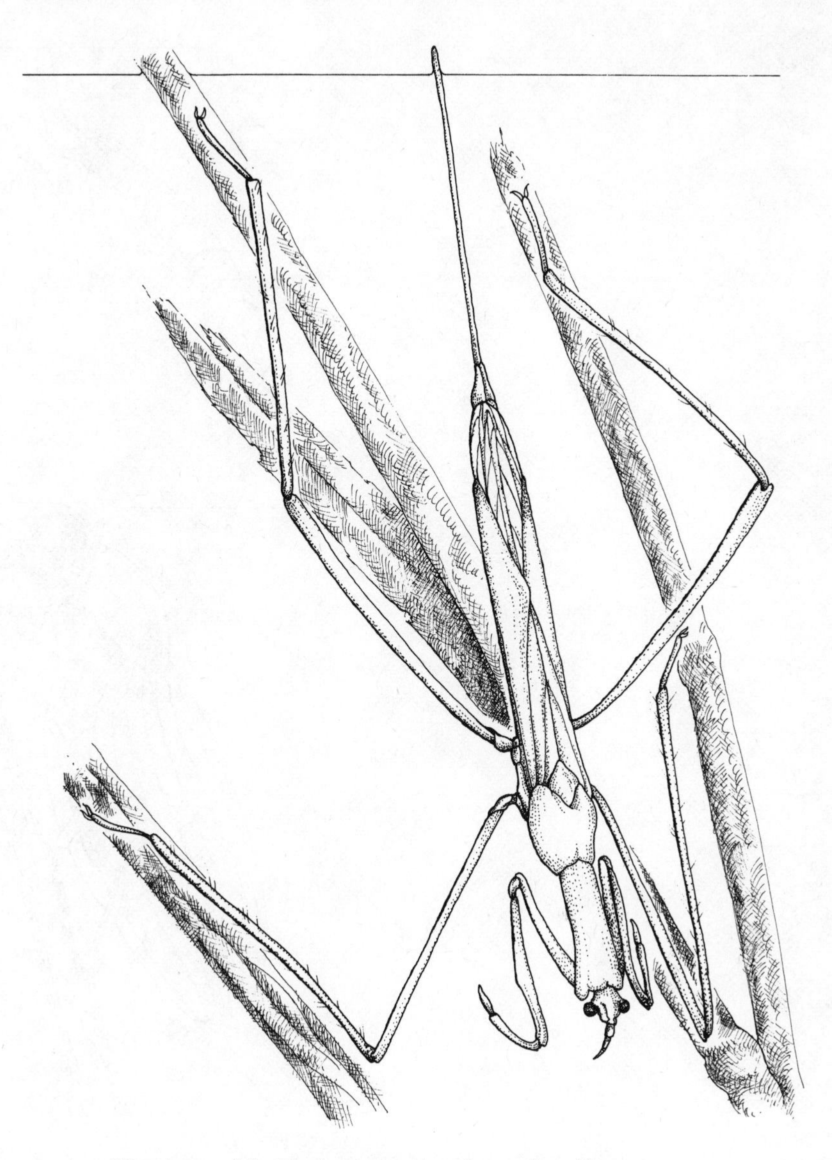

**FIGURE 8-3.** A water scorpion (Hemiptera: Nepidae) penetrating the surface film with its siphon or breathing tube bearing a terminal spiracle.

the insect uses oxygen from the bubble there is a corresponding increase in the bubble's nitrogen content. This alters the state of equilibrium between the gases in the bubble and those dissolved in the water. For equilibrium to be restored, nitrogen diffuses out of the bubble and oxygen diffuses into the bubble from the surrounding water. Because oxygen diffuses from water to air three times more rapidly than nitrogen diffuses from air into water, the bubble takes a considerable amount of time to dissipate. Temperature affects both the movement of the gases and the rate of metabolism of the insect, so at low temperatures the bubble may last many days, whereas at high temperatures the physical gill may be ineffective.

Some aquatic insects with open systems are able to remain submerged for most of their lives because of specialized structures that hold a permanent film of air on the surface of their bodies. Such a film provides an extensive

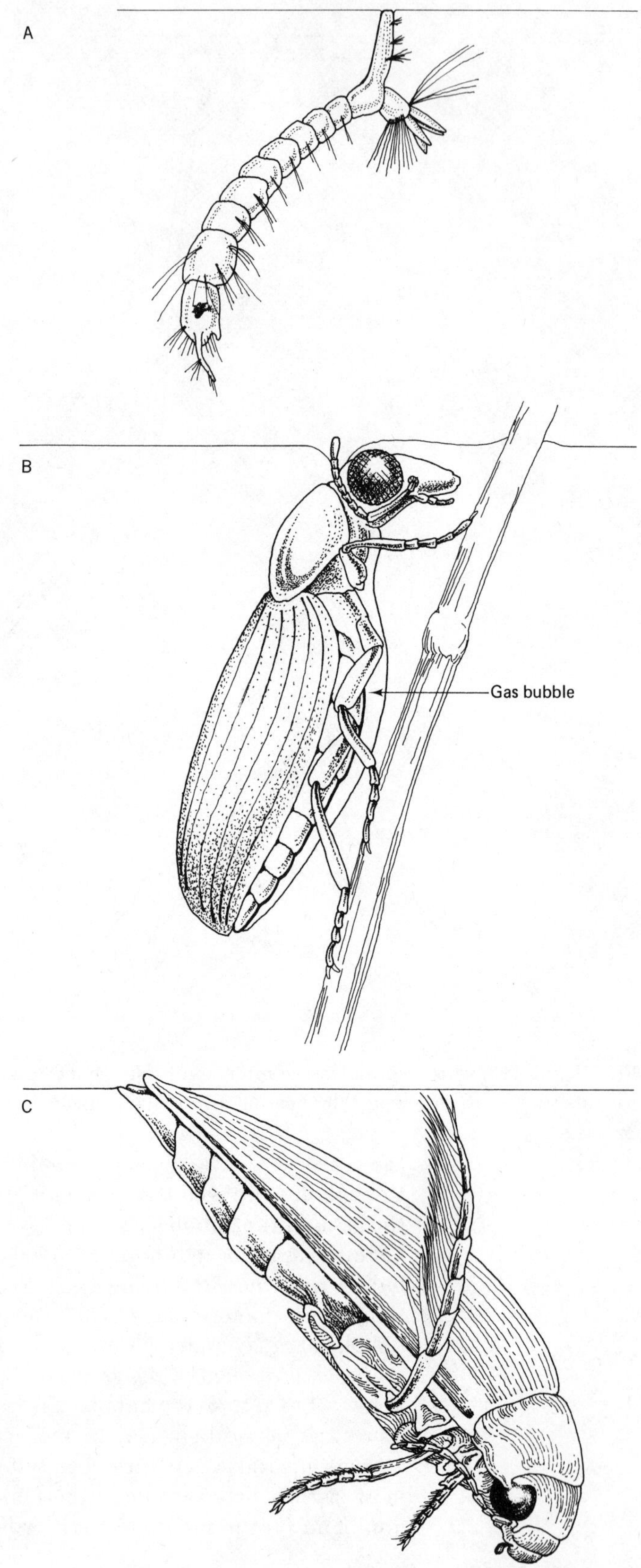


**FIGURE 8-4.** **A.** A mosquito larva at the surface of the water showing hairs associated with the breathing structure, **B.** Hydrophilid beetle replenishing its bubble by breaking the surface film with its antennae, **C.** Dytiscid beetle replenishing its sub-elytral air-store by penetrating the surface film with the tip of the abdomen and elytra.

area of air-water interface for gas exchange. This film, called a **plastron,** serves as a gas exchange mechanism rather than as a simple air store. The plastron is held in place by a dense pile of hydrofuge hairs that are thickened at the base to prevent their collapse. The hairs are also either sloped or bent at the tip to reduce an inflow of water. The pilosity (hairiness) can reach a density of two million hairs per square centimeter and may be able to stand a pressure of four atmospheres. Consequently, a plastron is permanent and does not dissipate as a bubble does. The spiracles open into the plastron, and there may be an arrangement of pathways through to hairs called **rosettes** that facilitate the flow of oxygen from the plastron to the tracheal openings.

A number of aquatic insects obtain oxygen by penetrating the tissue of aquatic plants, thereby extracting the air trapped between the cells. Most such insects live in the mud, where oxygen is in short supply, and pierce the plant tissue with a toothed siphon that has a spiracle near its tip.

The aquatic insects with closed tracheal systems depend entirely upon the diffusion of dissolved oxygen from the water into their tracheae or hemolymph. There are a number of factors that affect the amount of dissolved oxygen likely to be present in a body of water, so understanding these factors is necessary to an understanding of the distribution of aquatic insects and of the composition of the insect fauna of different aquatic communities.

Atmospheric oxygen enters the aquatic environment either by diffusion or by being carried and mixed in physically. Diffusion is a continuous process, although it may be curtailed periodically by a covering of ice. Diffusion is also slow; oxygen entering the water in this way may penetrate to a depth of only about six meters in the course of a year. Obviously then, the oxygen content of a body of water is influenced greatly by the ratio between its surface area and depth. The oxygen must reach deeper water by patterns of circulation that are set in motion by temperature changes at the surface. However, atmospheric oxygen may also be carried into a body of water by precipitation or be churned in by surface turbulence caused by wind or by the currents of rivers and streams. Another major source of oxygen is that given off by photosynthesizing aquatic plants, but the over-all contributions of oxygen by this means are highly variable, depending upon the amount of aquatic vegetation and seasonal changes in photosynthesis.

How much of the oxygen that enters a system remains there depends upon a number of factors, among the most important of which are temperature and the amount used during the oxidation of organic material. Generally, the highest oxygen concentrations occur in cool, clean, turbulent streams, and the lowest in warm, stagnant ponds with a high organic content. We might expect, therefore, the species diversity of aquatic insects with a closed tracheal system to vary more, both seasonally and between aquatic environments, than the diversity of those with an open system.

The insects that rely solely on the diffusion of oxygen into their bodies have evolved various means that increase the process in relation to the type of environment they inhabit. In general, the adaptations involve an increase in the surface area across which the oxygen can diffuse into the body. One of the most common adaptations is the development of **tracheal gills.** These are thin evaginations of the integument that are well supplied with tracheae. Such structures can occur anywhere on the body, but are usually located on the sides or tip of the abdomen (Figure 8-5*A* and *B*). Other species have **spiracular gills,** consisting of filamentous outgrowths of cuticle adjacent to the spiracles

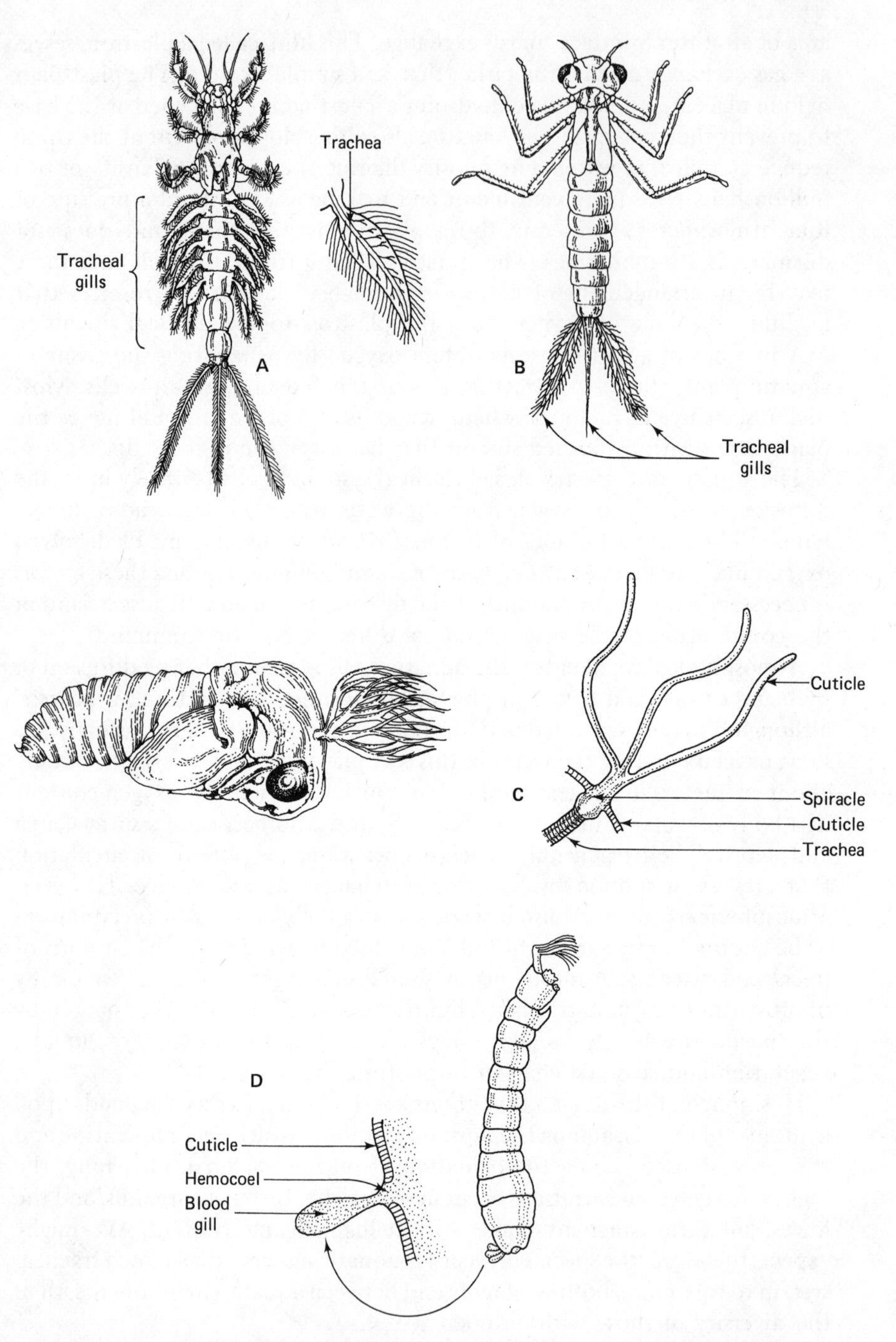


FIGURE 8-5. Various types of closed gas exchange systems found among aquatic insects. **A.** Lateral tracheal gills of a mayfly larva, **B.** Terminal tracheal gills of a damselfly larva, **C.** Spiracular gills of a black fly pupa, **D.** Blood gill of a midge larva. (C and D redrawn from Packard, 1898.)

that open directly into the corresponding trachea (Figure 8-5C). Still others have outward projecting pouches of cuticle filled with hemolymph, called **blood gills** (Figure 8-5*D*), that increase the area for direct diffusion of oxygen from the water into the hemolymph. All of these gill structures must be provided a fresh supply of water to assure an adequate supply of oxygen. If there is no water movement, the oxygen in the water adjacent to the gill becomes rapidly depleted and its replacement from the surrounding water is too slow to meet the insect's needs. Consequently, either the whole insect or its gill structures must be kept in motion. In this regard, it should be obvious that the surface of the gills must be kept free from an accumulation of debris that would curtail gas diffusion. In some mayfly young that live in silty environments, the first pair of abdominal gills are enlarged to form protective covers to reduce the fouling of the other gills with silt and debris. Other mayfly larvae harbor symbiotic chironomid larvae (Diptera) that keep the gills clean by feeding on the organic debris that collects on them.

### *Gas Exchange in Endoparasites*

Many insects, particularly some Hymneoptera and Diptera, spend their larval lives feeding within the tissues of other animals. From the standpoint of gas exchange, their environment is similar to that occupied by aquatic insects and presents the same alternatives for obtaining oxygei.. Many of them, especially those that are endoparasitic in other insects, are of small size and can acquire their oxygen by diffusion from the body fluid of their host. This may be facilitated by the development of various expansions of the body wall. In some, the hind gut is everted through the anus to form a caudal vesicle; others have tracheal gills like those of their aquatic relatives.

Larger, more active endoparasites with greater oxygen requirements have functional spiracles and obtain air from the host's gas exchange system or directly from the atmosphere. Most of the fly larvae that develop as endoparasites of vertebrates spend much of their time just below the skin and usually maintain contact with the atmosphere by way of small openings through which they project their posterior spiracles.

## General References

Chapman, R. F. 1969. The insects—structure and function. American Elsevier Publishing Co., Inc., New York, N. Y.

Edwards, G. A. 1963. Respiratory mechanisms. Pages 55–95 *in* K. D. Roeder, ed. Insect physiology. John Wiley & Sons, Inc., New York, N. Y.

Wigglesworth, V. B. 1972. Principles of insect physiology, 7th ed. Chapman and Hall Ltd., London.

# Reproduction, growth, and development

# 9

Even a casual observer cannot help but be impressed by the abundance of insects during certain times of the year and by how quickly they progress from scarcity during the winter to becoming common components of most ecosystems as the weather improves during the spring and summer. There is, of course, a complex interplay of physical and biological factors that determine how quickly insects of different kinds increase in number. These factors will be considered in a later chapter, but the sequence of events must begin with reproduction and the development of individuals with the capacity to leave offspring. In considering this subject, we will not be concerned with the production of eggs and sperm or the details of embryology. Instead, we will concentrate on the variety of reproductive adaptations and the significance of the different ways that insects develop after they leave the egg.

The entire reproductive process involves a complex interplay of behavior and physiology. The vast majority of insect species reproduce bisexually; that is, both males and females are involved. In most such species, the reproductive process consists of the following sequence of events:

MATE LOCATION: the mechanisms and behavior that bring potential mates together.
COURTSHIP: the behavioral patterns that lead to the pairing of a male and female.
COPULATION: the coupling of the male and female genital apparatus.
EJACULATION: the release of sperm by the male.
INSEMINATION: the introduction of sperm into the female reproductive system.
FERTILIZATION: the penetration of an egg (ovum) by sperm and the uniting of their nuclei.
EMBRYOGENESIS: the early stages of development within the egg.
OVIPOSITION: the deposition of a fertilized egg by the female.
ECLOSION: hatching, or the escape from the egg covering by a larval insect.

## *Modes of Reproduction*

Although most insects reproduce **bisexually,** reproduction in the absence of male gametes, called **parthenogenesis,** is a fairly common occurrence, and **hermaphroditism** (both sexes in one individual) is known in a few species. In bisexual species, the eggs are usually fertilized in the oviduct as they pass the opening of the duct leading from the sperm storage organ (**spermatheca**). However, fertilization may also occur in an ovariole and, in a few species, in the hemocoele. After fertilization the resulting progeny may then issue from the female in one of the following ways:

**Oviparity:** The eggs are laid following various degrees of embryonic development according to the species, but they do not hatch until sometime after oviposition. As embryological development and hatching are affected by several physical environmental factors, the young insects do not usually emerge unless conditions are favorable to their survival. They may hatch in a few days, remain in the egg stage all winter, or not hatch for two years. However, the longer the defenseless eggs are exposed, the more subject they are to mortality factors.

**Ovoviviparity:** Normally the eggs are retained within the body of the female until they hatch and then the young are released to the exterior, but sometimes the eggs hatch immediately after they are deposited. Ovoviviparity occurs in some species of most orders and has the advantage of reducing the time that the egg stage is exposed. However, the environmental conditions must be suitable for the survival of the young progeny when they are deposited, as they are normally less tolerant of adverse conditions than are eggs. Ovoviviparity is a particularly beneficial adaptation for species that utilize transitory food resources, such as animal excrement, that could otherwise deteriorate before the eggs could hatch.

**Viviparity:** There are several different forms of viviparity, but in general terms, the young are nourished for some period within the body of the female and released in a more advanced state of development than in ovoviviparity. In the viviparous roaches (Blattodea), earwigs (Dermaptera), and bugs (Hemiptera), the eggs are deficient in yolk, and the embryos are nourished in a brood pouch by way of a close association of special maternal tissue called a **pseudoplacenta** and the tissues of the embryo. In one group of flies (Cecidomyiidae), the larvae develop in the mother's body by feeding on her tissues, and in other dipterans, the larvae are nourished by the female and grow to full size in her vagina. They become pupae soon after they are deposited. The advantages and disadvantages of this type of reproduction are similar to those for ovoviviparity, except that in the extreme case in which the larvae pupate soon after they are deposited, no external source of larval food is necessary. Such insects, an example of which is the tsetse fly, are particularly difficult to control.

**Polyembryony:** The eggs are laid as in oviparity, but early embryonic development is modified so that a number of cell divisions produce small aggregations of cells, each one capable of becoming a larva. This mode of reproduction is found most commonly among hymenopterous parasites of caterpillars. Because most of these wasps are small, a caterpillar once located represents a large package of food resource, capable of supporting up to 3,000 offspring of the wasp. Consequently, relatively few caterpillars need to be found and successfully parasitized to maintain a high wasp population.

**Paedogenesis:** The term paedogenesis means reproduction by juveniles and describes a condition known for some time to occur in one species of beetle (Micromalthidae) and several genera of gall and fungus gnats (Cecidomyiidae). Ibrahim and Gad (1975) reported this phenomenon in a common flower fly (Syrphidae). The ovaries become functional in the larval stage and the eggs develop parthenogenetically. Ultimately, the body of the "parent larva" becomes filled with small daughter larvae that eventually break to the outside. Interestingly both the gnat and the micromalthid larvae feed on fungus in decaying wood, and as long as there is an ample supply of food, they continue to reproduce paedogenetically. But if conditions become crowded or the food supply grows short, there develop winged males and females, which leave the brood site and reproduce sexually. This strange means of reproduction would seem to be a rather good way to exploit favorable food conditions without "wasting" energy on the development of winged adults, unless needed. A modification of paedogenesis occurs in some aphids that begin to produce embryos prior to becoming adults, but the larvae are not deposited until the female matures. In several bugs and thrips, the ovaries are functional in the last larval stage and are inseminated at this time by mature males, but, as in aphids, the eggs are not laid until the females become fully developed.

**Parthenogenesis:** In most bisexual species, sex is determined by the sex chromosomes. Usually there is a pair of similar chromosomes (*XX*) in one sex and either a pair of dissimilar chromosomes (*XY*) or a single chromosome (*XO*) in the other. The sex with the *XX* pair is called **homogametic,** the one with the *XO* or *XY* pair, **heterogametic.** In humans, the male is always the heterogametic sex, whereas in insects, it may be either the male, or the female as in the Lepidoptera and Trichoptera. Sex is thus determined by the balance between genes for maleness and genes for femaleness following the recombination of genetic material at fertilization.

When an egg develops without being fertilized (a phenomenon known as parthenogenesis, which occurs quite commonly among insects), the sex of the offspring depends on the behavior of the chromosomes at the time of reduction division (**meiosis**). In general, eggs with one half the normal complement of chromosomes (**haploid**) produce males, whereas those with a full complement (**diploid**) produce females.

Species with populations comprised exclusively of females display **obligatory parthenogenesis;** that is, fertilization has been completely abolished. The females remain diploid in one of two ways. The reduction in chromosome number is compensated for by either a lack of meiosis or a premeiotic doubling of the chromosome number and a fusion of meiotic nuclei. Since there is no fertilization, there is no genetic recombination and all offspring are genetically like the parent. This means that evolution can proceed only by the accumulation of mutations in a single line; beneficial mutations that occur in separate individuals can never be combined. Any pronounced change in the environment could therefore eliminate all of the descendents of one genotype. In spite of this apparent disadvantage, several types of insects, particularly members of the order Homoptera, have obligatory parthenogenesis. They seem to be quite successful and able to withstand normal fluctuations in climate. But this is clearly and evolutionary cul-de-sac, and eventually their inability to adapt to environmental change will lead to extinction.

In species with both males and females, parthenogenesis is **facultative;** that is, it only occurs in the absence of fertilization. Among these species, females are produced from fertilized eggs or by diploid parthenogenesis, and haploid males are produced from unfertilized eggs. In the Hymenoptera, the females can lay either fertilized or unfertilized eggs by controlling the descent of sperm from the spermathecal duct with a sphincter. In a few rare cases, unfertilized females produce diploid females as the result of a few tetraploid ovarian cells producing diploid eggs by meiotic division.

The ecological significance of facultative parthenogenesis is substantial. In some parasitoid Hymenoptera, male offspring can develop in hosts too small for the development of females. Having located and "measured" a host, the female wasp can lay either a fertilized or an unfertilized egg, whichever is appropriate. In some species, the females may have a greater tolerance for certain environmental parameters than the males and can therefore temporarily expand the range of the species by reproducing parthenogenetically at the periphery where there are no males. This is referred to as **geographical parthenogenesis.**

Other insects, particularly aphids and other homopterans, display complex life histories involving cyclical parthenogenesis in which there is an alternation of sexual and parthenogenic generations. This combines the advantages of parthenogenesis when food is abundant with the evolutionary advantages of genetic recombination.

### *Eggs and Oviposition*

The initial stage of development of all insects is the egg. In oviparous species, it is released from the parent surrounded by a protective shell, or **chorion.** The chorion is a complex proteinaceous structure, produced by follicle cells in the ovaries. The outer surface often appears smooth but may be striated or sculptured to produce various patterns of ornamentation, some of which are strikingly beautiful (Figure 9-1). Near the anterior end of the egg there is a small opening, or **micropyle,** which allows entry of the sperm for fertilization. After a sperm enters an egg, a **fertilization membrane** is secreted that seals the micropyle and prevents the entry of additional sperm cells. In many terrestrial

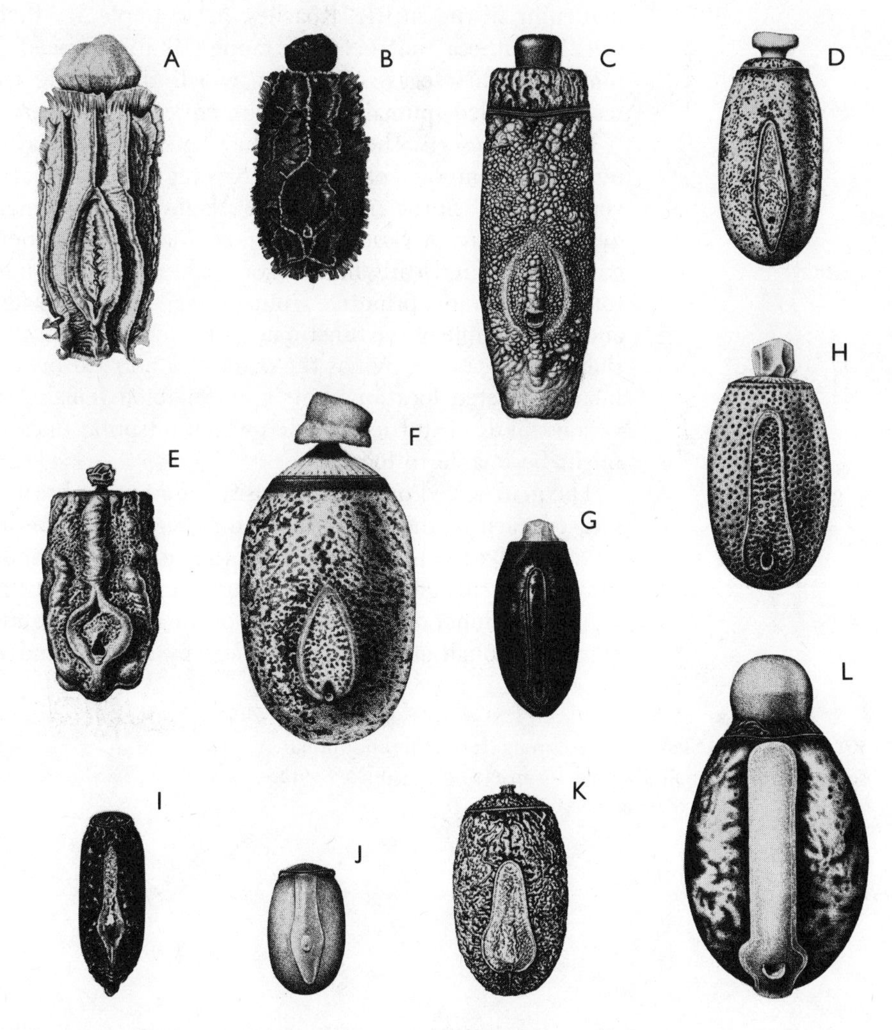


FIGURE 9-1. Beautifully ornamented eggs of Australian stick insects of the order Phasmatodea. (Drawings by F. Nanninga reproduced with permission of Melbourne University Press.)

species, the chorion has an inner meshwork comprised of vertical columns with spaces between. These spaces are filled with air, which is in contact with the atmosphere through numerous surface pores and thereby facilitates gas exchange. In some aquatic species, the outer chorion also traps a layer of air that functions as a plastron much as described in Chapter 8. Insect eggs are relatively large and much of their mass consists of yolk used as nourishment by the developing embryo.

In many insects such as the Lepidoptera, Coleoptera, and Diptera, the eggs are laid on the surface, so the female requires no specialized egg-laying apparatus. Sometimes the terminal segments of the abdomen form a telescoping tube, which facilitates placement of the eggs in shallow crevaces or even the penetration of some soft medium (Figure 9-2). Eggs laid out in the open are obviously exposed to unfavorable weather conditions and natural enemies, so in addition to being glued to the substrate by a sticky secretion, they are quite commonly covered by a protective material that either hardens as it dries or tans by a biochemical process similar to that which produces hardening of the cuticle. Roaches, for example, lay their eggs inside a capsulelike **öotheca** that becomes tanned as it is formed. Mantids and many moths lay their eggs in clusters, which the female covers with a frothy secretion called **spumaline** that subsequently hardens.

In many insects, the females have a special egg-laying device called an **ovipositor,** composed of three pair of integumentary shafts that arise from the ventral surface of the eighth and ninth abdominal segments. Although insect ovipositors have a varied appearance and serve a variety of functions, the components are clearly homologous. These uniquely insectan structures are found in the more primitive groups as well as in some advanced ones and are another example of the functional anatomy of the insects that contributed to their success. Not only has the ovipositor allowed insects to deposit eggs in more protected locations but it has enabled females to place the eggs in various kinds of nutrient media, which offspring that hatched in the open might be unable to locate.

The generalized ovipositor consists of a pair of basal structures, the **valvifers,** on each of the eighth and ninth segments. The **first valvifers** (eighth segment) give rise to a pair of shafts, the **first valvulae,** and the **second valvifers** (ninth segment) give rise to two pair, the **second** and **third valvulae** (Figure 9-3). The oviduct opens at the base of the ovipositor, and the eggs pass to the exterior through an egg canal formed by the first and second valvulae.

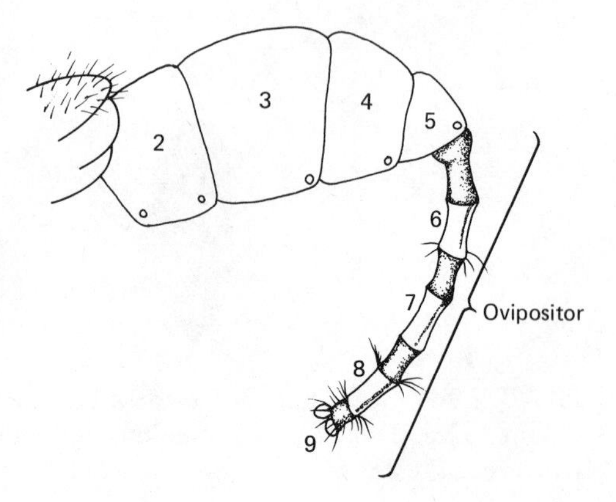


**FIGURE 9-2.** Slender telescoping terminal abdominal segments of a female fly used for inserting eggs into a soft medium. (Redrawn from West, 1951.)

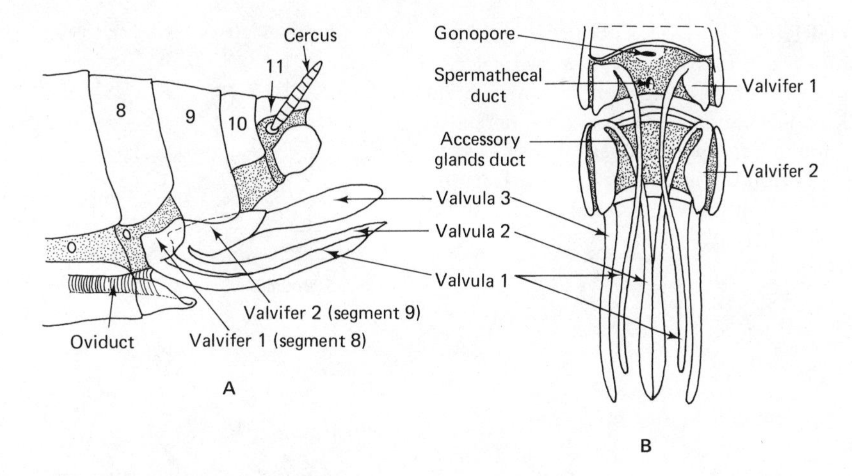


FIGURE 9-3. Diagram of the typical arrangement of ovipositor components which represent unique insectan structures that arise from the ventral surface of the eighth and ninth abdominal segments. (Adapted from Snodgrass, 1935.)

The shafts of the ovipositor have become highly modified in different groups, thereby permitting a variety of egg-laying possibilities. In the grasshoppers, the first and third valvulae form a pair of short, stout structures that are recurved at the tip and used for digging egg chambers in the soil (Figure 9-4*A*). In several groups, the shafts are toothed and may lock together to form a reciprocating cutting device used to penetrate the tissue of plants in order that the eggs can be deposited beneath the surface (Figure 9-4*B*).

The ovipositor reaches its maximum modification among different groups of the Hymenoptera. In the sawflies, it is short and broad and is used to place the eggs in plant tissue. In the parasitic families, the ovipositor is slender and greatly elongated, facilitating the deposition of eggs in or on hosts that might otherwise be out of reach. In the latter groups, the first and second valvulae form the egg tube, which is housed in a protective sheath (the third valvulae) that is lifted or coiled out of the way as the egg tube is inserted, with the hind coxae serving as a guide (Figure 9-4C). The eggs are elongate and often have very little yolk thereby facilitating passage through the slender egg tube. In the predaceous wasps and the bees and ants, the ovipositor has become a sting used either to narcotize their prey or as a defensive device (Figure 9-5). The stinging ovipositor forms a hollow needle for penetration and the injection of a venom, whereas the eggs pass from the genital chamber through an opening at the base of the sting.

## *Growth and Metamorphosis*

GROWTH. When insect eggs are laid, the stage of development of the embryo within varies from species to species. When development of the embryo is complete, the young insect may either hatch right away or remain in the egg until the environmental conditions are favorable. At the time of hatching the chorion is ruptured by muscular activity of the larva, which may also increase its volume by swallowing the air or fluid remaining in the egg. Many larvae have egg-bursting spines on different parts of the body, and sometimes the chorion has a special line of weakness along which the egg splits open.

The insect that emerges from the egg may bear a close resemblance to its

FIGURE 9-4. Variations in the form of insect ovipositors. A. Grasshopper, B. Katydid: lateral view with cut away section to show detail, C. Katydid: transverse section showing relative position of components, D. Parasitic wasp: note position of the hind coxae. (A, B, and C adapted from Snodgrass, 1935, D redrawn from Riley, 1888.)

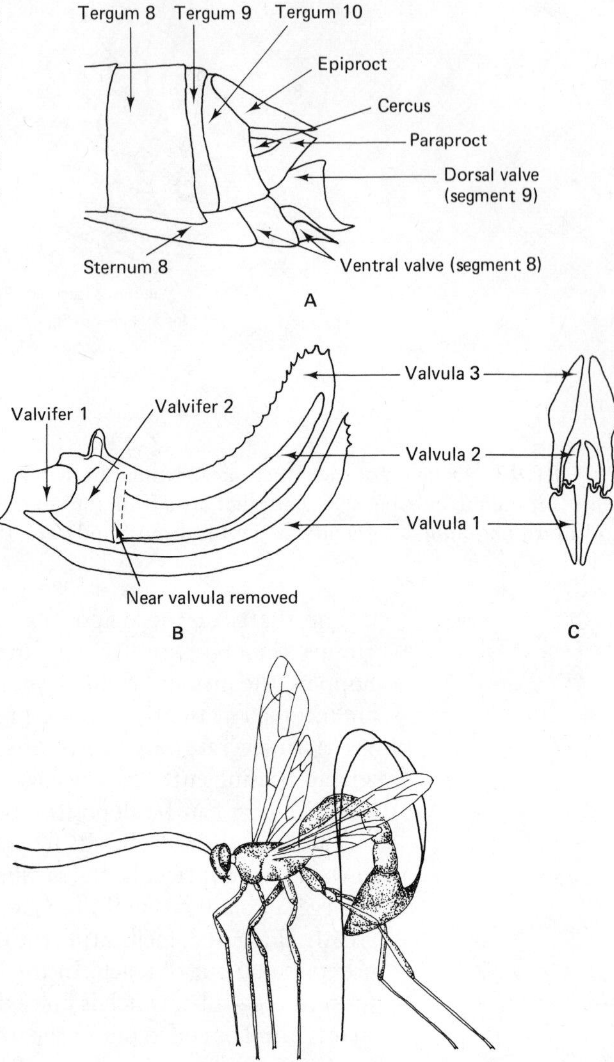


parents or look entirely different, depending upon the type of postembryonic development that will ensue. The name applied to the young insect that emerges is **larva** (pl. larvae), regardless of its appearance, although **nymph** is often used for those that look like their parents, and the term **naiad** is sometimes applied to the aquatic immatures of terrestrial or aerial species. Once the newly hatched immature begins to feed, it must cast off its original exoskeleton to allow for growth, as was discussed in Chapter 5. As a larva progresses toward the adult stage it must pass through a series of molts, each of which is normally accompanied by an increase in size and often by a slight change in appearance. Between each molt the larval insect is called an **instar,** and the time interval in a particular instar is called a **stadium;** the young insect that emerges from the egg is called the first instar, and following the first molt, the second instar, and so on. The number of larval instars varies considerably from a minimum of three in highly advanced forms to many in more primitive species. In some groups the number of instars is constant, whereas in others it varies even within species in relation to the availability of

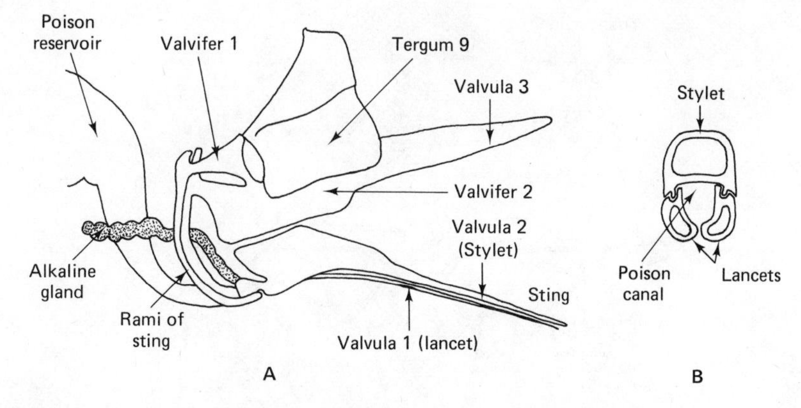


**FIGURE 9-5.** The ovipositor of a worker honey bee showing modification of the components into a sting. **A.** Lateral view, **B.** Transverse view. (Adapted from Snodgrass, 1935.)

food and variations in other conditions that affect development. In a few species, molts occur during periods of adversity that result in a reduction rather than an increase in size.

Growth, although apparently occurring in steps, occurs as a result of both cell division and an increase in the size of some cells. The rate and magnitude of growth among insects are quite astounding. The time for development to proceed from egg to adult may be as short as 48 hours for some flies under ideal conditions but may take years as in the cicadas. The magnitude of growth is also variable; a thousandfold increase in weight between the first instar and the adult is common, but in a few lepidopterans it may be 40 or 50 times that.

METAMORPHOSIS. The term **metamorphosis** means change of form and is used frequently to describe the postembryonic changes displayed by insects as they progress toward adulthood. Since all insects do not show the same degree of change during development, several different types of metamorphosis have been described. The Apterygota change very little in form during development and are said to be **ametabolous,** or lacking change. Among the winged insects, the more primitive groups, collectively called the **exopterygotes** because their wings develop externally, usually display a gradual change and are called **hemimetabolous.** The more advanced **endopterygotes,** the wings of which develop internally, show the most pronounced difference between larval and adult stages and are called **holometabolous.**

The life cycle of a typical hemimetabolous (**incomplete** or **gradual metamorphosis**) insect such as a grasshopper or bug consists of an egg, a series of larval (or nymphal) instars, and the adult. The nymphs resemble the adults in body form, including the presence of typical legs, antennae, and compound eyes but lack fully developed wings, gonads, and genitalia. In the course of development (Figure 9-6*A*), the wings appear as external pads that gradually increase in size but do not become fully developed and functional until the adult stage. (Except in the mayflies, which have a subimago mentioned earlier, all other winged insects do not molt again once they reach the winged stage.) Characteristically, the nymphs feed on the same food sources as the adults, so all postembryonic stages compete with one another. But this is not true in the Ephemeroptera, Odonata, and Plecoptera, which are aquatic as

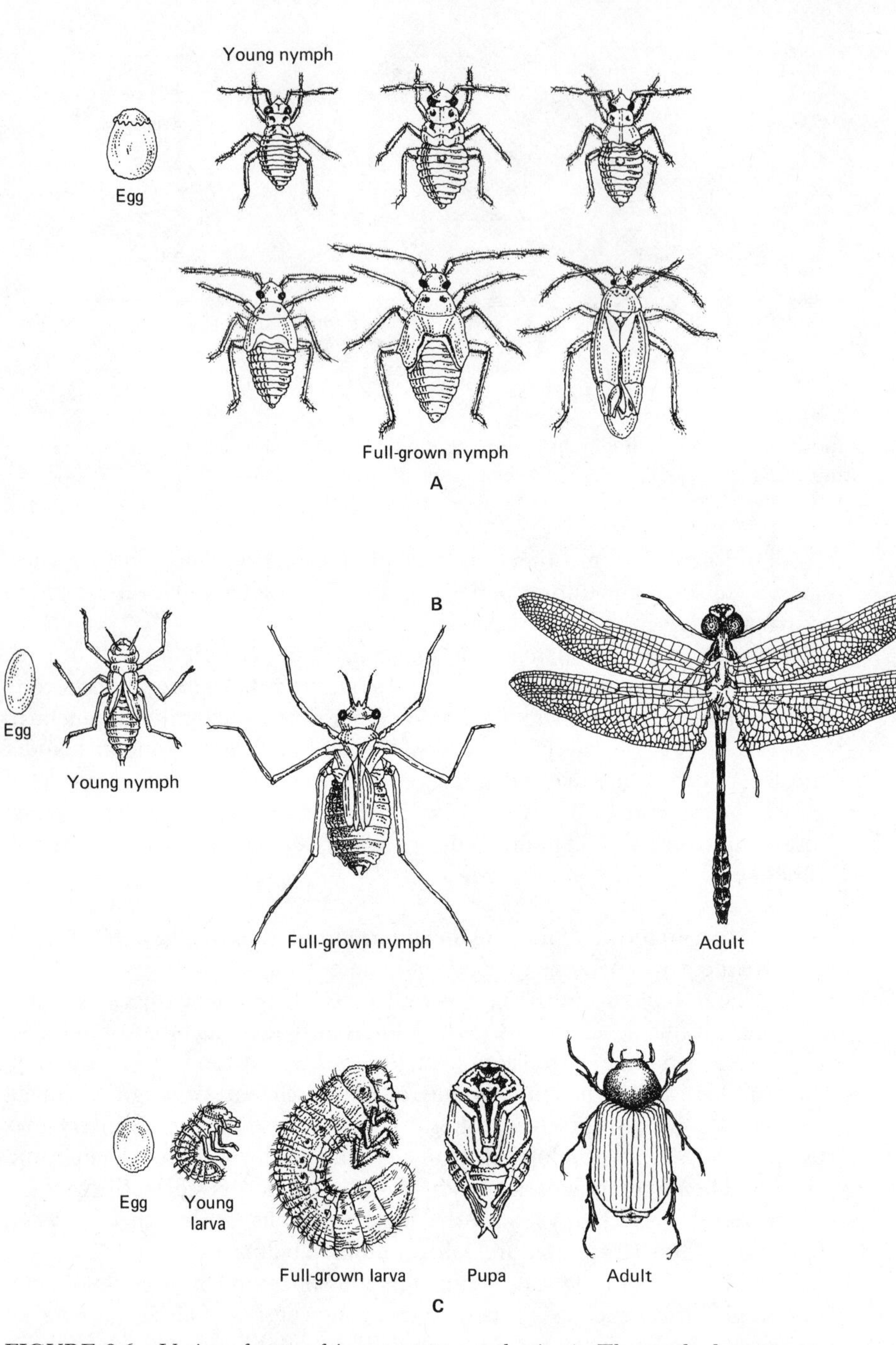


**FIGURE 9-6.** Various forms of insect metamorphosis. **A.** The gradual metamorphosis displayed by most exopterygote insects such as the members of the order Hemiptera, **B.** The gradual metamorphosis displayed by some aquatic exopterygotes like the dragonfly often involves some remarkable changes between the last larval stage and the adult, **C.** The complete metamorphosis of a beetle as representative of the changes that occur in the life history of most endopterygote insects.

immatures and terrestrial as adults. The aquatic nymphs (or naiads) of these groups are often quite dissimilar in appearance from the adult, but the development is typically gradual except for some rather remarkable changes that take place with the final molt (Figure 9-6*B*).

The rest of the insects have **complete metamorphosis.** Their life cycle consists of an egg, several larval stages, a transformation stage called the **pupa,** and the adult (Figure 9-6*C*). The larvae are usually clearly dissimilar to the adults and are adapted to distinctly different habitats and food sources (Figure 9-7). They lack compound eyes, have reduced antennae, and have

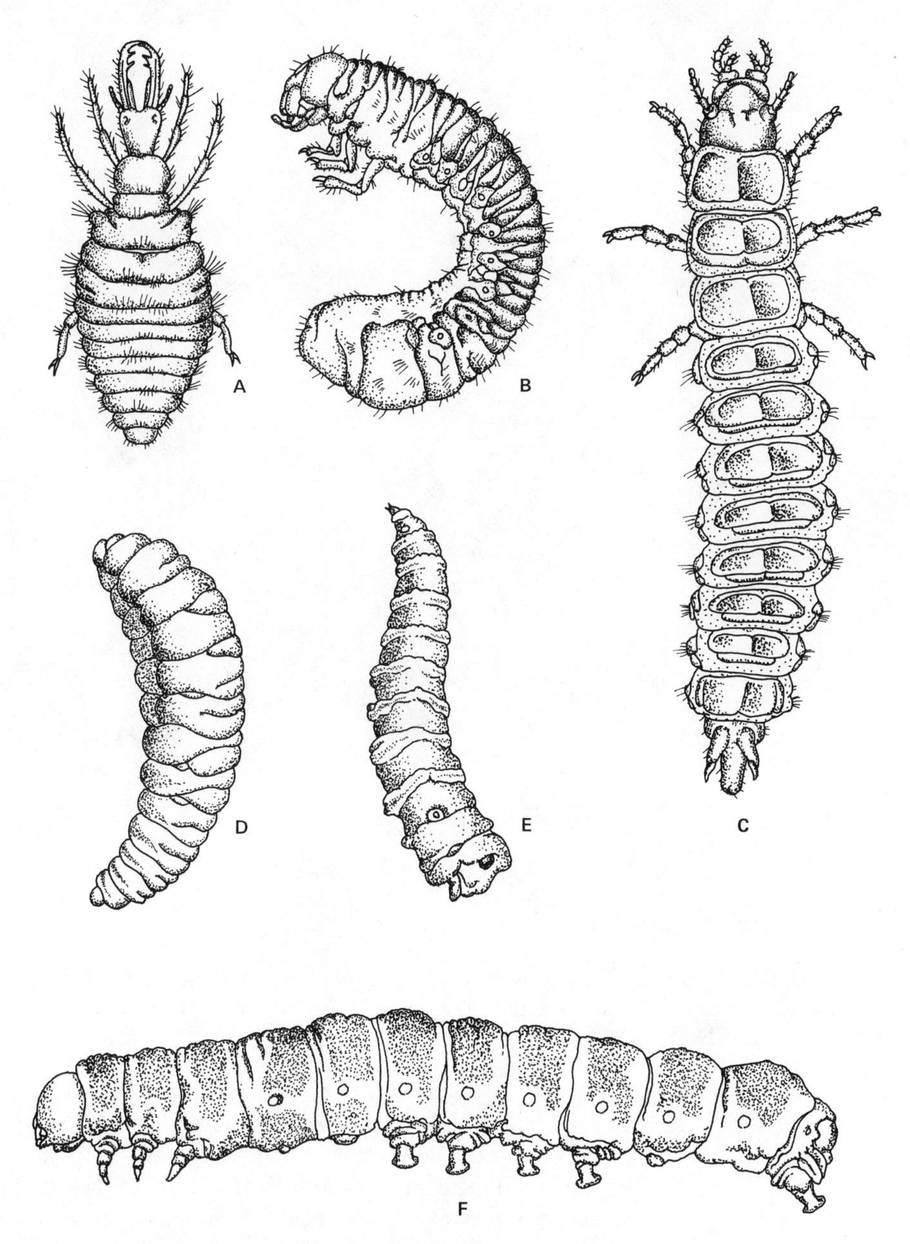


**FIGURE 9-7.** Some of the different types of endopterygote larvae. **A.** Ant lion (Neuroptera), **B.** Scarab beetle (Coleoptera), **C.** Carabid beetle (Coleoptera), **D.** Weevil (Coleoptera), **E.** Maggot (Diptera), **F.** Caterpillar (Lepidoptera).

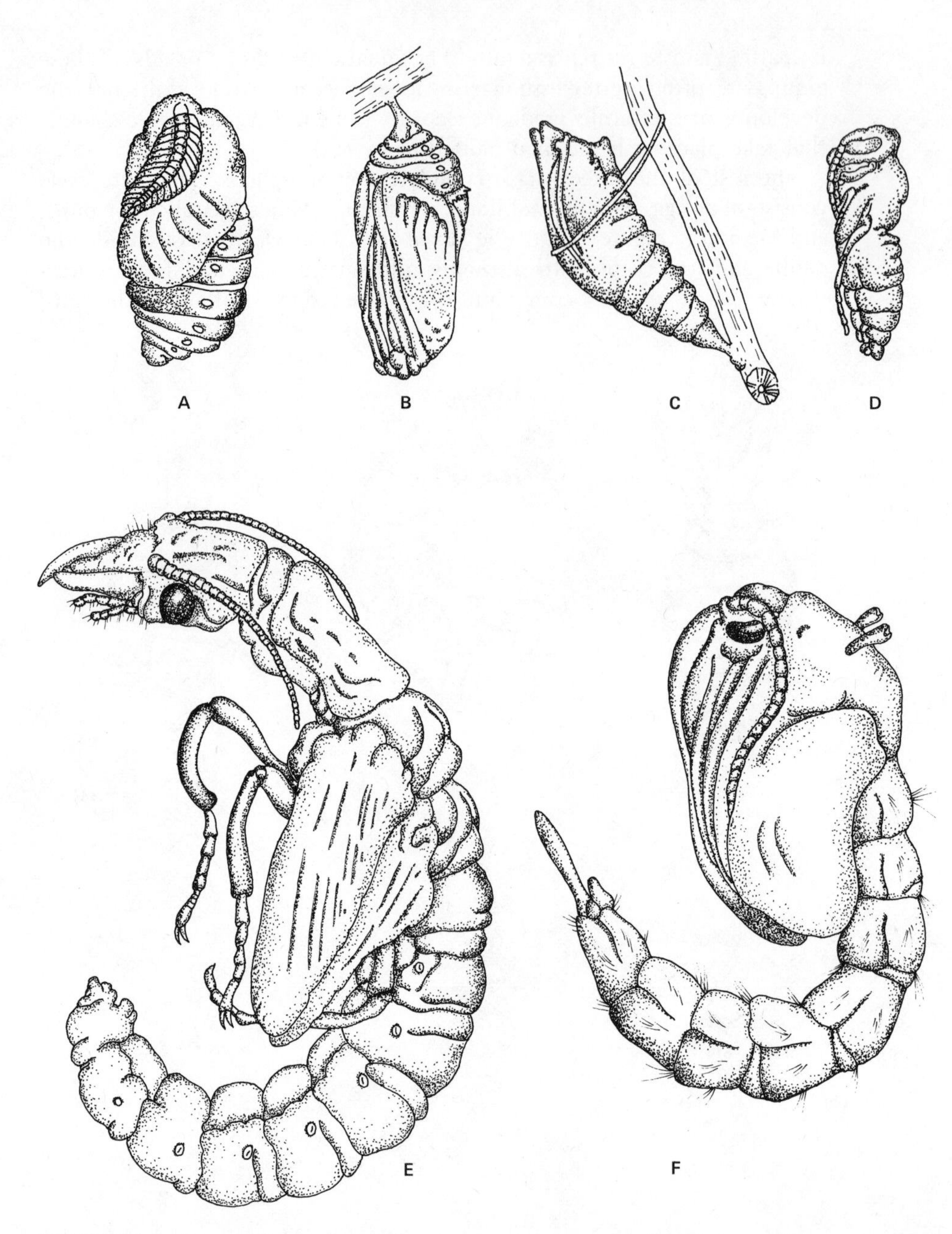


**FIGURE 9-8.** Some different forms of endopterygote pupae. **A, B** and **C.** Lepidoptera, **D.** Hymenoptera, **E.** Active pupa of a dobsonfly (Megaloptera), **F.** Active pupa of a mosquito (Diptera).

chewing or reduced mouth parts regardless of the type of mouth parts possessed by the adult stage. They may have both thoracic and abdominal legs, just thoracic legs, or no legs at all. The wings occur internally only as small buds of tissue. The almost total reorganization required to transform a larva to an adult takes place in the pupal stage (Figure 9-8), interposed between the last larval instar and the adult. The pupa is typically inactive, but there are a few exceptions to this. The pupal stage of some Megaloptera are weakly ambulatoral and have functional mandibles which they use to defend themselves (Figure 9-8*E*). The pupal stage of mosquitoes, called **tumblers,** have a pair of spiracular "horns" on the thorax that they project through the surface film for gas exchange; they also have specialized projections at the tip of the abdomen used to propel them through the water in a tumbling fashion (Figure 9-8*F*). Many inactive pupae are protected within a **cocoon** constructed of silk or silk-bound debris by the last larval instar. The higher Diptera pass the pupal stage within the last larval skin, which hardens to form a **puparium.**

Normally the larval instars of an endopterygote species are fundamentally similar and have the same life style, like the various instars of a leaf-feeding caterpillar. However, some species pass through one or more larval instars that are different from the rest, so there is even more change, or metamorphosis, in the course of their life history. This is called **hypermetamorphosis** and is most commonly encountered among those insects such as the Strepsiptera, which are parasitoids throughout most of their larval life; since the female does not deposit her eggs on the host, the first instar is an active hunting stage. Once the motile larva locates a host and attains a feeding position, it molts to a legless larva that spends the rest of its immature life in the host or host's nest (Figure 9-9).

It is interesting to consider how a phenomenon such as metamorphosis evolved. The immediate ancestors of insects and, indeed, the apterygotes display little or no change in form during development; yet, metamorphosis is a characteristic of all winged insects. Once wings evolved, why were they restricted to the adult? The answer to this question, even though conjecture, contributes to an understanding of metamorphosis. Growth, flight, and reproduction all have a high demand for energy; it would seem clearly advantageous to separate these processes, especially in terms of maximizing the utilization of cyclicly abundant food resources. Also, wings would be disadvantageous for actively growing larvae in many feeding situations and would create additional problems during molting. Once the possession of wings was relegated through evolution to the adult stage, some degree of metamorphosis became a requirement. The significance of wing possession as a contributing factor to the evolution of metamorphosis is illustrated further by the fact that members of the exopterygote orders that have become

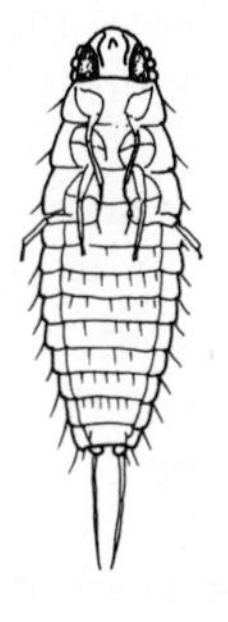


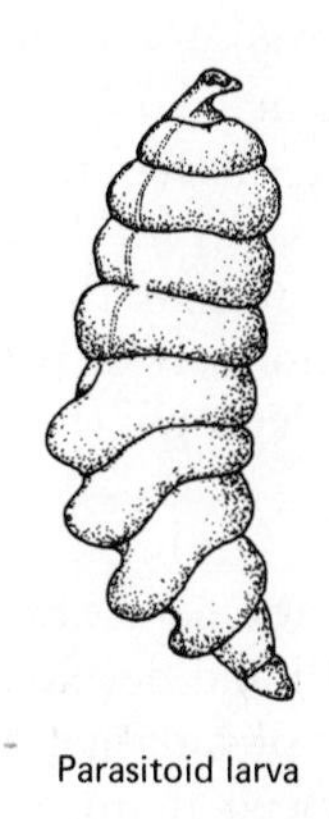


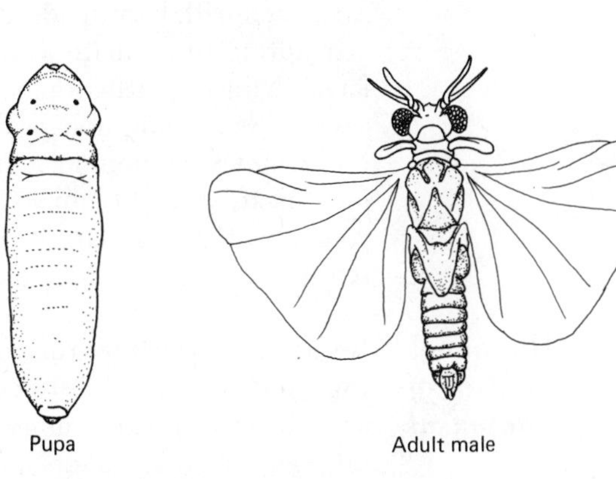


**FIGURE 9-9.** Life cycle of a twisted-winged parasite (Strepsiptera) showing hypermetamorphosis.

secondarily wingless develop in a manner resembling the apterygotes; that is, they are essentially ametabolous. In considering the endopterygotes, Hinton (1963) suggested that they require a pupal stage to accommodate the internal development of the wings and the flight muscles. He proposed that the wings are evaginated with the larval-pupal molt to permit the complete development of the flight muscles; an additional pupal-adult molt then became necessary to release the adult. In order for most endopterygote larvae to become adults, there must be a considerable amount of tissue reorganization, plus the breakdown of larval muscles (**histolysis**) that will not be needed in the adult. The pupal stage seems necessary to accommodate these changes.

However, the presence or absence of wings and how they develop does not provide the sole basis for metamorphosis among modern-day insects, and the pupal stage undoubtedly evolved independently several times. In some insects we observe rather marked changes in the appearance of immatures without a corresponding change in the appearance of the adults. This suggests two sets of genes, one governing larval characters and the other, adult characters. Thus, the two stages could evolve more or less independently, since changes beneficial to the larva need not benefit the adult, and vice versa. As beneficial adaptations accumulate independently, the two stages may diverge to the point where the life history consists of two quite different organisms with different modes of living; the larva becomes a feeding and growing stage and the adult, a dispersive and reproductive stage. The further apart the stages become, the greater the transformation necessary to convert one to the other, and this could require a special stage. The addition of such a stage has not occurred only among endopterygotes. For example, in the thrips, whiteflies, and scale insects, the feeding immature stages and the adults have diverged so markedly that a considerable amount of tissue reorganization is necessary for the completion of development. This is accomplished in one or more nonfeeding late larval instars that can quite properly be called pupae. So as Wigglesworth (1954) suggested, the larvae and adults of some insects may be essentially two organisms, the characteristics of which are expressed in sequence as a form of temporal polymorphism.

From a practical point of view, this divergence of immatures and adults compounds the problems of insect taxonomy. In some cases only the immatures or adults of a species are known; this is particularly true for the Trichoptera in which only about 15% of the described larvae have been associated with known adults (Denning, 1968). Frequently, entomologists in attempting to evaluate a situation, must make an identification based on larval characters rather than adult ones, and this requires an understanding of a second set of morphological features, especially for the endopterygotes. The immatures are commonly more voracious feeders than the adults, making identification of the immature stages a vital part of pest control. The two-volume set on the larvae of insects by Peterson (1960, 1962) is a useful reference.

### *Hormonal Control of Development*

Regardless of whether the young insect that emerges from the egg resembles the adult or is quite distinctive, its growth and development must proceed according to a predetermined plan in spite of variations in the complex array of environmental factors that tend to speed it up or slow it down. The neuro-endocrine system that provides the necessary coordination has been the subject of investigation for many years but has only recently been understood

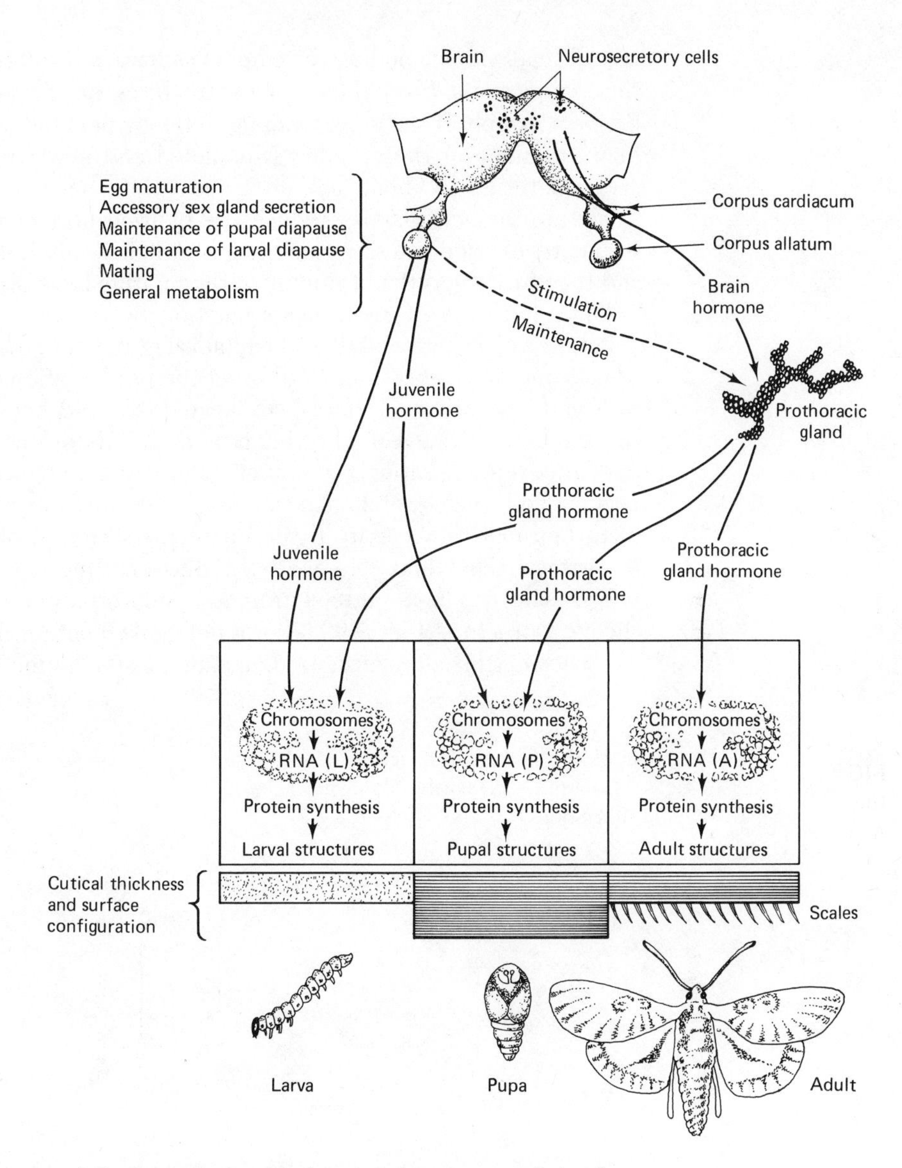


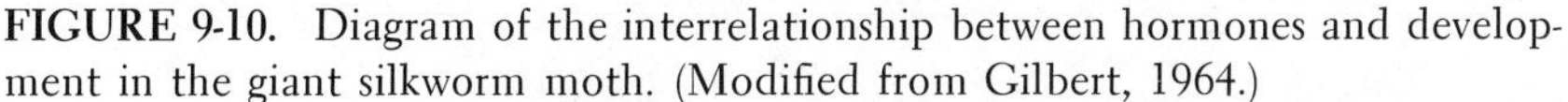

**FIGURE 9-10.** Diagram of the interrelationship between hormones and development in the giant silkworm moth. (Modified from Gilbert, 1964.)

and even now not completely so. There are three main organs involved in the regulation of development that are common to all insects. These are a group of **neurosecretory cells** in the brain, the small, white, glandular **corpora allata** just behind the brain, and some nondescript glandular tissue usually in the thorax called the **prothoracic glands** (Figure 9-10).

The neurosecretory cells cyclically release the **brain hormone** in response to various internal and external environmental stimuli received by a variety of sensory receptors; the brain hormone, which has been shown to mediate the production of the other hormones, therefore provides the bridge between the environment and the hormones that directly affect development. The brain hormone flows along inside the axons of some modified nerve cells that extend from the site of hormone production in the **pars intercerebralis** to a pair of small storage bodies, the **corpora cadiaca,** which often lie adjacent to the corpora allata. From these storage structures the hormone is released into the hemolymph. If the nerves leading to the corpora cadiaca are severed, the brain hormone in those bodies is depleted and newly produced hormone accumulates on the brain side of the cut where new corpora develop.

The prothoracic glands produce the molting hormone **ecdysone,** which stimulates division and secretion by the epithelial cells underlying the cuticle and thereby triggers the beginning of the molting cycle discussed in Chapter 5. The mediation of ecdysone production by the brain hormone can be demonstrated experimentally. The removal of the brain results in a cessation of molting, but it is resumed following the implantation of an active brain from another individual. If a larva is ligated (tied off) between the head and the thorax and between the thorax and the abdomen, and one brain is implanted into the thorax and another into the abdomen, the thorax will begin to molt but the abdomen will not. These experiments indicate that the brain hormone is necessary to stimulate the secretion of ecdysone by the prothoracic glands but cannot itself induce molting.

Ecdysone was first purified from the silkworm moth in 1954, but its chemical structure (Figure 9-11*A*) was not worked out until 1965. Since then, it has been artificially synthesized, making possible numerous studies on the

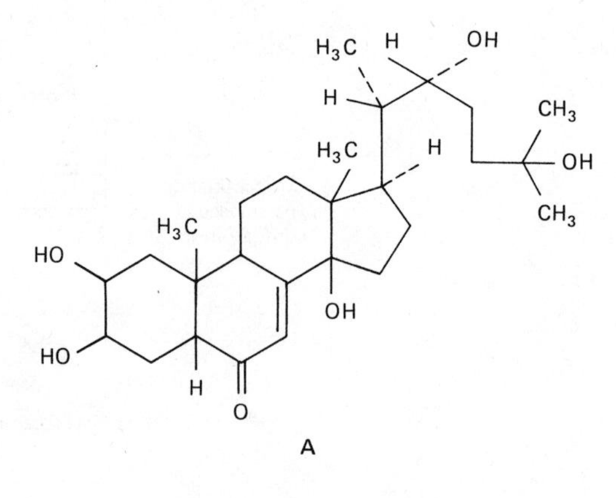


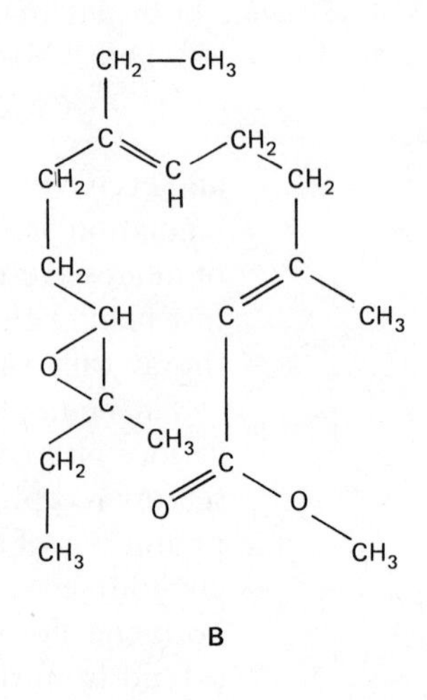


**FIGURE 9-11.** Diagrams of the chemical structure of the principal insect hormones presently known. **A.** Ecdysone, **B.** Juvenile hormone.

hormonal induction of molting. One of the most important findings to result from this work proved the error of an earlier contention that cell differentiation at the time of the molt is controlled by the concentration of ecdysone present.

Whether a larva molts to another larger larva or proceeds with the development of adult characters is now known to depend on the concentration in the hemolymph of the corpora allata secretion, **juvenile hormone.** This substance was not purified and identified until 1967 (Figure 9-11*B*). Its production by the corpora allata seems also to be mediated by the brain hormone. The presence of juvenile hormone in the hemolymph suppresses the expression of adult characteristics so that the juvenile morphology is retained. In the exoptygotes, growth is apparently accompanied by a progressive lowering of the juvenile hormone concentration in the hemolymph. As a result, the nymphs become a little more adultlike in appearance with each molt until at the time of the final molt juvenile hormone is absent and an adult emerges. The juvenile hormone effect can be demonstrated by removing the corpora allata from a young nymph and implanting it in a last instar nymph. The nymph from which the corpora were removed will develop precociously to a small adult, whereas the one which receives the young corpora allata, and therefore juvenile hormone, molts to a giant nymph instead of an adult.

In the endopterygotes, the immatures have larval tissues plus groups of cells called **imaginal discs** with the potential to become adult structures. As long as juvenile hormone is present, the growth of the imaginal discs is inhibited. Prior to the larval-pupal molt, juvenile hormone production ceases and its concentration in the blood declines, thereby permitting development of the imaginal discs and altering the metabolism of larval tissues. During the pupal stage, the juvenile hormone concentration drops to a negligible level and the final molt results in the emergence of a full-fledged adult. In the adult, the corpora allata become reactivated, and the juvenile hormone that is produced stimulates development of the ovaries and the production of yolk for the eggs.

To summarize then, the neurosecretory cells of the pars intercerebralis produce a brain hormone that serves as a mediator between the environment and the other hormonal systems. When stimulated by the presence of brain hormone in the blood, the prothoracic glands secrete ecdysone, which acts upon the epidermis to initiate a molt. The result of the molt is dependent on the concentration of juvenile hormone in the blood (Figure 9-12).

Before the purification and identification of the juvenile hormone, several compounds, which upon later comparison were found to be structurally similar to juvenile hormone, had been shown to interfere with normal insect development. In the early 1960s, an "accident of science" led to the discovery of a substance present in paper made from balsam fir. This "paper factor" disrupted the normal development of the bug *Pyrrhocous apterus,* resulting in an extra giant nymphal stage that was unable to molt to an adult (Slama and Williams, 1966). More recently a number of such compounds have been found to be quite commonly distributed throughout the plant kingdom, where they probably function, at least partially, as a defensive mechanism against insect herbivores. Ferns, for example, have quite high concentrations of these substances and are not fed upon by many insects. Scientists have also discovered that these **juvenile hormone mimics** do not usually affect all of the

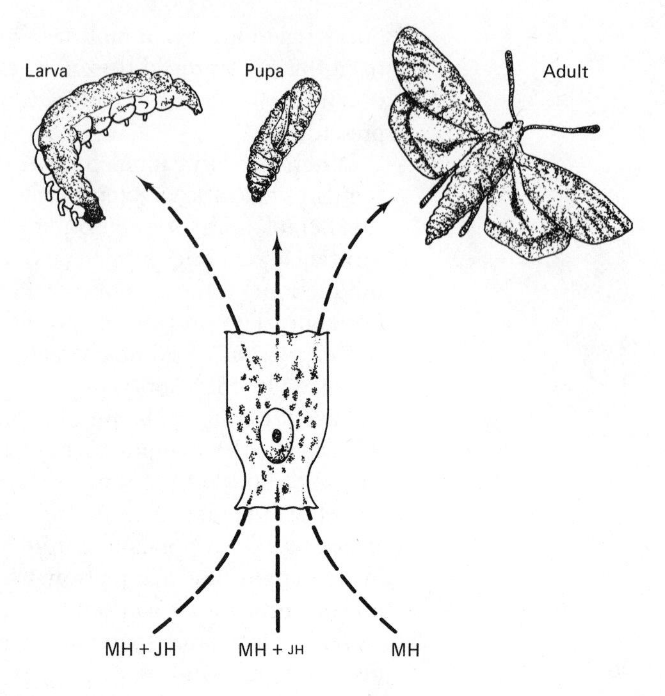


**FIGURE 9-12.** A schematic diagram of the role of hormones in complete metamorphosis. MH-molting hormone, JH-juvenile hormone. (Redrawn from Wigglesworth, 1968.)

different kinds of insects they are tested on, suggesting that some may have a specific action only on certain insects. Juvenile hormone analogs are now being tested as insecticides by several chemical research and development companies. They show a high level of activity against a variety of insects, particularly homopterans and the larvae of mosquitoes. Recently several such compounds, now called **third generation insecticides** (Williams, 1967), have been produced commercially for insect pest control (see Chapter 24). How far we can go from here will depend only upon our motivation and imagination.

Some insects will continue to reproduce generation after generation as long as conditions are favorable. This is true of some generalized feeders in the tropics and subtropics and also for many household and storage pests that have become adapted to the constantly favorable climate we have created in our buildings. In nature, insects with this type of life history may manage a variable number of generations in a given year, continuing to reproduce until the weather becomes unfavorable or food becomes scarce. Such insects are said to be **multivoltine.** Other insects have rather fixed developmental patterns, consisting perhaps of a single generation each year (**univoltine**). Still others take two or more years to develop, even though protracted favorable periods occur from time to time. If reproduction and development does not proceed continuously, a portion of the life cycle must be spent in some form of dormancy, during which development and metabolism are suppressed. Many terms and descriptions have been applied to the inactive states of insects (see Mansingh, 1971), but it is useful to distinguish between two basic types. The simplest form, **quiescence,** is a temporary arrest of development in response to some form of environmental adversity, most commonly a drop in temperature. When conditions become favorable again, development resumes more or less immediately. The second type of developmental arrest, **diapause,** is induced by some mechanism in advance of the onset of adverse conditions and persists until the adversity is safely past, regardless of whether or not there are brief returns to favorable conditions.

Insects may enter either quiescence or diapause in any stage of development. A single species may become quiescent in several stages during a single generation, but diapause is usually stage-specific for a given species. In some species, several generations may pass before conditions occur that induce diapause; this is called **facultative diapause** and is usually induced by a pronounced change in an environmental variable such as a drop in temperature or the drying up of a pond. In many species, diapause occurs in every generation regardless of environmental conditions and is called **obligatory diapause.** This form of developmental arrest is usually controlled by photoperiod; as the length of a day varies only slightly from year to year for any given location, it provides a safe and reliable regulatory stimulus. A more detailed treatment of diapause can be found in Wigglesworth (1954) or Lees (1955).

Dormancy, particularly diapause, is a highly significant adaptation that has enabled insects to cope with the fluctuating conditions that are characteristic of so many terrestrial environments. A dormant stage with little or no need for food can tide a species over periods of severe adversity or permit them to complete their development during two or more brief favorable periods separated by times when the environment is inhospitable. In this way, insects in the arctic and the subarctic can reproduce and develop during the short summer and survive the extreme cold of the long winter beneath the snow or ice. Likewise, desert insects can develop during or immediately following the rainy season and then survive in a dormant state through the hot, dry season.

The synchronization of insects and their natural enemies in hostile environments can be quite amazing. For example, the saturnid moth, *Hemiluca electra,* of southern California, flies during warm sunny weather in October and lays its eggs on the dead flower stalks of flat-topped buckwheat. The eggs remain dormant in the field until January when they hatch after the winter rains have stimulated the growth of their host plant. The larvae feed voraciously during winter and then pupate and enter an obligatory pupal diapause in April or May. The pupae remain just beneath the soil surface throughout the summer in cocoons of silk-bound grains of dirt. During this period, the host plant dries up and soil temperatures reach 50°C. The next generation of moths emerges in October to begin the cycle again. A small hymenopterous parasite of the *Hemileuca* larvae, *Apanteles electra,* lay eggs in the caterpillars during March and April. The parasite larvae complete their development quite rapidly and pupate at the same time as *Hemileuca* but do not emerge at the same time as the *Hemileuca* moths. They remain in their pupal diapause until caterpillars of the next generation are available as hosts the following spring.

Sometimes the physical environment is suitable for a long period of time, but a single source of food is abundant only for a shorter period. Many insects, particularly aphids, have evolved complex life histories that enable them to take advantage of such a situation. Their annual cycle may involve dormancy, a series of generations that reproduce by parthenogenetic viviparity, a generation that reproduces by sexual oviparity, both winged and wingless adults, and an alternation between two or more host plants. A relatively simple aphid life history is illustrated in Figure 9-13.

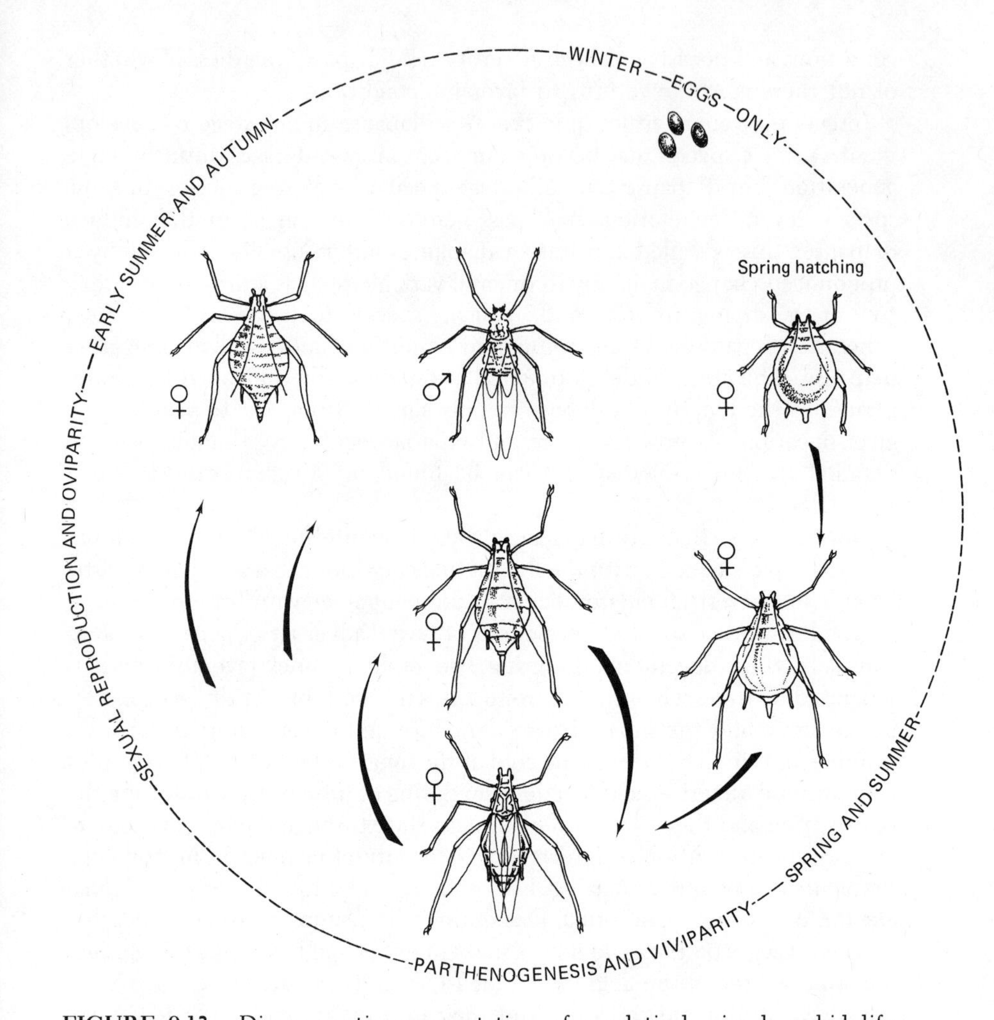


**FIGURE 9-13.** Diagrammatic representation of a relatively simple aphid life history. (Based on Lees in Kennedy, 1961.)

# References Cited

DENNING, D. G. 1968. Trichoptera. Pages 237–270 *in* R. L. Usinger, ed. Aquatic insects of California. University of California Press, Berkeley, Calif.

HINTON, H. E. 1963. The origin and function of the pupal stage. *Proc. Roy. Entomol. Soc.* Lond. A, 38: 77–85.

IBRAHIM, A. I., and A. M. GAD. 1975. The occurrence of paedogenesis in *Eristalis* larvae (Diptera: Syrphidae). *J. Med. Entomol.* 12: 268.

KENNEDY, J. S. ed. 1961. Insect polymorphism. Symposium No. 1, Royal. ent. Soc. London.

LEES, A. D. 1955. The physiology of diapause in arthropods. Cambridge University Press, Cambridge, England.

MANSINGH, AJAI. 1971. Physiological classification of dormancies in insects. *Can. Entomol.* 103: 983–1009.

PETERSON, ALVAH. 1960. Larvae of insects. Part II. Coleoptera, Diptera, Neuroptera, Siphonaptera, Mecoptera, Trichoptera. Published by the author, Columbus, Ohio.

PETERSON, ALVAH. 1962. Larvae of insects. Part I. Lepidoptera and Hymenoptera. Published by the author, Columbus, Ohio.

SLAMA, K., and C. M. WILLIAMS. 1966. The juvenile hormone. V. The sensitivity of the bug, *Pyrrhocoris apterus,* to a hormonally active factor in American paper pulp. *Biol. Bull.* 130: 235–246.

WIGGLESWORTH, V. B. 1954. The physiology of insect metamorphosis. Cambridge University Press, Cambridge, England.

WILLIAMS, C. M. 1967. Third-generation pesticides. *Sci. Amer.* 217(1): 13–17.

# Perceiving the environment 10

All animals have the ability to obtain information about their environment and to respond to that information in an organized manner. Obtaining the information, **sensory perception,** is the function of the central nervous system, assisted by a variety of sense organs or **sensilla** and the associated nerves that transmit the information. The active response to the various stimuli, **behavior,** results from nerve impulses transmitted to the appropriate affector organs. In this chapter we will review the variety and structure of insect sensilla and the nature of the stimuli they perceive. In the following chapters, we will examine some of the more important aspects of insect behavior that result from an integrated response to the environmental information perceived.

We have noted several times that insects have a rather hard exoskeleton, composed of a number of specialized layers that provide protection against water loss, physical damage, and the invasion of the soft tissues by harmful substances and microbes. If this body covering were unbroken except for the joints that allow for movement, the insect nervous system would be somewhat isolated from the external environment, much like a man in a full suit of armor. However, a suit of armor has several "windows" through which the man inside can see and hear. For insects, the sensilla scattered over the surface of the body form the "windows" through which the nervous system maintains contact with the outside world.

The sensilla arise from special cells in the epidermis. They usually consist of one or more **receptor cells,** each of which connects to the central nervous system by way of a nerve axon. Associated with the receptor cells are the **accessory cells,** which form the cuticular structures that either protect the sensory cells or are acted upon in some way by the environmental stimulus. The receptor cells are usually bipolar neurons that detect the stimulus and generate the nerve impulse which is transmitted to the central nervous system. There are also special internal multipolar sensory neurons associated with the muscles and internal organs that monitor certain aspects of the internal environment, but they will not be considered here.

Each type of sensillum is designed to detect a specific kind of energy or stimulus. One group perceives mechanical energy (**mechanoreceptors**) generated by physical forces such as molecular vibrations in the form of sound waves or various types of pressure. Another group perceives the chemical energy of various molecules (**chemoreceptors**), and a third group detects electromagnetic waves mainly in the form of light (**photoreceptors**) and heat (**thermoreceptors**). Regardless of the form of energy that stimulates a receptor, the electrical impulses transmitted along the nerve pathways are all the same. The central nervous system can identify the nature of each stimulus on the basis of the kind of sensory receptor that initiated the signal and the portion of the nervous system in which the signals are processed.

## *Mechanoreception*

The most common sensilla found on the body of insects involve a hairlike cuticular outgrowth called a seta (pl. setae) or a structure derived from a seta. As noted in Chapter 5, these sensory structures are produced by a trichogen that forms the hair or its derivative and by a tormogen that produces the basal membranous socket in which the seta moves (Figure 10-1*A*). The movement of the hair in its socket changes the pressure distribution across its basal membrane; this is detected by a push or pull on one end of the receptor cell. The receptor cell then transduces the mechanical energy into an electrical impulse. Although the tactile hair is the most common form of mechanoreceptor, it is frequently modified to improve the efficiency with which certain kinds of mechanical stimulation can be detected. The hair may be replaced by a domelike structure (Figure 10-1*B*) or there may be a group of receptor cells lying beneath a cuticular plate. These types of receptors that are called **campaniform** and **placoid,** respectively, are constructed so that when the cuticle is flexed or bent, the tip of the receptor cell(s) attached to the sensory structure detects the movement. These structures are reviewed by McIver (1975).

Dethier (1963) classifies the hair sensilla as either **pressure sensitive** or **velocity sensitive.** Other workers call these **tonic** or **phasic,** respectively. Tonic sensilla respond slowly to a static force like pressure by sending a steady flow of impulses to the central nervous system. These sensilla detect, among other things, contact with another object. Phasic sensilla adapt quickly and respond primarily to the velocity of bending rather than a final static position. These

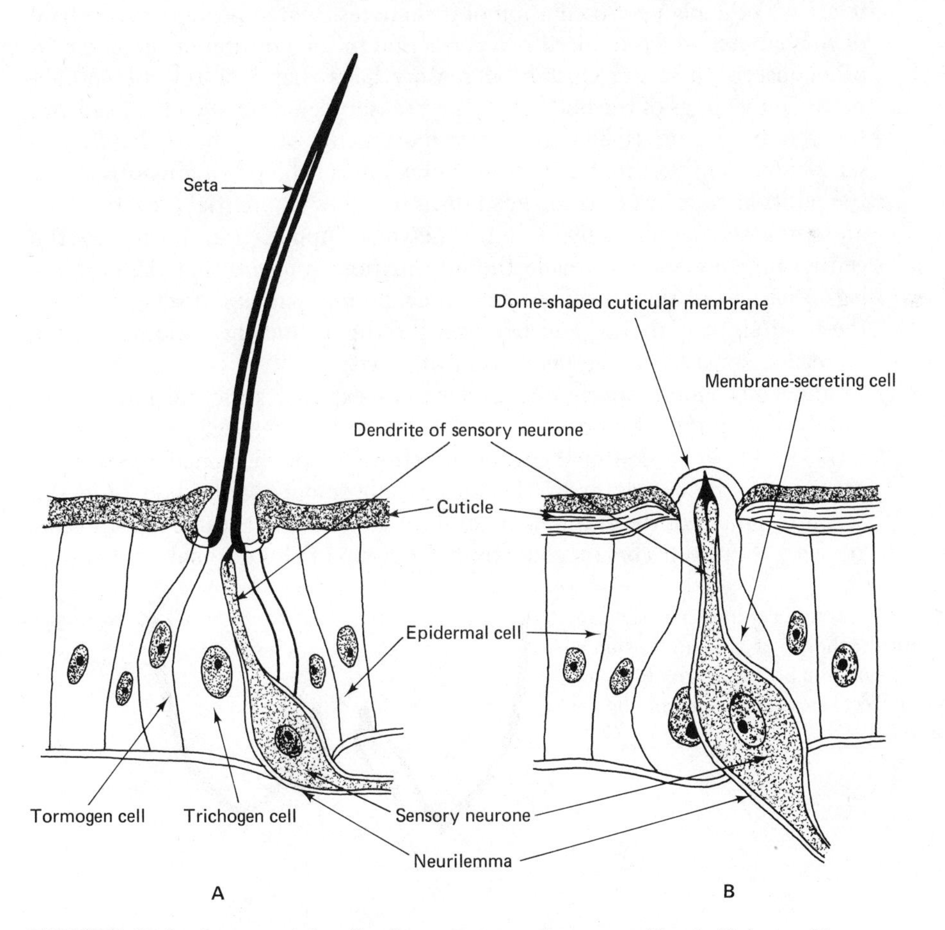


**FIGURE 10-1.** Structural details of two common insect sensilla. **A.** Hair sensillum, **B.** Dome sensillum. (Redrawn from Snodgrass, 1935.)

sensilla detect dynamic forces such as those generated by the flow of air or water.

Mechanoreceptors are located over much of the surface of the body. They are not distributed evenly, however, and are often concentrated in large numbers in specific places such as on the legs, antennae, and mouth parts that frequently come in contact with other surfaces. Many hair sensilla are extremely sensitive and can even detect a slight movement of nearby objects through perception of the air currents thus produced. A number of insects have tactile hairs on their face that detect the movement of air over the head during flight; some cease flying when these hairs are not adequately stimulated. In addition to detecting movements in their environment that may indicate some form of danger, insects also use vibrations as a means of locating prey. Water striders, for example, can detect the concentric pattern of ripples generated when an insect accidently drops into the water and can follow the ripples to their source in search of the potential meal.

Another important function of mechanoreceptors is to provide the insect with information about its position in its environment and the position of different parts of the body relative to one another. The groups of sensilla that perform this function are called **proprioreceptors.** In the discussion of flight (Chapter 7), we reviewed how a change in the plane of the body of a fly relative to the plane of oscillation of its halteres created pressure at the base of the halteres which resulted in a correction in the position of the body. In other insects there are clusters of sensory hairs that monitor the contact between two parts of the body. Ants have sensory setae between the head and the thorax (Figure 10-2) that detect movements of the head. Patches of sensory hairs on the front of the prothorax inform the praying mantis about the position of its head. As the head turns to follow a potential prey, the hairs are stimulated accordingly, and the nervous impulses transmitted to the central nervous system provide the information to assure that the grasping legs reach out in the direction of the prey. Similar patches of setae between the head and the thorax, and between the thorax and the abdomen of the honeybee are used in the perception of gravity.

Some mechanoreceptors do not have an external structure but consist simply of a bundle of bipolar neurons stretched between two integumentary surfaces. These **chordotonal organs** are stretched or shortened by external pressure on the cuticle or by the various movements of the body. On the second segment of the antenna of adults of all true insects there is a structure known as **Johnston's organ,** consisting of a group of chordotonal sensilla that

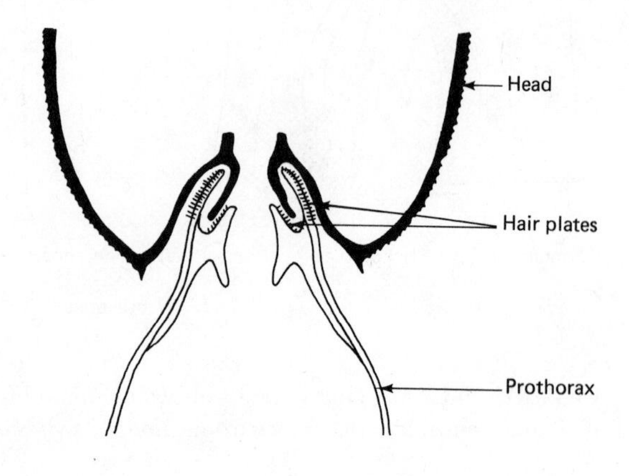


**FIGURE 10-2.** A diagram of a transverse section through the head and prothorax of the ant *Formica polyctena,* showing the position of the sensory hairs on the prothorax that detect movement of the head. (Redrawn from Markl, 1962.)

respond to movements of the antennal flagellum. In the fly *Calliphora,* Johnston's organ serves as a flight speed indicator by detecting the degree to which the air flowing past the head causes the third antennal segment to rotate relative to the second segment.

Johnston's organ also perceives information important in the orientation of aquatic insects. When a back swimmer is in its normal upside down position, the flagellum of the antenna is deflected away from the head by a small air bubble; when the back swimmer is the wrong way up, the antennae shift and the Johnston's organs perceive the new position of the body. Displacement of the antennal flagella caused by ripples on the surface of the water enables whirligig beetles (Gyrinidae) to avoid other individuals, and by detecting "echoes" of their own ripples, whirligig beetles avoid collisions with larger objects.

Johnston's organ and other special groups of chordotonal sensilla are also used as sound perceptors. Sound is a series of waves of mass movement of air molecules resulting from an alternation of compression and rarefaction. We tend to consider such waves passing through the air as sound which we perceive with our ears, and waves passing through a solid as vibrations that we detect quite differently. In insects there is really no difference in the way these signals are perceived by the receptor cells of chordotonal sensilla, although the structure of the organs that contain them is quite variable. Those receptors that perceive what we would consider to be vibrations are usually located in the proximal part of the tibiae and detect vibrations transmitted through the substrate to the legs. On the other hand, any structure such as a hair that is sensitive to movements of the air can become sensitive to the sound waves travelling through the air. Accordingly, Johnston's organ has become a hearing organ in some insects. Male mosquitoes locate their mates by detecting the sound of their wing beat. They have elaborate Johnston's organs which perceive vibrations of the flagellum caused by the sound waves that impinge upon their numerous antennal hairs (Figure 10-3).

The hearing organs of most insects consist of a membrane, or **tympanum,** backed by an air space so that it is free to vibrate at the frequency of the sound waves that impinge upon it. From two to 1,500 chordontonal sensilla associated with the tympanic membrane thereby detect the sound. Tympanic organs of this general construction are found in a variety of insects. In the Orthoptera, they are found on the front tibiae of crickets and long-horned grasshoppers and on the first abdominal segment of grasshoppers. In the Hemiptera, a few aquatic families have tympanic organs on the mesothorax, whereas the terrestrial cicadas have them on the first abdominal segment. Among the Lepidoptera, they occur on the abdomen of geometrids and pyralids and on the metathorax of noctuids.

### *Chemoreception*

The perception of chemicals is of great importance to insects. The detection of chemicals provides insects with a means of avoiding harmful substances and assists them in locating and identifying their mates, locating oviposition sites and nests, and recognizing their food. Although chemoreception has been divided into a general chemical sense, a sense of smell (**olfaction**), and a sense of taste (**gustation**), all chemicals are perceived in basically the same manner. That is, the chemical molecules come in direct contact with the receptor organs in order that the "potential energy existing in the mutual attraction

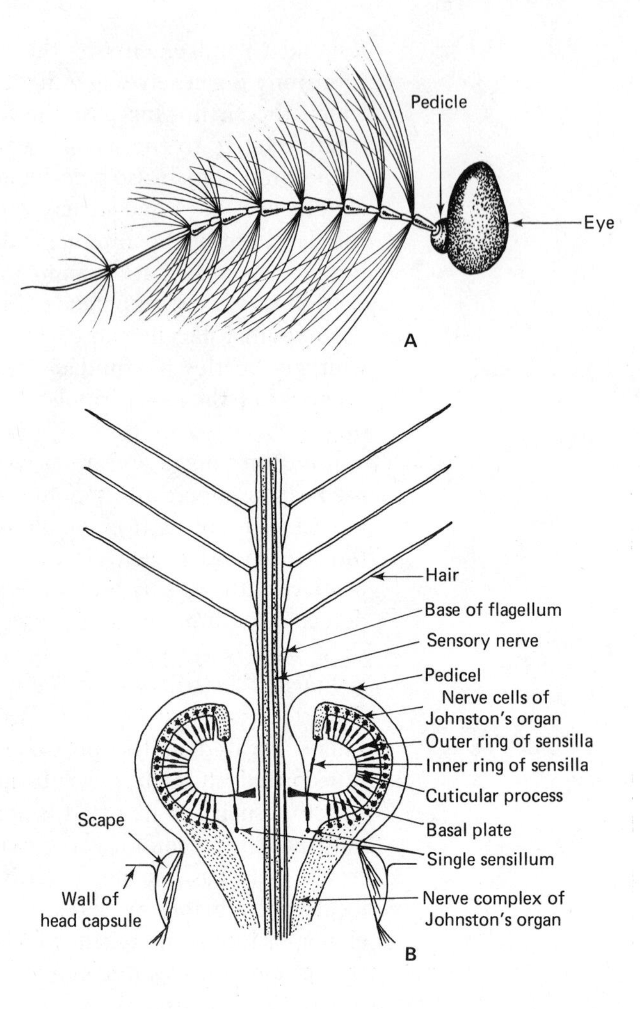


**FIGURE 10-3.** Johnston's organ. **A.** Hairy antenna of a male mosquito showing the swollen pedicel that houses Johnston's organ. **B.** A diagrammatic section through Johnston's organ. (Adapted from Chapman, 1969.)

and repulsion of the particles making up the atoms" can be detected (Dethier, 1963). Although we often say that insects can detect chemicals from a distance or by direct contact with them, we are really referring to their ability to detect the source of the chemical because in all cases the molecules must impinge upon a receptor before they can be sensed.

The fact that the chemical molecules must deliver their energy directly to the sensory neurons requires a different form of receptor organ than those used to detect mechanical stimuli, in which the sensory neurons are attached to a structure that responds to vibrations or changes in pressure. Chemoreceptors are generally characterized by having very fine nerve endings exposed to the environment through minute openings in the cuticle (Figure 10-4). Those that perceive chemicals in a gaseous state in relatively low concentrations (usually referred to as smell, or olfaction) normally have a large number of sensory neurons (Figure 10-4*B*). Each neuron responds to either a range of compounds or some special compound of importance. Those that perceive substances in a high concentration, in solution or a liquid state, have a smaller number of sensory neurons, each responding to a group of compounds in a manner we refer to as taste, or gustation (Figure 10-4C). Although the mechanism of detection is the same in each case, the olfactory receptors

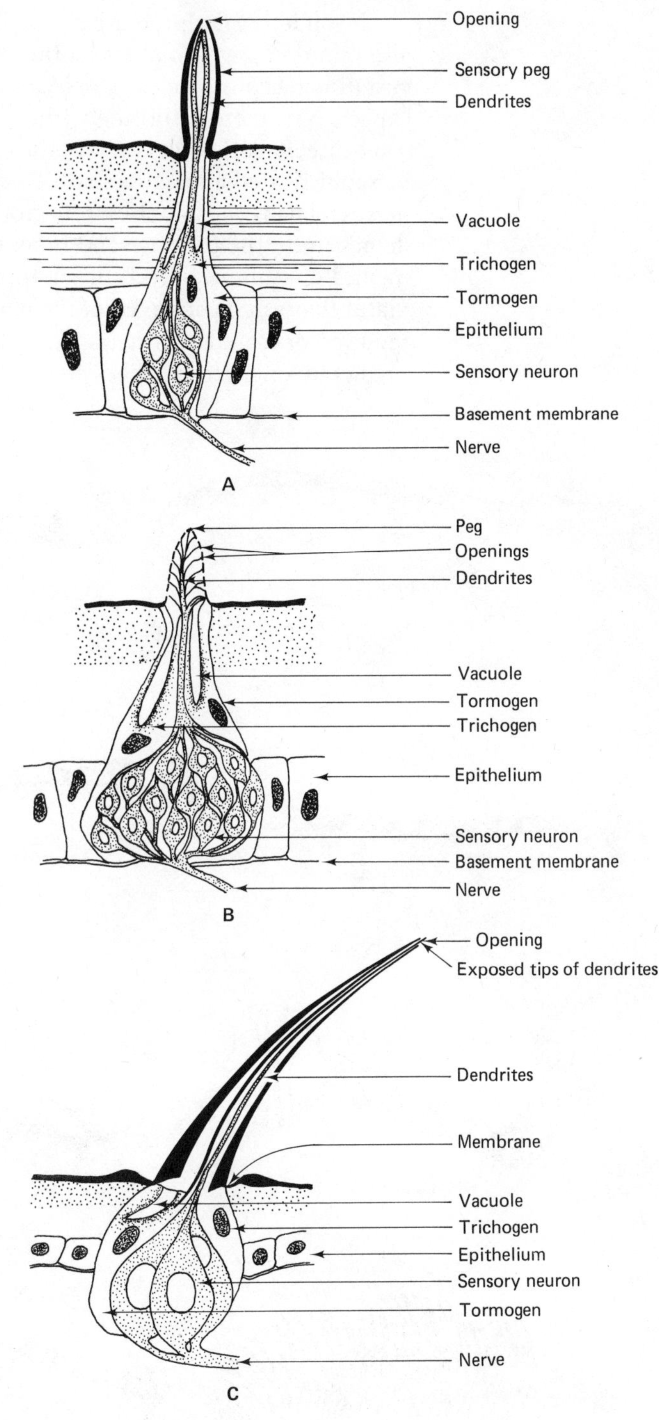


**FIGURE 10-4.** Insect chemoreceptors. **A.** Diagram of a thick-walled peg that functions in general chemical sensory perception, **B.** Diagram of a thin-walled peg typical of those that function as olfactory receptors, **C.** Diagram of a typical gustatory receptor. (After Slifer and others, 1959.)

enable the insect to detect chemicals in the air at some distance from their source, whereas the gustatory receptors are better designed for the identification of chemicals at their source. The receptors that provide the insects with a general chemical sense are apparently less discriminating than either the smell or taste receptors and initiate an avoidance response to irritating substances when they occur in high concentrations.

The olfactory receptors are most commonly found on the antennae but are also found in large numbers on the mouth parts, particularly the palps. The gustatory receptors are extremely abundant on the mouth parts, as one might expect, but, more surprisingly, they occur in large numbers on the tarsi of some insects. Although the response to chemical stimulation is now measured by sophisticated apparatus that detects changes in the electrical potential of nerves relaying the impulses from groups of chemoreceptors, it is quite easy to demonstrate the presence and function of these organs without such equipment. For example, insects often display an orientation to certain foods or to mates that can be observed in the laboratory. However, if their antennae are removed or coated with an impermeable substance, they no longer display the predicted orientation. The presence of taste receptors on the tarsi of a

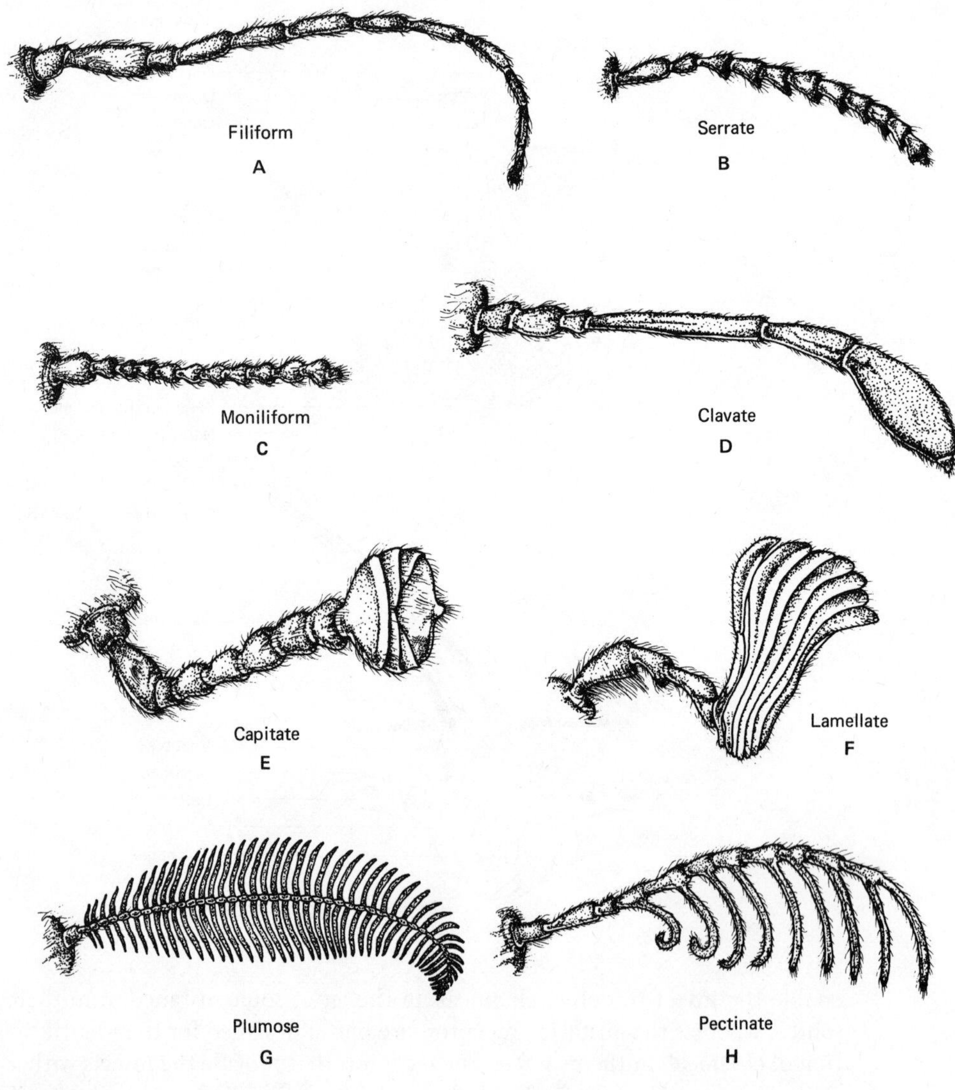


**FIGURE 10-5.** Different configurations of insect antennae showing various ways the sensilla-bearing surface is increased. A terminal club composed of a series of plates or lamellae typical of the antennae of scarab beetles is shown in F.

honeybee can be demonstrated by bringing the tarsi in contact with a sugar solution and noting that the bee will immediately extend its proboscis in search of food.

For a response to occur, a number of molecules must impinge upon receptors for that particular class of compound. This is facilitated by the presence of large numbers of receptor organs and the presence of the several neurons and nerve endings in each one. In various insects, the surface area that supports the receptor organs is often increased which often accounts for the elaborate configuration of the antennae (Figure 10-5). Such a relationship is particularly well illustrated by the terminal antennal clubs of scarab beetles, which are composed of a series of lamellae; when searching for food, the antennae are held up with the plates spread apart, thereby exposing more sensilla.

### *Hygroreception*

The perception of moisture in the air, **hygroreception,** is not very well understood for insects, even though there is ample behavioral evidence to support it. The detection of water is a form of chemoreception in the sense that water is a chemical compound but, at least in some cases, it may be perceived more like a mechanical stimulus. Many animal hairs are known to straighten and curl in relation to the wetness of the environment, which is why horsehair is used as the sensitive element in hygrographs and why women often complain about losing their curl on damp days. If insect hairs were to respond to moisture in a similar manner, it would change the distribution of tension across the membrane of the hair's socket and stimulate the associated sensory neuron. The relative humidity also affects the drying power of the air and, consequently, the amount of evaporational cooling that may be detected by thermoreceptors. In the human louse, *Pediculus humanus,* which is particularly sensitive to changes in the moisture content of the air, the antennae bear tufts of hair that absorb water relative to the amount present in the adjacent atmosphere.

### *Photoreception*

The ability to perceive that part of the electromagnetic spectrum referred to as light is **photoreception.** We tend to think of light reception as vision because the stimulation of our light-sensitive organs usually results in the formation of an image. However, many animals are unable to perceive images and respond only to the presence or absence of a light stimulus. The insects as a group are able to perceive light versus darkness, photoperiod, light intensity, the plane of polarization, movement, form, patterns, and some colors. Which of these light-related forms of environmental information an individual insect perceives depends upon the type of receptors it has.

Some insects do not have identifiable light receptors but respond to light striking or penetrating the cuticle in a manner referred to as a **dermal photo response.** The maggots of some higher Diptera have groups of photosensitive cells near the anterior end of their bodies which result simply in an orientation and movement away from the source of stimulation. However, most insects have well-developed photoreceptors in the form of simple eyes and compound eyes.

The **compound eyes** of insects are unique structures which appear to have arisen coincidently with the evolution of flight. They are not present among the apterygotes, and although the fossil record does not reveal their evolutionary origin, it would be hard to imagine a flying insect that lacked the ability to perceive the kinds of light-related environmental information already mentioned. Most adult insects have a pair of compound eyes, one on each side of the head, which tend to bulge outward and provide a wide field of vision. However, the organs have been secondarily lost in the adults of some parasitic, sedentary, and cave-dwelling forms for which flight and vision are less important. Compound eyes are also a characteristic of the immatures of most exopterygotes, whereas they are not present among the larvae of endopterygotes.

Each compound eye consists of an aggregation of individual sensory units called **ommatidia** (Figure 10-6). Each ommatidium consists of a group of structures that gather the light and a receptor apparatus that receives the light and transforms it into the electrical energy that is transmitted as a nerve impulse.

The light-gathering component of each ommatidium consists of an outer cuticular lens, or **cornea,** which is transparent and more or less biconvex. The lenses combine to form the surface of the compound eye, each making up an area known as a **facet.** Immediately beneath the **cornea** is a hard, clear structure called the **crystalline cone,** surrounded by a group of **corneal pigment cells.**

The light-receptor apparatus which lies beneath the light-gathering structures consists of six to eight **retinula cells,** each of which is connected through an axon to the optic nerve that leads to the brain. The retinula cells also combine to produce a central, compound rodlike sensory unit called a **rhabdom.** The retinula cells and rhabdom are surrounded by 12 to 18 **secondary pigment cells,** which isolate each ommatidium from neighboring ones to various degrees.

The number of ommatidia that make up the compound eyes varies greatly from species to species and is believed to be directly correlated with visual acuity. When only a few are present, the facets are round and separated from

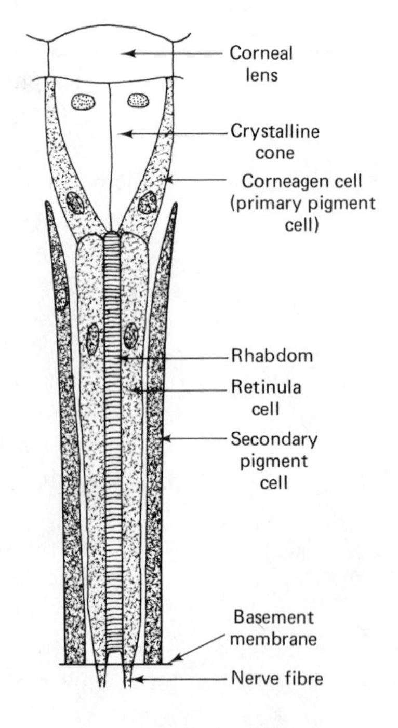


**FIGURE 10-6.** Diagram showing the structural details of a single light-sensitive unit called an ommatidium.

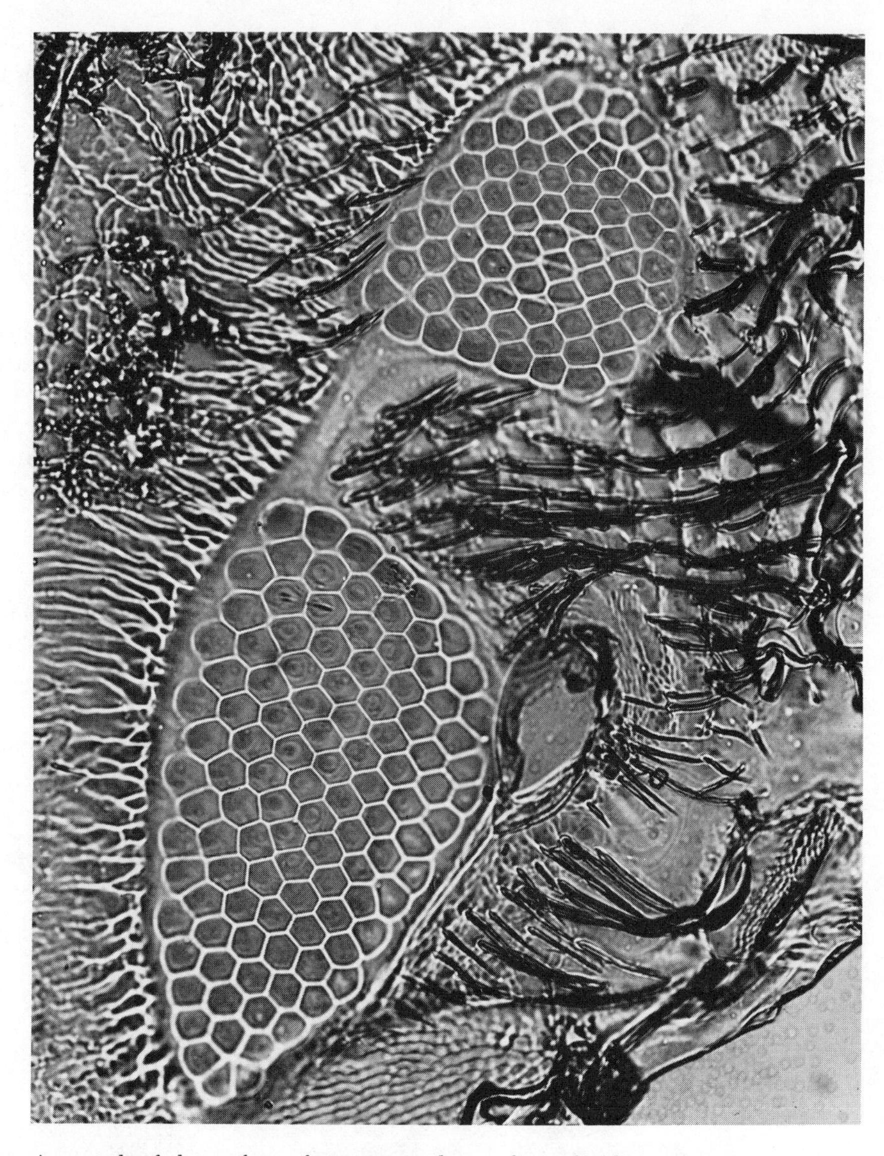

**FIGURE 10-7.** Photomicrograph of the surface of a compound eye of a scolytid beetle showing the hexagonal facets. (Courtesy of Canadian Forestry Service, Victoria, B.C.)

each other by narrow areas of pigmented cuticle, whereas when they are numerous, they are more or less hexogonal and closely packed (Figure 10-7). Each ommatidium perceives that part of the visual field from which light waves fall upon its rhabdom. The individual images received by the ommatidia are then combined to produce a mosaic image, which has often been likened to our viewing the environment through a bundle of drinking straws.

However, the amount of the environment perceived by each ommatidium is dependent upon the area from which light waves are received, and we now recognize two basic types of compound eye that are different in this respect. In the **apposition eye** (Figure 10-8*A*), the corneal and secondary pigment cells optically isolate the ommatidia so that a minimum of light passes from one ommatidium to another. In the **superposition eye,** the retinal rods, or rhabdoms, are short and there is some movement of the shielding pigment. At higher light intensities, the pigment is distributed so as to isolate the ommatidia (Figure 10-8*B*), in which case the eye is said to be *light-adapted* and functions as an apposition eye. At low light intensities, the pigment moves to the outward part of the secondary pigment cells so that some light passes obliquely from ommatidium to ommatidium without being absorbed (Figure 10-8C), and the eye is said to be *dark-adapted;* this is of considerable advantage to insects that are active during times when little light is available.

It has been theorized that the apposition eye or the light-adapted superposition eye forms a more highly resolved image because more light from a limited portion of the visual field strikes each individual rhabdom, thereby forming an over-all image composed of well-delineated, intense subimages. On the other hand, the dark-adapted superposition eye gives up some of the power of resolution in order that sufficient light can be received by each rhabdom to induce stimulation, but there is overlap of the images received by adjacent ommatidia. However, all workers do not agree with this interpretation.

Because of their position on the head and because the curved surface of the eyes projects to some extent, insects have a very extensive visual field. In some, the visual field extends through 240° horizontally and 360° vertically, and the fields of the two eyes overlap to provide binocular vision above, below, and in front of the head (Figure 10-9).

Because our eyes are completely different structurally from the eyes of

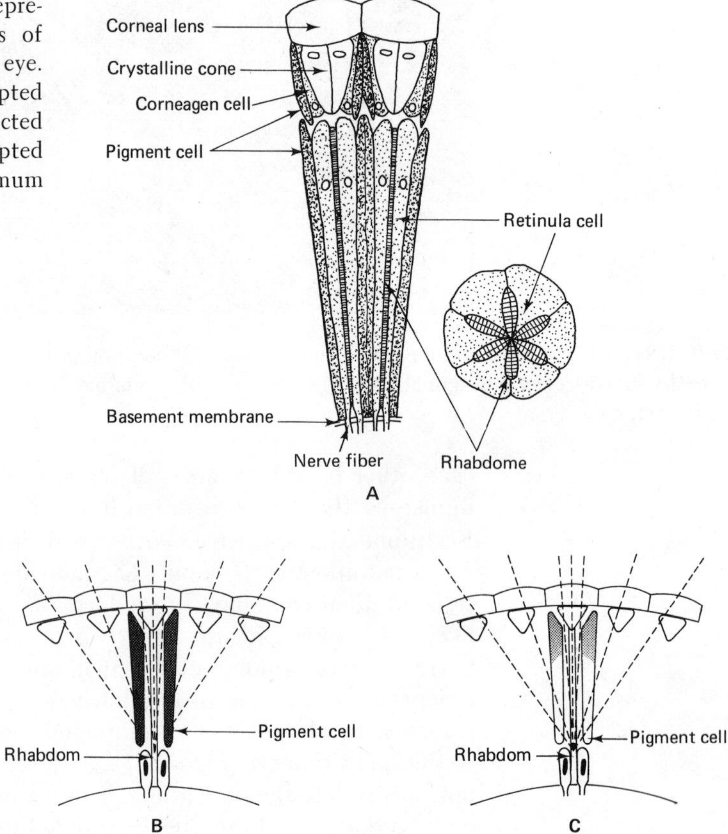


**FIGURE 10-8.** Diagrammatic representation of various arrangements of pigment cells in the compound eye. **A.** An apposition eye, **B.** A light-adapted superposition eye showing restricted light penetration, **C.** A dark-adapted superposition eye showing maximum light penetration.

**FIGURE 10-9.** Diagrammatic representation of the field of vision subtended by the compound eyes of a preying mantis. **A.** Dorsal view, **B.** Front view.

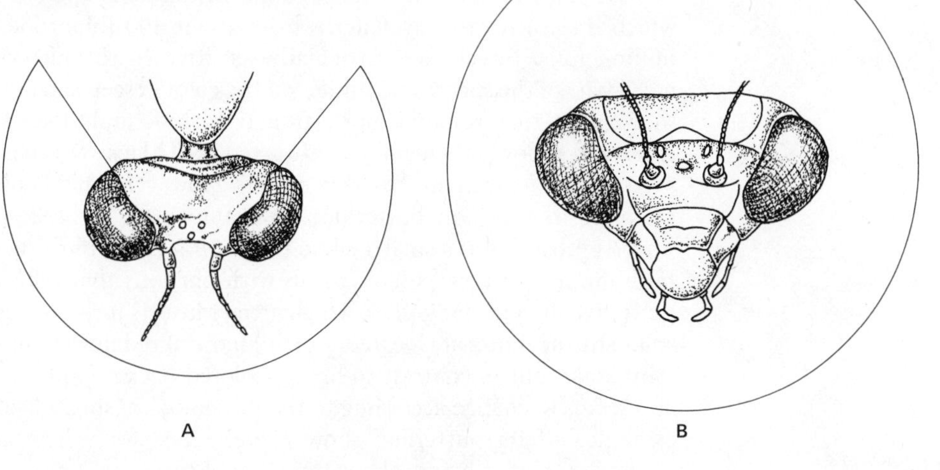


insects, we cannot develop a device that would enable us to see what an insect sees. Consequently, we develop our understanding of the insect's visual perception from an interpretation of behavioral evidence and the results of experiments. Much of the work related to insect vision has been conducted with honeybees because they can be trained to repeat behavior patterns through the offering of a reward in the form of sugar and water. Experiments have shown that bees can be trained to associate food with certain types of pattern that contrast with a uniform background color. However, they seem unable to distinguish between shapes such as triangles, squares, and circles that we differentiate readily. They can, however, distinguish solid forms from figures broken up into contrasting pieces (Figure 10-10). Studies have also shown that bees respond more readily to broken patterns that produce a high frequency of change in retinal stimulation. Wigglesworth (1965) suggested that the divided nature of flowers and also flowers moving in the wind produce flickering images that the bees associate with the presence of nectar. Hence, their ready response to a broken pattern that produces a flicker effect is very useful.

Insects have also been shown to be capable of perceiving color. Although

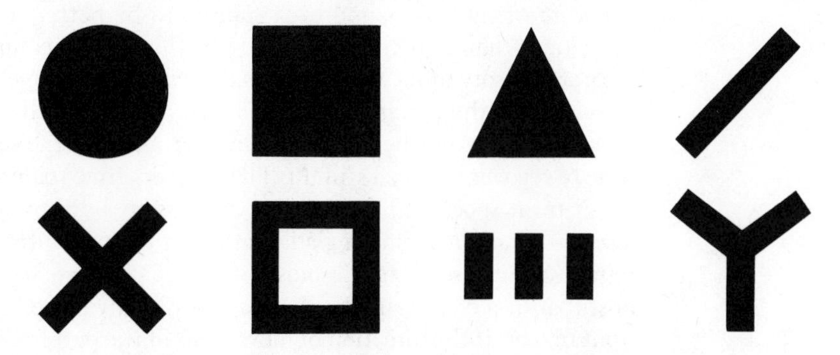

**FIGURE 10-10.** Symbols used to study pattern discrimination by the honey bee *Apis mellifera*. The upper row contains shapes not distinguished from each other, but which are distinguished readily from those in the lower row. (From Hertz, 1929.)

there is considerable variation in terms of the wavelengths perceived by different insects, as a group they are sensitive to wavelengths from about 240 (ultraviolet) to 700 (yellow-orange) millimicrons, compared to the human eye, which is sensitive to wavelengths from about 400 (blue-violet) to 800 (red) millimicrons. Insects are particularly sensitive to the ultraviolet and blue-green part of the spectrum. Studies on the color perception of honeybees have shown that they respond more strongly to blue and violet than to yellow-green and yellow, and seem to confuse red with black. This raises the question as to why so many plants have red flowers. However, not all pollinating species are blind to red; some butterflies, for example, are highly responsive to long wavelengths. Red is also attractive to nectar-feeding birds such as hummingbirds. Some red flowers have petals with portions that reflect ultraviolet in patterns called **nectar guides,** which attract insects into the base of the flower where the nectaries are located. Furthermore, if red flowers appear black, they must stand out in contrast to lighter-colored backgrounds. Color preference also seems to change according to the physiological state of some insects. For example, alfalfa butterflies show a preference for yellow and blue while feeding on nectar before they attain sexual maturity, but once mature, they favor the green color of leaves on which they lay their eggs.

When light waves from the sun enter the earth's atmosphere, they collide with gas molecules and dust particles. These collisions eliminate some of the planes of wave vibration until many of the waves vibrate in only one plane. Such waves are said to be polarized. The portion that becomes polarized varies in different parts of the sky but is generally greatest at an angle of 90° to the sun. For example, the plane of maximum polarization is directly overhead when the sun is just above the horizon. Many insects are able to detect polarized light and can therefore determine the position of the sun even when it is obscured from view. This was demonstrated by von Frisch (1950) to enable honeybees to communicate the position of food sources relative to the sun even on partly cloudy days. Other workers have suggested that polarized light orientation enables insects to maintain a steady course across open, unfamiliar territory when sun navigation or orientation by landmarks is not possible. Wellington (1974) indicated the importance of polarized light in a study that showed that insects often became inactive or restricted their range of movement during midday when polarized light was absent from the overhead sky.

Movement perception is also of considerable importance to insects, and the structure of the compound eyes appears to be better adapted to detecting movement than to perceiving images. As the eyes are composed of numerous, separate sensory units, movement is perceived as a source of stimulation that travels across the eye and activates a series of ommatidia in sequence. Rapid movement may not be detected, however, because the sensory units require time to recover between stimuli. The recovery time required between stimuli varies from species to species but appears to be shorter for rapidly flying insects—a rather fascinating adaptation to the perception of rapidly passing terrain. Some insects such as mosquitoes will continue to fly on a fixed mount in still air if a cross-striped pattern is continually passed beneath them from front to rear. If the direction of movement of the pattern is reversed, however, they will stop beating their wings even if a stream of air is directed backward over their heads. This suggests that the direction of pattern movement seems to provide an indication of flight speed relative to the ground.

Insects must also be able to judge distance for a variety of reasons, but this is particularly important for insects that capture prey. Distance perception requires simultaneous stimulation of similar groups of ommatidia in the two eyes. This can be demonstrated easily by damaging or masking one eye. However, insect eyes are fixed relative to each other and cannot determine distance by converging on a fixed point as is possible in vertebrates. The distance to an object is determined by its position in the visual field of the two eyes and the point of intersection of projected axes of the simultaneously stimulated ommatidia. The accompanying diagram (Figure 10-11) shows that a prey within reach of the prehensile mask of a dragonfly nymph would be perceived only by the ommatidia located in a particular part of the eyes, whereas other groups of ommatidia would perceive the prey to be out of reach. Maldonado and Barros-Pita (1970) have shown that in the praying mantis the image of a prey must fall into the region of high-resolution, called the **fovea,** before it will strike.

Many insects have **dorsal ocelli** in addition to compound eyes, whereas the larvae of endopterygotes never have compound eyes and usually have **lateral ocelli,** called **stemmata,** instead. The structure of stemmata is quite variable. Some, like those of the larvae of the Lepidoptera, are rather similar to a single ommatidium, as previously described; each consists of a cornea, crystalline cone, and a group of retinal cells that form a single rhabdom (Figure 10-12*A*). In other insects, there is a single cornea that overlies a group of retinal cells and several rhabdoms (Figure 10-12*B*). Regardless of their structure, stemmata function like the ommatidia of compound eyes and are believed to perceive images, form, color, and movement. Clearly, a small number of sensory units would not permit the formation of an elaborate mosaic image, but insects that have these light-sensitive organs are believed to be capable of scanning their environment by swinging their heads back and forth.

Dorsal ocelli are similar in structure to the second type of stemmata mentioned previously (Figure 10-12*C*). Typically, there are three separate ocelli located in a triangular pattern on the anterior part of the vertex of the head; however, there may be two or none. Consequently, dorsal ocelli do not perceive form as far as we know. Their main function is to detect changes in light intensity so as to provide a general stimulation of the nervous system that makes the insect more responsive to other environmental stimuli (Dethier, 1963).

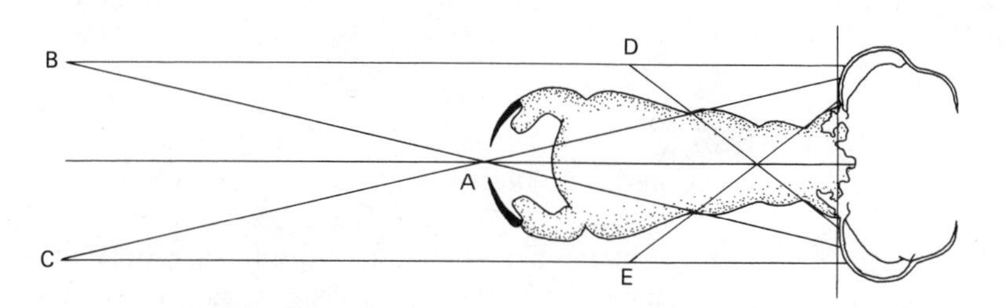


**FIGURE 10-11.** One conceptual analysis of depth perception by an insect. The series of straight lines represent the visual axes of selected ommatidia of a dragonfly nymph. Only prey at the intersection of visual axes in the vicinity of point A would be within reach of the extended prehensile labial mask. (Redrawn from Baldus, 1926.)

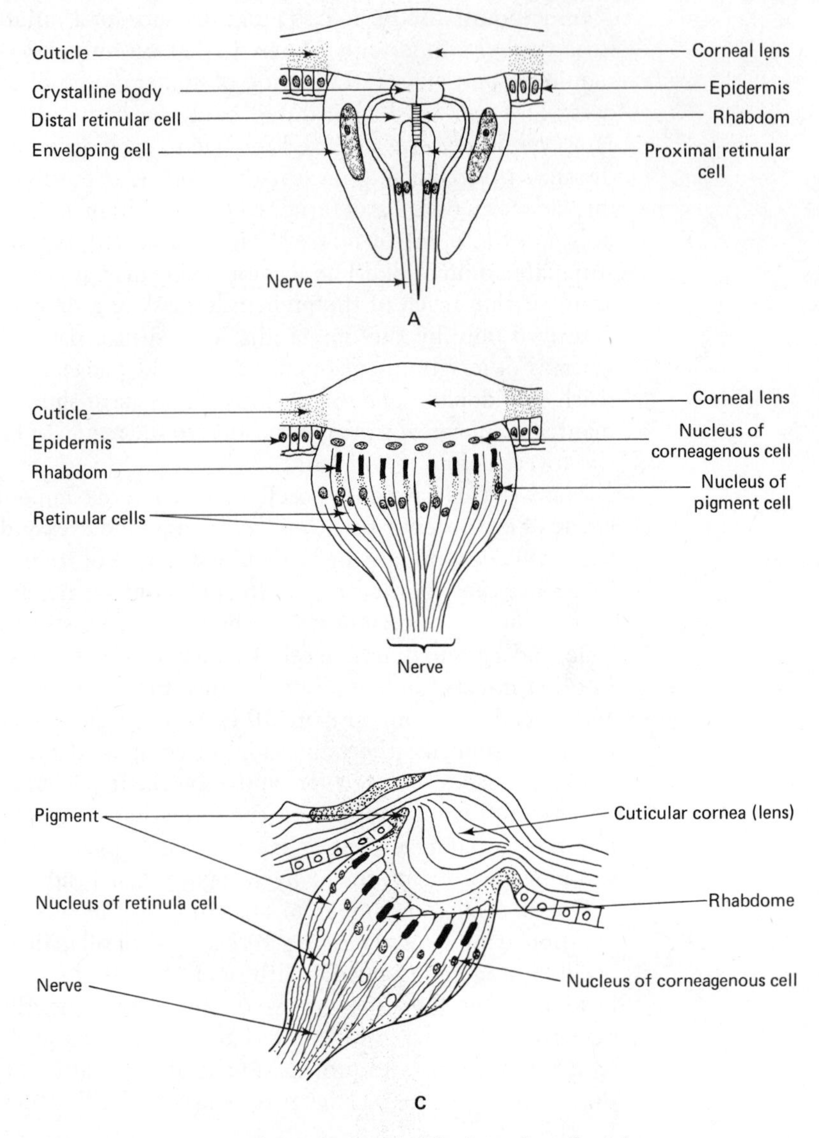


FIGURE 10-12. Diagrams of different types of simple eyes. **A.** The lateral ocellus of a caterpillar showing the single centrally located rhabdom, **B.** The lateral ocellus of a sawfly larva showing the absence of the crystalline body and larger number of rhabdoms. **C.** The dorsal ocellus of a hemipteran.

## *Thermoreception*

Because insects are cold-blooded animals, almost all of their activities and functions are regulated or affected by the temperature of their environment. It is not surprising then that insects are often able to detect rather small changes in temperature and to display a variety of temperature responses, including a tendency to aggregate in a particular zone of a temperature gradient. In addition to displaying what we might call a general temperature sense, some insects use temperature for specific purposes, such as the location of hosts. Many parasites of warm-blooded animals apparently orient to their hosts by following the temperature gradient surrounding the host's body. For example, the bedbug, *Cimex* (Hemiptera), is acutely sensitive to temperature and will leave a feverish person in favor of a bedfellow with a normal body temperature.

Despite all of the observed evidence for temperature sensitivity among insects, we have little well-substantiated information about the nature of temperature receptors. In the blood-sucking bug, *Rhodnius*, groups of thick-walled setae on the antennae are presumed to be temperature sensilla. One of

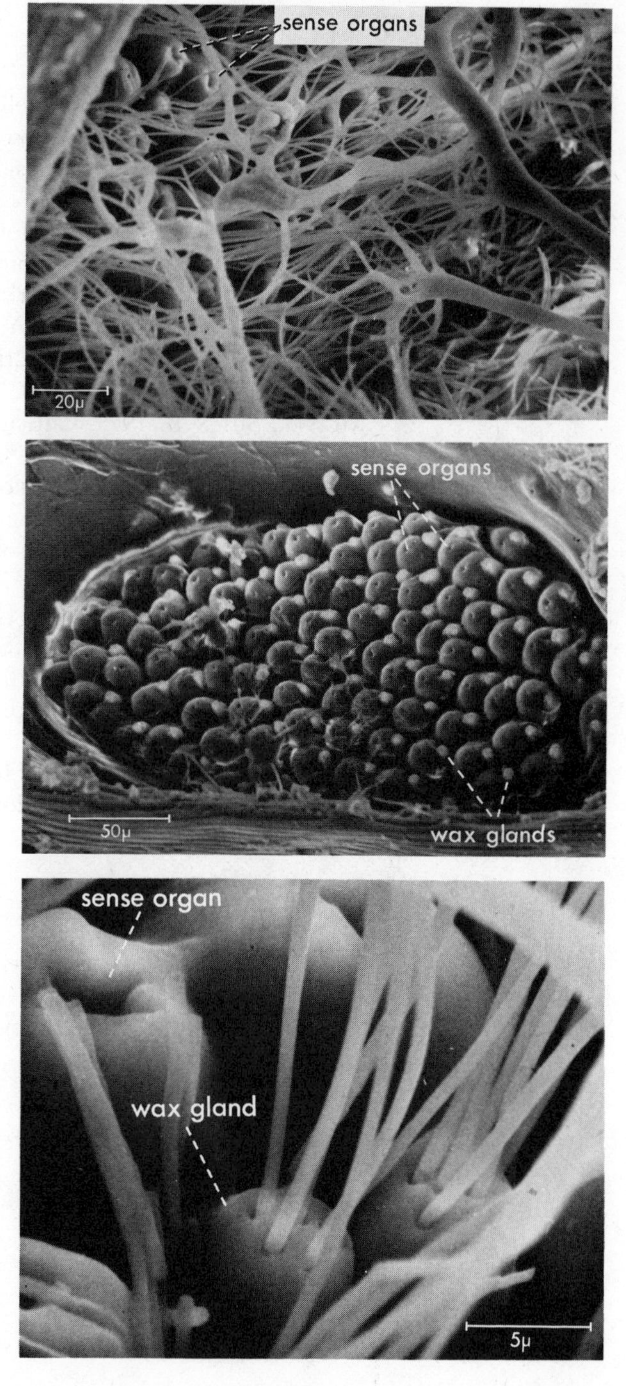


**FIGURE 10-13.** Scanning electron micrographs of an infrared sensory pit of the buprestid beetle *Melanophila acuminata*. **A.** An area of the pit showing mainly the wax filaments covering the sense organs, **B.** The sensory pit with most of the wax removed, **C.** A sense organ and adjacent wax glands extruding wax filaments. (Photographs courtesy of William G. Evans.)

the most exciting recent additions to our understanding of sensory perception by insects (Evans, 1975) involves the discovery of special infrared radiation detectors on the underside of the mesothorax of the buprestid beetle, *Melanophila* (Figure 10-13). These beetles oviposit in the bark of trees scorched by forest fires while the surface is still quite hot. Their infrared detection devices enable them to scan several miles of countryside while they are in flight and in search of trees being damaged by fire.

## References Cited


BALDUS, K. 1926. Experimentelle Untersuchungen über die Entfernungslokalisation der Libellen (Aeschna cyanea). Z. vergl. Physiol. 3: 375–505.

DETHIER, V. G. 1963. The physiology of insect senses. John Wiley & Sons, Inc., New York, N. Y.

EVANS, W. G. 1975. Wax secretion in the infrared sensory pit of *Melanophila acuminata* (Coleoptera: Buprestidae). *Quaest. Entomol.* 11: 587–589.

FRISCH, K. von. 1950. Bees, their vision, chemical senses and language. Cornell University Press, Ithaca, N. Y.

HERTZ, M. 1929. Die Organization des optischen Feldes bei der Biene I. Z. vergl. Physiol. 8: 693–748.

McIVER, SUSAN B. 1975. Cuticular mechanoreceptors of arthropods. *Ann. Rev. Entomol.* 20: 381–397.

MALDONADO, H., and J. C. BARROS-PITA. 1970. A fovea in the praying mantis. I. Estimation of the catching distance. *Z. vergl. Physiol.* 67: 58-78.

MARKL, H. 1962. Schweresinnesorgane bei Ameisen und anderen Hymenopteren. Z. vergl. Physiol. 44: 475–569.

SLIFER, E. H., J. J. PRESTAGE and H. W. BEAMS. 1959. The chemoreceptors and other sense organs on the antennal flagellum of the grasshopper (Orthoptera: Acrididae). J. Morph. 105: 145–191.

WELLINGTON, W. G. 1974. A special light to steer by. *Nat. Hist.* December 47–52.

WIGGLESWORTH, V. B. 1965. The principles of insect physiology, 6th ed. Methuen & Co., Ltd., London.


## General References


CHAPMAN, R. F. 1969. The insects—structure and function. American Elsevier Publishing Company, Inc., New York, N. Y.

DETHIER, V. G. 1963. The physiology of insect senses. John Wiley & Sons, Inc., New York, N. Y.

# Behavior

*In the foregoing six chapters we have examined the fundamentals of insect structure and function. On the basis of this examination, we have gained some appreciation of the complexity of insects and of how readily they have been able to adapt to the new situations that have arisen in the course of their evolution. However, adapting structurally or physiologically is only part of the evolutionary progress. Obviously, a functional capability is of little advantage unless it is put to the best possible use through an integrated system of responses combined with actions based on experience to form patterns of behavior.*

*To the average observer insect behavior appears to be rather stereotyped in that each species always seems to do the same thing under the same circumstances. For example, if we move a finger towards a settled fly, we can predict that it will take flight, or if we lift a stone, most of the insects beneath it will run quickly toward and under an adjacent object. These seemingly simple behavioral responses to specific stimuli are preprogrammed into the central nervous system of each species and are just as much a part of its inherited make-up as structural characteristics. Thus, the first time a specific response is needed, it can be performed. We call this kind of behavior* **instinct,** *which is distinguished from learned behavior on the basis that no prior experience is necessary.*

*All animals display instinctive behavior, but in ourselves its importance is subordinated to behavior patterns that we develop through experience and practice. Because our lives are dominated by learned behavior and experiences that arouse our emotions, we often tend to think of the behavior of other organisms in human terms. This tendency to attach human feelings and emotions to other animals is called* **anthropomorphism,** *which is a tempting pitfall when observing insects because some of their behavior patterns, particularly those displayed by social species, often parallel our own behavior. However, the assignment of human characteristics to insect behavior should be avoided because it tends to obscure the fact that these patterns of behavior are adaptations which have developed as a result of natural selection.*

*The capacity to learn is largely dependent upon the number of nerve cells that comprise the central nervous system. How well the inherent capacity to learn can be exploited depends upon the time available to experience events to which alternate responses are possible. Insects have relatively few nerve cells in their central nervous system, and most of them do not live for a very long time. Most of the nerve cells must be utilized to perform a rather fixed set of behavioral responses that will be required to assure survival and procreation under normal circumstances.*

*In order to survive, insects often have to respond extremely rapidly. In order to do this, they must be able to rapidly transmit information monitored by various sensory receptors to the motor mechanism that will produce the response. In the course of their evolution, insects have sacrificed the refinement of their perception and learning ability in favor of the ability to respond quickly to common stimuli* (Wigglesworth, 1968).

*The usefulness of learned behavior among many insects is also diluted by the fact that their life history patterns require distinctly different behavior patterns in different developmental stages. In those species which display a high level of immature-adult divergence, two or more very different behavior patterns have evolved. Behavior learned in one stage would serve little purpose in a subsequent stage. Nevertheless, insects can benefit greatly from the ability to learn, and experiments show that some have a rather remarkable ability to do so.*

*In the chapters that follow we will examine some of the basic types of behavioral responses of insects and how these are integrated into patterns of behavior that serve various biological functions. Although the biological functions of behavior are frequently related and all lead ultimately to reproduction and to the survival of the species, some of the more important aspects of behavior will be considered in separate chapters.*

# Basic responses and patterns of behavior

# 11

There are two basic kinds of behavior, that which is inherited and that which is learned. Inherited, or **innate**, behavior, which is often referred to as **instinct,** consists largely of more or less predictable responses or sequences of responses to the different types of stimuli identified in Chapter 10. Learned behavior is not inherited but is acquired as a result of the experiences of each individual. Consequently, members of the same species share inherited patterns of response and a capacity to learn but may behave differently because of different experiences involving various stimuli in their environment. Sometimes the two kinds of behavior become blended together, often making it difficult to distinguish between what has been inherited and what has been learned.

## *Innate Behavior*

The simplest form of innate behavior is the reflex. The rapid involuntary removal of your finger from a hot surface is a typical reflex reaction. This simple form of reflex involves a nerve cell or neuron with receptor dendrites that are associated with a sensory receptor and an axon that runs to and terminates in the central nervous system. Excitation of the receptor dendrites results in a nerve impulse that travels along the axon to its end. The impulse is then passed to an association neuron through a synapse and then through another synapse to a motor neuron. The motor neuron transmits an impulse along its axon to a muscle fiber that contracts in response to the stimulus received. This simplification, which is often called a **reflex arc,** is presented diagrammatically in Figure 11-1. Reflexes may result in the rapid movement of one part of the body or a change in the position or attitude of the entire body. Those reflexes that produce a rapid contraction of a specific muscle or group of muscles are called **phasic reflexes.** Those that activate slow-acting muscles, are set off by tonic receptors (e.g., stretch receptors), and help to maintain the position of the whole body are called **tonic reflexes.**

Reflexes vary in complexity and usually occur in groups or sequences that are coordinated within the ganglia of the central nervous system. Many of these complex groups of reflexes result in behavior patterns that are associated with orientation either in space or to or away from sources of stimulation. The maintenance of the position of the body in space, which we tend to take for granted, is sometimes called **primary orientation.** Most studies of individual behavior are concerned with **secondary orientation,** or the directional response to various stimuli. It should be pointed out, however, that the movement toward or away from a source of external stimulation is often influenced by internal stimuli related to the physiological state of the individual. The same individual may orient toward a stimulus at one time and away from it at another.

The classification of animal orientation has become complex, and the use of many specialized terms varies from one behaviorist to another. Although much new information about orientation has been gained in recent years, the descriptions provided by Fraenkel and Gunn (1961) are among the best available. The following three basic types of orientation are widely accepted:

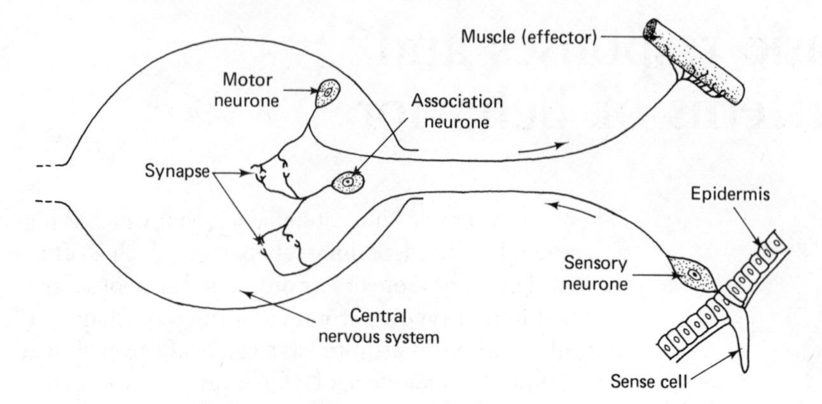


**FIGURE 11-1.** Diagram of a simple reflex arc comprised of a sensory, an association, and a motor neuron.

(1) **kineses,** random or undirected locomotor reactions initiated by a stimulus the intensity of which governs the rate of movement, (2) **taxes,** directed movement toward (positive) or away from (negative) a source of stimulation along a line that runs through the long axis of the body, and (3) **transverse orientations,** directed movements in which the long axis of the body is oriented at a fixed angle relative to the source of the stimulus.

The simplest type of kinesis is called an **orthokinesis.** This is an activity response to a stimulus. For example, daytime active insects will often remain motionless in the dark. If the light intensity is gradually increased, a point is reached at which the insect starts to move. This is the threshold for that stimulus under the existing conditions. If the light intensity (strength of the stimulus) is increased further, the rate of movement of the insect will increase as well. In the case of other stimuli such as moisture or special chemicals, which are less evenly distributed than light, the inactivation associated with the absence of the stimulus may produce aggregations of individuals.

More complex, undirected responses, which result in a behavior pattern comprised of a change in the frequency of random turns in the presence of unfavorable conditions, are known as **klinokineses.** In the presence of a strong danger signal, an insect will react with an instantaneous response called an **avoidance reaction.** When the stimulus is more diffuse, the response may consist of a series of more frequent random turns, which continue until the insect enters a more favorable zone characterized by an absence of the stimulus. Such responses are important in that they help to prevent entrapment in unfavorable situations. For example, if an insect displayed an orthokinetic response to cool air, it might simply move deeper into a cold zone and become immobilized before it could escape. On the other hand, the turns that characterize a klinokinetic response would probably result in the insect leaving the cold zone. The same kind of response to a favorable stimulus such as the smell of food would lead an insect to a specific location (Figure 11-2).

The directional nature of taxes involves a more complex system of sensory receptors that are capable of determining the direction from which the stimulus is coming. In some situations, the insect can compare the intensity of a stimulus in the immediate vicinity by making alternate movements from side to side. A caterpillar orienting toward a single light source often will

**FIGURE 11-2.** A hypothetical track of an insect orienting toward the center of a concentration gradient of a diffuse stimulus such as odor. The track is straight when the intensity of the stimulus is constant or increasing, but random turns occur when the intensity of the stimulus declines. (Adapted from Wigglesworth, 1968.)

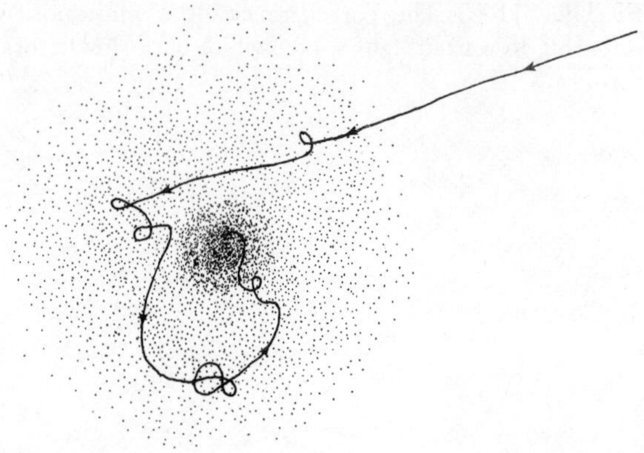

swing the front portion of its body from side to side and turn its track toward the side receiving the highest stimulation (Figure 11-3). Often insects have symmetrically located sensory structures, such as those on the antennae, which allow the intensity of stimulation on both sides of the body to be monitored simultaneously. When the track deviates from a direct route to or from the source of stimulation, the sensory receptor on one side of the body receives more stimulation, so a turn occurs which will restore sensory equilibrium to the pair of receptors. This can be demonstrated easily by placing a beetle on a piece of paper in front of a point source of light. If the paper is rotated to the right after the beetle has begun to walk toward the light, the beetle will

**FIGURE 11-3.** Orientation of a caterpillar toward a light source achieved by balancing the stimulus received by the lateral ocelli as the head is swung from side to side.

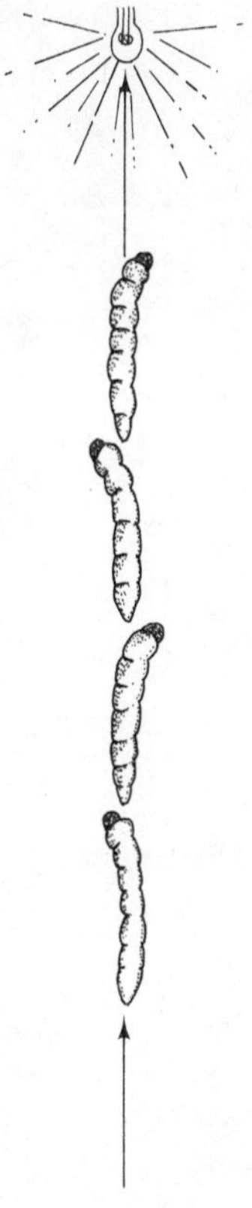

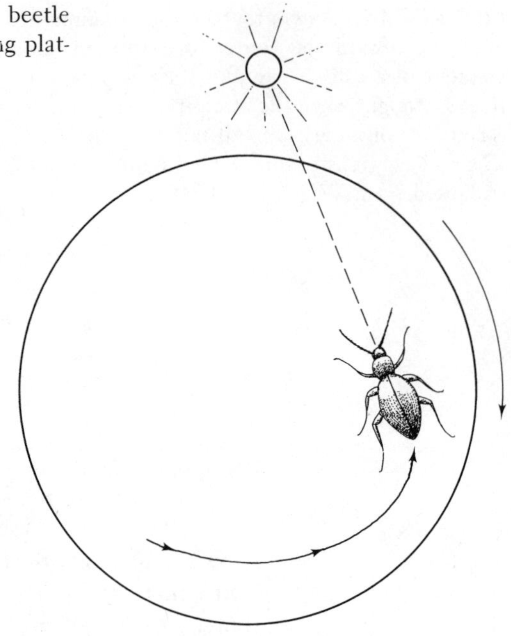

**FIGURE 11-4.** The curved track of a photopositive beetle orienting to a fixed light source while on a slowly rotating platform.

make a compensatory turn to the left (Figure 11-4). Similarly, when one eye is masked, the insect will continually turn toward the uncovered eye.

These orientation movements or taxes are usually named according to the kind of stimulus to which they are a response. A response to light would be a **phototaxis**; to an air current, an **amenotaxis**; to gravity, a **geotaxis**; and so on. If the response is toward the source of the stimulus, we say it is **positive,** and if it is away from the source, we say it is **negative.**

Transverse orientations result in an alignment of the body at a fixed angle relative to the direction of the source of a stimulus, but they do not necessarily involve locomotion. The dorsal (ventral) light reaction, displayed by many insects, is an example in which locomotion is not usually involved. Yet, it is important as a means of maintaining primary orientation. The dorsal (ventral) light reaction is one means by which free-swimming aquatic insects maintain their normal position; back swimmers use a ventral light reaction, whereas waterboatmen use a dorsal light reaction.

The **light-compass reaction** is a transverse orientation in which locomotion is frequently involved. As the insect orients and moves at a fixed angle relative to the light source, it becomes an important aid to navigation. This kind of behavior was observed first in ants, which use it as a means of returning to their nest after a foraging trip. A simple experiment can be conducted with many ants to show that they use sun-compass orientation. Simply place a small black container over an ant returning to its nest, and leave it there for a period of time long enough for the position of the sun to change substantially. When the ant is released, it will continue its trip back to the nest along a new course, displaced by an angle equal to the change in the position of the sun (Figure 11-5). Similarly, if the image of the sun is blocked from the view of an insect navigating by a sun-compass reaction and a mirror is situated so that the insect sees a reflected image in a new position, it will change direction. The new track will be oriented at the same angle relative to the reflected image as the original track was to the real sun.

FIGURE 11-5. Course followed by an ant using sun compass navigation. After spending two hours in a dark box at point *y* the ant will change its track direction by an angle (*x*) equivalent to the angle through which the sun has passed during the period of captivity.

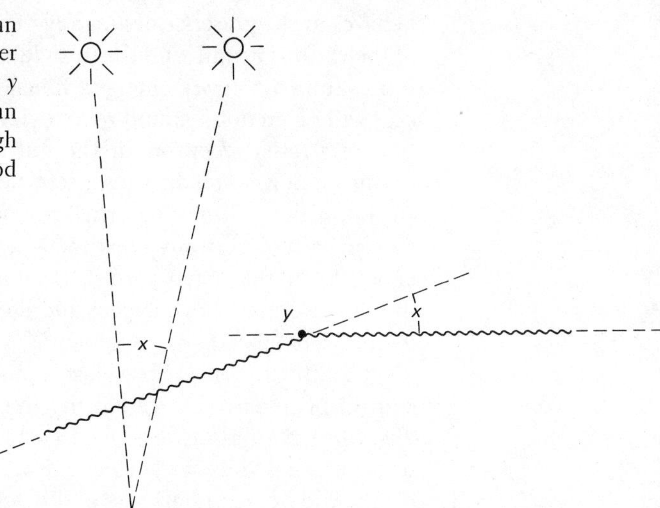


The sun, moon, and stars make particularly good navigational reference points because they are so distant that an insect can maintain a constant orientation angle and travel in a straight line for a long distance. If the light source is close, its angle of incidence on the retina changes after the insect has travelled in a straight line for only a short distance. The insect can only maintain its transverse orientation if it continually turns toward the source. This will place the insect on a spiral course that will ultimately end at the light itself (Figure 11-6). It is in this way that moths are thought to fly into the flame of a candle; yet, they may only be displaying an innate pattern of behavior adapted to more distant sources of stimulation.

Many more examples of different responses to a wide variety of environmental stimuli could be presented, but such a presentation would serve little purpose. Under normal circumstances, insects are not exposed to one dominant stimulus as they are in many laboratory experiments used to gain an understanding of a specific response. Instead, they are exposed to a variety of stimuli to which they must respond in an integrated way. How an insect

FIGURE 11-6. Diagram of the course followed by a moth as it approaches a fixed light source. The orientation to the light source is maintained at a constant angle and the moth flies into the light source along a "logarithmic spiral."

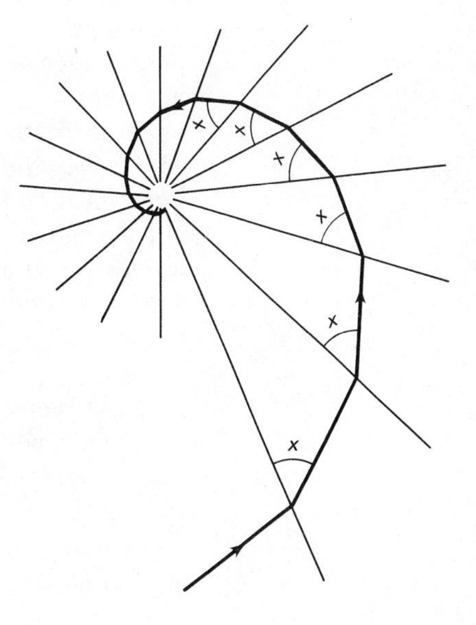

behaves in the presence of an array of stimuli depends on the relative strength of the various stimuli and the individual's internal physiological environment. For example, a newly emerged female with ovaries that do not contain ripe eggs will often not respond to an oviposition stimulus. Similarly, the stimulation of stretch receptors in the gut caused by the presence of food often inhibits a normal feeding resonse; when the ingested food has been digested, the inhibition is lost and normal responses to food are displayed. Many of our present interpretations of animal behavior are based on observations and experiments that have not taken some of these factors into account. Consequently, we must both expect and be receptive to many new interpretations in the years ahead.

Often the inhibition or release of normal patterns of response are under the control of inherited, internal rhythms which induce behavioral periodicity. The diurnal rhythms displayed by many insects provide ample evidence of some internal "biological clock." For example, cockroaches are usually active during the hours of darkness and inactive during daylight hours. However, if the time of light and dark periods are transposed, the roaches may still be active at about the same time even though it is light. The innate rhythm may persist for several days before any adjustment to the new light-dark cycle occurs. When a population of some species is reared under artificially constant conditions, the individuals may not display similar patterns of periodicity. However, if a regular time cue is introduced, the activities of all the individuals will become synchronized. For example, some mosquitoes reared under constant light display no rhythm in their egg-laying, but when a regular, short dark period is introduced, the females begin to lay a batch of eggs every 24 hours (Corbet, 1966).

In most species we find a great deal of compatibility in the response to different stimuli. For example, insects that respond negatively to gravity often respond positively to light, whereas those that live in the soil often display a negative phototaxis, a positive geotaxis, hygrokenesis, and a positive thigmotaxis (response to touch or contact with other surfaces).

Nevertheless, in the performance of everyday functions there must be a considerable amount of coordination between different responses. Often a sequence of occurrences involves stimuli that inhibit one response while setting up the next response in a chain of reflexes. Even in a rather simple process such as feeding, a well-coordinated sequence of responses may be necessary. For example, blood-sucking flies will orient toward a host in response to the host's silhouette, odor, moisture, and temperature; as it approaches the host, the fly may extend its proboscis in response to an intensification of the same stimuli; a probing response may then be stimulated by odor and the actual feeding by taste; feeding would stop when the fly was disturbed by a threat stimulus (perhaps some avoidance reaction on the part of the host) or when stretch receptors associated with the crop inhibited the normal feeding response.

There are, of course, other patterns of behavior that involve more complex successions of responses and even a progression through several successions at the same time. The nest-building and provisioning activities of solitary bees and wasps provide excellent examples of such complex sequences. The solitary leaf cutter bees construct a series of larval cells in pre-existing holes of appropriate size in some woody material. A female bee must first locate such a hole that is not already occupied. She must then leave the nest site in search of leaves or petals of an appropriate type from which to construct the first larval cell (less turgid leaves or the petals of certain flowers are favored). The cells are constructed of oval pieces of leaf or petal, cut to precision with the mandibles, and carried separately to the nest. When the first cell is complete, it is provisioned with 7 to 12 loads of pollen, topped with a load of nectar. An egg is then laid on the store of food, and the cell is capped with several discs of plant material. One such cell may take from one to four days to construct and provision, depending on the weather and the availability of resources. A series of such cells will be made sequentially; the number will depend upon the depth of the nest tunnel (Figure 11-7). Each step in this process involves different stimuli, plus navigation to and from the nest. Each act must be done in the appropriate sequence; if the work of a female is interrupted, she begins again exactly where she left off.

The nest-building behavior of the solitary wasp *Ammophila campestris* is even more remarkable (Figure 11-8). The sequence begins with the construction of a vertical burrow in the ground, which upon completion is sealed with several appropriately sized pebbles. The female wasp then leaves the nest site in search of a caterpillar to serve as food for her larva. When she finds a caterpillar, the wasp stings it in successive segments to assure that it is paralyzed but alive and carries the immobilized prey back to the nest. The small stones blocking the entrance are removed so that the caterpillar can be dragged into the burrow, after which a single egg is laid upon it. The wasp then leaves the burrow and seals it with the same pebbles selected earlier. What is even more amazing than the behavioral sequence itself is the fact that *A. campestris* may have several such nests in different stages of development and can switch from one to another without any apparent confusion (Baerends, 1959).

Most workers in the field consider the use of tools to be an advanced level of behavior, and some insist that the use of tools implies intelligence. If that is so, there are a few intelligent insects. After an *Ammophila* female has sealed a burrow with pebbles, she attempts to eliminate all trace of the entrance by smoothing out the soil over it. This is often accomplished with a small, smooth pebble held in the mandibles and moved back and forth with short, quick strokes of the head.

An ant found in India constructs its nests by sewing leaves together along their edges. The worker ants work cooperatively to draw the edges of adjacent leaves together. When the leaves are in position, one worker seizes one of its own larvae in its mandibles. As the larva produces silk from its mouth, it is moved back and forth between the leaves until they are "stitched" together (Jeannel, 1960) (see Figure 15-3).

Nest building in itself involves complex patterns of behavior performed by

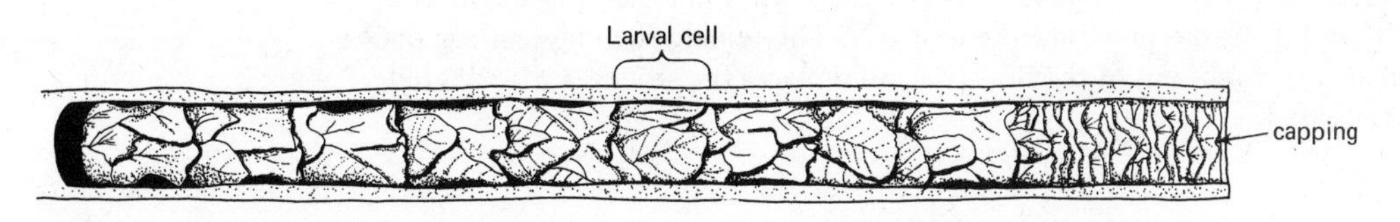


**FIGURE 11-7.** Diagram of a nest of the leaf-cutter bee *Megachile rotundata*, showing a series of nine larval cells and a capping of leaf pieces. (Redrawn from Stephen, Bohart, and Torchio, 1969.)

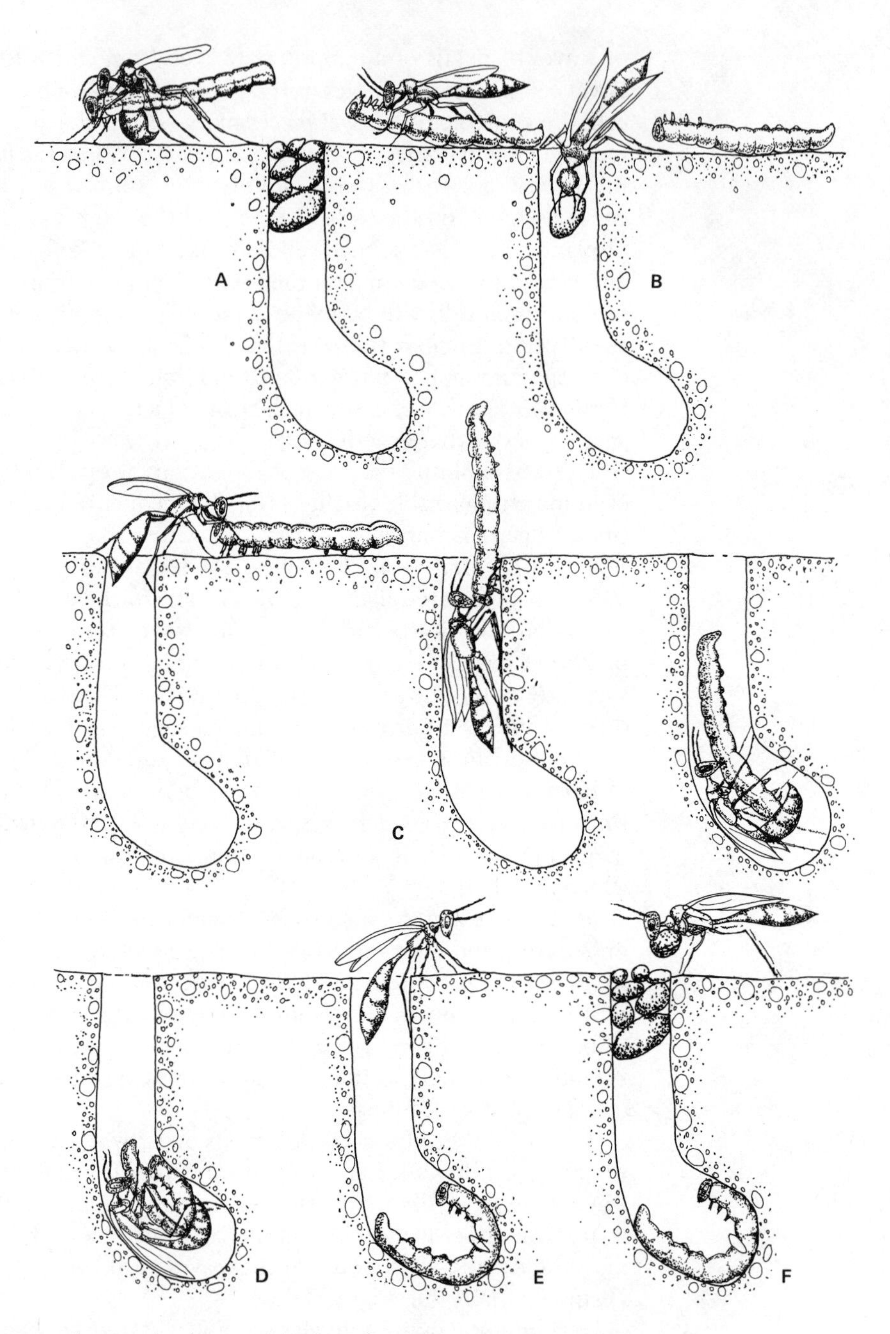


FIGURE 11-8. The successive acts in the provisioning of a previously constructed nest by the solitary wasp *Ammophila campestris*. After returning to the nest with a paralyzed caterpillar (*A*) the female removes the stones she has placed in the entrance (*B*) and drags the prey into the nest (C). The female then lays an egg on the caterpillar (*D*) climbs out of the nest (*E*) and replaces the stones (*F*). (Adapted from Baerends, 1941.)

a surprisingly large number of insects (see Chapter 16). But there would be little value to nests constructed for subsequent use if the builder could not find its way back to them. The establishment of territories and nests requires yet another behavioral capability, namely, navigation or homing ability. Probably no single group has attracted more interest in this regard than the nest-building Hymenoptera. Earlier we made brief mention of light-compass orientation as one navigational method used by ants. The fact that ants find their way while crawling rather than flying affords them numerous navigational possibilities. In addition to the light-compass, ants use visual landmarks, odors both on and off the trail, feel, and perhaps even some measure of distance travelled. What is even more interesting about these back-and-forth movements is that the insect must have some way of knowing which way it is travelling and be able to reverse some cues on the return trip. If the position of the sun is used to determine the direction of travel away from the nest, the insect must reverse its position along with the position of other landmarks on its return. Apparently some ants can lay a scent (pheromone) trail that is polarized, or they detect chemical differences between the materials to each side of the trail and are able to reverse them on the way home.

Flying insects cannot lay trails, but they too can use sun navigation in combination with other visual cues present in the landscape. When solitary bees and wasps have established a nest site, they often take one or two short flights around the nest, apparently to reconnoiter the nearby terrain. Honeybees clearly show these preliminary orientation flights whenever a hive is moved to a new location.

Insects seem to become familiar with the landmarks of their territory and move along a predictable course day after day depending on the condition of the sky. Wellington (1974) showed that when there is sufficient polarization overhead (Figure 11-9) insects are able to traverse open, featureless spaces with ease. At such times, in fact, both crawling and flying insects will set their courses across such spaces by referring to the plane of polarization, even when there are numerous suitable landmarks. For varying periods during the middle of summer days, there will be little or no polarization near the zenith, partly because the zone of maximal polarization shifts nearer the horizon during the middle of the day (Figure 11-9) and partly because the unpolarized glare around the high midday sun attenuates the remaining polarization still further. Insects then either reduce their amount of activity or restrict their travel to areas where they have learned the landmarks.

Obviously, the use of landmarks as navigational aids cannot be just a matter of instinct. An insect that uses landmarks must be able to remember them and learn how to use them in a sequence which will help it find its way. Only the capacity to remember or learn can be inherited; the learning must come from experience.

Several workers have claimed to have demonstrated that insects are capable of a classical conditioned response of the type discovered by the Russian behaviorist Pavlov. But the experiments conducted to date have produced inconclusive results. On the other hand, there is considerable evidence that insects, particularly ants, can learn their way through a maze placed between the nest and a source of food. Extreme care must be exercised in the conduct of this type of experiment, however, because of the variety of cues that ants are capable of using. Nevertheless, under well-controlled conditions insects

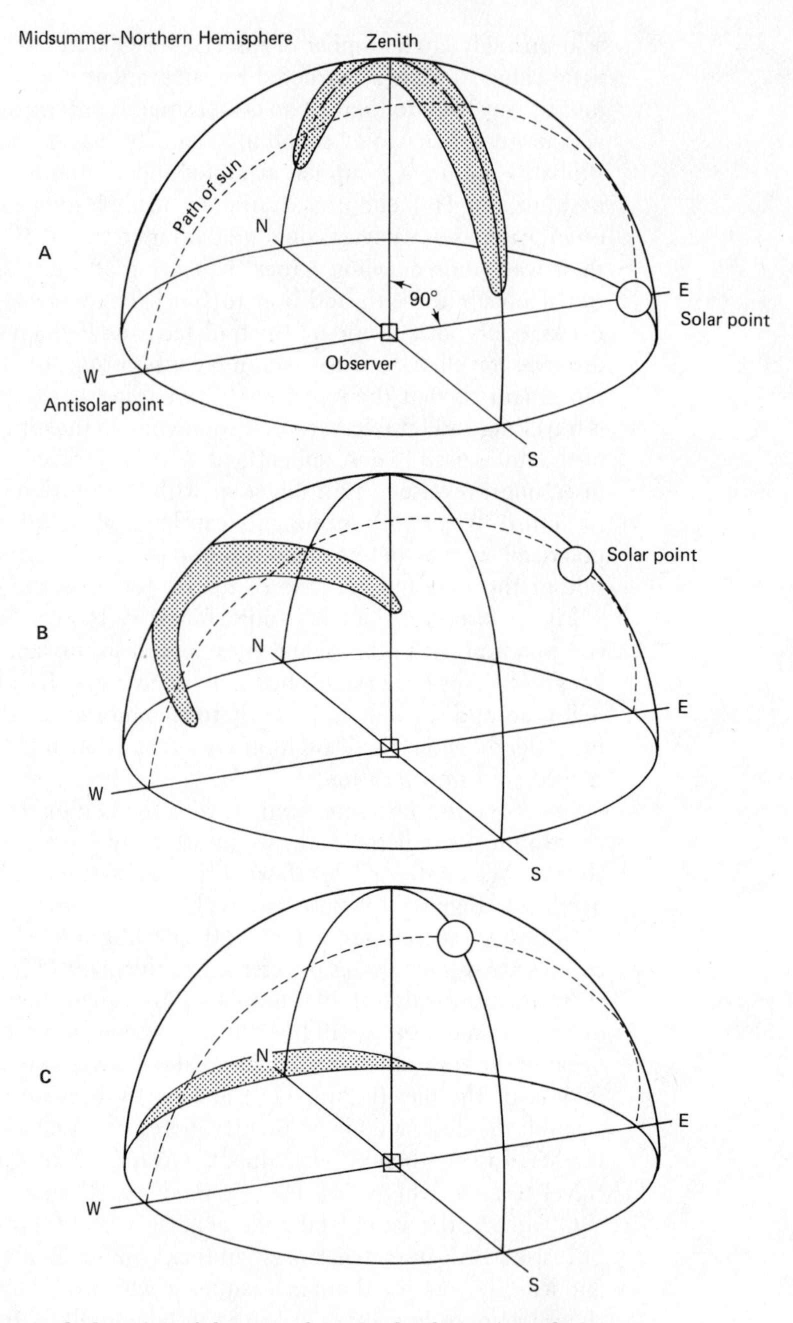


FIGURE 11-9. Changes in the location of the area of maximal polarization relative to the position of the sun and the observer. **A.** When the sun is near the horizon, the area of maximum polarization is directly overhead. **B.** About midmorning when the sun is about 45° above the horizon, the area of maximum polarization drops below the zenith to the northwest. C. When the sun is overhead at noon the area of maximum polarization is just above the northern horizon. (Reprinted, with permission, from W. G. Wellington, *A Special Light to Steer By.* Drawing by H. Peter Loewer, *Natural History Magazine,* December, 1974. © The American Museum of Natural History, 1974.)

can find their way through a maze with fewer and fewer wrong turns on successive trials, as shown in Figure 11-10.

A number of simple experiments that have been conducted to demonstrate the color sense and chemical sense of bees can be repeated as a demonstration of learning ability. A feeding platform is simply covered with colored squares of cardboard and a small glass dish is placed on each color. If a 50% sugar syrup is added to one dish and distilled water is added to all other dishes, the bees will aggregate at the sugar dish. If the dish containing sugar syrup is switched with a dish of water placed on another color, the bees will continue to return to the color where they had previously found the syrup.

If a few individual bees are marked when they first come to the color with the sugar syrup, it is possible to study the learning ability of these bees. Simply note the color of the cardboard square on which the bee initially lands each time it returns to the feeding station. Then move the color for which the reward is offered to new locations. In this way it is possible to develop a learning curve much like that presented for the ants moving through a maze. The more experience each marked bee gets, the fewer will be its landings on the wrong color.

Bees have also been shown to be capable of learning some appreciation of time. When bees are presented with food at an artificial feeding station at a

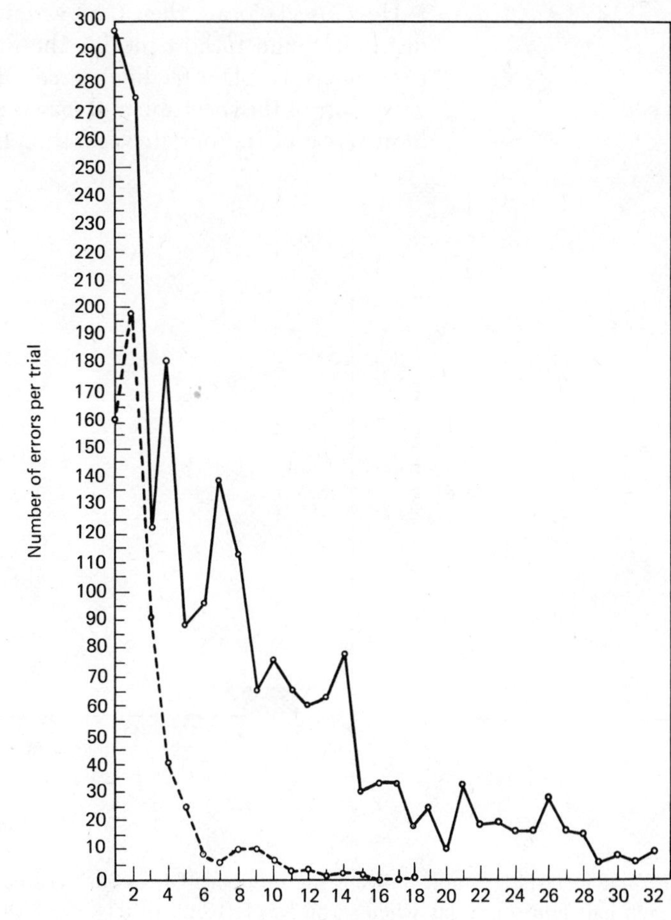


**FIGURE 11-10.** The learning curve for the ant *Formica pallide-fulva* in a six-point maze (solid line) compared to that for hooded rats (broken line). (Adapted from Schneirla, *in* Roeder, 1953.)

particular time for several consecutive days, they learn what time to visit the station. If food is subsequently made available all day, the bees will only visit the feeding station at the time food has been there previously (Figure 11-11) until they learn the new situation. Under natural conditions, this time memory is believed to be important in enabling the bees to visit certain flowers only during those hours when they make their pollen and nectar available (Wigglesworth, 1965).

In the foregoing pages, we have reviewed a range of behavior patterns from simple, innate reflexes to much more complex patterns, some of which involve a mixture of instinctive and learned behavior. All of these patterns serve a relatively few basic biological functions, which enable each species to develop, spread throughout its range, and reproduce. In the next few pages and in the chapters that follow, some of the important aspects of these basic functions will be examined in more detail.

The location and acquisition of food is without doubt of fundamental importance to all organisms. Dethier (1966) described feeding behavior per se as "a complex and interacting sequence of responses to a variety of stimuli culminating in ingestion and repletion." He outlined the basic sequence of events involved as: ". . . locomotion bringing the insect to its food, cessation of locomotion on arrival, biting or its equivalent (probing, sucking, etc.), continued feeding, termination of feeding."

How insects locate their food varies widely in relation to how the food is distributed and to how specific the insect is in its food preferences. Many phytophagous (plant-feeding) insects are attracted to the yellow or yellow-green part of the spectrum and may use color or other forms of visual cues to locate areas of appropriate vegetation from a distance. Likewise, many biting

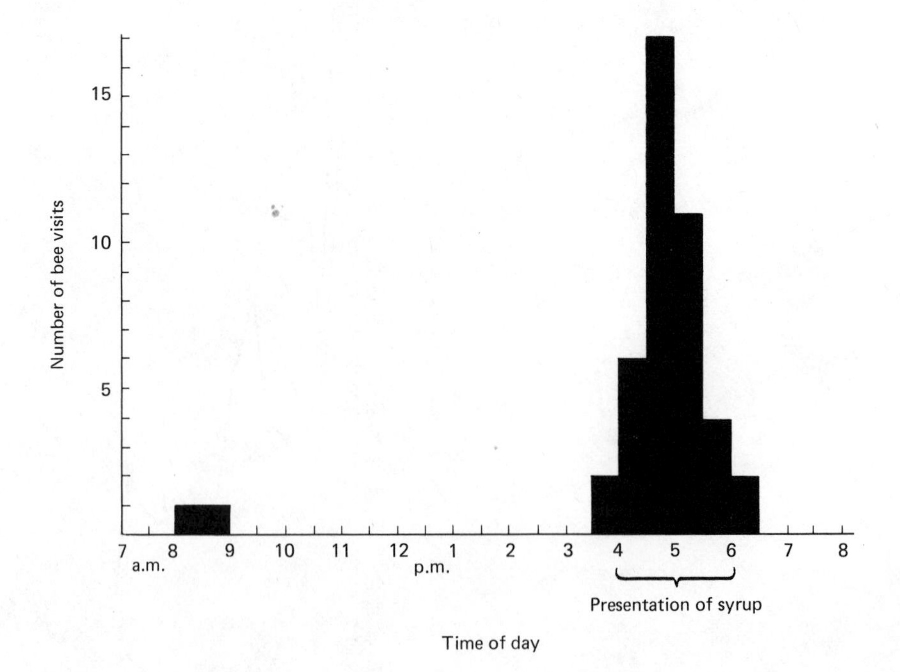


FIGURE 11-11. Graph showing the number of bees that visited a feeding place each half hour on a day when syrup was left out all day. Over the previous three weeks syrup was put out only between 4 and 6 p.m. (From Beling, 1929.)

flies orient toward certain silhouettes; black flies prefer silhouettes that generally resemble a large four-footed animal. Honeybees orient toward the color and the movement of flowers. In the case of parasites in search of concealed hosts, the initial response may be to a group of stimuli that simply characterize the host's environment. In some insects—carion feeders. for example—the source of food may be located at a considerable distance by way of olfactory cues.

Once the insect is in the general vicinity of food, stimuli which result in more discriminating behavior become dominant. Phytophagous insects may be able to locate their specific food plant on the basis of **token stimuli,** which may be attractive, yet of no nutritive value. After the host species has been located, there may still be some selection of one plant versus another on the basis of physical characteristics that are discerned by tactile senses or of nutritive characteristics that are tested by probing or preliminary biting.

In the vicinity of their hosts, mosquitoes use a wide range of stimuli to make the final feeding selection; they have been shown to respond positively to various volatile components of perspiration, carbon dioxide, moisture evaporating from the surface of the skin, and even convection currents set in motion by the warmth of the host's body. Aquatic insect predators sometimes locate their food by responding to the ripples produced by the swimming or struggling of potential prey (Carthy, 1965). Similarly, some lice (Malophaga) apparently locate their duck hosts by perceiving the ripples they make while swimming.

However, much of the food utilized by insects involves other living organisms. This leads to the development of a dynamic system of attack and counterattack on the part of the interacting species, which is discussed in Chapter 18. Inasmuch as insects are the eaten as frequently as they are the eaters, we should expect they too must have evolved patterns of behavior that provide means of escape and defense. Some of these aspects are considered in Chapter 14.

The behavioral process of food location and feeding is relatively simple compared to the sequence of events that eventually leads to procreation. From a purely behavioral point of view, we can argue that the sequence of events that leads ultimately to the leaving of offspring begins with the emergence of the young adults. In many species, the young adults migrate either before mating or soon thereafter. This migration is an important part of a species' behavior both in a biological and an applied sense, so it will be treated separately in the next chapter.

Once the individuals of a population have become separated temporally and spatially as a result of emergence and dispersal, it is necessary that sexually mature individuals be able to locate each other. When males and females come together, there is often an evolutionary advantage to behavior patterns that lead to selected pairings and mating and the elimination of weaker individuals. Consequently, once the sexes locate each other and come together, they may go through some elaborate form of courtship behavior. In both mate location and courtship there is a strong emphasis on communication, which, as we will see in Chapter 13, involves the production of and response to sound, light, chemical messengers, tactile stimulation, and visual displays.

Oviposition, like food location and feeding, may involve a sequence of behaviors which leads the female to the appropriate habitat and hence to the finite position where the eggs must be laid. In many species the survival of the young depends upon the ability of the mother to select the proper host upon which to lay her eggs. Yet, she may do this flawlessly with no prior experience. In the case of generalized feeders or species in which the first instar larvae can survive long enough to find food for themselves, the eggs may be laid indiscriminately. Some butterflies, for example, drop their eggs while in flight over mixed vegetation. On the other hand, highly specific hymenopterous parasites engage in elaborate search procedures, which culminate in the placement of each egg, with the help of their highly sensitive ovipositor, in exactly the right part of the host's body.

Depending on the species, a female insect might engage in a considerable amount of parental care. Whether or not insects care for their young or defend territories cannot be used as a measure of whether their behavior is advanced, since it is entirely instinctive. Some of the older groups such as dragonflies, termites, and earwigs display protective behavior. Some of the defensive methods employed by insects will be considered in Chapter 14, but the phenomenon of defense, like other forms of behavior, reaches a peak of complexity among the social insects, the evolution and behavior of which will be discussed in Chapter 16.

## References Cited


BAERENDS, G. P. 1941. Fortplanzungsverhalten und Orientierung der Grabvespe *Ammophila campestris.* Jur. Tidjschr. Entomol. 84: 68–75.

BAERENDS, G. P. 1959. Ethological studies of insect behavior. *Ann. Rev. Entomol.* 4: 207-234.

BELING, I. 1929. Uber das Zeitgedachtnis der Bienen. Z. vergl. Physiol. 9: 259–338.

CARTHY, J. D. 1965. The behavior of arthropods. W. H. Freeman and Company, Publishers, San Francisco.

CORBET, P. S. 1966. The role of rhythms in insect behavior. Pages 13–28 *in* P. T. Haskell, ed., Insect behavior, Royal Entomological Society of London, Symposium 3.

DETHIER, V. G. 1966. Feeding behavior. Pages 46-58 *in* P. T. Haskell, ed, Insect behavior, Royal Entomological Society of London, Symposium 3.

FRAENKEL, G. S. and D. L. GUNN. 1961. The orientation of animals. Dover Publications, Inc., New York, N. Y.

JEANNEL, R. 1960. Introduction to entomology. Hutchinson & Co., Ltd., London.

ROEDER, K. D. ed. 1953. Insect physiology. John Wiley and Sons, Inc., New York, N. Y.

SCHNEIRLA, T. C. 1953. Modifiability in insect behavior. Pages 723–747 *in* K. D. Roeder, ed. Insect physiology. John Wiley and Sons, Inc., New York, N. Y.

STEPHEN, W. P., G. E. BOHART and P. F. TORCHIO. 1969. The biology and external morphology of bees. Agric. Exp. Sta., Oregon State Univ., Corvallis, Ore.

WELLINGTON, W. G. 1974. A special light to steer by. Nat. Hist. (4)47–52.

WIGGLESWORTH, V. G. 1965. The principles of insect physiology, 6th ed. Methuen & Co., Ltd., London.

# Dispersal and migration 12

The high level of motility displayed by insects, particularly that which results from their ability to fly, has long been a subject of considerable fascination to man. But in spite of the attention that insect flight has attracted, the formulation of widely accepted generalities has been elusive. One of the major controversies has centered around the use of the terms in the title of this chapter and whether or not they are behaviorally distinctive. Both terms are useful, but frequently they are used interchangeably or without reference to the basic behavior patterns associated with different kinds of active or passive movement.

Any change of location can be called displacement. It may come about as a result of the accidental passive transport of insects by wind currents, water currents, **phoresy** (in association with another organism), by active locomotion, or by active locomotion that leads to passive transport. In spite of the frequency with which passive transport, particularly the wind, becomes involved in the movement of insects, there is relatively little displacement that is completely accidental. Small insects are not blown to inhospitable habitats as frequently as one might expect. However, many insects are highly susceptible to being blown about by the wind and so must have behavioral mechanisms that make them less susceptible to displacement accidents. On the other hand, the wind provides an inexhaustible source of external energy that insects have been able to exploit through the evolution of specific patterns of behavior.

The movements of insects can best be separated into those often referred to as **trivial** and those which are truly **migratory.** Trivial movements tend to be local and lacking in directionality of a type that leads to predictable displacement. A butterfly fluttering from plant to plant in a meadow and pausing to feed on the nectar of flowers, or to lay eggs on suitable host plants, is a typical example of trivial movement. The insect tends to change direction frequently and traverse a familiar territory rather than to fly in a more or less straight line over new territory. We cannot describe trivial movements as random, because they may involve a variety of responses to an array of stimuli. It is during this kind of activity that insects often locate food, mates, or oviposition sites. The initial phase of these selective behavior patterns often involves the detection of host odors or pheromones which are unevenly distributed as odor plumes. The trivial movements result in the interception of these cues which may then lead to some oriented movements such as a positive chemotaxis or klinokinesis (see Chapter 11). The degree of displacement that results from trivial movements may be quite substantial or rather small, as illustrated in Figure 12-1. Usually insects involved in trivial flights remain within the boundary layer (the relatively thin layer of the atmosphere immediately above the substrate where friction retards air movement and creates turbulence) and within a rather local area. Trivial flight is also characterized by numerous pauses during which insects feed, lay eggs, engage in mating behavior, or simply rest.

Migration, on the other hand, involves behavioral patterns that lead to a

FIGURE 12-1. Diagram of a hypothetical track of an adult insect engaged in trivial flight. Such flights are usually interrupted by numerous brief stops (dots). The ultimate displacement between the starting point and termination point (broken line) is usually considerably less than the total distance traversed (solid line).

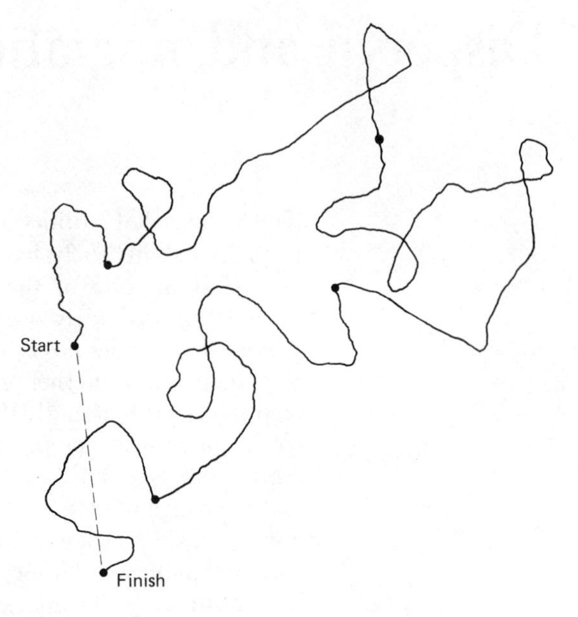


departure from one habitat and movement to another. In the case of aphids, it may result in movement from one host plant to another quite nearby; for other species, it may result in travel over hundreds of miles. Frequently migration involves a behavioral pattern that leads to an escape from the boundary layer into horizontally moving air at higher elevations, the speed of which often exceeds the species' intrinsic flight velocity. Migratory movements are also more sustained than trivial movements and, in most cases, are characterized by a general straightening out of the track (Kennedy, 1974). Insects that migrate largely under their own power appear to be motivated by an inner drive that results in the suppression of their responsiveness to appetitive stimuli such as food or mate odors, which would normally distract them during trivial flight.

Much of the past confusion in this area has arisen from the fact that many workers have considered migration to be a regular back-and-forth movement between two habitats or two parts of a species' range. These workers usually use the term dispersal to describe a unidirectional spread of a population to new or less populated areas. This distinction seems to be a carry-over from our basic understanding of bird behavior, a phenomenon that most students unfortunately encounter before reflecting on the movement of insects. There is a limit to the distance that most insects can migrate and return because they are relatively short-lived as adults. Nevertheless, there are some species that migrate and return (remigrate) in a highly predictable manner.

The key to understanding migration, therefore, seems to be not whether there is a regular back-and-forth pattern of movement, but rather that the travel has become an integral part of the species' life history as opposed to being a random event. Kennedy (1975) made the latest of a series of attempts to clarify the subject of insect migration. Kennedy contends rightfully that if an insect species must travel from one habitat to another as a regular and vital part of its biology, specific behavioral adaptations must have evolved that made such a life-style possible. Insect migration, therefore, can simply be defined as **adaptive travel.**

The interpretation of migration as back-and-forth movement, leads to such movement being considered as migration when that is not always the case. This basic confusion has lead some authors to believe that the regular travel between feeding and oviposition sites within the same habitat constitutes migratory behavior. Certainly such movements are behaviorally different from the trivial flight of the butterfly in a meadow, described earlier, in that the flight path may be straight and the displacement quite predictable. But this kind of behavior usually occurs between well-defined parts of the habitat and tends to concentrate the adults and their offspring rather than disperse them (Schneider, 1962). These back-and-forth movements, then, involve responses to feeding and reproductive stimuli not displayed by migrants and result in the fulfillment of the same basic functions as do trivial movements.

Some of the best examples of regular back-and-forth movements between parts of a habitat can be drawn from the group of scarab beetles known as chafers. The adults emerge from the soil of pasture land where pupation occurred and remain in the grass until mating has taken place. The inseminated females then fly to nearby woodland where they feed on the leaves of deciduous trees and develop their eggs. They then fly back to the pasture to oviposit in the soil. This results in the repopulation of previous breeding sites rather than the distribution of the population to new areas.

Many observers have noted that not all individuals in a population migrate; some remain in the vicinity of their breeding site. When the electric buck moth *Hemileuca electra* (Saturniidae) emerges in the fall, some of the females are so heavy with eggs that they are capable of only short, weak flights from plant to plant and consequently lay their eggs in the immediate vicinity of their pupation site. Others emerge, fly upward out of the boundary layer, and leave the area completely. In fact, these moths are recognized for their ability to fly strongly well above the ground on windy days. Whether the behavioral differences displayed by individual moths is under genetic control or is the result of nutrition and developmental history, remains unknown. In my work on the flight behavior of the bark beetle *Dendroctonus pseudotsugae* (Scolytidae), I found a correlation between fat content and the tendency to display migratory behavior (Atkins, 1967). Hagen (1962) suggests that it is possible to identify migrant convergent lady beetles heading for hibernation sites by squeezing them to see if they contain much fat. Perhaps nutrition alone can provide the differences in behavior that would contribute to the exploitation of both local and distant habitats. However, we should not rule out the possibility and selective advantages of genetic differences.

Migration serves a completely different function than does trivial flight behavior, regardless of how structured the latter may appear to be. Migration provides an organized method for responding to local adversity, aids in the occupation of new breeding sites, serves to even the distribution of the population, and, in the process, assures a high degree of genetic mixing.

Although insects are often observed to migrate coincidently with crowding or some deterioration of the quality or quantity of their food supply, the actual causes of migration remain in doubt for most species. So many species seem to migrate as a matter of course even though they are not crowded and their habitat appears to be suitable for continued utilization. The present view is that there are several environmental stimuli which precede impending adversity and to which insects have become adapted in the course of their evolution. These stimuli influence development and cause some individuals to become migrants.

The problem is one of reconciling the patterns of observed behavior with the actual physiological mechanisms and the general logic of the outcome. For example, some insects respond to a shortage of food with increased locomotion, but the locomotion may continue long after the food shortage has passed. When a migration occurs late in the season as is often the case among grasshoppers, it is difficult to determine which of several factors serves as the stimulant. Crowding, a reduction in the quality and quantity of vegetation, higher temperatures, drought, and a decline in day length are all interrelated phenomena.

From an evolutionary point of view, it would seem that a mechanism that integrates the environmental cues with the insect's developmental physiology would have the strongest selective advantage. Experimental evidence suggests that this is accomplished by way of the environmental effects on the endocrine system. For example, there is an established relationship between crowding and the size and activity of the prothoracic glands. The prothoracic glands of crowded locusts are small at emergence and disappear a few days after the final molt, whereas in isolated individuals they are large and persist until sexual maturation. Crowded individuals engage in more active locomotion than isolated individuals, and when the latter are injected with heamolymph from crowded locusts, they too become active (Haskell and Moorhouse, 1963). Other relationships exist between environmental stimuli and activity of the corpus allatum and the development of ovaries. Poor-quality food and declining photoperiod will both result in retardation of ovarian development, a condition usually encountered among migrants. Thus, there is a considerable degree of coordination between the development of the flight apparatus and the ovaries in relation to various environmental changes. A generalized diagram of the more important interrelationships is provided in Figure 12-2.

Because of the incompatability of a well-developed flight mechanism and fully developed ovaries, migration tends to occur when ovarian development is attenuated. The balance between the flight and the reproductive system may vary according to the environmental conditions, as suggested previously, but in many species this balance is a regular part of development. In quite a large number of species migratory flights occur soon after emergence. The extent and duration of such flights may vary greatly from individual to

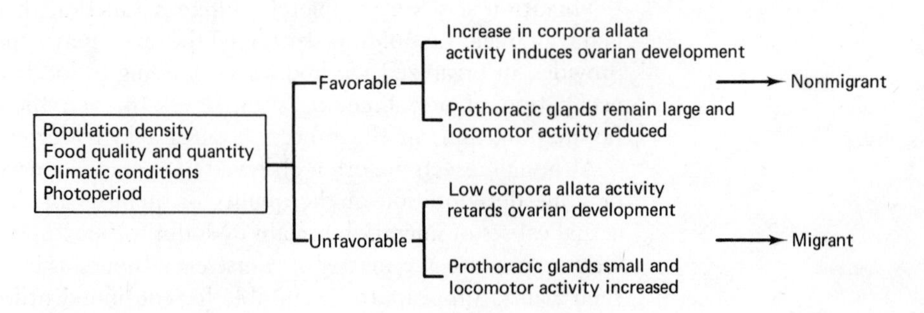


FIGURE 12-2. The interrelationships between environmental conditions, hormonal activity, and locomotion that lead to the induction of nonmigratory or migratory behavior.

individual depending on their developmental histories, but they occur nonetheless. During these early flights, many insects are clearly nonresponsive to appetitive stimuli, as demonstrated by the fact that older individuals tend to dominate the catches of baited traps. The nonresponsiveness of young migratory individuals has been demonstrated in studies of the coddling moth (Geier, 1960), the screwworm (Crystal, 1964), the Douglas fir beetle (Atkins, 1966), and other insects.

Such flights early in development result in the rapid disappearance of many individuals from their breeding site. For insects that inhabit transitory environments, this type of behavior is required so they can keep pace with changes in the location of necessary resources. Other insects go through a period of adult dormancy in a habitat some distance from their breeding sites. These insects migrate to such sites following emergence, enter dormancy, and then migrate back to the breeding area later.

The distance between the breeding range and the site where dormancy is passed may be considerable, as it is for the convergent lady beetle *Hippodamia convergens* (Coccinellidae). The young beetles leave their feeding territory in the central valley of California during May and June and fly many miles to hibernation sites on the western slopes of the Sierra Nevada. In February and March of the following year, the same beetles fly back to the valley to breed.

### Classes of Migrations

In order to clarify the diversity of insect migratory activity, which both behaviorally and ecologically ranges from something akin to trivial movement within the breeding area to travel over very long distances, Johnson (1969) provided a classification of migration consisting of the following three broad classes:

- Class I. Adults with a life span limited to one season that emigrate from their breeding site, disperse, oviposit, and die.
- Class II. Insects that emigrate from their breeding sites to a feeding site and then return after oögenesis to oviposit in the former or new breeding sites.
- Class III. Insects that emigrate from their breeding site to a hibernation or aestivation site from which they return the following year to oviposit in the original area.

Class I migration may be extremely variable in terms of distance or duration but usually occurs soon after emergence and before the gonads have matured. Johnson recognized five subtypes of this class of migration, but a pair of examples should give an adequate impression of the variation that exists. A rather simple illustration is provided by the flights of ants and termites, in which winged individuals that are produced seasonally become the founders of new colonies. They leave the nest soon after emergence and fly weakly with the wind. As the wind changes, they spread in various directions within and beyond the area where breeding is possible. Relatively few new nests need to be established to keep the area adequately populated, so large numerical losses during migration can be sustained.

Although seemingly different and certainly of greater magnitude, the migration of the desert locust *Schistocerca gregaria* (Acrididae) is also an example of Class I migration. In tropical and subtropical areas, the desert locust must move from breeding areas that are deteriorating because of drought to new areas that are receiving rain and have green vegetation. These breeding areas are often widely separated and tend to change location because of variations in the pattern of seasonal rainfall. The migratory behavior displayed by the desert locust has evolved to take advantage of the seasonal winds as an aid to traversing the long distances between breeding habitats. Individual flying locusts seem to orient in such a way as to maintain the cohesiveness of the swarm rather than to fly in a particular direction relative to the ground or the wind. The swarm as a whole is displaced downwind and thus to zones of convergence where the associated rainfall produces a suitable breeding site.

The pattern of Class II migrations is illustrated by many insects that display a high degree of larval-adult divergence in terms of habitat or food preference. Female mosquitoes, for example, must often leave the area of the larval habitat from which they emerged so as to obtain a blood meal necessary for egg development. The larvae are filter-feeders in aquatic habitats, whereas the females often search for vertebrate hosts some distance from the water. After feeding and oögenesis, they return to aquatic habitats to oviposit.

Class III migration varies considerably from species to species. At one extreme, the young adults emerge from their breeding site and spread in all directions throughout the available breeding territory until the onset of unfavorable conditions. The insects then enter diapause within the breeding area and then migrate again some months later prior to oviposition. The spruce beetle *Dendroctonus obesus* (Scolytidae) migrates only a short distance to diapausing sites in the bark at the base of spruce trees but makes a longer migration in the spring prior to attacking new trees where their brood is reared.

The convergent lady beetle, mentioned earlier in this chapter, engages in a long migratory flight from the breeding area in the valley to hibernation sites in the mountains. Equally long migratory flights return them to their breeding territory the following spring.

Some of the longest migratory flights known are undertaken by insects that fly both before and after they overwinter. The monarch butterflies *Danaus plexippus* (Danaidae) in North America migrate hundreds of miles to a rather well-defined overwintering area from which they return to breeding areas the following spring. The vast distance covered by monarch populations has made it difficult to determine exactly how far individuals fly during each of the phases of the migration. The breeding range of these butterflies is extensive in a north-south direction so there is considerable variation in the latitude at which they develop. The adults that result from the last generation in Canada migrate to regularly used overwintering sites in California, Mexico, and Florida. The butterflies are active on warm days during the winter but engage in only trivial flights in the immediate vicinity of the hibernation site. In the spring the overwintered adults fly northward to new breeding areas; some are believed to terminate their migration early and establish spatially intermediate generations, but others are thought to return directly to their northernmost breeding areas.

### *The Adaptive Nature of Migrations*

The energy required to sustain flight for an extended period is substantial. The intermittent feeding that characterizes trivial flight and provides the day-to-day energy requirements of an individual is not a part of normal migratory behavior. Although a large amount of fat is stored during larval feeding for use in the adult stage, there are limits to the intrinsic capacity of insects to fly. Some insects have a truly remarkable ability to sustain flight; I have often obtained eight hours of continuous flight from bark beetles tethered on a rotating flight mill (Atkins, 1961). However, most insects are not capable of very high flight velocity (see Table 7-1), and their ability to progress against the wind is rather limited. On the other hand, the wind is an inexhaustible supply of external energy that can be exploited. In recent times, entomologists have realized that insects are rarely blown about at random but have evolved patterns of behavior that enable them to exploit the wind in the fulfillment of their migratory needs.

The general pattern of atmospheric circulation is governed by the interaction of large pressure cells variously positioned according to the seasons. Near the ground these regional circulation patterns are modified by local climatic and topographic effects. The frequency with which local winds shift direction would make their general exploitation risky in that this would lead to enormous numerical losses when insects are carried to inhospitable habitats. Insects then must be strong enough to combat the wind, restrict their flight to periods of relative calm, or fly with winds that would normally carry them to favorable habitats.

Some insects such as the painted lady *Vanessa cardui* (Nymphalidae) are well known for their strong directional flights. In southern California during early spring, mass flights of adult painted ladies can be seen heading northward from breeding sites in Mexico. Abbott (1961) reported that these butterflies fly in a northerly direction regardless of the direction of the wind. In 1973 I observed a large mass migration in the San Diego area that lasted several days. One day, the butterflies flew northward against a steady wind of about 15 mph, and on another day, the same flight path crossed an easterly wind of similar velocity.

In contrast, other strong flying insects tend to fly with the wind during migrations and are displaced long distances as a result. Spruce budworm moths are reasonably good fliers but are also displaced considerable distances on the wind. The moths fly within the forest throughout the day, but in the evening as the light intensity declines, they tend to fly at treetop level rather than beneath the forest canopy. While flying above the trees, the moths are subject to convectional currents and transport on the prevailing wind. During the day the moths are also stimulated to fly upward to the top of the canopy by the decline in light intensity and change in barometric pressure associated with approaching thunderstorms. Moths flying above the canopy are subject to carriage aloft by convectional currents in the storms and then become swept along with them for long distances. The moths are then deposited elsewhere, sometimes more than a hundred miles from their source, when the conventional activity of the storm declines (Greenbank, 1963).

Many insects that are weaker fliers also have specific behavior patterns that clearly enhance their opportunity to migrate to specific areas with help from the wind. One such example that has been well documented involves the prediapause and postdiapause migrations of *H. convergens* (Hagen, 1962). In years when aphid populations are high, large numbers of young adult lady beetles emerge from fields in the lowland valleys during May and June. The general reduction in the abundance of aphids because of prior feeding by the beetle larvae leads to the departure of the young adults. About this time, large numbers of lady beetles are often observed in the vicinity of aggregation sites in the mountains. The beetles apparently leave the fields by way of vertical take-off flights on warm, calm mornings. These vertical flights, assisted by convectional currents, continue upward to a temperature ceiling of 11° to 13°C, which curtails flight. The nonflying beetles are believed to fall into warmer air that permits a resumption of their upward flight. This alternating pattern of upward flight and falling produces oscillations of movement that may have an amplitude of up to 1000 feet (305 meters). During these oscillations the beetles are carried on horizontal, westerly winds toward the mountains, where the beetles are deposited in the zone of intersection between the temperature flight ceiling and the ground level (Figure 12-3*A*).

Warm days during February and March in the Sierra Nevada are associated with a high pressure system over southern Idaho that produces northeasterly winds aloft. The aggregation sites on the western side of the mountains are protected from these upper-level winds, but they experience convectional currents and warm upslope winds from the valley. When the temperature of the aggregation sites rises, the beetles break dormancy, fly upward, and eventually engage the winds aloft that carry them back to the valley. The beetles again oscillate up and down in the vicinity of the temperature flight ceiling as before but are forced to the ground as the temperature ceiling declines late in the day (Figure 12-3*B*).

The importance of adaptive flight behavior in relation to variable environmental conditions is illustrated well by the migrations of the beet leafhopper *Circulifer tenellus* (Cicadellidae) in the San Joaquin Valley of California, summarized by Cook (1967). The leafhoppers overwinter and pass their first generation predominantly on wild vegetation at the southern end of the valley. In the spring, the prevailing northwesterly winds enter the valley in the area of San Francisco and are deflected southward through the valley; this is in the opposite direction of the spring leafhopper migrations. The leafhoppers are capable of flight speeds in the order of only 2 mph and tend to fly when the prevailing winds have abated. This frequently occurs in the late afternoon when the warm air in the valley rises and is replaced by a downpouring of air from the surrounding hills. This creates a northwesterly flow across the breeding site that carries the leafhoppers already in flight well out into the valley where sugar beets are cultivated. As a result, the spring migration is against the prevailing wind. The combination of leafhopper flight behavior and wind patterns at other times throughout the season allows the beetles to move throughout the valley and ultimately end at their hibernation sites in the southern hills in the autumn.

The numerical losses that occur during insect migrations are clearly high but the benefits that accrue must more than compensate for the in-flight population attrition. Insects that utilize scattered temporary habitats must produce enough progeny to ensure that a few survive the migration to new habitats. The fact that the habitats of such species are scattered unevenly throughout their range must have been a strong selective force that favored a strategy in which migration and habitat location are the functions of the winged adult. Yet, this seems inefficient in respect to all the resources consumed during the development of the high percentage of adults that die during the movement between habitats. On the surface it would seem to be more efficient for dispersal losses to be absorbed early in the life history, but this occurs in only a relatively few species of insects. As one might expect,

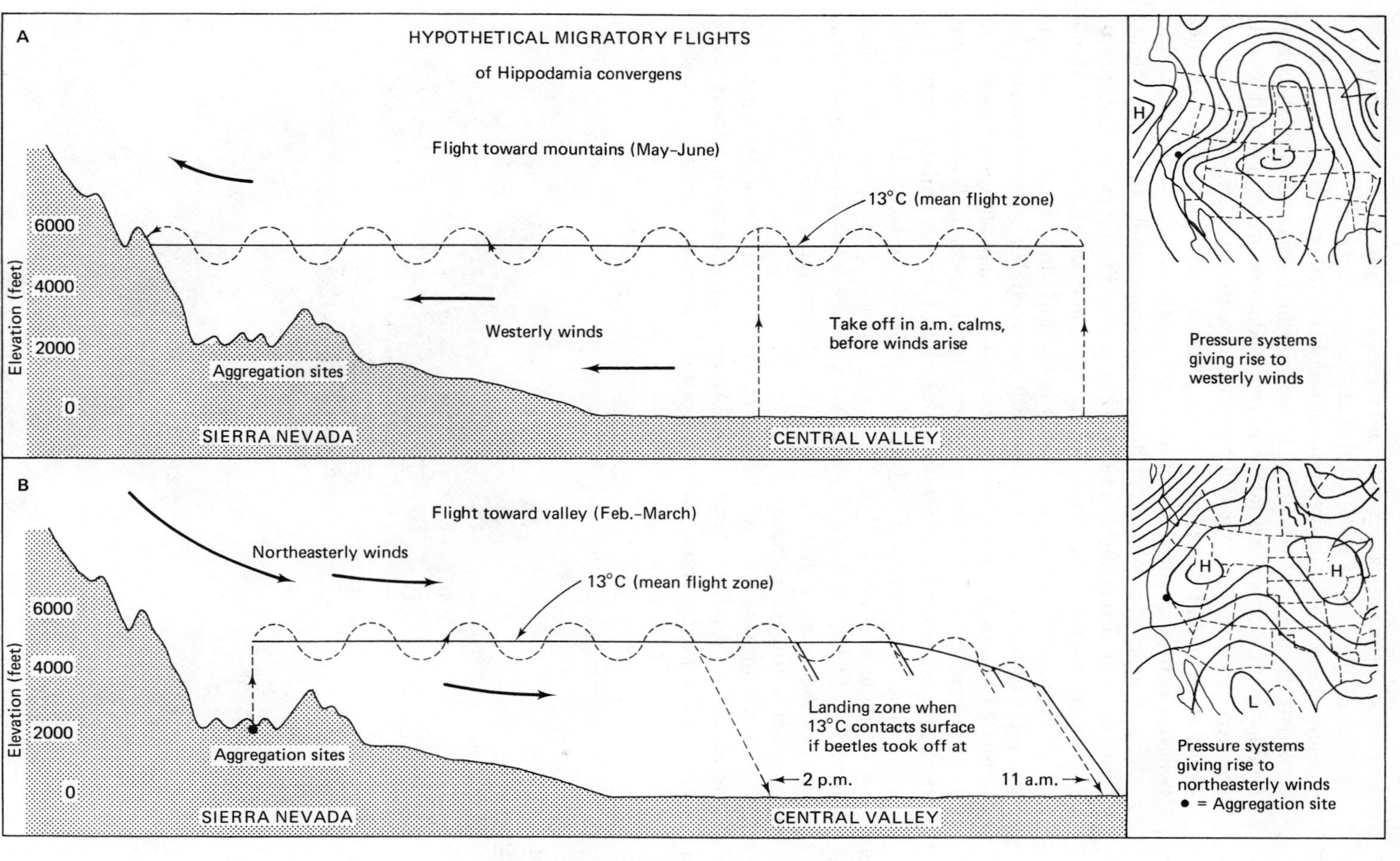


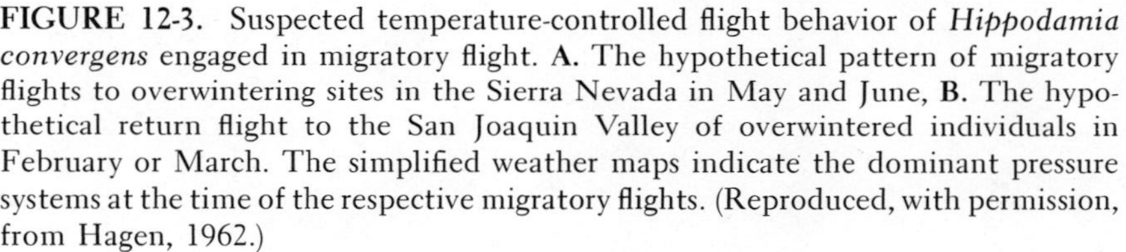


**FIGURE 12-3.** Suspected temperature-controlled flight behavior of *Hippodamia convergens* engaged in migratory flight. **A.** The hypothetical pattern of migratory flights to overwintering sites in the Sierra Nevada in May and June, **B.** The hypothetical return flight to the San Joaquin Valley of overwintered individuals in February or March. The simplified weather maps indicate the dominant pressure systems at the time of the respective migratory flights. (Reproduced, with permission, from Hagen, 1962.)

those that do migrate as juveniles are rather generalized feeders or users of a somewhat specific but widely distributed food source.

Those species that migrate as immatures have little capacity to navigate or terminate their travel. In order to migrate and maximize the success of establishment in a new habitat, appropriate adaptive behavior would seem to be a necessity. Studies of the behavior that leads to and terminates the passive dispersal of young caterpillars and first instar homopterans reveal that their behavior is truly adaptive and therefore migratory.

Migration by young larvae of the tussock moth family Liparidae is quite common. The first instar larvae of the gypsy moth display a rather precise pattern of behavior that leads to their trsnsport on the wind. McManus (1973) reported that most eggs hatch in the early morning, independent of the ambient temperature and humidity. Between 0800 and 1000 hours, and again, but to a lesser extent, in the afternoon, the small caterpillars ascend the trees and move out to the tips of the branches. Between 1300 and 1500 hours, the larvae drop on strands of silk. At this time of the day both the horizontal and vertical air currents typicallly reach their greatest velocity. The suspended larvae swing back and forth until the wind is strong enough to break them loose, at which time they sail on the wind to new habitats. The long lateral setae of the larvae (Figure 12-4) increase their buoyancy and appear to be an adaptation to passive transport.

The balsam woolly aphid *Adelges piceae* (Chermidae or Adelgidae) was accidentally introduced into North America on European nursery stock. Although a winged form occurs in Europe, winged adults have never been

FIGURE 12-4. Young larvae of the gypsy moth showing the long lateral setae believed to improve their passive transport by the wind. (Photograph courtesy USDA.)

encountered under field conditions on this side of the Atlantic. Nevertheless, this pest of fir forests spread out widely from several introduction sites on both the east and west coasts of North America.

The first nymphal stage, commonly called a crawler, is the only stage in which any significant displacement is possible. The eggs are commonly entangled in strands of wax (the "wool") produced by the female, and all of the other stages are passed with the sucking mouth parts embedded in the tissue of a host tree. Cards that are treated with a sticky substance and placed throughout an area of infested trees trap large numbers of crawlers and provide ample evidence of their movement on air currents within the forest.

A detailed study of crawler responses to different environmental stimuli provided no evidence of predictable oriented behavior (Atkins and Hall, 1969). However, the over-all level of crawler activity increased with an increase in both temperature and light intensity. When a high population of crawlers became active at one time, they would let go of the substrate and drop from their host. Greenbank (1970) showed that up to 80% of the crawlers drop from the host tree on which they originate.

The tiny crawlers have a free-fall velocity of approximately 30 cm/sec, so they can be blown about by air currents as low as one or two miles per hour. The catch data from the sticky traps indicated considerable aerial transport of crawlers of a magnitude that would assure the displacement of many individuals from tree to tree. However, most fir forests are characterized by an interlocking canopy of branches, which probably acts as an effective barrier to displacement out of the forest, and the dense shade inhibits the development of convectional currents. Along the forest edge or around small openings where the sun penetrates the foliage, both higher temperatures and light intensities occur, particularly after midday. The related high level of activity results in a peak of crawler dropping at this time. Once airborne, the tiny insects are carried by the gentle outflow of air from the forest margin, and then encounter convectional currents generated by the warmer surfaces in the open (Figure 12-5).

Newly hatched crawlers are the most active and can remain sufficiently vigorous to establish themselves on new hosts for two or three days at 17° to 21°C and midrange humidities. The dropping behavior of newly hatched crawlers in relation to temperature and light levels that are coincident with good conditions for passive displacement is clearly adaptive.

### *Over-all Significance*

Insects, like most organisms, derive both genetic and ecological benefits from being able to move from one place to another. Such behavior increases the mixing of the gene pool and hastens the spread of beneficial mutations. Ecologically, migration enables a species to vacate crowded areas or habitats where the requisites for life are deteriorating, in favor of sparcely populated areas or habitats with an abundance of appropriate resources. As is usually the case with evolution, the persistence of a trait depends on whether or not it contributes to the survival of the species. The benefits of migration, therefore, must outweigh the associated physiological costs and numerical losses. Obviously, all insect species do not have the same need to migrate. Those with more or less permanent breeding habitats may be able to achieve adequate displacement by way of the general diffusion that accompanies trivial flight. At the other extreme, species that occupy very temporary or transitory habitats need to migrate regularly.

**FIGURE 12-5.** Diagram of air currents associated with a forest margin as a result of the greater heating of the ground in cleared areas.

In order that the genetic and ecological benefits can be obtained, substantial losses in the form of energy expended and of mortality must be borne. Apparently, these costs and benefits have been favorably balanced in the course of evolution. The evolution of adaptive behavior patterns has proved to be one way to maximize the migratory gains and to minimize the attendant losses. In fact, it would be surprising if many species of insects could survive long if they had to rely on some haphazard means of moving from one habitat to another. Even so, there would seem to be a limited number of behavioral strategies that can be exploited, even though there may be an infinite number of minor variations in the details of such behavior.

The large-scale population changes that accompany the departure or arrival of migrants are of considerable practical importance in pest evaluation and control. However, detailed studies of pest migrations are difficult and time-consuming. Perhaps if we focus our attention on the broad similarities in the adaptive behavior associated with different patterns of displacement we will be able to identify groups of insects that move between habitats in a similar way. This would be a giant step forward in our efforts to understand and manage insect populations.

## References Cited


Abbot, C. H. 1951. A quantitative study of the migration of the painted lady butterfly, *Vanessa cardui* L. *Ecology*. 32: 155–171.

Atkins, M. D. 1961. A study of the flight of the Douglas-fir beetle *Dendroctonus pseudotsugae* Hopk. (Coleoptera: Scolytidae) III Flight capacity. *Can. Entomol.* 93: 467–474.

Atkins, M. D. 1966. Laboratory studies on the behavior of the Douglas-fir beetle *Dendroctonus pseudotsugae* Hopk. *Can. Entomol.* 98: 953-991.

Atkins, M. D. and A. A. Hall. 1969. The effect of light and temperature on the activity of the crawlers of the balsam woolly aphid. *Can. Entomol.* 101: 481-488.

Cook, W. C. 1967. Life history, host plants and migrations of the beet leafhopper in the western United States. *Tech. Bull. U. S. Dept. Agric.*, No. 1365.

Crystal, M. M. 1964. Observations on the role of light, temperature, age and sex in the response of screw-worms to attractants. *J. Econ. Entomol.* 57: 324-325.

Geier, P. W. 1960. Physiological age of codling moth females *Cydia pomonella* (L.) caught in bait and light traps. *Nature*, London. 185: 709.

Greenbank, D. O. 1963. The analysis of moth survival and dispersal in the unsprayed area. Pages 87–99 *in* R. F. Morris ed. The dynamics of epidemic spruce budworm populations. *Mem. Entomol. Soc. Can.* 31.

Greenbank, D. O. 1970. Climate and the ecology of the balsam woolly aphid. *Can. Entomol.* 102: 546-578.

Hagen, K. S. 1962. Biology and ecology of predaceous Coccinellidae. *Ann. Rev. Entomol.* 7: 289-326.

Haskell, P. T. and J. E. Moorhouse. 1963. A blood-borne factor influencing the activity of the central nervous system of the desert locust. *Nature*, London. 197: 56–58.

Johnson, C. G. 1969. Migration and dispersal of insects by flight. Methuen & Co., Ltd., London.

Kennedy, John S. 1975. Insect dispersal. Pages 103–119 *in* David Pimentel ed. Insects science and society. Academic Press, Inc., New York.

Schneider, F. 1962. Dispersal and migration. *Ann. Rev. Entomol.* 7: 223-242.

# Insect communication 13

With the exception of those species such as the desert locust that migrate en masse, the migratory behavior of insects typically leads to the dispersion rather than aggregation of individuals. Although dispersive behavior is ecologically important, as we have seen, it introduces problems associated with the postmigratory separation of individuals. If mating has yet to occur, the separated individuals must have some means of coming into contact with each other. Filling this need and the production of aggregations for purposes of resource exploitation, overcoming host resistance by mass colonization, and defense are the major roles of communication among insects.

Insects employ tactile, visual, auditory, and chemical methods of communication, and in many species a combination of methods is used in the direction of behavior to fulfill a single biological function. For example, a special odor or sound may be produced to bring a number of scattered individuals together for the purpose of reproducing. Once together, pairing may result from recognition and courtship behavior involving visual cues. Once paired, the male may induce the female to copulate by releasing a volatile aphrodisiac or by exciting her tactilely. Because so many of the behavioral sequences engaged in by insects involve more than one method of communication, it is easier to organize a discussion of the subject according to the basic methods rather than the biological functions they serve.

## *Chemical Communication*

Chemical communication among insects involves a variety of compounds, collectively called **pheromones.** Pheromones can be likened to hormones in that they are of a specific composition, produced by special glands, to be released at specific times. Furthermore, they inhibit or stimulate specific biological functions. Whereas hormones coordinate the physiological and behavioral processes *within* the individual, pheromones coordinate the physiological and behavioral activities *between* individuals of the population. The rapidly expanding field of pheromone research has been reviewed by a number of authors, including Karlson and Butenandt (1959), Butler (1967), and Shorey (1973).

Insect pheromones can be variously grouped according to the kind of behavior or activity they coordinate. Some pheromones function over long distances as sex attractants or as population aggregators, whereas others function over only short distances as is true of some sex stimulants. Consequently, pheromones can be important in stimulating reproductive behavior, producing aggregations at sources of food, stimulating mass defensive tactics, regulating population density, and so on. Among social insects, pheromones are particularly important in coordinating colony members to perform all of the activities necessary for the survival of the colony as a whole.

Sex attractants are employed by a wide variety of insects (see Jacobson, 1965) but probably have been most widely studied among the butterflies and the moths. In many species, the female produces a pheromone that attracts males to her location. Usually the scent is released only during those periods of the day when the males are active and responsive but continues to be released throughout that period until mating has occurred. In species that mate several times, the female will release her pheromone for several days, thus ensuring the receipt of sperm from several males. In the moths, the pheromone molecules are perceived by thousands of olfactory receptors located on the large plumose antennae of the males. When the antennal receptors of a resting male intercept the scent of a "calling" female, the male will often respond by vibrating its wings and then taking flight. The male will fly into the odor-bearing air stream until he locates the female or can no longer perceive her scent. If the male loses the scent, he will engage in a random flight which increases his chances of intercepting the trail again. In this way, widely separated males and females are effectively brought together; in one experiment, more than one quarter of the released males of the saturniid moth *Arctias selene* were able to locate caged females about seven miles away.

Chemical communication between mates is particularly important for bisexual species in which the females are flightless. Such is the case with a number of moths belonging to the families Saturniidae and Lasiocampidae, in which the females are either extremely sluggish or have short, nonfunctional wings. In some of the tussock moths, for example, a newly emerged female sits on her empty cocoon, releases her pheromone that attracts a male, mates, and then lays her eggs in the immediate vicinity; the young larvae then migrate as described in Chapter 12.

Sex pheromones not only serve to bring males and females together but they may also cause some sexual excitation that facilitates copulation. Sometimes a low concentration of a pheromone will only agitate the male, but a higher concentration will stimulate oriented flight, and an even higher concentration will result in courtship and copulation (see Traynier, 1968). In some species the females need a little stimulation as well, so the males produce a scent that acts as an aphrodisiac. For example, the male of the grayling butterfly *Eumenis semele* has a patch of special scent scales on the upper surface of the forewings which are exposed to the female by his spreading the wings during courtship (Figure 13-1).

Some insects release a mixture of chemicals that attract both sexes. This is common among the bark and ambrosia beetles, which use pheromones not only to attract members of the opposite sex but also to create an aggregation of individuals large enough to collectively overcome host resistance or, in the case of susceptible hosts, to maximize the utilization of contagiously distributed resources. In the genus *Dendroctonus*, the females initiate the attack of the host tree and subsequently release a sex-aggregating pheromone. In those genera such as *Ips*, in which the brood gallery is started by the male, the male produces the pheromone.

As more beetles are attracted to the host and attack it, more pheromone is released, strengthening the stimulus. Obviously, such a procedure cannot go on indefinitely. As the host material approaches an optimal level of occupancy, the attraction of new individuals slows down. Subsequently, the fully occupied host becomes unattractive as the result of a change in the pheromones produced. The production of the attractant compounds ceases, and an antiaggregation pheromone, which acts as a deterrent to late arrivals, is emitted.

Many ant species lay chemical trails which assist them in traversing the distance between their nests and sources of food they have located. Ant trails

FIGURE 13-1. Chemical communication in the grayling butterfly *Eumenis semele.* **A.** The position of the scent patches cross hatched on the dorsal surface of the male's forewings, **B.** The male bowing to a female during courtship to expose the scent patches which the female contacts with her antennae. (Redrawn from Chapman, 1969, after Tinbergen, 1951.)

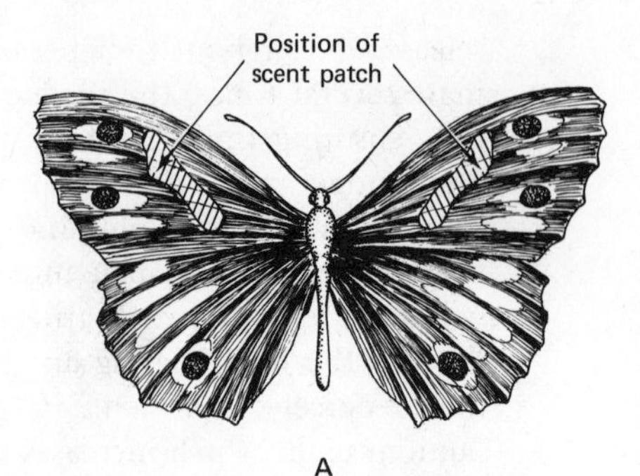


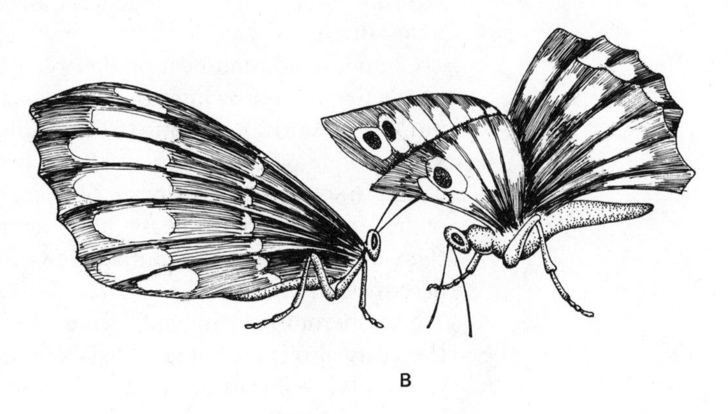


are frequently marked by the release of small droplets of pheromone at regular intervals along the travelled route. In addition to marking the way, the released pheromones serve to recruit other individuals to gather food from the same source. As numerous ants travel back and forth over the same route, leaving spots of pheromone as they go, the spots blend together to form a continuous trail. The fact that ant trails involve chemical communication can be readily demonstrated by rather simple experiments involving the disruption of an established trail by replacing a portion of it with unmarked soil. It is also possible to create artificial trails with a crude extract of the ant's **Dufour's gland** where the pheromone is produced (see Figure 13-5).

Termites mark their trails with a pheromone produced in special glands located in their abdomens. In the primitive species, an odor trail is used to recruit workers to damaged portions of nests that need to be repaired. In some more complex species, however, the same pheromone is used to recruit workers to food gathering, as in the ants.

The social termites, ants and bees, also use chemical communication to alert members of a colony to some danger and to recruit individuals to the colony's defense. These and other aspects of chemical communication in the honeybee will be discussed later.

### *Audio Communication*

Sound is also quite an effective means of communicating over both short and long distances. Sound is perceived as vibrations by specialized mechanoreceptors, which may be grouped together into "ears," one type of which is illustrated in Figure 13-2. Sounds produced by insects can be (1) a by-product of another activity, (2) the product of striking the substrate with some part of the body, (3) the result of rubbing two body surfaces together, (4) the result of a vibrating membrane, or (5) the result of a pulsating air stream.

The sound produced by the beating wings is used by some insects to locate other individuals and to bring the sexes together. A swarm of small flies is clearly audible to other individuals in the area. Males of the mosquito *Aedes aegypti* are attracted to the flight tone of sexually mature females, whereas immature females have a different tone to which the males do not respond.

A variety of insects produce sound by striking the substrate with some part of their body. Some grasshoppers strike the ground with their hind tibiae. Because both sexes engage in the activity, it probably serves to bring individuals together by way of a response to low-frequency vibrations through the substrate. Individuals of the termite *Zootermopsis* rock back and forth on

**FIGURE 13-2.** The tympanal organ of a katydid (Tettigoniidae: Orthoptera). (*Right.*) Close up photograph of the openings to the organ on a front tibia. (*Below.*) Diagram of a transverse section through the organ showing structural details.

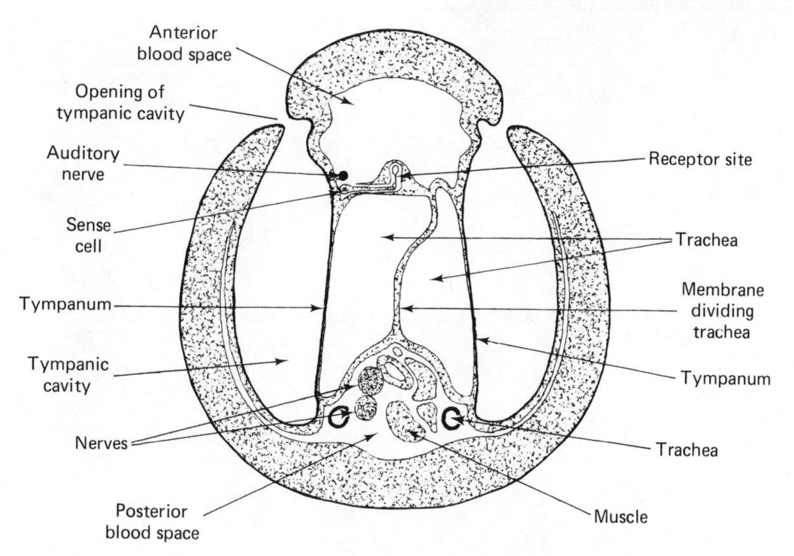

their middle legs so that the mandibles tap against the floor of the tunnel. This behavior is initiated by a disturbance near one part of the nest and warns other members of the colony to move elsewhere (Howse, 1964).

**Stridulation** is the term applied to sound production by the rubbing together of two body surfaces. The mechanism, which is used by many different insects, can be likened to running one's thumb nail down the teeth of a comb. Usually a ridgelike structure (**the scraper**) on one part of the body is moved back and forth over a ridged surface (**the file**) on an adjacent part. Variations of this type of system are common among the grasshoppers, crickets, katydids, true bugs, and beetles. In the grasshoppers, a row of pegs along the inside of the hind femora (Figure 13-3) is rubbed back and forth over the edge of the parchmentlike forewings. In katydids and crickets, the cubital vein of each forewing is toothed. When the wings are folded at rest, a ridge near the base of the left wing overlaps the file on the right wing. When the katydids and crickets sing, their wings are repetitively opened and closed part way, the sound being produced by a vibration of the adjacent wing membrane as the wings are closed.

The stridulation of orthoptera has been studied in considerable detail. Some of these insects produce five distinctly different songs used for calling, courtship, copulation, aggression, and alarm (Chapman, 1969), but the common calls of crickets and katydids that break the silence of summer evenings serve to draw the females to their singing suitors.

Beetles use many different parts of their well-sclerotized bodies as stridulatory organs, but the elytra are most commonly involved. Male bark beetles of the genus *Dendrotonus* have an abdominal scraper that is moved back and forth across an elytral file to produce a clearly audible chirping sound. The function of the sound is not understood but commonly occurs when the male is sitting in the entrance to a female's gallery while the host tree is under attack.

The production of sound by a muscle-driven membrane, or **tymbal,** occurs among a few moths and true bugs. The most thoroughly studied mechanism

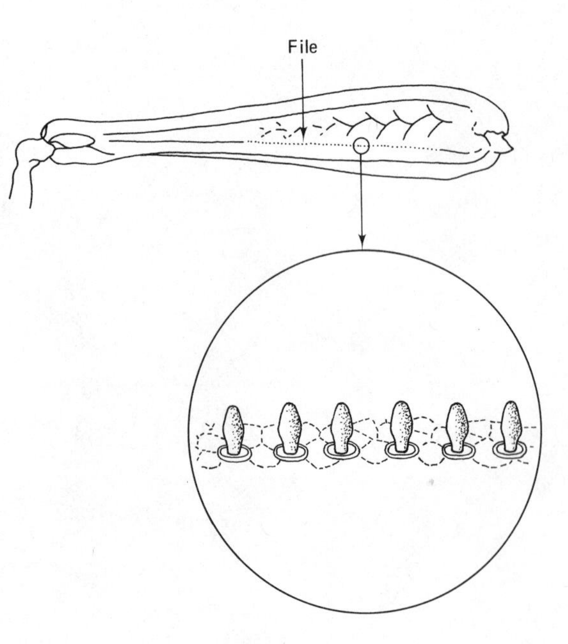


**FIGURE 13-3.** Inner surface of the hind femur of a grasshopper with an accompanying enlargement of a portion of the file. (Redrawn from Roscow, 1963.)

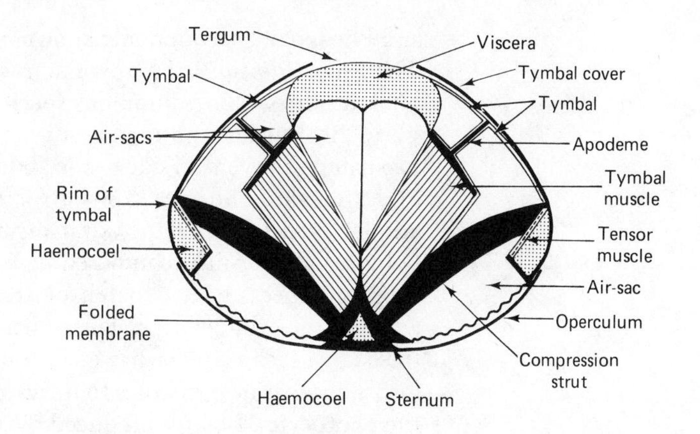


FIGURE 13-4. Diagram of a transverse section of the first abdominal segment of a cicada showing the structures concerned with sound production. (Redrawn from Chapman, 1969, after Pringle, 1954.)

of this type is that of a well-known insect chorister, the cicada. In these insects, the first abdominal segment is highly modified for sound production (Figure 13-4). There are a pair of dorsolateral structures that resemble drumheads; each consist of a thin disc of cuticle, supported by a thicker cuticular rim. These discs form the tymbals which are protected by tymbal covers composed of regular cuticle. Each tymbal gives rise to an apodeme on its inner surface to which the tymbal muscles are attached. When the muscles contract and cause an inward buckling of the tymbals there is an associated click. When the muscles relax, the tymbals return to their normal shape and produce a second click. The succession of clicks associated with a cycle of rapid muscle contraction and relaxation produces the familiar trill of the cicada. North American cicadas have one song which stimulates species' specific aggregations and a second song used for courtship.

The only known case of sound produced by a pulsed airstream occurs in the moth genus *Acherontia,* which draws air in through its proboscis by dilation of the pharynx. The in-rushing air causes a flaplike epipharynx to vibrate and to produce a pulsed airstream and related low-pitched sound. This is followed by a high-pitched whistle as the air is expelled.

### *Visual Communication*

Visual communication among insects is common and highly variable. Unlike chemical and sound communication, visual communication requires a direct line of sight between individuals and is only effective over relatively short distances. Vision can be important as a means of bringing individuals together and also in the sometimes elaborate patterns of courtship behavior.

The females of mayflies, caddisflies, and midges are attracted visually to swarms of males engaged in dance-like up-and-down flights in compact groups. Male butterflies frequently pursue appropriately colored females and can be persuaded to chase pieces of colored paper moved in a manner that imitates a female's flight. Once butterflies have paired, visual displays become an important part of the courtship behavior without which copulation may not occur.

The most spectacular visual displays encountered among the insects involve luminescence (see Lloyd, 1971). In some insects, luminescence is incidental and serves no specific communicative function. Some collembolans, for example, give off light as a by-product of their metabolism, as do many bacteria; insect larvae infected by such bacteria may also appear luminescent.

The most brightly luminous insects are beetles of the families Lampyridae and Phengodidae, commonly known, respectively, as fireflies and glowworms. The genus *Photinus* (Lampyridae), which is common through parts of the United States, is one luminous species which has been studied quite intensively. Both males and females have light-producing organs; the males have two pair on the ventral side of the abdomen, compared to a smaller single pair in the females. The larvae also have a single pair on their abdomens, but these are not progenitors of the adult organs.

Each light organ is composed of 7,000 to 8,000 large cells, called **photocytes,** grouped under a patch of transparent cuticle. The photocytes are arranged in cylinders through which tracheae and nerves pass. The main trachea in each cylinder has numerous side branches, each of which divides into a number of tracheoles that carry oxygen into the immediate vicinity of the photocytes. Light is produced by the oxidation of a compound known as **luciferin** in the presence of an enzyme called **luciferase.** The complex biochemical process that occurs within the light organs is believed to be triggered by the acetylcholine released by an activated nerve ending. The acetylcholine is believed to react with ATP and coenzyme A to produce pyrophosphate. The pyrophosphate is thought to remove the inhibition of the enzyme luciferase, enabling oxidation of the luciferin to take place (Chapman, 1969).

There is quite a long delay between nervous stimulation and the production of light which suggests that the acetycholine must diffuse into the photocytes or that the intermediate steps occur quite slowly. The flash of light produced by each cylinder of photocytes is quite short, but the lack of coordination between cylinders may result in the entire organ producing a relatively long flash. In *Photinus,* the flash given off by an entire light organ lasts several hundred milliseconds, and flashes occur at regular intervals. In some species the frequency of flashing increases with temperature.

Perhaps the most intriguing aspect of insect-produced light is its great efficiency. The light falls entirely within the visible portion of the spectrum and almost all of the radiant energy given off is light rather than heat. The energy emitted by *Photinus,* for example, is only about 10% heat, compared to a standard light bulb that gives off about 95% of its radiant energy as heat. Scientists are obviously interested in similar chemical light-producing systems as a possible means of reducing energy consumption, but perhaps even more intriguing than the efficiency of insect light production is the fact that the luciferin is not destroyed by the light-producing reaction. Apparently luciferin passes from a high-energy state to a low-energy state with the release of radiant energy but no change occurs in its molecular structure.

Both sexes of the North American firefly emit flashes of light. The males flash during flight, and the females perch and flash back at males that come close to them. After several flashes have been exchanged, the male orients toward the female, and mating ensues (McElroy, 1965). In some of the Asian tropical species, the females are flightless, so their flashing light clearly facilitates mate location by the males. Since the males flash as well, the situation would seem to be one of confusion, but this is overcome to some extent by adjacent males flashing synchronously. This phenomenon spreads until large numbers of males flash in unison.

The larvae of some fungus gnats of the genus *Bolitophila* have light-producing organs formed from enlarged ends of their Malpighian tubules. The colonial larvae construct a snare consisting of strands of silk supplied with tiny beads of sticky material. At night the larvae come out and emit a steady glow, which apparently attracts small insects into their trap.

***Tactile Communication***

Communication by touch obviously can occur only after other means of communication bring individuals together. Nevertheless, insects engage in rather elaborate forms of contact behavior for courtship and sexual stimulation. In *Drosophila*, for example, the male is attracted to the female visually, but final species recognition results from the male tapping the female with his forelegs. In many species the female is not immediately receptive to the male's copulatory advances and will only allow the male to mount after some appropriate foreplay. In a few insects, the male appeases the female with an offer of food and then copulates with her while she eats his offering. Male scorpion flies of the genus *Panorpa* (Mecoptera) secrete droplets of saliva which harden and serve as a snack for the female during copulation. The males of other species appease their mates with seeds or nonedible objects such as a brightly colored petal.

The appeasement behavior of males would seem to guard against exciting a female to respond to a mate as a potential attacker. In the mantids, the female responds to the careful approach of the male by grasping him with her forelegs and then proceeds to remove his head with her mandibles during copulation. The outcome, however, is more vigorous copulatory activity resulting from the severing of the male's subesophageal ganglion.

# The Integration of Communication Methods

In the foregoing discussion, the four main methods of communication used by insects have been treated separately. However, we must realize that they are not used independently. Each has its special attributes. Odors, for instance, are carried great distances on the wind and, if detectable in low concentrations, can provide a way by which insects beyond the range of sight or hearing can communicate effectively. At closer range, other forms of relaying information may be more appropriate. Furthermore, biological functions are frequently fulfilled by way of complex behavior patterns composed of a sequence of responses. Such is often the case with those functions served by communication.

For mating to occur, individuals may have to come together over considerable distances. The danger of attack by natural enemies is considerably lessened when individuals can communicate while they remain concealed. Odor and sound would seem to serve this strategy well. Once individuals have been able to gain proximity, visual cues would become useful as a means of species recognition. Finally, tactile stimulation can serve to provide the excitation required before the female submits to copulation.

A behavioral sequence of the type just described involving several types of communication is relatively simple compared to the communication that must take place among the subsocial and social insects. In truly social species, the survival of entire colonies depends upon the integration of the relatively simple individual patterns of communication and response into a coordinated system of mass behavioral phenomena. An examination of communication

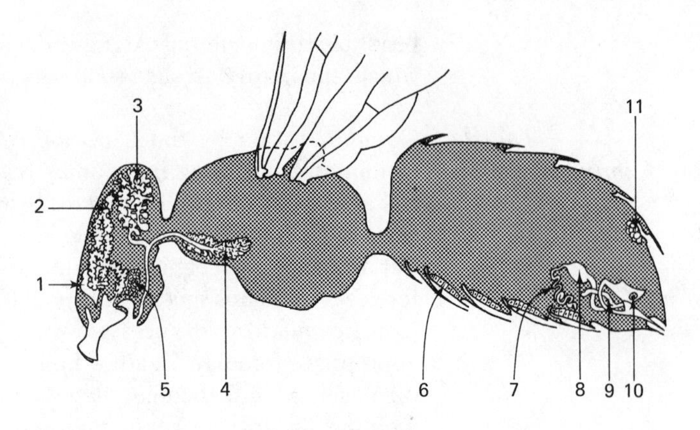


FIGURE 13-5. Chemical producing glands of the worker honey bee *Apis mellifera*. *1.* mandibular glands, *2.* hypopharyngeal gland, *3.* head labial gland, *4.* thoracic labial gland, *5.* hypostomal gland, *6.* wax glands, *7.* poison gland, *8.* vesicle of the poison gland, *9.* Dufour's gland, *10.* Koschevnikov's gland, *11.* Nasanov's gland. (Redrawn from *The Insect Societies*, Wilson, 1971, with permission of the author and Harvard University Press.)

within a colony of honeybees therefore seems to be a fitting way to complete this chapter.

The coordination of life and behavior within a beehive involves communication of each of the types discussed. Since social bees return to their nest at regular intervals, mechanisms for long-distance communication have been replaced largely, but not totally, by elaborate communication in and around the nest site. Although visual, sound, and tactile communication occurs between members of a hive, chemical communication predominates. In addition to chemical communication, as discussed previously, bees exploit chemical signals in the form of special substances in the food that is exchanged between individuals. (The glands of a honeybee that produce behavior-mediating chemicals are shown in Figure 13-5). The following discussion, therefore, will be organized according to the functions served by communication rather than by the method employed. The discussion will of necessity be shorter than the subject justifies, but interested students are referred to the outstanding treatment by E. O. Wilson (1971) for further details.

Wilson divides social communication functionally into alarm and assembly, recruitment, recognition, food exchange, grooming, and group effects. A brief examination of the kinds of communication used in the regulation and coordination of these functions in but a single species should provide some perspective of the complex nature of this subject over the entire range of social insects.

### *Alarm and Assembly*

Alarm and assembly can be used for both defensive and foraging purposes. An initial sting by a honeybee may provoke other bees in the vicinity to become aggressive. Because the worker's sting is barbed, it catches in the victim's skin, and as the bee attempts to fly away, the poison gland and Dufour's gland are often left behind. The act of stinging releases the compound isoamyl acetate, produced by secretory cells that line the sting pouch (Shearer and Boch, 1965). The highly volatile substance attracts other bees to the source, and some subsequent stinging may occur. However, it does not stimulate the release of the substance by other bees not engaged in stinging, so as the danger passes and the alarm pheromone dissipates, behavior returns to normal.

Much more important in honeybee behavior is the chemical communication that leads to other forms of assembly. Worker bees release a mixture of chemicals produced by their **Nasanov glands.** The substance is often released by bees located near the hive entrance, during swarming, and when new food sources are first located. Individuals that have been isolated from their nest mates for a period of time will also release the scent as a means of reestablishing contact.

The scent continuously laid down around the entrance to the hive assists the foraging bees to locate the nest on their return. However, this hive odor does not serve to distinguish one colony from another, as once thought. During swarming, the Nasanov substance stimulates the assembly of workers that ultimately leads to the familiar cluster of bees around their queen.

Another assembly pheromone consists of a group of chemicals, commonly called **queen substance**, produced by the mandibular glands of the queen. This secretion and another from **Koschevnikov's gland** are at least partially responsible for the formation of the cluster of "court bees" that constantly surrounds the queen. Also, when a colony swarms, the workers are attracted to the queen in flight and follow her trail of evaporating pheromone to the settling site located by the scout bees. Once the queen settles, she releases another mandibular gland secretion that tends to settle the nearby workers. A first group of settled workers dispense their own Nasanov secretion, which stimulates the rest of the swarm to cluster.

***Recruitment***

Recruitment involves the gathering together of nest mates at a particular place for purposes of applying a joint effort to a specific task, such as nest construction or food retrieval. The most elementary form of recruitment communication in the honeybee involves the recognition of food sources from the scent that adheres to the bodies of foraging bees and the nectar they regurgitate upon return to the hive. When a particular food source is abundant and near the hive, this simple form of communication is quite adequate. In fact, investigators have increased pollination by training the bees on sugar syrup tainted with the odor of the crop (von Frisch, 1967).

Ants forage for food on the ground so are able to leave well-marked trails to recruit their nest mates to the task of exploiting a food source. Bees, on the other hand, forage on the wing, so they must have some other means of communicating the exact location of food to their colleagues. Some bees do lay odor trails, however. The South American genus *Trigona*, having established a course from the hive to a food source, will stop every few meters on a homeward flight, depositing a droplet of mandibular gland secretion at each point. Other bees then follow the odor trail (Lindauer, 1971). The honeybee employs a rudimentary form of odor trail, used for short-distance orientation in the immediate vicinity of the hive. The trail is laid down by workers returning to the hive over a short distance on foot; the chemical is referred to as the "footprint" pheromone.

The honeybee communicates the distance and the direction to good foraging sites by an elaborate method of communication, widely known as the "bee dance" (von Frisch, 1967). Different races behave slightly differently, but usually if the source of pollen or nectar is fairly close to the hive, returning workers perform the **round dance** (Figure 13-6). As indicated in the illustration, a worker that has just returned from a successful foraging trip penetrates the hive to where there are other field bees and engages in an excited circular pattern of running. Other field bees follow and in the course of the dance pick

FIGURE 13-6. The round dance of the scout honey bee that is thought to communicate the presence of a nectar source near the hive. (Adapted from Karl von Frisch: Bees: Their vision, chemical senses, and language. © 1950, 1971 by Cornell University. Used by permission of Cornell University Press.)

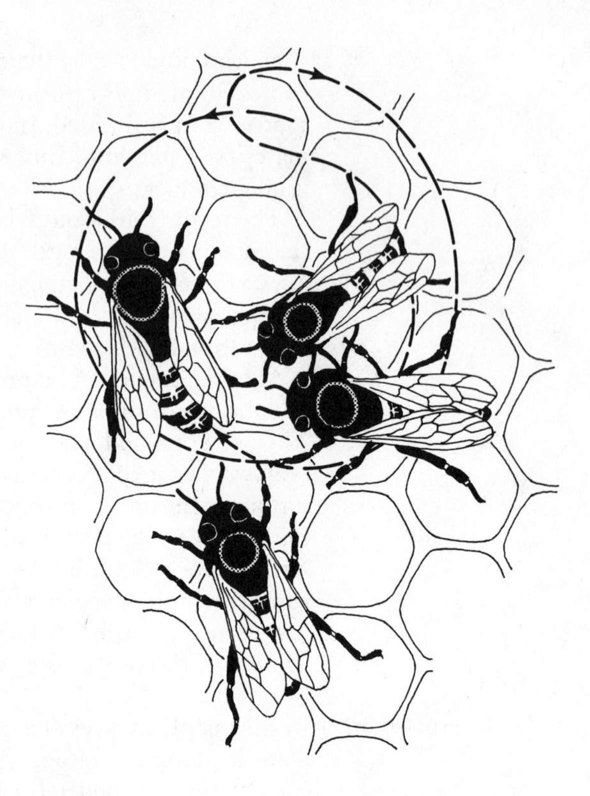

up odor information about the source of food. This simple form of the bee dance recruits other workers to search for pollen or nectar of a certain kind in the immediate vicinity of the hive.

If the food source is more than 50 meters or so from the hive, the information transmitted by the round dance would be of little use. As the distance to the foraging site increases, the round dance becomes modified by the incorporation into it of a straight run, during which the performing worker waggles its abdomen from side to side (Figure 13-7). The new pattern of behavior that results is called the **waggle dance** (Figure 13-8). After much painstaking study, von Frisch was able to decode this remarkable means of communication.

When a scout bee returns to the hive after having made several successful trips to and from a forage site a considerable distance away, she penetrates deep into the hive, regurgitates nectar from her crop, and begins to perform the waggle dance on one of the vertical honeycombs. The sun is used as the key reference point for the communication of direction. If the forage site is located on a line between the hive and the sun, the straight run of the dance

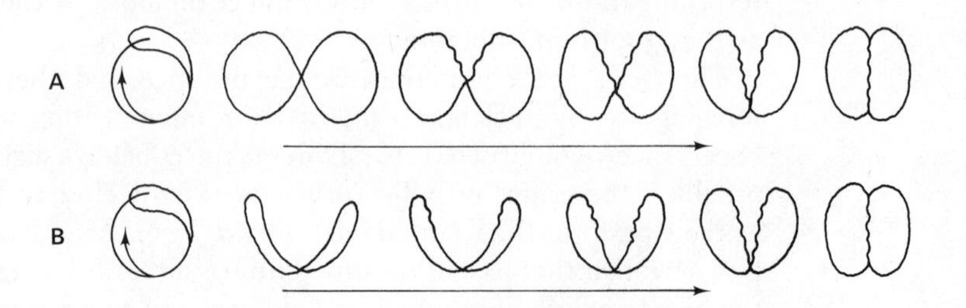


FIGURE 13-7. The changes in the round dance that occur as the distance between the hive and the nectar source increases. (Adapted from von Frisch, 1967.)

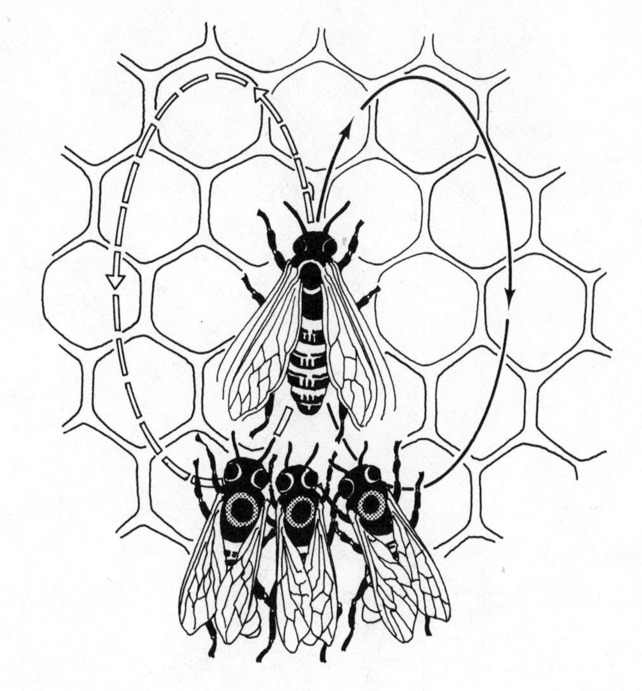

**FIGURE 13-8.** The waggle dance of the scout honey bee that is believed to communicate both the direction and the distance to nectar sources away from the hive (see text for details); (Adapted from Karl von Frisch; Bees: Their vision, chemical senses, and language. © 1950, 1971 by Cornell University. Used by permission of Cornell University Press.)

will be up the comb. If the direction is away from the sun, the straight run will be down the comb. The straight run is followed by a circle to the right, another straight run followed by a circle to the left, and so on. Likewise, if the forage site is located 30° to the right of the sun, the straight run of the dance will be performed at an angle 30° to the right of the vertical (Figure 13-9). The compound eyes of the honeybee are so sensitive to ultraviolet radiation that they can detect the position of the sun and communicate the direction to forage sites even on lightly overcast days.

The communication of the direction to more distant food sources is highly beneficial, but foraging efficiency would clearly be enhanced by some communication of distance as well. This information is provided by the duration of the straight run and the speed with which the dance is performed. During the straight run, the dancing bee waggles her abdomen from side to side with a frequency of about 13 to 15 vibrations per second, at the same time producing an audible buzzing sound by vibrating her wings. By observing the dance of scout bees returning from feeding platforms placed at different distances from the hive, von Frisch was able to decode the dance language. The further the food from the hive, the longer was the duration of the wagging portion of the dance and the fewer were the number of complete cycles of the dance per unit of time.

Apparently, the duration of the straight run is not based on the absolute distance to the foraging site but on the energy that must be expended to get there. If the trip involves a flight up a steep slope or against the wind, the straight run performed in the hive will be longer. Furthermore, different genetic races of honeybee have incorporated variations into their dance that have been referred to as dialects. For example, the Carniolan race has the most rapid dance tempo, and that of the Italian race is the slowest. This means that an Italian worker would interpret the shorter straight run of a dancing Carniolan worker to mean that the forage site is closer to the hive

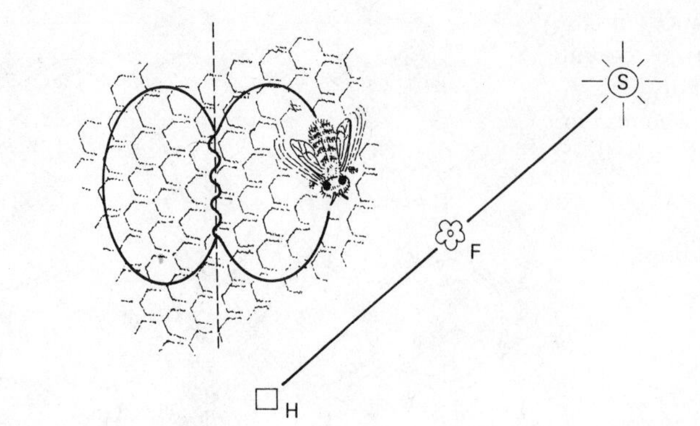


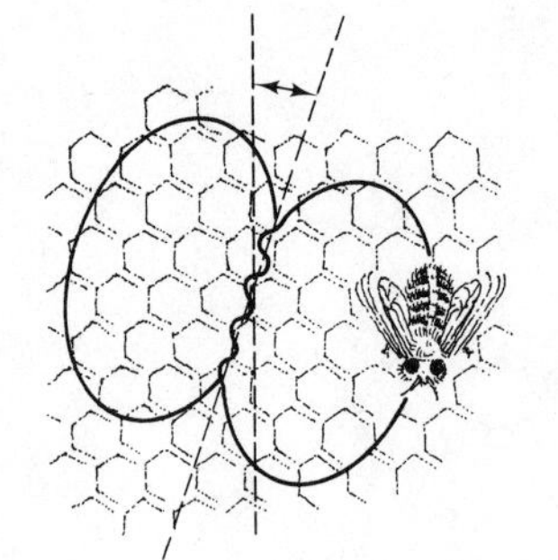

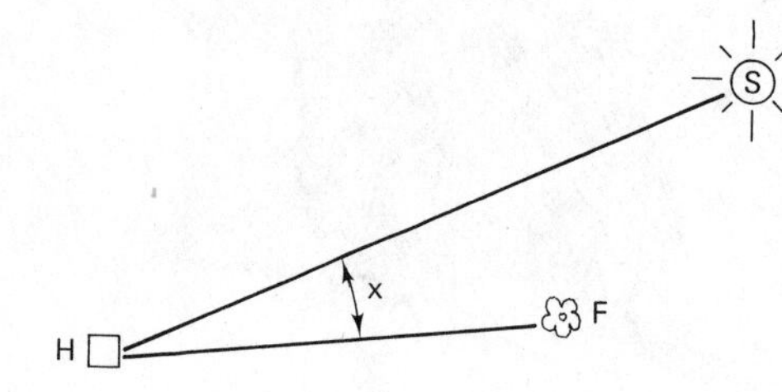


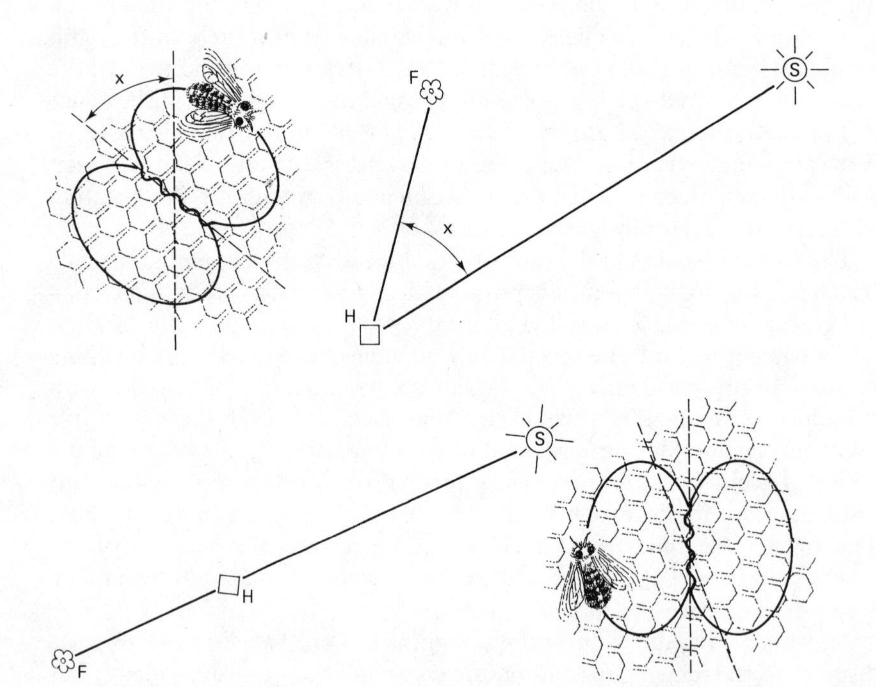


FIGURE 13-9. Diagrams showing the orientation of the straight run of the waggle dance relative to the position of the hive (H), forage site (F), and the sun (S).

than it actually is. Experiments with colonies consisting of a mixture of the two strains revealed that Carniolan workers consistently overestimated the distance from Italian dances, and vice versa (von Frisch, 1967).

The waggle dance is a truly fascinating example of insect communication, but it is even more amazing than meets the eye. As Wilson (1971) so clearly stated, it is ". . . a signal constructed from a ritualized and miniaturized imitation of the journey that the signalling bee has taken in the past and upon which some of its sister bees are about to embark." The workers in the hive are able to rehearse their flight in miniature before they set out, much as we do when we trace a proposed plane trip on a map.

In 1967, Wenner and Johnson challenged von Frisch's interpretation of the bee dance as the major means of communication for foraging. They contended that the bees could adequately achieve recruitment with the odor of the nectar brought to the hive and the recruitment pheromone left by foragers in the field (Wenner et al, 1969). This stimulated additional research, the results of which indicate that the two hypotheses, although in conflict are not mutually exclusive, as pointed out by Gould (1975). Gould concluded that von Frisch and Wenner were simply examining different parts of the same process. The elaborate dance language is of great value in the rapid recruitment of foragers to remote, isolated patches of food, but when an extensive supply of food is available, recruits can use odor alone. As Gould pointed out, the honeybee evolved in areas where a means of exploiting spatially and temporally discontinuous food sources that were also sought by other species would have been a significant advantage.

Honeybees engage in other dance variations that have yet to be studied as thoroughly as the waggle dance. One called the buzzing run initiates swarming. During the warm part of a suitable day, one or several workers begin running through the colony in an excited zig-zag pattern, vibrating their wings and abdomens as they go. The sound produced is quite distinctive and may be part of the total pattern of communication. The dance is highly contagious, and soon many workers join in. Within minutes, the bees near the hive entrance rush out and are followed by a mass of workers and the absconding queen. Wilson believes that this form of communication may be unique in that the initial signal stimulates others to produce the same signal in the form of a chain reaction.

### Recognition

Recognition is an extremely important part of the communication system of social groups. In a bee colony, for example, there is so much activity involving the comings and goings of the field workers that an inability to distinguish friend from foe could be a major problem. The recognition of nest mates seems to be very casual, but intruders of other species are quickly and violently expelled. The reaction of honeybees to individuals of the same species but from a different hive seems to vary considerably in relation to the possible significance of the intrusion. During periods when the nectar flow is strong, the accidental arrival of a field bee from another colony will probably be accepted. But when nectar is scarce, strangers are repelled and robber bees meet with violent rejection. Each colony does not have an inherently distinctive odor as once thought, but each colony does become recognizable by the odor of the pollen and nectar the bees are collecting. During a major nectar flow when all the hives are working a common food source, there is probably little difference between them. On the other hand, during hard times each hive may have a very distinctive odor.

The recognition of castes is also important among social insects. The queen honeybee is clearly distinguished from all other members of the colony and is treated with apparent respect. The workers not only recognize the queen on the basis of the odor substances she secretes but are also able to determine when she is becoming reproductively ineffective.

After the old queen leaves a colony with a swarm, the first virgin queen to emerge communicates audially with other young queens which have not yet emerged. If the first queen to emerge killed all the others, the colony could be left queenless by an accident on her nuptial flight. By producing a chirping sound called "piping," the first queen delays the emergence of her sisters, who chirp back from within their cells. This has been interpreted as a means of delaying the mortal combat that must occur whenever there are two queens in a colony.

### *Group Effects*

Group effects in the honeybee are obvious from the incredible control that a viable queen exerts over the functioning of her colony. As long as a healthy queen is present to give off the appropriate signals, the workers contribute unerringly to her needs and those of her offspring and the colony as a whole. If a queen is removed, the organized behavior of the workers gives way to disorganized restlessness in a matter of minutes. Within a few hours, workers will draw several worker cells into enlarged cells suitable for rearing a replacement queen. After a few days, some of the workers experience ovarian development and actually lay infertile eggs which develop into drones. The presence of "queen substance" apparently acts to inhibit queen cell construction and reproduction, probably by suppressing the activity of the corpora allata of the workers. This raises the question of how the workers are stimulated to produce young queens to facilitate the reproduction of the colony by swarming. Butler (1960) found the amount of "queen substance" produced by queens during the normal swarming season to be about one quarter of that of queens of nonswarming colonies. Thus the queen exerts her control over all individuals that comprise a colony; yet, she in turn is influenced in her behavior and egg laying by the workers that tend to her needs.

## References Cited


Butler, C. G. 1960. The significance of queen substance in swarming and supersedure in honey bee (*Apis mellifera* L.) colonies. *Proceedings of the Roy. Entomol. Soc.*, London (A) 35: 129–132.

Butler, C. G. 1967. Insect pheromones. *Biological Reviews* 42: 42–87.

Chapman, R. F. 1969. The insects: Structure and function. American Elsevier Publishing Co., Inc., New York, N. Y.

Frisch, K. von. 1967. The dance language and orientation of bees (tr. Chadwick). Belknap Press of Harvard University Press, Cambridge, Mass.

Frisch, K. von. 1971. Bees, their vision, chemical senses, and language. Cornell University Press, Ithaca, N. Y.

Gould, J. L. 1975. Honeybee recruitment: The dance language controversy. *Science* 189: 685–693.

Howse, P. E. 1964. The significance of the sound produced by the termite Zootermopsis angusticollis (Hagen). *Animal Behaviour* 12: 284-300.

Jacobson, M. 1965. Insect sex attractants. John Wiley & Sons, Inc., New York, N. Y.

Karlson, P. and A. Butenandt. 1959. Pheromones (ectohormones) in insects. *Annual Review of Entomol.* 4: 39-58.

Lindauer, M. 1971. Communication among social bees. Harvard University Press, Cambridge, Mass.

Lloyd, J. L. 1971. Bioluminescent communication in insects. *Ann. Rev. Entomol.* 16: 97-122.

Roscow, J. M. 1963. The structure, development and variation of the stridulatory file of *Stenobothrus* lineatus (Panzer) (Orthoptera, Acrididae). *Proc. R. ent. Soc. Lond.* A. 38: 194-199.

Traynier, M. M. 1968. Sex attraction in the Mediterranean flour moth, *Anagasta kühniella:* Location of the female by the male. *Can. Entomol.* 100: 5-10.

Wenner, A. M. and D. L. Johnson. 1967. [Reply to K. von Frisch, *Science* 158: 1072-1076]. *Science* 158: 1076-1077.

Wenner, A. M., P. H. Wells, and D. L. Johnson. 1969. Honeybee recruitment to food sources: Olfaction or language? *Science* 164: 84-86.

Wilson, E. O. 1971. The insect societies. The Belknap Press of Harvard University Press, Cambridge, Mass.

# Offense and defense

# 14

The subject of insect offense and defense could be considered appropriately in several places within this book, because it involves a combination of morphological, physiological, and behavioral adaptations that have developed in response to a variety of selective pressures. I have chosen to discuss it here because virtually all of the structural and chemical mechanisms that have evolved have been accompanied by the development of appropriate patterns of behavior that make them effective. Protective coloration would be useless if an animal associated itself with an inappropriately colored background.

To humans insects may seem more offensive than they really are, because we see ourselves under attack by insects such as biting flies in quest of a blood meal or bees and wasps whose nests we have disturbed. Some insects are extremely aggressive in their acquisition of food, as are many others in defense of their nests. But this is not offense in the strict sense. In fact, when it comes down to a conflict between members of the same species or between different species using the same resource, insects are generally rather passive. Species that are so strongly solitary that they engage in mortal combat upon contact are rare. Even the grappling between a pair of males for a single female, as occurs among sphecid wasps, or the territorial defense displayed by dragonfly males seems largely ritualistic.

Such is not the case when it comes to defense. The mechanisms that have been evolved by insects for their own protection are truly incredible. The fantastic attention to detail that nature seems to have paid in the evolution of defensive traits provides ample evidence of the intensity and the variety of selective pressures to which the insects have been subjected. The principal mechanisms of defense can be classified as predominantly behavioral, structural, chemical, or colorational. However, in most instances insects employ two or more of these methods, either in sequence or in combination. Normally there is such a high degree of integration between the different mechanisms, particularly behavior, that to separate them as I have done throughout much of the following discussion can be justified only on the grounds of convenience. The subject of mimicry in the broad sense cannot be considered under any single heading, so will be mentioned several times.

## Behavioral Defense

As already indicated, behavior is an important component of all forms of defense, but there are some insects that rely almost entirely on behavior for protection. Perhaps the most straightforward approach to defense is to flee; this is extremely effective among insects because of their small size and their ability to accelerate very rapidly. Many leaf hoppers (Cicadellidae) move rapidly sideways to the opposite side of the branch they are on and simply attempt to stay out of sight. Jumping or taking flight are two particularly effective means of escape, as anyone who has attempted to capture a fly or a grasshopper with the hands will readily testify. Another effective form of escape consists of reflex dropping. Behavior of this type is displayed by a number of kinds of beetles, especially weevils, and a variety of caterpillars. Geometrid larvae often drop from their food plant on a strand of silk when disturbed and then reel themselves back up when the danger has passed. Other caterpillars, like those of the California tussock moth, just drop freely to the underlying vegetation. The opposite of rapid movement is, obviously, to remain motionless. Remaining motionless reduces visibility or gives the appearance of being inanimate. Numerous insects adopt a posture that feigns death (thanatosis) as a defense against natural enemies that prefer live prey. A death posture functions particularly effectively in combination with rapid movement. If an adversary is persistent, the surprise change from a lifeless pose to a sudden evasive movement introduces an element of surprise which effectively confuses the pursuer.

Another form of mainly behavioral defense involves the adoption of a threatening posture as in the katydid *Neobarettia spinosa* (Figure 14-1). Such poses can be quite effective in frightening adversaries sufficiently that they will at least cease to be aggressive. Threatening postures are often enhanced by bizarre structural features that make the insect appear grotesque, or by patches of bright color which when displayed suddenly change an attacker's search image. Some threatening postures actually mimic the defensive behavior of other species that use an additional mechanism as their first line of defense. This can be seen among nonstinging species that engage in the abdominal thrusting behavior characteristic of stinging species. One of the most fascinating threatening postures is adopted by the larvae of the Brazilian sphingid *Leucorrhampha*, which at rest blend well with their normal background. When disturbed, the larva twists its body to expose a pattern of scalelike markings and then sways back and forth in the manner of a small snake (Figure 14-2) (Jeannel, 1960).

**FIGURE 14-1.** The katydid *Neobarettia spinosa* in its threatening posture. (Reproduced, with permission, from T. J. Cohn, 1965.)

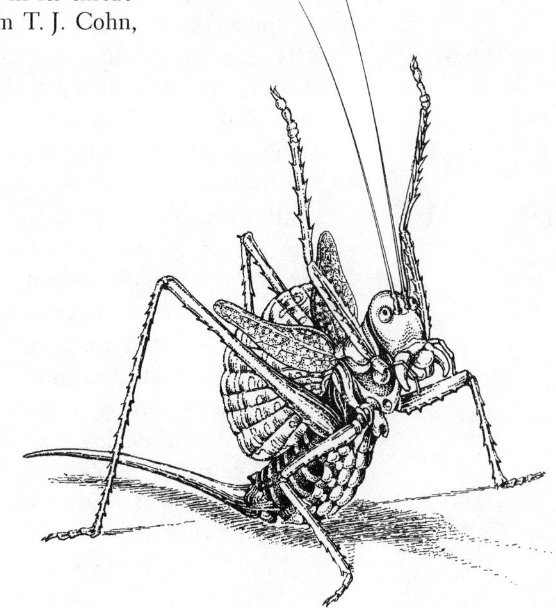

**FIGURE 14-2.** The caterpillar of *Leucorrhampha ornata* assuming a defensive posture in which it resembles a small snake. (Redrawn from Jeannel, 1960.)

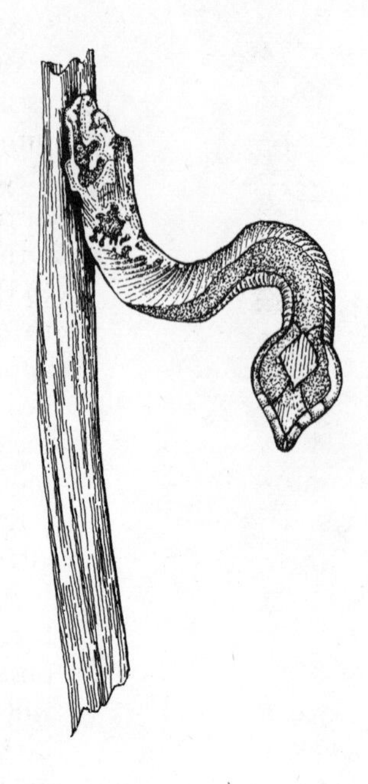

## Structural Defense

In some insects the integument becomes so heavily sclerotized that it provides adequate protection against almost any form of natural attack, including the beaks of insectivorous birds. This is especially true of beetles, among which the ironclad beetles (Tenebrionidae) and some weevils (Curculionidae) are well known by collectors because of the difficulty of penetrating them with mounting pins. In some ants, the front of the well-sclerotized head capsule of the soldier caste is flattened and can be used to plug temporary openings in the nest against intruders (Figure 14-3).

The mouth parts of insects are particularly useful in defense, as are the modified cerci that form the forceps at the tip of the abdomen of earwigs; the latter become proficient at picking up adversaries such as ants and throwing them to one side with a flick of the abdomen. Insects do, however, have a variety of specialized defensive structures, mainly in the form of spurs and modified setae. The leg spines of insects are primarily an aid to locomotion but become effective weapons when the legs are raked across a victim. The adults of tiger moths (Arctiidae) have poison glands associated with their leg spines that greatly increase their effectiveness. Caterpillars belonging to several families have a cloak of hairlike setae that make them unpalatable to a variety of insectivores and increase the difficulty of oviposition upon them by parasites. Short fragments of these setae break off and readily enter the skin, causing irritation and festering similar to that caused by a bark sliver. In some species, however, the setae are provided with a toxin secreted by associated epithelial gland cells. These urticating hairs are capable of causing serious irritation in humans. Precautions to avoid inhalation of hair particles must be taken by laboratory workers whenever large numbers of pubescent caterpillars are reared. Following a recent outbreak of the Douglas-fir tussock moth in the western United States, members of salvage logging crews suffered severe nose and eye irritation caused by the disturbance of a large residue of urticating hairs on the caste larval skins that remained in the forest.

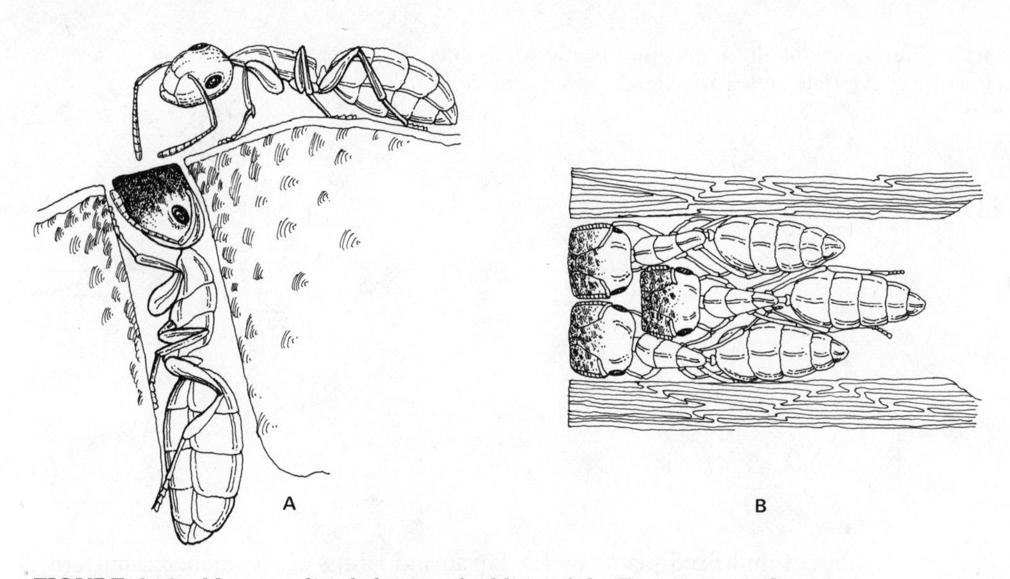


**FIGURE 14-3.** Nest-guarding behavior of soldiers of the European ant *Camponotus truncatus*. **A.** A lone soldier blocking a small opening, **B.** A group of soldiers blocking a large entrance hole. (Redrawn from Wilson, 1971, after Szabó-Patay.)

Numerous insects have evolved structural modifications which make them similar in appearance to dangerous species, as in the case of the snakelike caterpillar, or to some insignificant object in their environment. The latter, which is simply structural camouflage, will be considered as a component of cryptic form and coloration.

# Chemical Defense

Arthropods in general and insects in particular display the widest diversity of chemical defenses of any group of terrestrial animals. Chemical defense among the insects can be divided broadly into the use of compounds referred to as **venoms** that are injected by way of mouth parts or some special stinging apparatus, and the use of odoriferous or repugnatorial substances produced by special integumentary glands. Arthropod venoms were reviewed by Beard (1963), whereas other forms of chemical defense were reviewed by Roth and Eisner (1962) and Eisner (1970).

For the most part, the defensive chemicals are produced by multicellular structures comprised of glandular epithelium and a saclike reservoir with a cuticular lining in which the secretion is stored. The defensive compound is usually discharged by muscular or hydrostatic compression of the reservoir or by its evagination. Several staphylinid beetles have tubular glands which under pressure of the hemolymph turn inside out like fingers of a glove (Figure 14-4).

Nonvenomous defensive secretions are highly variable. Roth and Eisner (1962) and Eisner and Meinwald (1966) listed numerous defensive compounds that are used singly or in combination by arthropods. Not all such compounds are insect-made. Often they are simply special chemical constituents of the diet that when sequestered by an insect render it unpalatable. Probably the best-known example is the sequestration of the cardiac glyco-

FIGURE 14-4. Ventral view of the staphylinid beetle *Malachius bipustulatus* showing evertible defensive glands. (Redrawn from Jeannel, 1960.)

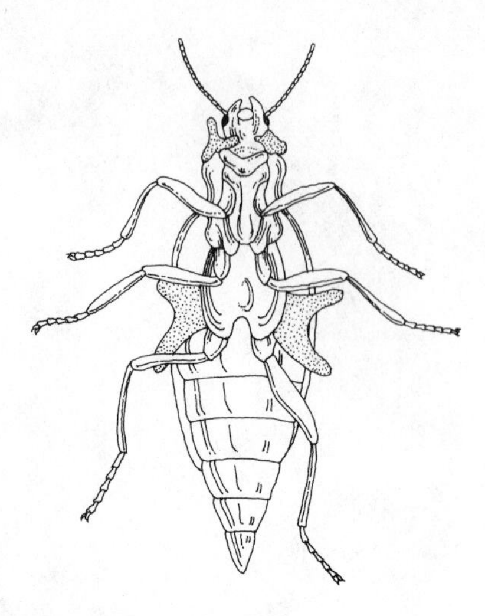

cides of milkweed plants by the larvae and adults of the monarch butterfly. Other insects, lepidopterans in particular, are rendered undesirable as prey by various other plant-derived compounds that become incorporated into their tissues. Obviously, the potential predators of such insects have no way of knowing that these prey are unpalatable unless they try them and learn that they should be avoided. Apparently the learning of predators is enhanced if distastefulness is associated with a specific visual image. In the course of these predator-prey relationships, many unpalatable species have evolved patterns of bright-colored markings often referred to as **warning** or **aposmatic coloration.** This will be discussed briefly under colorational defense later in this chapter.

In some species, the distasteful components of an insect's food are actually concentrated for release against an adversary before it takes a trial taste. The larvae of some butterflies have a thoracic pouchlike structure called an **osmeterium,** where odiferous plant components accumulate and from which they are expelled when the pouch is everted in response to a disturbance. One well-known example is the larva of the anis swallowtail which upon disturbance everts a bright orange-red osmeterium and its contents of volatile anis oil. Recently, Roth and Eisner discovered that a man-made herbicide known to be repellent to ants is sequestered by a species of grasshopper when it feeds on sprayed foliage; the herbicide is incorporated into the froth the grasshoppers release for defense.

Defensive chemicals that are not injected can be separated broadly as being odoriferous or repugnatorial. Numerous insects, including roaches, earwigs, several families of beetles, and several families of bugs, have rather characteristic odors that our own sense organs readily detect as undesirable. Many of these compounds are not released merely in the presence of danger but are released constantly to the surface of the integument from the epithelial glands that produce them. Anyone who has handled the large black "stink beetles" (Tenebrionidae) or whirligig beetles (Gyrinidae) knows how long these odors can linger. Indeed, they may cause an entire insect collection to smell for months after the specimens have been pinned and dried. One species of ant produces citral, a compound closely related to citronella, commonly used in commercial insect repellents.

Most of the chemicals used in insect defense function largely as repellents that deter attackers, but in higher concentrations some of these substances are highly toxic. Some chemicals, however, seem to be primarily toxic. The meloids are often referred to as blister beetles because a compound called **cantharidin** in their integument is a powerful mucous membrane irritant and vesicant that was once thought to be useful as a human aphrodisiac. Another compound, pederin, produced by staphylinid beetles of the genus *Paederus* produces a dermatitis. A saponin compound that is extracted from a beetle and used by Kalahari bushmen to coat their arrows is a powerful paralytic. The formicine ants derive their name from the fact that they are stingless and rely on formic acid secretions in their attack on other organisms and in colony defense. The formic acid is applied topically as a spray that these ants can propel up to 30 centimeters or subcutaneously through wounds caused by their mandibles.

Perhaps the most often described example of chemical defense encountered among the insects is that of the famous bombardier beetles of the carabid genus *Brachinus*. These small black beetles derive their name from the audible report that accompanies the explosive emission of their defensive secretion. By raising and rotating the tip of its abdomen, a bombardier beetle can accurately aim its "cannon" at an adversary. The internal defensive apparatus consists of a reservoir into which hydroquinones and hydrogen peroxide are secreted by associated glands. The two compounds pass posteriorly into a cuticular chamber where they come in contact with a catalytic enzyme (Figure 14-5). The reaction that occurs when the three compounds are mixed together results in a sudden liberation of oxygen that expels a visible cloud of quinone at the enemy (Eisner and Meinwald, 1966).

In the broad sense, venoms include all forms of toxic compounds that are injected. This definition would therefore encompass the chemicals associated with urticating hairs and spines as already mentioned, salivary secretions injected by the mouth parts, plus the venoms of aculeate hymenopterans.

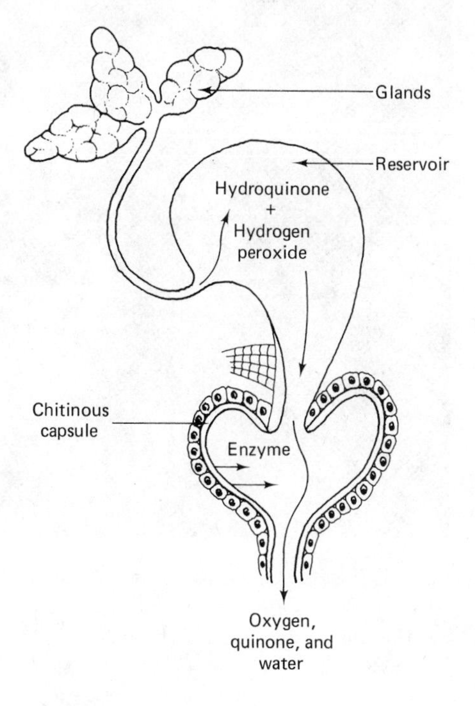


**FIGURE 14-5.** Defensive system of the bombardier beetle *Brachinus*. (Adapted from Schildknecht and Holoubek, 1961.)

Unlike nonvenomous compounds, venoms are usually ineffectual when applied topically. However, when injected into an organism, venoms induce their characteristic response. Some venoms such as the salivary secretions of many predaceous insects have a general toxicity because of their primary function as aids to digestion. These tend to degrade or liquify the surrounding tissue in an irreversible manner. In some blood-feeding species, such as mosquitoes, these secretions contain an anticoagulant that facilitates feeding by retarding clotting.

The venoms produced by the sting glands of aculeate hymenopterans may have evolved from the paralyzing venoms employed by the parasitic wasps. Some paralyzing venoms have a fairly widespread effect, but others are specific to only a few host species. Such specificity appears to be attributable to the chemical nature of the compounds rather than to the host preference of the parasite. Furthermore, many venoms are unstable and produce only a temporary paralysis that accommodates oviposition, whereas others produce permanent paralysis. The venoms of social bees, wasps, and ants are used mainly for the defense of the colony. The chemical nature of insect venoms is highly variable and only a few have been adequately studied from a biochemical standpoint.

The social insects often enhance their defensive capability through coordinated group behavior. Angered bees and wasps will often respond to danger

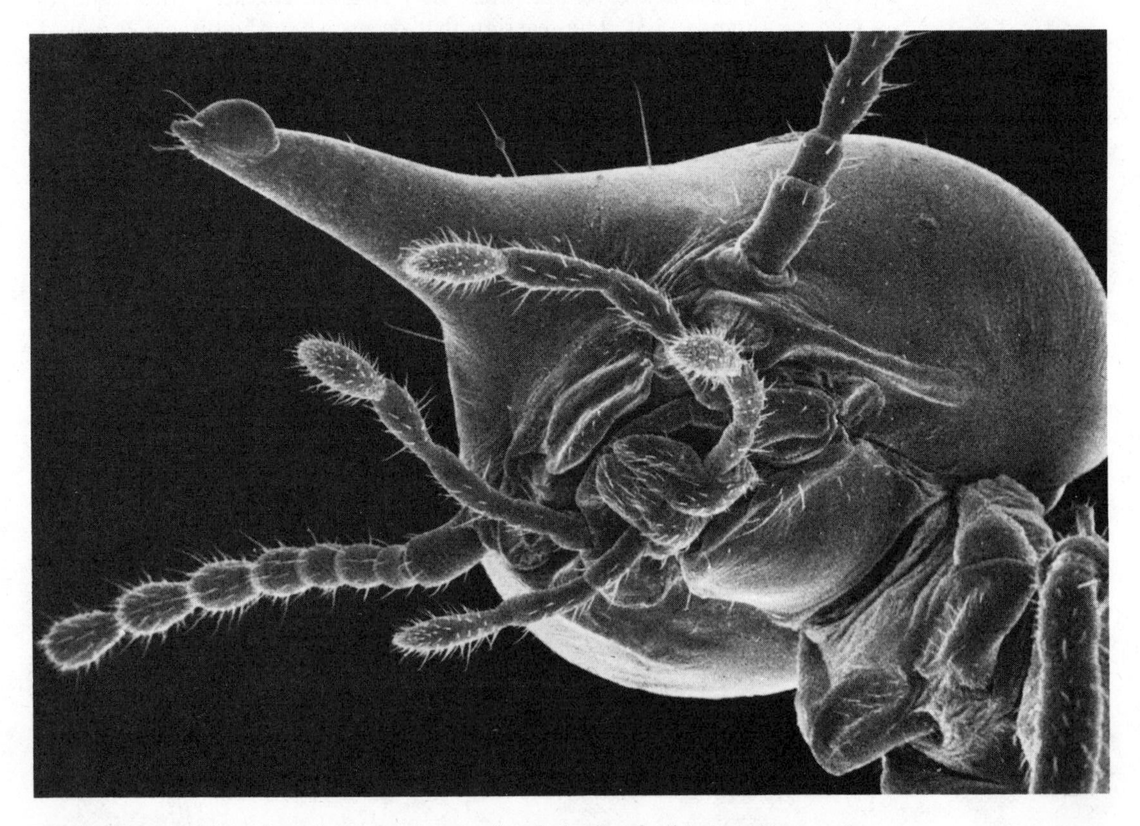

**FIGURE 14-6.** Scanning electron micrograph of the head of a termite nasute *Nasutitermes exitiosus* viewed from below. Note the droplet of defensive secretion on the tip of the snout. (Photograph by courtesy of Thomas Eisner, Cornell University.)

with a rush of defenders from the nest; group stinging behavior may then result from a response to an alarm pheromone released in association with the venom. Some ants will encircle a portion of food and discharge a defensive secretion outward to ward off members of other foraging species. In some termite species, specialized soldiers known as nasuti (Figure 14-6) eject a sticky frontal gland secretion from a projection of the head that resembles a nozzle. When workers are travelling back and forth over exposed trails, the nasuti line up along the sides of the pathways with their snouts pointing outward to provide protection to the foragers. A group of nasuti in a cooperative effort will also encircle an adversary and discharge the secretion that hardens into a sticky thread. By swinging their heads from side to side, they can throw loops into the threads and quickly entangle an enemy much larger than themselves.

## Colorational Defense

Defensive or, more correctly, protective color patterns can be grouped under cryptic coloration, flash patterns, warning signals, and mimicry.

### *Cryptic Coloration*

Cryptic coloration is simply camouflage. Crypsis can result from either blending into a featureless background or looking like a particular object that forms a common component of the environment. The success of this form of protection is highly dependent on the closeness of the match between the color and pattern of a species and that of its habitat background, but it also depends on the existence of appropriate behavioral mechanisms. A cryptic individual must select a suitable background on which to rest and must either remain motionless or move in an appropriate fashion, such as the gentle swaying motion displayed by some leaflike species. For example, the mottled color pattern of the grasshopper genus *Philippiacris* from the rocky desert areas of central Chile ranges from a light sandy beige to a dark gray-green. I collected a number of these interesting creatures (Figure 14-7) and invariably found them to rest among stones that matched their color almost perfectly.

Certainly one of the truly fascinating aspects of protective coloration in general, and of cryptic patterns in particular, is the level of perfection that has evolved in so many species, as some of the following examples will demonstrate. The study of industrial melanism discussed in Chapter 4 indicates the survival value of camouflage and how quickly changes in a color pattern can occur when driven by the selective pressure of predation. The long evolutionary history of insects has certainly provided ample opportunity for a high degree of refinement in their protective color patterns.

Protective coloration also provides numerous examples of similar patterns having evolved in totally unrelated groups that occupy similar habitats where they are subjected to the same kind of selective pressures. The structural similarity between caterpillars and the larvae of sawflies is quite remarkable, but the similarity in the color patterns of these groups is truly amazing. Members of both groups that feed on the needles of coniferous trees are often the green color of the needles and bear similar longitudinal light green stripes that resemble the midrib and characteristic lighting pattern of these elongate leaves.

Several basic color patterns are employed that help insects to blend well

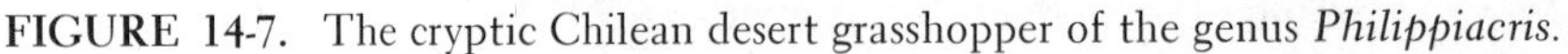

**FIGURE 14-7.** The cryptic Chilean desert grasshopper of the genus *Philippiacris*.

into featureless backgrounds. **Disruptive coloration** tends to break up the outline of the body with light and dark markings that suit a mottled background or create the impression of a random pattern of sunlight and shadow (Figure 14-8). No matter how well an organism's color matches its background, the insect will reveal its position readily if it casts a shadow. **Shadow elimination** by compressing the body and wings tightly against the substrate therefore becomes an important behavioral aspect of camouflage. Even when an organism is colored to match its background, predators with stereoscopic vision can detect the presence of a prey as the result of the three-dimensional image created by differential lighting. This is overcome quite well by patterns of **counter-shading.** A solid colored caterpillar, for instance, would be highly visible on a leaf of similar color because of the higher level of light reflectance from its dorsal surface compared to that of its lateral surfaces. If an individual is more darkly colored along the middorsal line and there is a gradation to lighter color along the sides, it has the effect of flattening out the visual image perceived by a predator.

Only a few insects are capable of **adaptive color change** such as that displayed by chameleons and some fishes. Some locusts can vary their color from nearly white through yellow and brown to almost black and thereby match the color of their backgrounds. The change is not as rapid as in some vertebrates but comes about gradually, supposedly in response to the light intensity reflected from the background. Similarly, the color patterns developed by the pupae of certain lepidopterans are determined in advance by the background color perceived in the larval stage (Wigglesworth, 1964). Several genera of stick insects, however, display a chromatic rhythm such that they are paler by day than by night. This change is the result of the movement of

FIGURE 14-8. Disruptive coloration of moth wings.

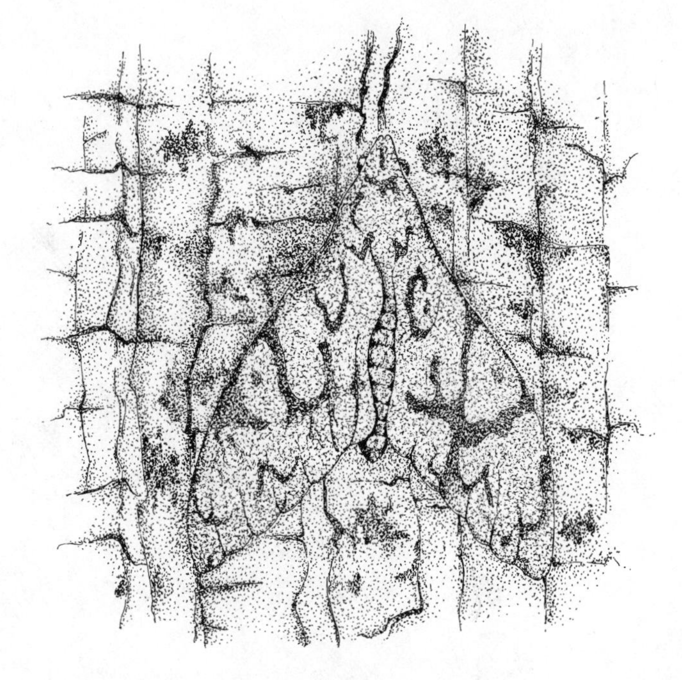

pigment granules in the epithelial cells, which is under the control of hormones.

Not all insects live in habitats with a generally featureless background into which they can blend. In habitats characterized by open space, where an individual might frequently form a silhouette against a rather uniform color field such as the sky, better protection is afforded by the resemblance of a commonly occurring object. This similarity to inanimate objects is usually referred to as **homomorphism** (shape), **homochromism** (color), or **homotypism** (form and color), whereas mimicry is more properly reserved for the resemblance to another animal.

A common kind of homotypism displayed by insects involves the resemblance to leaves, twigs, and floral parts. Leaf look-alikes, including katydids, phasmids, and butterflies, present a wide variety of examples which illustrate the incredible perfection that evolution can produce. In this regard, one of the interesting aspects of leaf resemblance emerges when a series of species with activity periods that span the passage of the seasons is examined relative to the changing condition of the leaves among which they hide. The wing veins as well as markings on the wing surface simulate the appropriate pattern of leaf venation and configuration. Insects that are active during the spring, when the leaves are fresh and unmarred, are green and look like perfect new leaves. Those that are active later in the year may resemble tattered older leaves, discolored and damaged through the course of time. These patterns are achieved with irregular wing margins and clear areas that resemble perforations, combined with blotches of color that resemble physical damage or splashes of bird droppings (Figure 14-9).

Larval stages also resemble inanimate objects. The caterpillars of some sphingid moths have counter-shading and diagonal markings which make the individual look like a rolled-up leaf, complete with stalk provided by a thick terminal spine (Figure 14-10). Geometrid larvae, or loopers, often resemble

FIGURE 14-9. The protective resemblance to a discolored leaf of the Indian butterfly *Kallima inachus*. (Redrawn from Jeannel, 1960.)

twigs, complete with budlike swellings and appropriate bark markings (Figure 14-11).

One cannot examine the cryptic coloration and behavior of insects without wondering how it is possible for such perfect copies of noninsectan objects to have evolved and how much protection they afford the species that have evolved them. If they do not occur by chance, why are all species not

FIGURE 14-10. The protective resemblance of a sphingid caterpillar to rolled leaves. (Redrawn from Jeannel, 1960.)

FIGURE 14-11. The protective resemblance of a geometrid larva to twigs of its host plant. (Redrawn from Jeannel, 1960.)

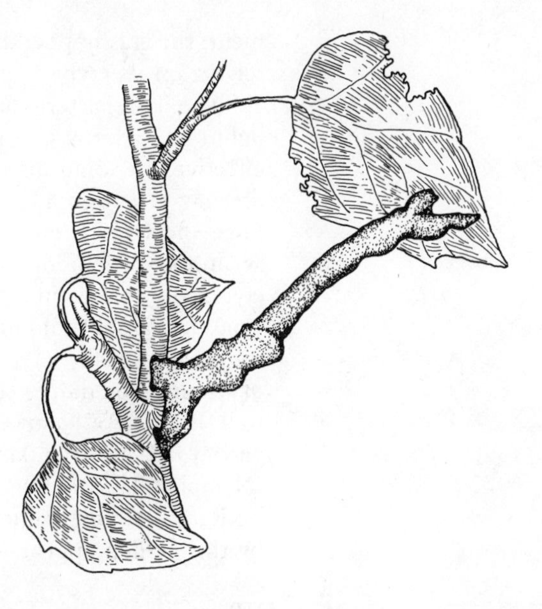

protected in this manner? Crypsis is just one of a number of defensive strategies, and it is best suited for those species that are sedentary, diurnal, and live in well-lighted habitats. Furthermore, as perfect as camouflage may be, it can never provide complete protection. For one thing, a predaceous species must improve its search ability in order to survive. If we recognize a coevolution between the hunter and the hunted, the members of the latter group that most perfectly blend into their surroundings will always have the best chance to survive predation and will consequently leave proportionately more offspring than the variants that are more easily seen. There is a sizable body of experimental evidence that illustrates the protective value of a slight resemblance to a noninsectan object, even under cage conditions where the search effort of a predator is greatly reduced. Since some apparently poor crypsis reduces predation at least a little, we can see how there would be enough selection to drive evolution toward better and better crypsis. Thus, after countless generations, we see an accumulation of inherited characteristics which in some cases border on cryptic perfection. The fact that some of the details cannot be seen clearly without the aid of a magnifying glass, an objection sometimes raised by skeptics, is a concern that stems from an anthropocentric view of the natural world.

### *Flash Patterns*

Even the best camouflage does not fool all of the predators all of the time, so many species have evolved a second line of colorational defense, referred to as flash patterns. The basis of the effectiveness of flash patterns is to induce a rapid change of the search image of the pursuing species. This is accomplished by quickly changing from cryptic to conspicuous, and back again. The efficacy of the flash pattern strategy is amply demonstrated by grasshoppers with brightly colored hind wings. When resting among the vegetation, many grasshoppers are well camouflaged but may be found with a little careful searching. As you approach close enough to make a capture, your quarry will often take flight, during which the colorful hind wings are exposed. This changes your search image, and by the time you have made a mental adjustment, the grasshopper drops out of sight amongst the vegetation, once again relying on its cryptic coloration for concealment.

Some flash patterns are apparently more effective than others. Concentric light and dark rings, usually referred to as **eye spots,** seem to be particularly effective. In some insects, eye spots are displayed perhaps as a means of drawing attention away from some vital part (Figure 14-12), but they seem more effective in deterring the attack of a predator when displayed suddenly. A number of lepidopterans such as the eyed hawk moths (Saturniidae) are cryptically colored in their normal resting position, but when disturbed, they expose a pair of prominent eye spots on their hind wings (Figure 14-13). Eye spots may actually serve a dual purpose. They are certainly an effective means of altering a predator's search image, but they may induce a fright response as well. Blest (1957) showed that birds which had not had any learning experience with such markings regularly drew away from peacock butterflies (Nymphalidae) that suddenly displayed their eye spots. The fact that the birds in Blest's experiment drew away suggests that they were at least startled by the sudden appearance of an ominous pair of eyes.

### *Warning Coloration*

Whereas some insects are so perfectly camouflaged, others seem to flaunt themselves with gaudy displays of color and ornamentation. Most brightly colored insects are adequately protected by special defensive structures, chemicals, or by the fact they are unpalatable. However, this is not always the case, as illustrated by those species that resemble flowers as a form of cryptic coloration. The extremely showy color patterns of many foul-tasting butterflies and beetles provide an obvious association between the color pattern and distastefulness that is readily learned by vertebrate predators while they are young. With periodic reinforcement of the learned association, experienced predators generally avoid prey displaying similar patterns. Likewise, the alternating bands of colors presented by the abdomens of stinging hymenopterans provide a specific image that can be remembered by a predator that has been stung. One can demonstrate the effectiveness of warning coloration quite simply. If an inexperienced frog is fed a bumblebee from which the

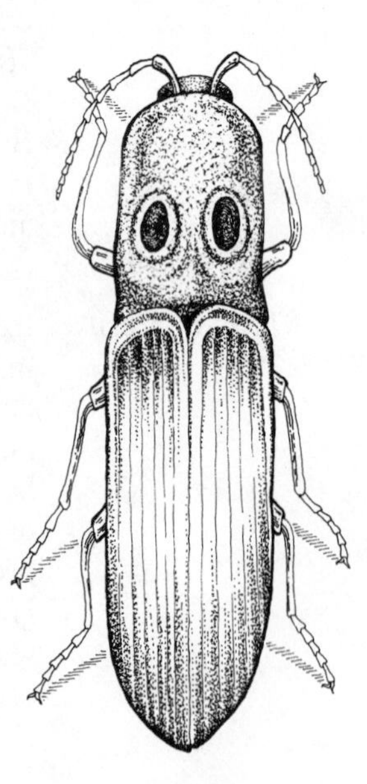

**FIGURE 14-12.** Elatarid beetle with a pair of prominent eye-spots on the hard pronotum.

FIGURE 14-13. The defensive display of eye-spots on the hind wings by the South American peacock butterfly. (Redrawn from *Mimicry* by Wolfgang Wickler. © W. Wickler, 1968. Used with permission of McGraw-Hill Book Company.)

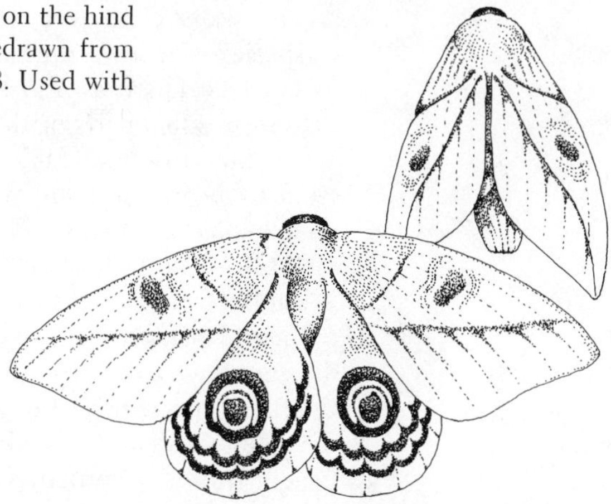

sting has been removed, it will subsequently take an intact bee. However, once having been stung, the frog will refuse subsequent offerings of similar appearance.

***Mimicry***

The concept of mimicry was first proposed more than 100 years ago by the English naturalist Henry W. Bates. On the basis of observations he made on an expedition to the Amazon basin, Bates proposed that species with overlapping ranges (sympatric) could evolve a resemblance if the common color pattern was beneficial to both species. Bates had observed that some palatable species had similar color patterns to sympatric unpalatable species. He concluded that the unpalatable species (**the models**) were employing warning coloration and the palatable species (**the mimics**) derive benefit from the fact that they are not attacked by predators that have learned the association between the color and bad taste.

The above concept is known as **Batesian mimicry** and has been the subject of controversy ever since it was proposed. It seems to rule out the possibility that species come to resemble each other as a result of parallelism. However, it does not seem necessary to consider every case of resemblance as a case of mimicry. The predominance of black and red patterns in the wings of South American butterflies compared to a predominance of black and green patterns among African and black and blue patterns among Indian species is well known. We might expect, therefore, to encounter more similarity in the appearance of groups of species within each of these geographic regions than between them.

Opponents of the concept of mimicry argue that the gradual accumulation of traits that seems to function suitably in the development of crypsis cannot operate effectively in the development of mimicry. If a mimic survives because it looks like a model that predators have learned to reject, it has been questioned whether a species could gain sufficient protection from early minor changes for them to be selected. We must remember, however, that many mimetic patterns function while the species are in motion so perhaps do not need to be as highly refined as those employed in crypsis; camouflage must withstand the careful search of a predator often very close at hand. It can be counterargued that such changes occurred as the result of major mutations or had their beginning in convergence. No matter which side of the argument one chooses to take, it is a matter of fact that there are many pairs of species that look alike in which one is well defended and the other is defenseless. There are also a number of species that retain their original color pattern in some parts of their range but in other areas are clearly similar in appearance to unpalatable species with which they overlap. Finally, there seem to be just too many examples of mimicry in which details of the resemblance are extraordinary and result in some obvious benefit to the mimic to pass it all off as an accident of parallelism or convergence.

For Batesian mimicry to remain effective, it is obvious that there must be a numerical balance between the model and the mimic such that a predator does not have more experiences with palatable than with unpalatable prey of the same appearance. For a mimetic pattern to evolve, a palatable species must be considerably less abundant than the model and remain so in order to benefit from the similarity. This raises an interesting question as to how the development of a mimetic pattern by a palatable species affects the level of predation on the unpalatable species. Certainly the learning of young predators would be slowed by encounters with palatable prey. This could increase the predation pressure on the model and lead to the selection of divergent color patterns. Viewed in this way, the gradual development of mimicry seems completely possible, as both model and mimic could evolve together rather than only the mimic having to undergo a large-scale change.

The naturalist Müller noticed that in some groups of mimetic species all of the individuals are unpalatable. Consequently, when two or more unpalatable sympatric species have a similar appearance, we call it **Müllerian mimicry.** At least in theory, it would seem to be beneficial for a group of species to adopt a common warning signal rather than different ones. A predator would have to learn to recognize only one pattern instead of several separately. Assuming that the predator requires a number of bad experiences to learn the appearance of an inedible prey, a group of inedible species could share the number of individuals that must be sacrificed to educate the predator.

Various workers have argued that closely related species may look alike because of a common ancestry. For example, two genera of black and yellow striped wasps probably inherited the pattern from a similarly colored ancestor. In such a case, we cannot consider one to be the model and the other the mimic since, in a strict sense, there is no mimicry at all. However, this technicality does not seem as important as the more fundamental question as to whether Müllerian mimicry is effective, which it apparently is. For example, in South America there is a group of Müllerian mimics consisting of four different families. In addition to the Müllerian species, there are two species of edible Batesian mimics, and all of them are avoided by predators (Wickler, 1968).

Mimicry is commonly thought of as similarity in color patterns because of the influence of the early workers who mainly studied butterflies. However, it is obvious that color similarity alone would not be as effective as similarity in over-all appearance; there are in fact many examples of insects from different orders that resemble each other in both form and color. One of the most diverse groups that forms a so-called mimicry ring involves a cerambycid beetle, a hemipteran, a butterfly, and a spider-wasp that all bear a rather striking similarity to a highly unpalatable soft-shelled beetle (Figure 14-14).

There are also many examples of mimicry that rely mainly on form in that both model and mimic are monochromic. In some cases, the benefit that

FIGURE 14-14. A mimicry-ring consisting of the distasteful soft-shelled beetle *Lycus rostratus* (A), a cerambycid beetle (B), a lygaeid bug (C), a butterfly (D), and a pompilid wasp (E). (Redrawn from *Mimicry* by Wolfgang Wickler. © W. Wickler, 1968. Used with permission of McGraw-Hill Book Company.)

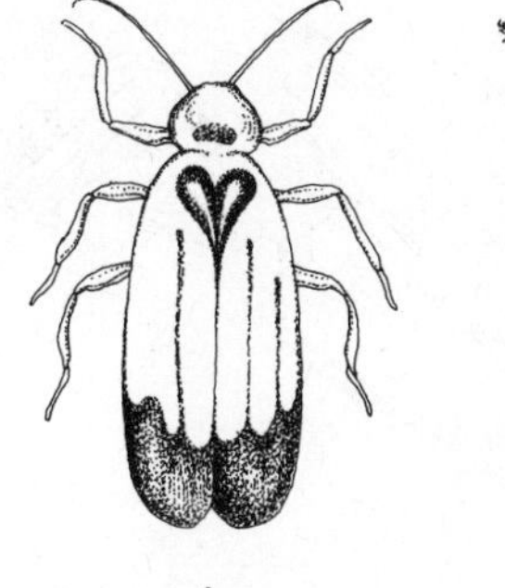


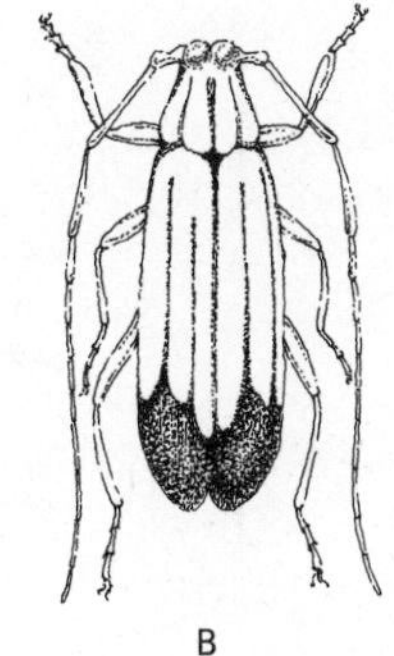


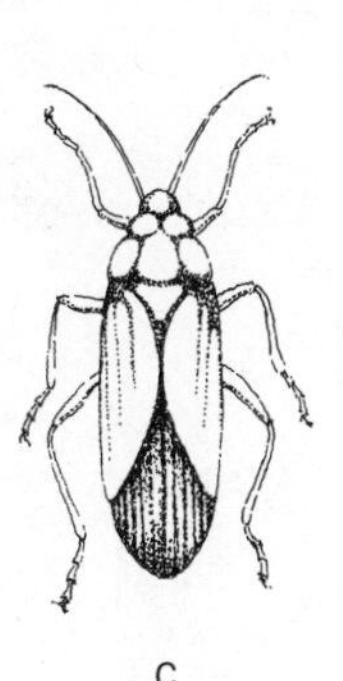


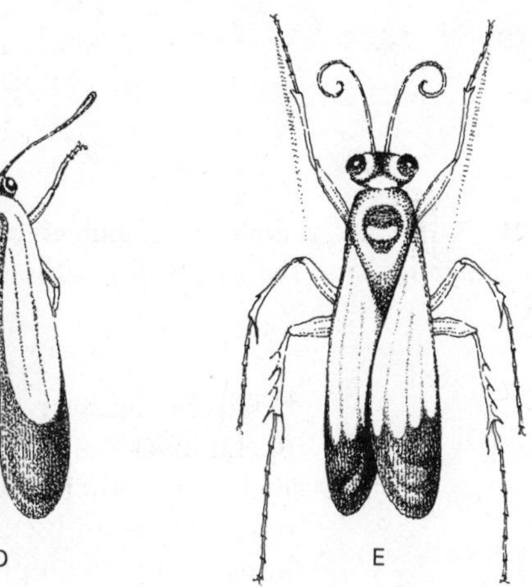

FIGURE 14-15. A defenseless, monochromatic, cerambycid mimic (left) of a foul smelling tenebrionid model (right) with which it is sympatric.

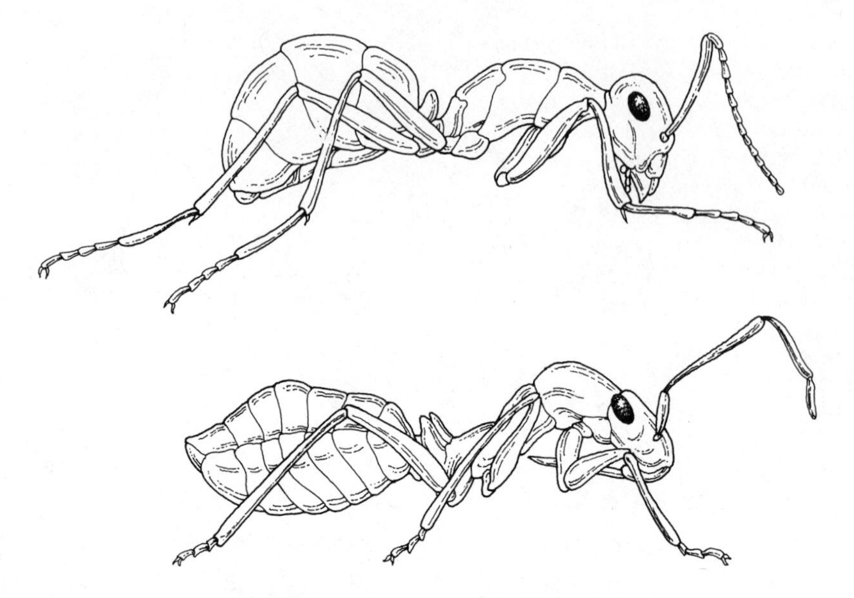

**FIGURE 14-16.** An example of ant mimicry. (*Above.*) The ant model. (*Below.*) A small bug mimic. Note the wing stubs of the bug resemble the node on the ant's pedicle.

befalls the mimic is obvious, as in the case of the defenseless long-horn beetles (Cerambycidae) that closely resemble foul-smelling tenebrionids (Figure 14-15). In other cases, as with many ant mimics, the significance of the resemblance is obscure. In spite of their chemical defense, ants are eaten by birds, reptiles, amphibians, and mammals, so it seems unlikely that ant mimics gain any protection against predation simply through resemblance. It may help some species to mix with ants and remain unnoticed, even to the extent of becoming permanent residents in their nests, but there are many successful ant guests that do not look like their hosts. Perhaps the answer lies in the fact that because ants are generally numerous, to be able to move among them without being conspicuous simply increases the chance of survival. Whatever the basis for it, the similarity between ants and totally unrelated insects, such as the small hemipteran illustrated in Figure 14-16, can be truly remarkable.

# References Cited


Beard, Raimon L. 1963. Insect toxins and venoms. *Annual Review of Entomol.* 8: 1-18.

Blest, A. D. 1957. The function of eyespot patterns in the Lepidoptera. *Behaviour* 11: 209–256.

Eisner, T. 1970. Chemical defense against predation in arthropods. *In:* Chemical ecology (E. Sondheimer and J. B. Simeone, eds.). Academic Press, Inc., New York, N. Y.

Eisner, T. and Y. C. Meinwald. 1966. Defensive secretions of arthropods. *Science* 153: 1341–1350.

Jeannel, R. 1960. Introduction to entomology. (Trans. by Harold Oldroyd). Hutchinson & Co., Ltd., London.

Roth, L. M. and T. Eisner. 1962. Chemical defenses of arthropods. *Annual Review of Entomol.* 7: 107-136

Schildknecht, H. and K. Holoubek. 1961. Die Bombardierkäfer und ihre Explosionschemie V. Mitteilung über insekten Abwehrstoffe. *Angew. Chem.* 73(1): 1-6.

Wickler, Wolfgang. 1968. Mimicry. (Trans. by R. D. Martin). World University Library. McGraw-Hill Book Company, New York, N. Y.

Wigglesworth, V. B. 1964. The life of insects. The New American Library, Inc., New York, N. Y.

# Insect-made structures 15

In addition to gaining protection in the different ways described in Chapter 14, the majority of insects seek shelter of either a permanent or temporary kind. A positive thigmotaxes often leads to concealment under stones or in a variety of natural cracks and crevices. In addition, many insects engage in the construction of a wide variety of structures, some of which are as elaborate as any built by nonhuman species (see von Frisch, 1974).

Often insect dwellings are little more than a by-product of their feeding. Boring insects, for example, create a protective place in which to live as they tunnel through in the woody parts or leaves of plants. Others such as the ambrosia beetles excavate tunnels as a place in which to culture fungus and rear their young, without actually consuming any of the material they remove. Still other insects utilize vegetative parts that become distorted because of their feeding or which are actually tied together into protective living places.

## Leaf-rollers and Leaf-tiers

The leaves of plants are used for protective shelters by a number of insects collectively referred to as **leaf-rollers** and **leaf-tiers**; some of these species are well known as economic pests (Metcalf and Flint, 1939). Leaves are often distorted by the juice-feeding of insects such as thrips and aphids; the associated chemical and mechanical stimulation may alter the structure of the leaf tissue and thereby create a sheltered living place. The typical leaf curling or folding that results creates protected habitats wherever two surfaces overlap. However, leaf rolling and tying often involve some deliberate cutting and folding on the part of an insect or the tying of several surfaces together with strands of silk. A simplistic form of this behavior is displayed by those caterpillars that bind a group of conifer needles into a tube which they use as a protective retreat.

The larvae of leaf-rolling lepidopterans on deciduous plants create folds or tubes as shelters from which they emerge to feed; such tubes are used as resting places during larval life and later serve as molting and pupation sites. The creation of these abodes may involve no more than pulling one half of a leaf over on top of the other half or rolling a leaf into a tube fastened with silk (Figure 15-1). In other species, however, the larvae cut the leaf in a very precise way to facilitate the formation of a shelter with a specific appearance (Figure 15-2). Some ants also construct nests by sewing leaves together as a coordinated group activity that involves the use of a silk-producing larva as a needle and thread (Figure 15-3).

## Insect-caused Galls

Almost everyone is familiar with localized plant deformities called galls. Many of these abnormal structures are inconspicuous and nondescript, but fre-

FIGURE 15-1. Grape leaves folded and tied by the larvae of the grape leaf folder (Lepidoptera).

quently they are large, brightly colored, and highly ornamented. Plant galls are caused by various organisms, including fungi and nematodes, but most are caused by insects and mites. The gall structure itself can only form from plant parts with active meristematic tissue, so galls most commonly arise from buds or developing leaves. The gall is the result of abnormal cell multiplication, so it cannot develop on leaves that have ceased active growth. The stimulus for gall formation is not well understood. A gall usually originates at the site where an egg has been inserted into the plant tissue, but it does not usually begin to develop until after the egg has hatched. This suggests that some stimulation must be produced by a larval secretion or the physical irritation associated with the larva's feeding.

Gall formation may involve a mutual relationship between the insect and the plant. The developing larvae either eat the inner tissue of the gall they produce or suck out the liquid contents of the gall cells. In the process, the insects produce an enzyme that converts starch to sugar in the same way plant

FIGURE 15-2. The specific pattern of leaf cutting and rolling that is characteristic of the basswood leaf roller (Lepidoptera).

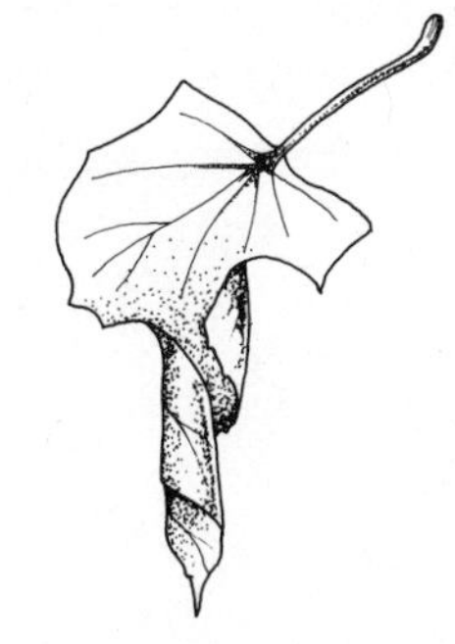

FIGURE 15-3. Workers of the ant *Oecophylla smaragdina* drawing together leaves which are tied in place with silk produced by the larva held in the mandibles of the worker on the right. (Drawing by F. Nanninga, reproduced from *Insects of Australia* with permission of Melbourne University Press.)

enzymes make the same conversion. The excess supply of nutrient material may stimulate the protoplasm and cause additional plant cell division. Thus both the plant and the insect may derive nutrition from the association.

Gall-forming insects are scattered throughout the orders Thysanoptera, Homoptera, Coleoptera, Lepidoptera, Hymenoptera, and Diptera, but the vast majority of galls are produced by cecidomyiid flies and tiny wasps of the family Cynipidae. Gall insects attack buds, leaves, flowers, stems, and roots; the result is an ideal living place complete with an abundance of food. In addition to the insects that stimulate the development of galls, many other species use abandoned galls as temporary living quarters.

The galls caused by some insects are so consistent in their appearance that they are more easily identified than are the insects that inhabit them. Some have numerous spines or other surface configurations that do not seem to bear any relationship to the seemingly featureless surface from which they develop. A small sample of insect galls is illustrated in Figure 15-4. Keys to the galls of some major plant families are provided by Felt (1940).

Galls have long been known to contain useful chemicals and, in some cases, to be suitable as food. Tannic acid, used in the tanning of leather and in the dyeing of wool and of skins, is a common constituent of oak galls. Several oak galls caused by cynipid wasps contain a high-grade pigment employed in the

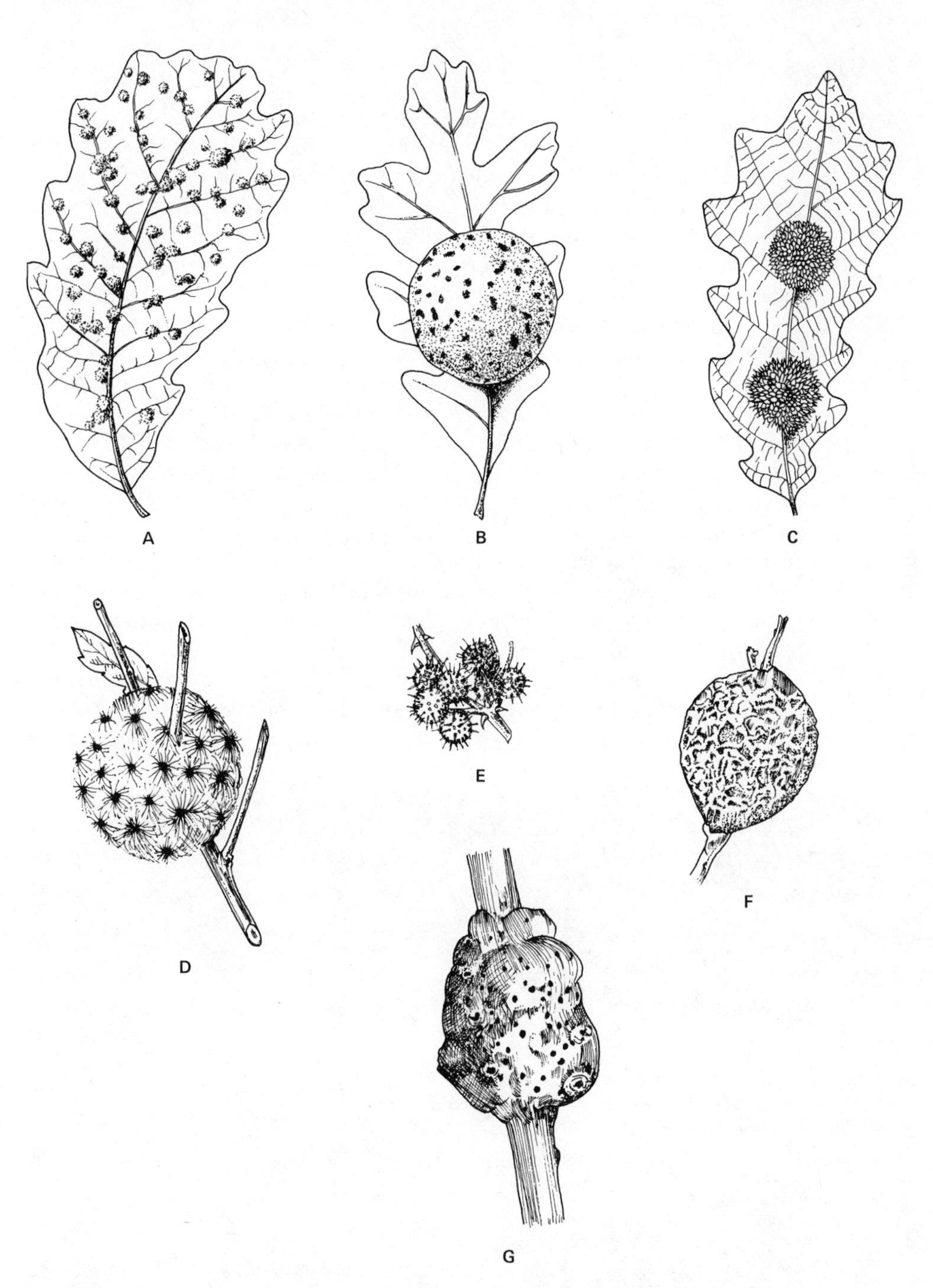


FIGURE 15-4. A sample of insect galls showing the great variability in their form.

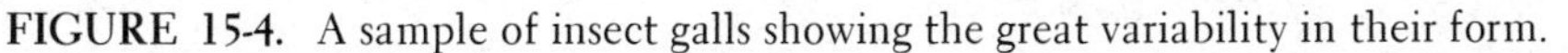

manufacture of top-quality permanent inks used in the printing of government documents and currency. The Eurasian Aleppo gall, or ink marble, caused by *Cynips gallaetinctoriae,* is most highly valued for its pigment. Galls also contain strong astringents and other chemicals used in the past for pharmaceutical purposes.

## Protective Cases

The eggs, larvae, and pupae of insects are frequently protected by partial or complete cases. Roaches and some primitive termites, for example, deposit the eggs in protective pods called oothecae, whereas other insects such as mantids and a variety of moths cover their eggs with a coating of frothy spumaline that hardens soon after it has been deposited. The pupal stage is often passed in a silken cocoon or a cocoon made from earthen material or plant debris bound together with silk or a sticky secretion. In the more highly evolved groups of flies, the pupal stage is passed inside the last larval skin, which becomes tanned to form a leathery puparium. All of these structures provide some protection against parasitism and have a modifying effect on the physical environment during otherwise vulnerable stages of development. Case-making behavior and its evolution are discussed by Frost (1959).

Insects of practically all orders construct protective cases of varying complexity, although the practice is most common among those with complete metamorphosis. In the broadest sense, a case can be considered as any extra covering that is intentionally formed by an insect, regardless of whether it is comprised entirely of a body secretion or largely of foreign material. Thus, the frothy secretion of the spittle bugs (Figure 15-5), the covering of wax threads of the wooly aphids, and the shells of the scale insects (Figure 15-6) can be considered, along with silken cocoons, to be extraorganismic protective cases.

Some insects gain protection simply by piling debris on their backs. The larvae of the datura beetle (Chrysomelidae) carry about accumulations of

**FIGURE 15-5.** The typical protective secretion produced by a spittle bug (Cercopidae: Homoptera).

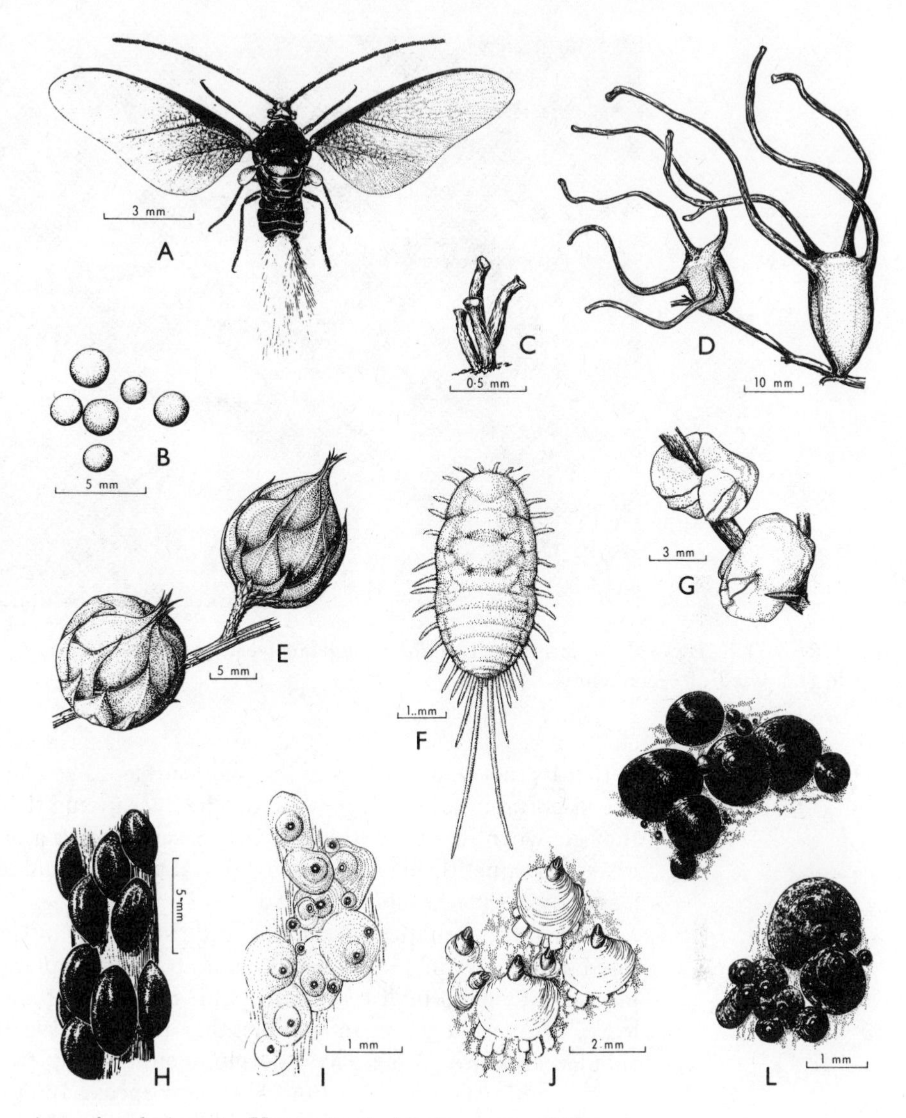


**FIGURE 15-6.** A variety of scale insects (Homoptera) showing some of the protective structures they form. **A.** Winged male, **C, D, E.** Galls, **H** through **L.** Protective covers of females. (Drawings by S. Curtis, reproduced from *Insects of Australia* with permission of Melbourne University Press.)

their own sticky fecal material, which contains a high concentration of plant alkaloids that most predators find unpalatable (Figure 15-7). The masked hunter (Reduviidae) carries a load of lint and dust stuck to its dorsal surface as an effective means of camouflage. The predaceous larvae of some brown lacewings (Hemerobiidae) are called trash carriers because they pile the remains of their victims upon their backs.

The larval lepidopterans are the predominant group among the terrestrial case-makers. Some fasten their cases to fixed objects; others construct portable cases that they carry about. Often the case of the last larval instar is sealed to serve as a pupation site. The coleophorids construct tubelike cases of silk, leaf material, and feces that cover their entire body with exception of the head capsule. When feeding, the larva is attached to the plant by its mouth parts

FIGURE 15-7. Larva of the datura beetle (Chrysomelidae) with a defensive dorsal shield composed of excrement.

so that its case stands out from the plant surface. The bagworms (Psychidae) live in portable cases and move about freely, carrying their abodes wherever they go; when it is time to pupate, the case becomes a pupation site. Their cases are frequently made from twigs or leaf pieces bound together with silk in a species-specific fashion (Figure 15-8).

Among the aquatic insects, the caddisflies are by far the best-known case-builders. As in the related Lepidoptera, trichopterans construct both permanent and portable cases from a wide range of materials, including bits of leaves, twigs, sand grains, and pebbles that are bound together with silk into a tubular structure. When larval development is complete, the case is firmly secured and sealed for pupation. For many species the case is so distinctive that it can be used for purposes of identification (Figure 15-9).

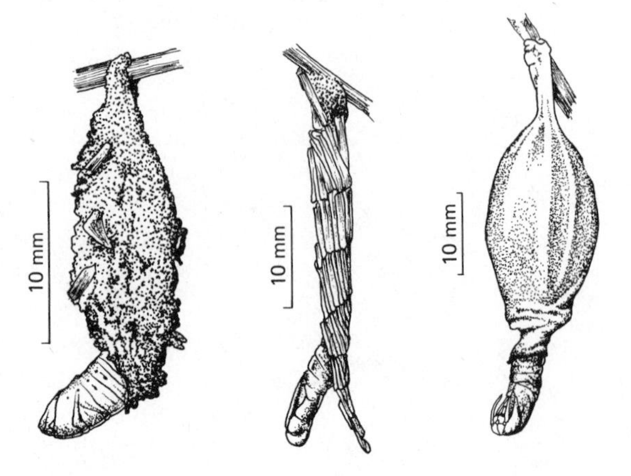


FIGURE 15-8. Several different forms of portable cases constructed by bagworms, the larvae of psychid moths. (Redrawn from Nanninga, 1970, *Insects of Australia*.)

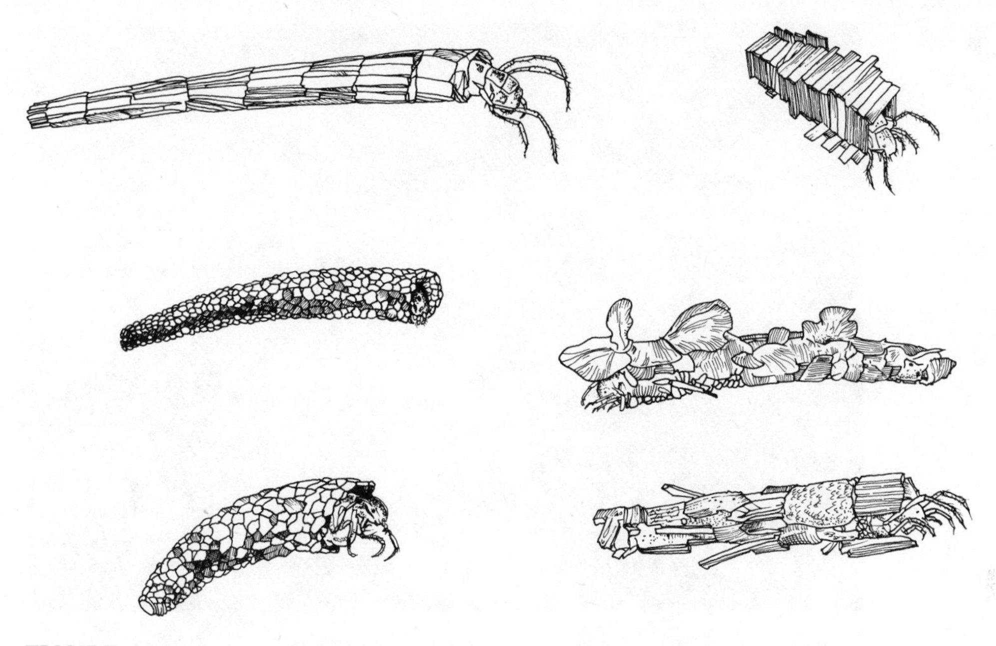

**FIGURE 15-9.** Cases constructed of different materials by the larvae of caddisflies (Trichoptera). (Redrawn from Quick, 1970, *Insects of Australia*.)

# Insect Nests

Only a few insects, such as the tent caterpillars, build communal abodes as a cooperative larval effort (Figure 15-10), but insects representing almost every major order construct nests in which the female lays her eggs. In some species, the female may do no more than build the nest and provision it, but in many presocial species the young are cared for in the nest until they are able to fend for themselves. Such nests may be no more than a burrow dug in the soil, but often they are quite elaborate structures that require highly integrated patterns of behavior for their construction. The nest-building behavior of presocial insects results in a wide range of structures, which can be as enjoyable to collect and to observe as the insects that make them. A few selected nests constructed by hymenopterans are illustrated in Figure 15-11 as examples.

The most elaborate insect-built structures are the nests of some social species. The nests of wood-boring termites consist of nothing more than a network of tunnels, but the foraging and fungus-growing termites construct elaborate aboveground nests from a mixture of soil, saliva, and feces. The gigantic mounds of tropical termites often have distinctive forms that are characteristic of particular species (Figure 15-12). In addition to providing protection, termite nests are often constructed in such a way as to provide a well-regulated interior microclimate. *Amitermes* orient their nests in such a way as to maximize solar warming in the morning and evening (Figure 15-12 *A*, *B*). Microclimatic control is most highly developed in the fungus-growing genus *Macrotermes.* Its members construct a series of passageways through which air flows according to its own density, such that temperature and carbon dioxide conditions favorable to the growth of fungus are maintained

**FIGURE 15-10.** Tent constructed by the larvae of the western tent caterpillar. (Photograph courtesy of W. G. Wellington.)

in the culture chamber. The basic structure of the nest of *Macrotermes natalensis* of Africa and a diagram of its air conditioning system are shown in Figure 15-13.

The nests of ants are highly variable, both in terms of location and structure. However, those of species that develop large colonies are usually entirely in the soil, either under a rock partially imbedded in the soil or beneath a mound of plant debris. The soil itself provides an effective buffer against fluctuations in temperature and moisture. A large rock set partially in the soil can serve as an efficient thermoregulatory aid, as does varying the depth of the brood chamber according to the seasons. According to Wilson (1971), the most advanced microclimatic control has been attained by mound-building species. Although ant mounds do not seem to have the air-conditioning structure of the termite mounds, they are much more than an accidental accumulation of material. The soil of such mounds often has a different texture and higher organic content than the surrounding soil, and the surface is often thatched with a specific kind of material that may shed water, increase the absorption of solar radiation (because of color and texture), or provide insulation. The regular conical shape of such mounds seems to be related to improving the heating of the nest when the sun is low in the sky.

Generally speaking, we are more familiar with the nests of bees and wasps than those of ants and termites. Probably no nest structure is better known than the wax comb of the domestic honeybee even though it has long been a practice to provide the bees at least the outer shell of their hive. In the wild state, *A. mellifera* utilizes existing nest sites and may form combs with no more shelter than that provided by some surrounding foliage. However, the familiar wax comb, consisting of numerous hexagonal cells, is usually constructed in a confined space such as a hollow tree. Any cracks and crevices are

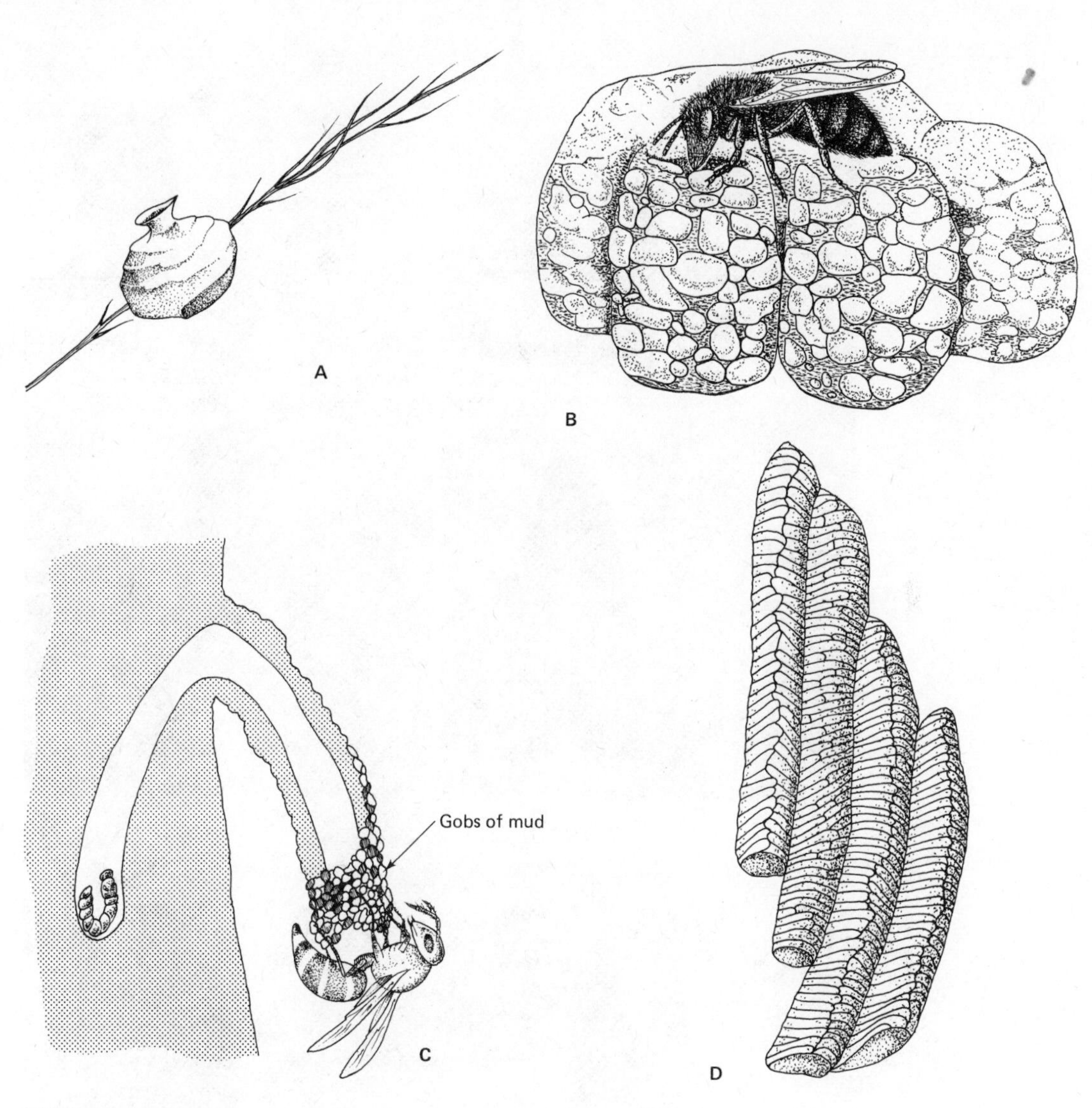


**FIGURE 15-11.** Some examples of nests constructed by solitary hymenopterans. **A.** Jug-like nest of the potter wasp *Eumenes,* **B.** Group of larval cells constructed by the mason bee *Chalicodoma,* **C.** Section through the nest of the solitary wasp *Oplomerus* showing the storage tunnel and the down-curved entrance tube, **D.** Nest of the pipe organ mud dauber *Trypoxylon.*

usually sealed with a resinous material called propalis. The combs are suspended vertically such that there is a rather precise space ($\frac{1}{4}$ to $\frac{3}{8}$ inch) between the surface of completed combs. Each typical worker cell is a very definite size and fits together with those neighboring it in a very precise way (Figure 15-14). If a queen encounters a large or irregular cell, she will deposit there a haploid egg that will become a drone. On the other hand, workers will feed the larvae developing in the cuplike base of queen cells on a diet of royal jelly. Thus the structural details of the nest as a whole as well as the individual cells seem to have become an integral part of the species' social organization.

The nest of a honeybee provides protection in a general way but also forms a highly defensible site for brood rearing, food storage, and survival over the

**FIGURE 15-12.** Mounds of some Australian termites. **A** and **B.** North-south and east-west aspects of mounds of *Amitermes meridionalis*, **C.** *Nasutitermes triodiae*, **D.** *Nasutitermes walkeri*, **E.** *Amitermes vitiosus*. (Reproduced from *Insects of Australia* with permission of Melbourne University Press.)

winter. In respect to brood rearing and overwintering, the ability to regulate nest temperature is highly beneficial. The precision with which honeybee colonies do this is truly remarkable, but it would not be possible if it were not for the type of nest site selected and the way it is modified and utilized. The actual thermoregulatory process, however, is accomplished by the bees themselves rather than physically, as in the termites. During the months when a brood is being reared, the brood chamber is maintained at a temperature of 34.5° to 35.5°C even though the outside air temperature may reach more

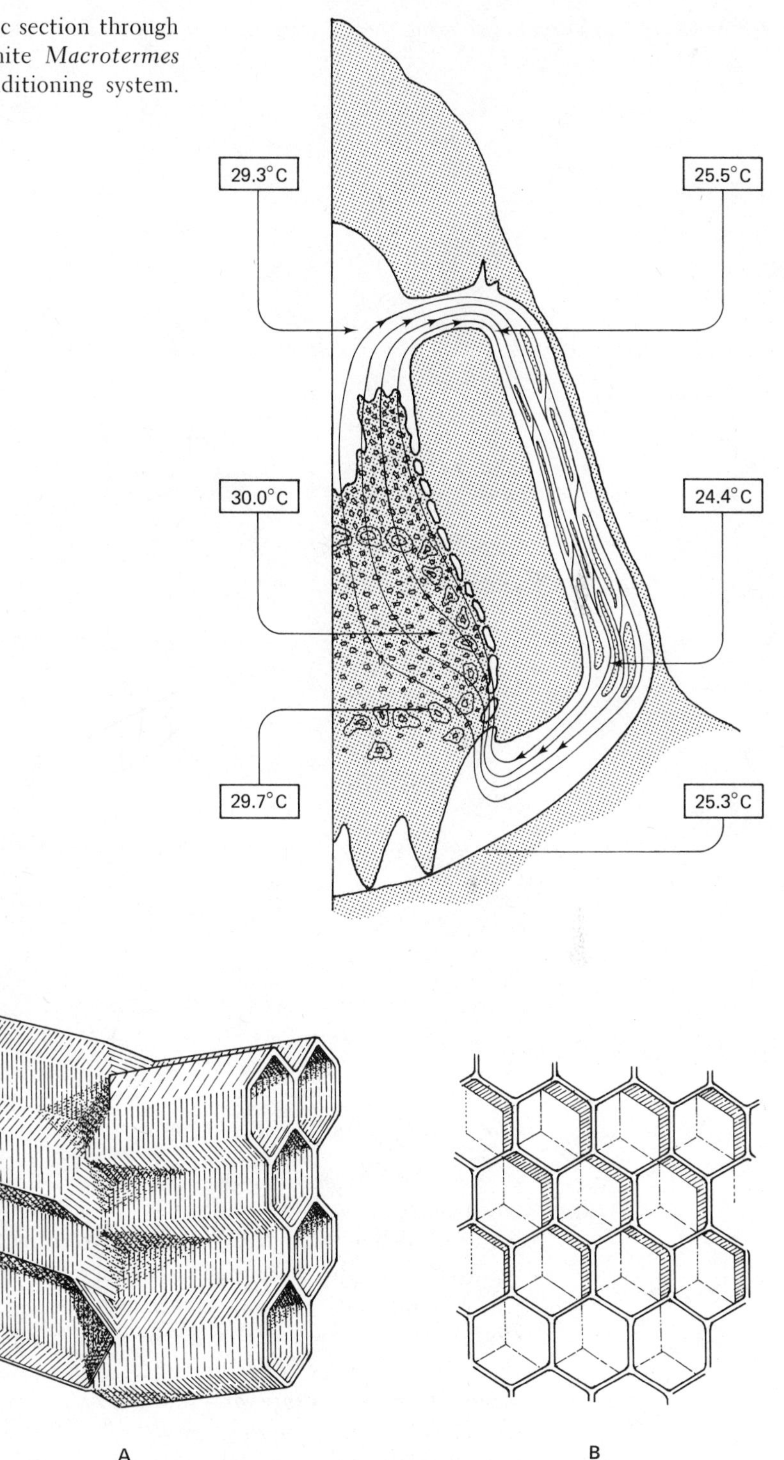


**FIGURE 15-13.** Diagrammatic section through a mound of the African termite *Macrotermes natalensis* showing the air-conditioning system. (Adapted from Lüscher, 1961.)

**FIGURE 15-14.** Diagrammatic details of the wax cells of the honey bee *Apis mellifera*. **A.** Lateral view with a few cells cut away to show the way the cells fit together, **B.** A group of cells viewed from the comb surface. (Drawing copyright © 1974 by Turid Hölldobler. Reproduced by permission of Harcourt Brace Jovanovich, Inc., from *Animal Architecture* by Karl and Otto von Frisch.)

FIGURE 15-15. Diagrams showing the steps in the construction of a paper wasp nest (Vespidae). (Adapted from Kemper and Döhring, 1967.)

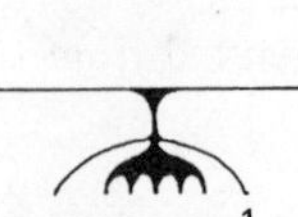


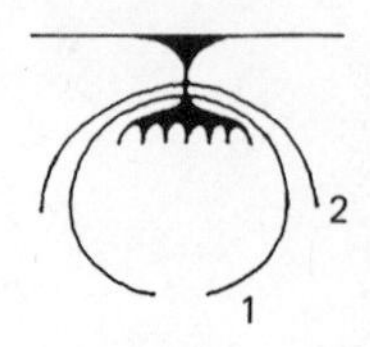


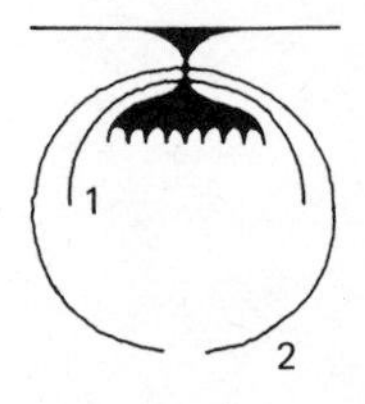


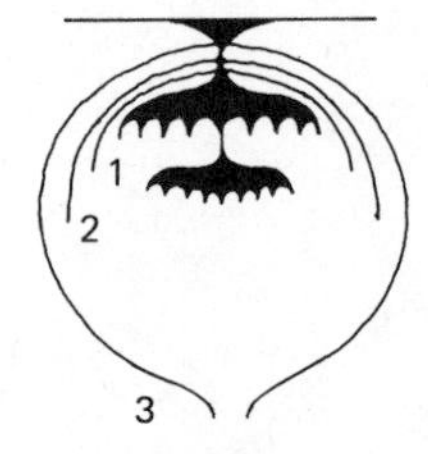


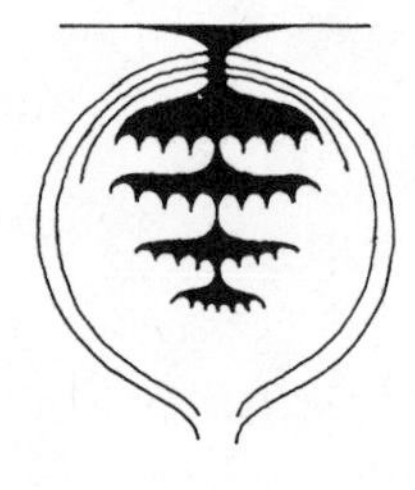

than 60°C. In the winter the cluster temperature ranges from 20° to 30°C and is never permitted to drop below 17°C, regardless of the outside temperature. When the outside temperature drops, the temperature in the hive is maintained by the metabolic heat generated by the workers, the behavior of which changes as the temperature declines. At first the workers form a loose cluster toward the center of the hive. As the temperature drops, the cluster tightens. The workers at the center consume small amounts of honey and generate heat by vibrating their muscles; those forming the outer layers of the cluster act as an insulating blanket. As time passes, the outer bees move toward the center of the cluster while those near the center move into the outer layers. During hot weather, the temperature of the brood area is maintained by a circulation of air in the hive created by workers that fan with their wings at the hive entrance. When this activity proves to be inadequate, other workers carry water into the hive and distribute it over the brood cells, which are then cooled as the water evaporates.

The more advanced social wasps construct paper combs made up of hexagonal cells similar in appearance to those of the honeybee, but the combs themselves are oriented horizontally instead of vertically. In addition, many of these wasps construct the outer walls of their nests of the same paperlike material used to build the combs. The beginning nest, constructed by the queen, consists of a small comb, made up of only a few cells and attached to an overhanging surface by a stalk; at first, there is no paper envelope. As the colony grows, the outer paper envelope is added and then is expanded to accommodate the addition of more brood cells to the comb. Still later, additional tiers of comb are suspended from the one above. More outer layers are added to enlarge the envelope as the smaller inner envelopes are torn down (Figure 15-15). As in the honeybee, the enclosed nest of social wasps provides a temperature environment that can be regulated to a considerable extent. Colonies of *Polistes, Polybia, Vespa,* and *Vespula* all use both fanning and water transport as means of cooling overheated nests (Wilson, 1971).

## References Cited


Felt, Ephraim Porter. 1940. Plant galls and gall makers. Comstock Publishing Company, Inc., Ithaca, N. Y.

Frisch, K. von. 1974. Animal architecture. A Helen and Kurt Wolff Book, Harcourt Brace Jovanovich, Inc., New York, N. Y.

Frost, S. W. 1959. Insect life and insect natural history. 2nd ed. Dover Publications, Inc., New York, N. Y.

Kemper, H. and Edith Döhring. 1967. Die sozialen Faltenwespen Mitteleuropas. Paul Parey, Berlin.

Lüscher, M. 1961. Air-conditioned termite nests. *Sci. Amer.* 205(1): 138–145.

Metcalf, C. L. and W. P. Flint. 1939. Destructive and useful insects, their habits and control. McGraw-Hill Book Company, New York, N. Y.

Wilson, Edward O. 1971. The insect societies. The Belknap Press of Harvard University Press, Cambridge, Mass.

# Social organization and behavior

# 16

Insects display a behavioral continuum with a strong solitary instinct at one extreme and a truly social (**eusocial**) instinct at the other. In insects such as the larvae of some megalopterans, the solitary instinct is so strong that mortal combat results whenever two individuals meet. On the other hand, colonies of eusocial insects may contain millions of individuals cooperating to the benefit of the group. As different as these extremes of behavior are, they are not mutually exclusive. In fact, gregarious and solitary behavior sometimes occur in sequence within the life history of a single species. Many caterpillars, for instance, are highly gregarious in their early instars and seem able to move about in an orderly fashion only when a certain individual leads the procession (Figure 16-1), yet in their last instar and as adults they become solitary.

Wilson (1971) suggests that the eusocial insects, which include the termites, ants, and the more highly organized wasps and bees, can be distinguished as a group by the possession of three characteristics: cooperation among individuals of the same species in the care of young, a division of labor in which more or less sterile individuals assist fecund individuals, and an overlap of generations so that offspring are able to assist their parents. On the basis of these criteria, Michener (1974) provided the following classification of insect social behavior:

*Solitary:* insects possess none of the three traits involving cooperative care of the young, division of labor, and assistance to parents.
*Subsocial:* the adults care for their own immature offspring for a period of time.
*Communal:* adults of the same generation use the same nest but do not cooperate in the care of young.
*Quasisocial:* adults of the same generation use the same nest and cooperate in the care of young.
*Semisocial:* a worker caste of the same generation as the reproductive caste cares for the young.
*Eusocial:* an overlap of generations so that offspring belonging to the worker caste assist their parents in the care of the young.

Different authors unfortunately tend to use these and other terms in a variety of ways which has led to some confusion. Consequently, the following discussion will be based on the broad view that any stage of development less than eusocial can be considered as **presocial** and the term **society** will be applied to any group of individuals organized in a cooperative manner.

Much has been written on the behavior of social insects, and it is perhaps a grave injustice to treat such a fascinating subject in the limited space available here. Nevertheless, a brief examination may at least serve to stimulate a more detailed study. If such proves to be the case, there is no better place to begin than with Edward O. Wilson's marvelous book, *The Insect Societies.*

## Presocial Behavior in Noneusocial Orders

Although eusociality is confined to the orders Isoptera (termites) and Hymenoptera (ants, some wasps and some bees), several other orders display

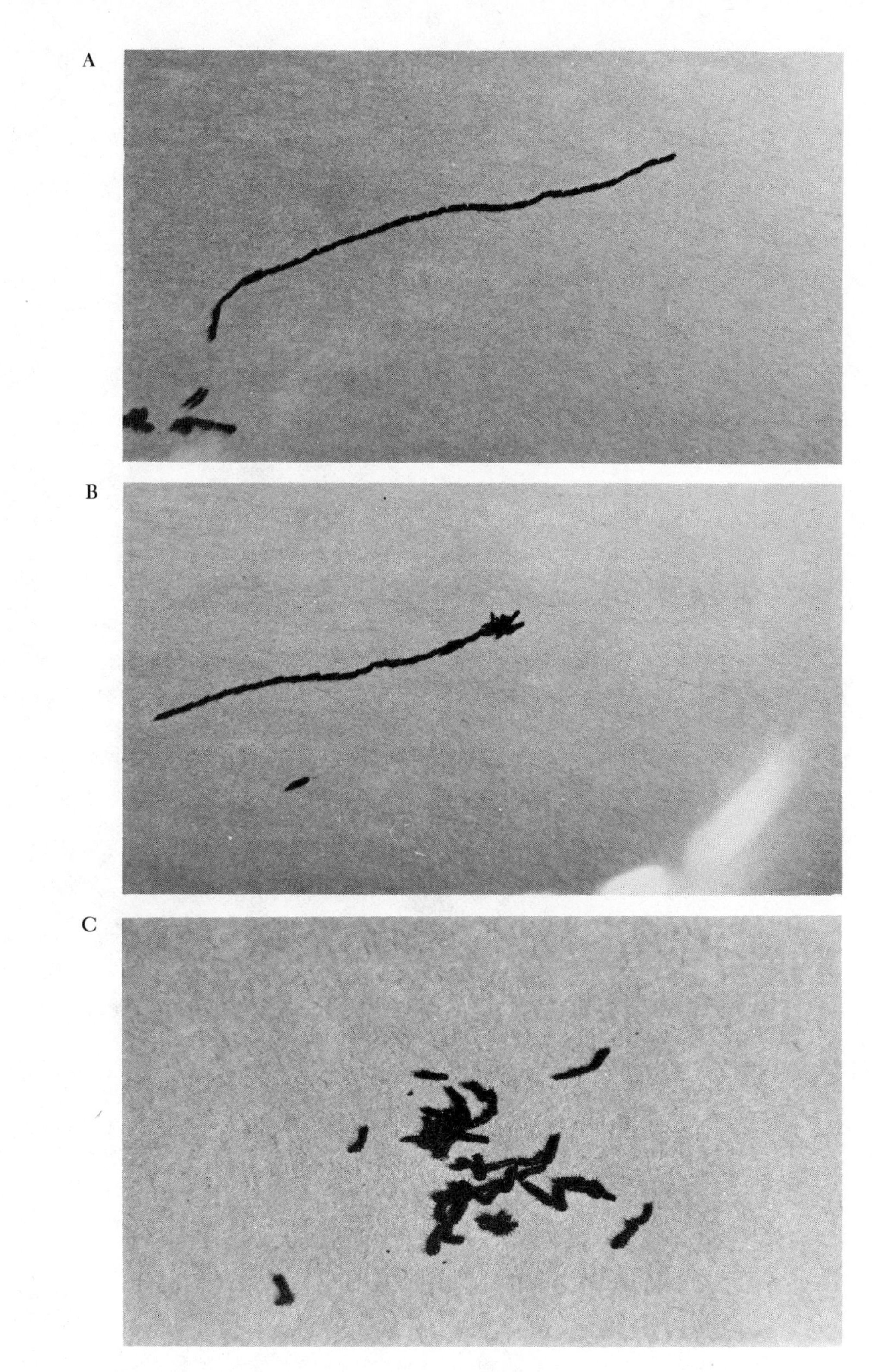


FIGURE 16-1. Processionary behavior of first instar *Hemileuca electra* larvae. **A.** A normal procession, **B.** Confusion following the removal of the leader (below the column) causing a cessation of forward progress, **C.** Complete disorientation of the group.

**FIGURE 16–2.** [OPPOSITE] The brood gallery of the Douglas-fir beetle at the interface of the wood and bark of the host tree showing the larval mines progressing away from the central gallery formed by the adults.

various degrees of presocial behavior, ranging from parental care to some division of labor of the type considered representative of a semisocial level of organization. Parental care would seem to have a distinct selective advantage over the abandonment of eggs immediately after they have been laid. However, parental care does not seem to have any generalized environmental prerequisite and is displayed both by solitary and gregarious species. Although gregarious by nature, female earwigs (Dermaptera) become solitary during the raising of their young. The female usually deposits a cluster of eggs in a preconstructed earthen cell, which she tends until the nymphs attain their third instar. After the young leave, the female abandons the cell but may return to it sometime later to deposit a second cluster of eggs. Hemipterans belonging to several families provide shelter with their bodies for their nymphs during early stages of development.

The web spinners (Embioptera) also guard their eggs until they hatch, but these insects are gregarious and live in loosely arranged colonies that consist of a system of interconnecting silken tubes. Even the first instar nymphs are capable of producing silk from their specialized foretarsi and contribute to an enlargement of the system of runways.

More elaborate forms of parental care and even some division of labor occur among the Coleoptera. Dung beetles (Scarabaeidae) show various degrees of complexity in the behavior associated with the care of their young that seem to suggest an evolutionary progression. In its simplest form, the brood care consists of the female fashioning a ball of dung that is placed in a shallow burrow and then oviposited upon; the female subsequently abandons the nest to excavate another burrow elsewhere. In other species, the male assists the female to form the dungball and may guard it while the female prepares the hole in which it will be buried; this represents a simple division of labor. In the genus *Copris*, the male and female cooperate in the excavation of the nest, plus the gathering and preparation of food for the larvae. In one species, the female guards and cleans the nest throughout the period of larval development. In a related species, both parents remain throughout the life of the larvae, but only the female cleans and maintains the nest (Wilson, 1971).

Fairly advanced levels of presocial behavior are displayed by the bark and ambrosia beetles (Scolytidae). In the Douglas-fir beetle *Dendroctonus pseudotsugae*, the female initiates a brood gallery by chewing through the outer bark of the host tree to the interface between the wood and the inner bark. Males and other females are attracted to the host in response to a pheromone released by the attacking female (see Chapter 13). Each female is joined by a male that initially sits in the gallery entrance and produces a chirping sound. The purpose of the chirping sound is not clearly understood but may serve to communicate the fact that pairing has occurred. The female then proceeds to construct a gallery oriented with the grain of the wood and lays small groups of eggs in niches cut in the side walls of the tunnel. As the main gallery is lengthened, the male keeps it clear of boring dust and excrement by forcing debris from the entrance with his legs and body. Eventually, the gallery opening is sealed with a plug of boring dust that presumably prevents the entrance of natural enemies. Once the entrance is sealed, the male cuts one or more ventilation holes through the bark. The larvae mine away from the main gallery (Figure 16-2), so there is little opportunity for contact between the female and her offspring, even though she often remains in the gallery throughout larval development.

The level of presocial organization in the ambrosia beetles is still further advanced. In the genus *Gnathotrichus*, it is the male that initiates the brood gallery and releases the aggregating pheromone. Once joined by a female, both parents cooperate in excavation of the main gallery, which is innoculated with fungal spores carried by the male in special structures called mycangia. The female lays eggs singly in cradles carved into the side walls of the main gallery and provides fungus for the larvae. The developing larvae feed on fungus that grows on the wood chips and the feces they produce while enlarging their niches. The larvae do not mine away from the main gallery as the bark beetles do but are located in such a way that they can be contacted and given food by their mother. The female remains throughout larval development, during which time she tends her brood and keeps the gallery system free of debris.

According to Wilson (1971), the most advanced form of parental care known among the Coleoptera is displayed by the burying beetle genus *Necrophorus* (Silphidae). Male and female cooperate in the excavation of an earthen burrow and in the preparation of the ball of carrion that is interred within. Once the larval food has been lowered into the burrow, the entrance is sealed from within. The eggs are laid in a depression in the carrion, but when the larvae hatch, they not only feed on the putrifying flesh but also receive a brown liquid passed directly to their mouths from the mouth of their mother. If the larvae are deprived of the food provided by their mother, they are unable to complete their development. In two other species, the male also participates in feeding the brood (Wilson, 1971). The social behavior of these beetles is also more advanced than most in that they display some degree of altruism. Small members of the same species may assist in the internment of the carcass and then depart without taking any further part in reproduction. The dominant pair then take over the food supply (Milne and Milne 1976).

# Social Behavior Among the Hymenoptera and the Termites

***Wasps*** The wasps are a particularly interesting group because they display all of the steps one might envision as significant in the evolutionary progression from truly solitary to truly social behavior. This fact has long attracted the attention of biologists interested in the development of eusocial systems; the most thoroughly constructed behavioral sequence has been provided by the prominent wasp biologist, Howard E. Evans (1958). In his schema, Evans provides a phylogenetic sequence based on both morphological changes and the sequence of related behavioral adaptations. The two clearly go hand in hand, but for simplicity, only the principal elements of Evans' behavioral sequence will be presented here (Table 16-1).

The aculeate (stinging) Hymenoptera are believed to have evolved from parasitic forms, and the behavior of some solitary wasps is scarcely different from that of many parasitoids. In a number of species, the female searches for

**TABLE 16-1** **Major Steps in Development of Eusocial Behavior of Wasps (The families that demonstrate each level of behavior are given in parentheses. Modified from Evans, 1958.)**

| Step of Behavioral Sequence | Comments |
|---|---|
| 1. Prey–egg (Pompilidae) | The female locates a prey, temporarily paralyzes it with her sting, lays eggs on the prey, and departs. The prey recovers and carries the wasp larvae which feed as external parasitoids. |
| 2. Prey–natural crevice–egg (Pompilidae, Sphecidae) | The female drags the paralyzed prey to an available protective crevice where it is left with an egg attached. The female thus provides young with a level of protection. |
| 3. Prey–nest–egg (Pompilidae, Sphecidae) | Female paralyzes a prey and then constructs a nest in which it is placed along with an egg. This is a slightly advanced level of parental care. |
| 4. Nest–prey–egg (Pompilidae, Sphecidae) | Same as Step 3 except the nest is constructed before the prey is captured. This introduces homing in that the female must return to a previously selected nest site. |
| 5. Nest–prey–egg–prey (Sphecidae) | Similar to Step 4, but the addition of more prey after egg is laid introduces mass provisioning as a more advanced form of parental care. |
| 6. Nest–prey–egg–prey–prey (Sphecidae, Eumenidae) | Instead of mass provisioning, the nest is provisioned progressively with fresh prey. This brings the female into contact with her developing offspring. In some species, the female remains in the nest when not provisioning, thereby reducing predation, and may also clean the nest of partially consumed food. |
| 7. Prey macerated by female (Eumenidae, Vespidae) | In the process of progressive provisioning, the fresh prey are macerated by the female and fed to the larvae. This brings the female into direct contact with her offspring and provides opportunity for trophallaxis and the transfer of pheromones. |
| 8. Female life prolonged and offspring remain with nest (Vespidae) | The prolonged female life results in overlap with the first generation offspring, which remain and lay eggs in cells they add to the nest. This results in small colonies consisting of the mother and a group of undifferentiated daughters. |
| 9. Trophallaxis and division of labor (Vespidae) | Mother and daughters cooperate in nest building and the care of young, but there is no permanent division of worker and egg-laying castes. Trophallaxis paves the way for queen dominance. |
| 10. Queen dominance (Vespidae) | The original offspring are all females that are incapable of producing their own female offspring, thus separating the reproductive and worker castes; intermediates may be common. |
| 11. Differential larval feeding (Vespidae) | Differential feeding of the larvae and trophallaxis leads to the production of a well-defined worker caste strongly differentiated from the queen, and a reduction in the number of intermediates. |

a larval prey, paralyzes it with her sting, lays an egg on the surface of the prey's body, and departs. Prey that are left out in the open are clearly subject to consumption by other organisms or to accidental destruction. The behavioral tendency to conceal the prey, either before or after an egg is laid upon it, would have obvious benefits. Progressive provisioning at least has the potential benefit of improving larval nutrition as well as of opening the way to trophallaxis and the possibility of parental control over development, as seen in the burying beetle. The ultimate development of colonies, in which a worker caste is dominated by a single reproductive, improves the efficiency of reproduction and the level of brood survival by way of the environmental control provided by the nest and its occupants.

The most familiar wasps belong to the genera *Polistes* (paper wasps), *Vespa* (hornets), and *Vespula* (yellow jackets). All of these are eusocial, but their colonies die out each fall and must be started anew by overwintered queens. Nests are typically started in the spring by a single female. However, founding *Polistes* queens are occasionally joined by other overwintered females that become subordinate to the foundress. The auxiliary females gather food and regurgitate it for the foundress; she overtly prevents her associates from laying eggs and eats any eggs they manage to deposit (Eberhard, 1969).

A single foundress usually constructs a small group of paper cells in which she rears the first group of larvae on macerated insects killed freshly each day. The first offspring are all workers, which upon emergence forage for food and add new cells to the nest. The queen then becomes little more than an egg-layer and only occasionally leaves the nest. The colony grows rapidly throughout the summer, each generation of workers increasing the size of the nest and the number of larvae that can be reared at one time.

Toward the end of summer, larger brood cells are constructed in which numerous new queens and males are reared. Soon thereafter, the founding queen dies and the population of workers declines rapidly. The virgin queens and males leave the nest and mate. The males soon die, but the inseminated females seek secluded crevices in which to hibernate before renewing the cycle the following spring.

Adult wasps feed to some extent on nectar and the juice of ripe fruit, but these substances are not stored as bees store honey, and the larval diet consists mainly of macerated insects rather than pollen. This fundamental difference in the food preferences of wasps and bees probably contributed to the failure of wasps to evolve the tendency to overwinter as intact colonies. The insects that comprise the wasp diet are not readily obtainable during cool weather and cannot be stored because they tend to putrify. Honey and pollen, on the other hand, store very well.

***Ants*** Until rather recently, the ancestry of the ants was uncertain, but an examination of the fossil *Sphecomyrma* suggests a close affinity to the nonsocial parasitic wasps of the family Tiphiidae. All living ants are eusocial, so it is not possible to trace an evolutionary progression of behavior for them as it is for the wasps. Wilson (1971) feels that the diversity of ant social behavior most likely had its beginning in the behavior of solitary, ground-dwelling, wasplike insects that divested themselves of their wings and constructed a subterranean nest in which they reared a small group of young, which were provided freshly killed prey.

Although we cannot observe in ants the numerous steps between solitary and eusocial behavior, the group is fascinating in terms of the almost endless variety of eusocial life-styles it exemplifies. The following paragraphs will provide a hint as to the scope of this variety. The interested reader should then refer to other sources, including the old but fascinating book *Ants, Their Structure, Development and Behavior* by W. M. Wheeler, and E. O. Wilson's excellent book, *The Insect Societies.*

All living species of ants have a caste system that generally parallels that of the eusocial wasps and bees and clearly reveals a division of labor. The principal castes consist of males, queens, and workers, the latter group being subdivided into different types in most species (Figure 16-3). Several genera also have castes intermediate between males and workers and between workers and queens, but these are generally less important than the main castes. Typical males are winged reproductives that contribute absolutely nothing to the labor of the colony, although they do groom other adults and are groomed in return. Queens are the colony founders and mothers. Within an established colony, their social behavior is restricted to grooming, as in the males, but in the establishment of new colonies they perform the full range of duties later performed by workers.

Workers are sterile females, and in the broad sense include both the laborers and defenders of the colony. There are often three subcastes of workers, based partially on size. The largest are the **majors** or **soldiers,** which are often considered to be a distinct caste because of the oversized (allometric) development of their heads and mandibles. The other workers are distinguished as being **media** if of intermediate size and **minor** if small. The degree of worker polymorphism varies greatly from species to species, but in species in which it is pronounced, a clear division of labor among the subcastes can be observed.

Perhaps most spectacular are the legionary or army ants of the humid tropical forests. These ants do not construct nests but form temporary clusters called bivouacs in the shelter of a fallen tree trunk or other such partially exposed location. The workers form a solid mass up to a meter across, consisting of layer upon layer of individuals linked together by their tarsal

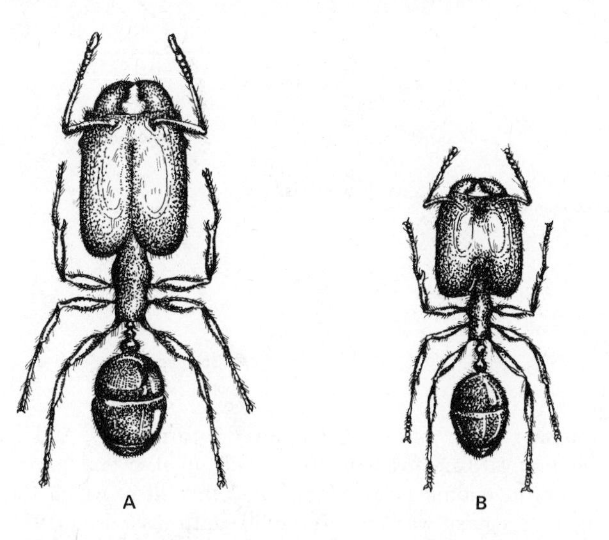


**FIGURE 16-3.** Different forms of the worker cast of ants. **A.** Major, **B.** Media, **C.** Minor (After Wheeler, 1910.)

claws. The queen, thousands of larvae and pupae, and, at certain times of the year, up to 1,000 males and a few virgin queens are located near the center of the ball of up to a half million workers.

The night is passed in a tight cluster, but at first light, the chains of workers become detached and a hoard of individuals begin to move away from the bivouac site in all directions. Soon one or more raiding columns form and move out in search of food. The workers lay a pheromone trail to guide those that follow, while the soldiers guard the column's flanks. Some species engage in column raids, whereas others employ a swarming tactic (Figure 16-4). But regardless of their method, the foraging ants form a number of changing columns behind the advancing hoard of workers that flush and capture a variety of mainly arthropod prey. The prey are stung to death, dismembered, and transported rearward as food for the developing larvae (Rettenmeyer, 1963). At the end of each day of marauding, the protective cluster is re-formed.

As the colony rapidly reduces the food supply in the immediate vicinity of its bivouac, it must move to a new location. However, in at least one species the change in bivouac site has been correlated with the reproductive cycle within the colony, which in turn has an effect on its food requirements. According to Schnierla (1971), the queen periodically undergoes rapid ovarian development which results in great distension of her abdomen. Over a period of several days, such a queen lays 100,000 to 300,000 eggs, which hatch about

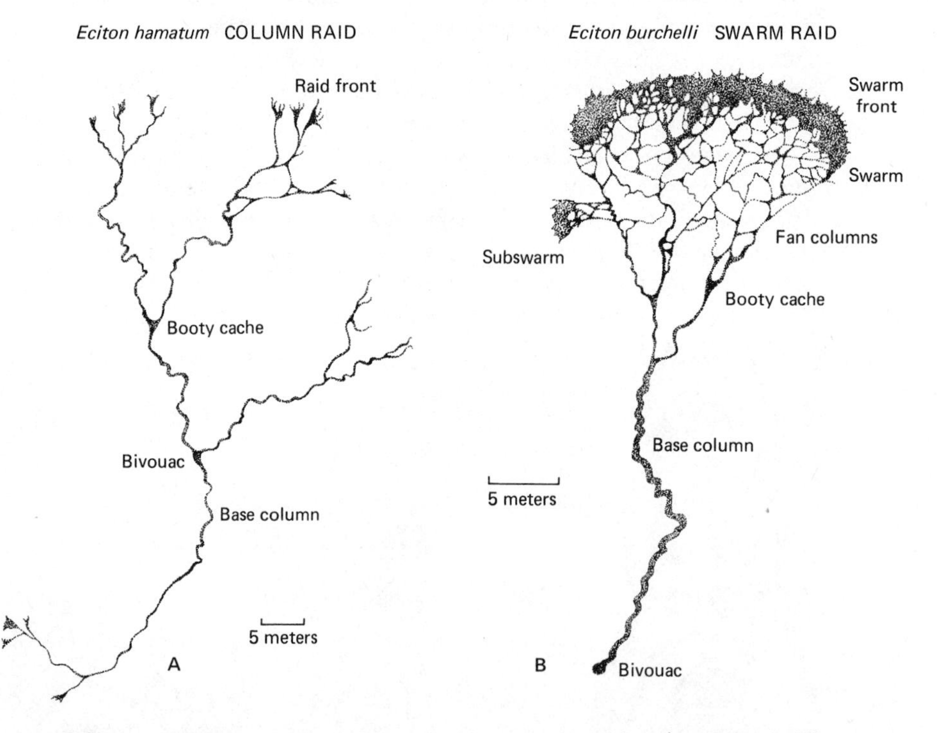


**FIGURE 16-4.** The two basic patterns of raiding employed by army ants. **A.** The column raid of *Eciton hamatum* with an advancing front made up of several narrow bands of workers, **B.** The swarm raiding tactic of *Eciton burchelli* in which the advancing front is made up of a mass of workers. (Reproduced, from Wilson, 1971, after Rettenmeyer, 1963, with permission of the author and Harvard University Press.)

three weeks later. During this **statary phase,** the larvae from the previous period of oviposition complete their pupal development and emerge as young workers. The addition of these new workers seems to stimulate the colony to increase the intensity of its daily foraging raids and change the location of its bivouac site each evening. This **nomadic phase** therefore coincides with the period during which a large number of larvae must be fed. Once these larvae complete their development and pupate, the nomadic behavior declines and the colony enters another reproductive period, during which a single bivouac site is utilized.

People living in the temperate zone are most familiar with the nest-building ants such as members of the genus *Myrmica* that usually nest in the ground under a large stone or the genus *Formica* that construct the familiar mounds of twigs and other plant debris. These ants are characterized by the production of winged reproductives during the summer and early fall. In some species the emergence of the reproductives produces quite spectacular swarms, during which mating occurs. The nuptial flight results in the ubiquitous spread of potential foundresses of new colonies. After a short time, the newly mated queen sheds her wings and excavates the beginning of a new nest in which she seals herself. In the spring, she lays a small group of eggs and tends the larvae until they attain adulthood and become the colony's initial group of workers. These first offspring may take more than a year to complete their development, but the queen never leaves, relying on her internal food reserves and flight muscles for her own sustenance as well as that of her young.

Over the succeeding years, the colony grows very slowly until enough workers are present for it to enter a period of accelerated growth. In areas characterized by cold winters, the colonies enter hibernation for several months, and no eggs are laid until warmer conditions prevail in the spring. In most species, several years pass before winged males and females are produced to leave and to establish new colonies in their turn. Colonies of many species contain only a few hundred individuals, whereas some contain many thousands; most nests contain only one queen, but some *Formica* nests have hundreds of active queens.

In spite of the general similarities displayed by the nest-building ants as far as the establishment and growth of new colonies is concerned, they show extensive adaptive radiation in other aspects of their behavior. Most display a high level of variability in their choice of prey. Like the army ants, they forage widely for almost any kind of small animals, especially arthropods, that are available. A single nest of the European red ant *Formica polyctena* is said to gather up to a kilogram of such food in the course of a single day. Chauvin (1967) concluded that the population of 300 million million red ants estimated to inhabit the Italian Alps would be capable of destroying 15,000 tons of insects a year. But not all ants have the catholic taste of *F. polyctena.* Some feed specifically on only certain kinds of arthropods or their eggs. Others, often called "harvester" or "agricultural" ants, subsist entirely on seeds and thereby become pests in some grain and grass growing regions. When protein rich foods are not needed for the rearing of larvae, many ants feed upon the nectar of flowers.

A number of ant species feed exclusively on the anal excretion of aphids, scale insects, and other homopterans. The so-called "honeydew" is rich in sugars and in free amino acids and forms an inexhaustible source of nutrient throughout the time when the producing insects are actively feeding. Some species of ants can be seen to stroke the cornicles of aphids to induce the flow of honeydew from the anus. Ants in their tending of homopterans can seriously interfere with the biological control programs directed against these insects by actively repelling parasites and predators. Some ants actually transport aphids or their eggs from their host plant in the fall to underground "barns" in which they are cared for over the winter. In the spring, the ants carry the aphids back to their host plant once again to "milk" them of their honeydew (Way, 1963).

Members of the New World tribe Attini are sophisticated fungus culturers. Many species in the group gather small pieces of green leaves or the petals of flowers to form underground beds on which to culture a specific variety of fungus. Members of the genus *Atta* are among the more serious pests of New World tropical agriculture as a result of their leaf collecting; in English, they are commonly referred to as leaf-cutting ants. Media workers carry the plant material piece by piece deep into their nests where it is licked, cut into small pieces, wet with an anal secretion, and formed into beds of moist pulp. The newly formed beds are then "planted" with fungal mycelia collected from established beds. The fungus grows rapidly and soon produces on the tips of the hypae small spheres which are fed to the ant larvae. The fungus beds are tended and harvested by smaller workers (minora) that never engage in the collection of leaf fragments. However, in *Atta cephalotes* these smaller workers may accompany the larger leaf-gatherers; they do not assist with the leaf cutting but ride back to the nest on the leaf portion, apparently warding off parasitic phorid flies with their mandibles and hind legs (Figure 16-5).

One of the more fascinating aspects of the fungus culture is how the ants are able to maintain a monoculture of their specific fungus in that abandoned beds are rapidly overgrown by alien fungi of various species. Weber (1957) discovered that the worker ants tending the fungus beds "weed" them of alien hypae with their mandibles. It has also been postulated, but not proven, that the ants employ fungicidal and bacteriacidal substances secreted by their salivary or anal glands. The cultured fungus is carried to each new nest by the foundress. Before departing on her nuptial flight, the virgin queen packs a small wad of mycelia into a cavity near the base of her labium. The wad is then deposited in the nest she excavates. The queen tends the initial

**FIGURE 16-5.** Media worker of the fungus-growing ant *Atta cephalotes* accompanied during leaf gathering by a minor worker which protects the media worker from parasitic phorid flies. (Redrawn from Eibl-Eibesfeldt and Eibl-Eibesfeldt, 1967.)

fungus garden but does not consume any herself. However, the first workers to emerge feed upon it and fertilize it with their fecal material.

Not all ants are as industrious as the aphid herders and fungus gardeners. The Amazon slaver ant *Polyergus lucidus* and the red ant *Formica sanguinea* are common species that make raids on other ant species to capture slaves. The raiders carry larvae and pupae back to their nests where they are raised to assume the duties of nest maintenance. *Polyergus lucidus* is an obligatory slave keeper as the workers cannot feed themselves. *F. sanguinea*, on the other hand, is a facultative slaver (Wilson, 1975).

**Bees** Bees, like wasps, display many degrees of presocial behavior. Within small groups of related species it is possible to identify numerous steps in the evolutionary progression from solitary to eusocial behavior. As mentioned earlier, a major difference between the bees and the wasps lies in their diets and the morphological adaptations related to food collection. Adult bees not only feed on nectar and honey but raise their young on pollen or a pollen-honey mixture rather than on macerated insects. Because both honey and nectar are readily storable, the larval cells of bees can be provisioned initially rather than progressively throughout development as in many wasps; this tends to reduce the occurrence of contact between adult bees and their developing offspring.

It is not possible to consider the natural history of bees in more than a superficial way in the space available. However, I will present a brief description of the principal evolutionary steps envisioned by Michener (1969) and then take a somewhat more detailed look at the eusocial organization of the bumblebees and the common, domesticated honeybee.

In the **solitary bees,** each female independently constructs one or more nests. Each nest contains a group of cells that are mass-provisioned with sufficient pollen and nectar to fulfill the needs of the larvae throughout their development. One egg is laid in each provisioned cell, which is then sealed and left. Such is the way of the leaf-cutter bee *Megachile rotundata* (see Chapter 19), which under wild conditions uses widely scattered, naturally occurring nest sites. However, some solitary bees form **nesting aggregations.** Although the nests and their entrances are independent, they may be close together, as is usually the case with alkali bees such as *Nomia melanderi,* which are also discussed in Chapter 19. Such aggregations are probably a response to the contagious distribution of suitable nesting sites, but they also require the development of a tolerance for other individuals of the same species which would seem to be a prerequisite to cooperative behavior.

Some bees, including other species of the genus *Nomia,* utilize composite nests, consisting of a common entrance but with cells constructed and provisioned independently by individual females. Michener calls these species **communal** and suggests there is economy of labor in both nest construction and defense, which also tends to be more persistent since at least one female is present much of the time.

The next important evolutionary step would involve **quasisocial** groups of bees, the females of which cooperate in the construction and guarding of the nest and the brood cells and in the provisioning of the cells. There would be no true division of labor in that all the females would engage in all of the necessary tasks, including the laying of eggs. However, behavior of this type has not yet been discovered among bees.

The final step before full-fledged social behavior would consist of a **semisocial** group in which cooperating females of the same generation display some division of labor in the sense that some females help with provisioning and other tasks but do not lay eggs. There is limited evidence that such behavior occurs in several species of bees, but more detailed study is needed.

On the basis of the criteria stated at the beginning of this chapter, members of five major groups of bees qualify as being eusocial. The tribe Apini contains the bumble bees, which have a relatively simple social life style, and the honeybees, which represent the most advanced of the social bees.

The bumblebees are primarily adapted to cooler climates. Like the eusocial wasps, they normally overwinter as queens that survive the cold months in the solitude of protected hibernation sites. In the spring, the queen emerges and sets out in search of a suitable subterranean nest site in the form of some pre-existing cavity. Within the nest she constructs a cuplike cell from wax that is secreted by the intersegmental glands of her abdomen. The cup is then provisioned with a ball of pollen on which she lays her first batch of eggs, and the cell is sealed. After the initial egg cell has been completed, the queen constructs a honeypot near the entrance to the nest and fills it with nectar. The first group of offspring are all workers that upon emergence assist the queen in producing more larval cells and feeding the new brood. The subsequent brood is either fed by the workers or is allowed to feed directly on a mass pollen store, depending on the species. By late summer, the colony consists of several hundred workers and produces males and new queens; the colony then abandons its nest. The virgin queens are met by waiting males, copulate, and then enter hibernation.

The social life of the honeybee *Apis mellifera* is far more complex, as was indicated by the discussion of its rather elaborate communication system. The honeybee originated in the tropics or subtropics but seems to have been preadapted, through its ability to regulate its nest temperature, to expand its range into regions of colder climate and still maintain perennial colonies. Entire colonies, instead of just fecund queens as in the wasps and bumblebees, overwinter. This not only permits the colonies to become much larger but permits and requires other changes as well.

Because honeybee queens do not initiate new colonies each year by their own labor, they do not require any of the capabilities of their worker caste. Consequently, they have become much more differentiated from the workers and serve solely the functions of reproduction. New colonies are founded through the process of colony division by swarming. The swarming process usually begins when a colony is strong and crowded. In response to some unknown stimulus, the queen reduces the level of queen substance she produces, and the workers respond by constructing a small number of queen cells. The larvae that hatch from the eggs laid in these larger cells are fed exclusively on royal jelly and complete their development in about two thirds of the time it takes workers and drones (males) to develop. When virgin queens are developing within the colony, the initial queen's egg production declines, and she is treated less hospitably until she is more or less forced to leave with a large retinue of older workers. Swarms containing the old queen are called **prime swarms** and may be followed a short time later by one or more **afterswarms** each containing a new queen, produced within the hive and having returned from her nuptial flight.

Swarms usually settle in a cluster similar to that of army ants at their bivouac. Once scout bees have communicated the location of a suitable nesting site, frequently in a hollow tree, the swarm moves there and immediately begins to construct new waxcombs for brood rearing and food storage (Figure 16-6). Since the queen does not have to do anything in the way of nest building, as in the case of foundress bumblebees, she can proceed with egg laying as soon as workers prepare some brood comb. Consequently, the new colony grows extremely rapidly.

Within a few days of the departure of the last swarm, the virgin left in the original colony goes on a series of mating flights and finally returns with sufficient sperm to last five to seven years, during which time she will lay hundreds of thousands of eggs. The brood left in the original nest is usually in all stages of development when the swarms leave, so the colony is replenished rapidly by a new complement of workers that help restore it to its original strength within a year. The rapid rate of population growth and the large colonies made possible by their perennial nature requires that the workers efficiently exploit the pollen and nectar sources within flying distance of their nest. This is made possible by the communication for recruitment, discussed in Chapter 13.

Drones are produced throughout the more favorable part of the year and are usually most abundant when the colonies are reproducing by swarming and therefore contain virgin queens that must be inseminated. Drones contribute absolutely no labor to the colony but are tolerated by the workers until late in the summer. As the supply of food in the field begins to decline in the fall, the workers become intolerant of the drones, driving them from the hive, and not permitting them to reenter.

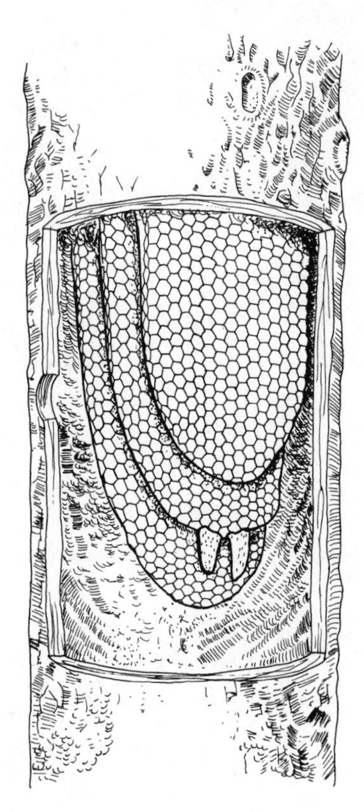

**FIGURE 16-6.** Combs of wild honeybees in a hollow tree. (Redrawn from Wilson, 1971, with permission of the author and Harvard University Press.) Each comb consists of a double layer of wax cells. When complete, the surfaces of adjacent combs are separated by a space of $\frac{1}{4}$ to $\frac{3}{8}$ of an inch. How the bees determine where to begin each comb remains unknown. The horizontal orientation of the combs in a new natural hive is usually the same as that of the hive from which the swarm emerged, and is apparently determined by magnetic field imprinting.

***Termites*** The order Isoptera is clearly more ancient and more primitive than the order Hymenoptera, yet all living termites are eusocial and have achieved a level of caste differentiation and social organization equal to the wasps, ants, and bees. The most primitive living termite, *Mastotermes darwiniensis,* seems closely related to the primitive, social, wood-eating roach *Crytocercus,* with which they share a number of intestinal protozoans that are essential to the digestion of the cellulose they consume in a wide variety of forms. Some termites also lay eggs in small groups that are protected by a covering similar to the oöthecae of roaches.

Termites lack cellulose-digesting enzymes so rely on their intestinal fauna of mutualistic flagellate protozoans, which must be passed continuously between individuals, to aid in the break-down of their food. The passage of gut symbionts from old to young individuals seems to have set the stage for the development of more advanced social behavior. It has been postulated, therefore, that the social order of termites began with trophallaxis and led to the addition of brood care, rather than the reverse, as in the social hymenoptera.

The natural history of termites is quite variable. The so-called dry wood termites that frequently cause damage to the dry, seasoned wood of buildings usually live in colonies consisting of only a few hundred individuals. Winged reproductives are produced seasonally, usually during the warmer months, leave the nest soon after emergence, and spread rather ubiquitously. When a female completes her flight, she divests herself of her wings and runs about in an excited fashion until joined by a male. After pairing, the king and queen engage in a behavior called **tandem running** in which the king closely follows his queen in search of a nest site. When the pair locate a suitable place to begin a nest, they take turns excavating an initial tunnel with a small chamber at the bottom. The king and queen seal themselves in by plugging the tunnel with chewed wood, and the queen proceeds to produce a small batch of eggs. When the young nymphs emerge, they are fed regurgitated food until they are able to feed upon the surrounding wood and thereby enlarge the nest. As the colony grows in the ensuing years, soldiers are produced that provide the colony with defense. When the colony is several years old, more alates are produced which found new colonies. The foundress of a colony may live for more than 10 years, but if her egg production declines or she is killed, her duties may be assumed by secondary queens.

Most of the more highly evolved termites are soil dwellers which construct conspicuous mound nests. Alate reproductives are again produced seasonally and often emerge from their nests at the beginning of the rainy season. After a short, feeble flight, the reproductives shed their wings and engage in nest founding behavior similar to that already described. The first brood consists of workers, but soldiers are produced later. In addition to workers, soldiers, and huge physogastric females, some colonies contain secondary and tertiary queens (Figure 16-7). As the number of individuals in the colony grows from year to year, the size of the mound is increased by the addition of layers of soil and excrement.

The mound-building termites have adapted to virtually every form of food with a high cellulose content. The food is gathered by the workers, which forage throughout a large area surrounding the nest by way of a system of subterranean tunnels or covered pathways; some even forage on the surface over well-trodden trails marked with pheromones. Termites of the subfamily

FIGURE 16-7. The five castes of the termite *Ameritermes hastatus* drawn to same scale. **A.** Worker, **B.** Soldier, **C.** Physogastric primary queen, **D.** Secondary queen, **E.** Tertiary queen. (Reproduced from Wilson, 1971, after Skaife, 1954, with permission of the author and Harvard University Press.)

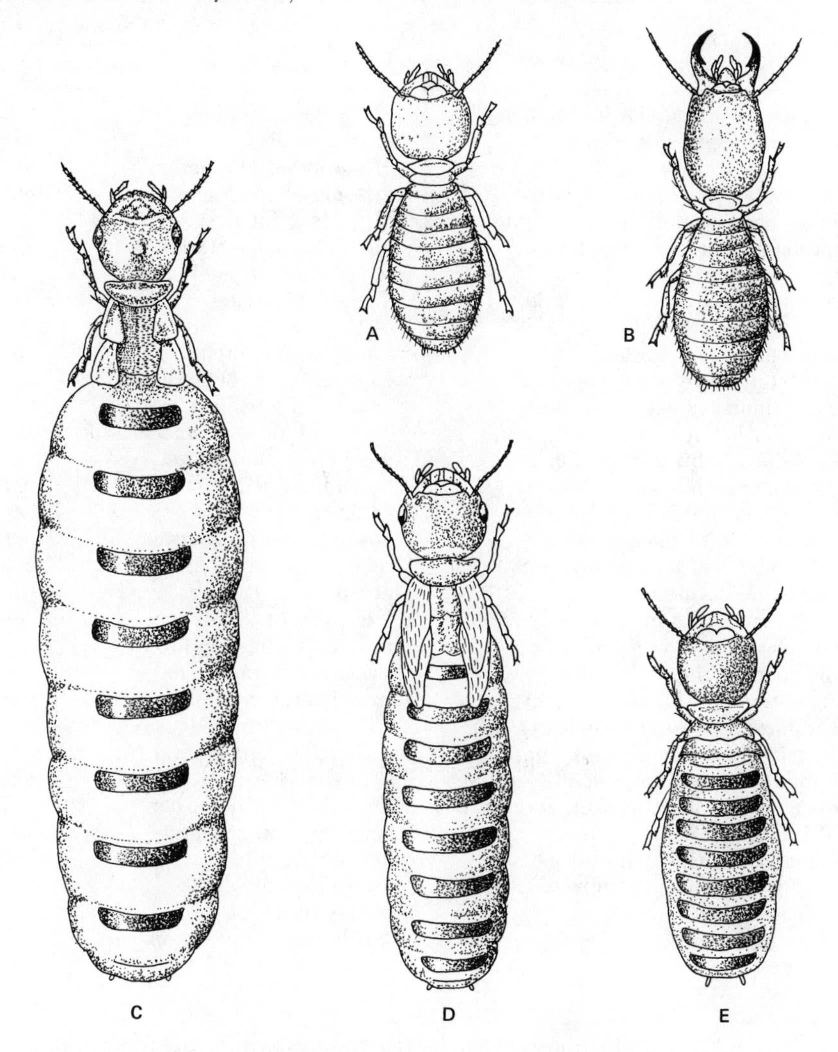


Macrotermitinae culture fungus on comblike structures built from their excrement. As in the ant genus *Atta*, the fungus beds of occupied nests are a pure culture. Rather than cropping the fungus in the manner of ants, termites consume the entire mass, including the excrement substratum. The colonies of fungus-growing termites may contain as many as two million individuals. The primary queen is capable of laying up to 30,000 eggs a day; in a life-span of up to 10 years, a single queen may produce tens of millions of offspring.

The evolution of the termites is far removed phylogenetically from that of the Hymenoptera; yet, the two groups have, by quite different routes, arrived at a complex level of social organization. There are some striking similarities in the social biology these distinct orders display, but there are some fundamental differences as well. The most important differences markedly affect the make-up of the colonies. In the termites, the queen cannot control the sex of her offspring, so the workers are both males and females (genetically).

**TABLE 16-2** **Basic Similarities and Differences in Social Biology Between Termites and Higher Social Hymenoptera (Wasps, Ants, Bees). Similarities Are the Result of Evolutionary Convergence. Reproduced from *The Insect Societies* by E. O. Wilson, 1971 with the author's permission.**

| Similarities | Differences | |
|---|---|---|
| | *Termites* | *Eusocial Hymenoptera* |
| 1. The castes are similar in number and kind, especially between termites and ants.<br>2. Trophallaxis (exchange of liquid food) occurs and is an important mechanism in social regulation.<br>3. Chemical trails are used in recruitment as in the ants, and the behavior of trail laying and following is closely similar.<br>4. Inhibitory caste pheromones exist, similar in action to those found in honeybees and ants.<br>5. Grooming between individuals occurs frequently and functions at least partially in the transmission of pheromones.<br>6. Nest odor and territoriality are of general occurrence.<br>7. Nest structure is of comparable complexity and, in a few members of the Termitidae (*e.g.*, *Apicotermes*, *Macrotermes*), of considerably greater complexity. Regulation of temperature and humidity within the nest operates at about the same level of precision.<br>8. Cannibalism is widespread in both groups (but not universal, at least not in the Hymenoptera). | 1. Caste determination in the lower termites is based primarily on pheromones; in some of the higher termites it involves sex, but the other factors remain unidentified.<br>2. The worker castes consist of both females and males.<br>3. Larvae and nymphs contribute to colony labor, at least in later instars.<br>4. There are no dominance hierarchies among individuals in the same colonies.<br>5. Social parasitism between species is almost wholly absent.<br>6. Exchange of liquid anal food occurs universally in the lower termites, and trophic eggs are unknown.<br>7. The primary reproductive male (the "king") stays with the queen after the nuptial flight, helps her construct the first nest, and fertilizes her intermittently as the colony develops; fertilization does not occur during the nuptial flight. | 1. Caste determination is based primarily on nutrition, although pheromones play a role in some cases.<br>2. The worker castes consist of females only.<br>3. The immature stages (larvae and pupae) are helpless and almost never contribute to colony labor.<br>4. Dominance hierarchies are commonplace but not universal.<br>5. Social parasitism between species is common and widespread.<br>6. Anal trophallaxis is rare, but trophic eggs are exchanged in many species of bees and ants.<br>7. The male fertilizes the queen during the nuptial flight and dies soon afterward without helping the queen in nest construction. |

Furthermore, after the first few instars, the termite immatures engage in work for the good of the colony. In the social Hymenoptera, the workers are all female, and the queen can regulate the production of males; the larvae are, in a sense, parasitic and contribute nothing to the well-being of the colony. These and other differences are summarized in Table 16-2 with the kind permission of E. O. Wilson.

# References Cited


Chauvin, Rémy. 1967. The world of an insect. (English tr. by H. Oldroyd). McGraw-Hill Book Company, New York, N. Y.

Eberhard, Mary Jane West. 1969. The social biology of polistine wasps. Miscellaneous publications, Museum of Zoology, University of Michigan. 140: 1-101.

Eibl-Eibesfeldt, I. and Eleonore Eibl-Eiblesfeldt. 1967. Das Parasitenabwehren der Minima-Arbeiterinnen der Blattschneider-Ameise (*Atta cephalotes*). *Zeit. für Tierpsych.* 24: 278–281.

Evans, H. E. 1958. The evolution of social life in wasps. Proceedings of the Tenth International Congress of Entomology, Montreal, 1956. 2: 449–457.

Michener, C. D. 1974. The social behavior of bees: A comparative study. The Belknap Press of Harvard University Press, Cambridge, Mass.

Milne, L. J., and Margery Milne. 1976. The social behavior of burying beetles. *Sci. Amer.* 235(2): 84–89.

Rettenmeyer, C. W. 1963. Behavioral studies of army ants. Kansas University *Sci. Bull.* 44: 281–465.

Schneirla, T. C. 1971. Army ants, a study in social organization. W. H. Freeman and Company, Publishers, San Francisco, Calif.

Skaife, S. H. 1954. Caste differentiation among termites. *Trans. Roy. Soc. South Africa* 34(2): 345–353.

Weber, N. A. 1957. Weeding as a factor in fungus culture by ants. *Anatomical Record.* 123: 638.

Way, M. J. 1963. Mutualism between ants and honeydew-producing Homoptera. *Ann. Rev. Entomol.* 8: 307–344.

Wheeler, W. M. 1910. Ants: Their structure, development and behavior. Columbia University Press, New York, N. Y.

Wilson, E. O. 1971. The insect societies. The Belknap Press of Harvard University Press, Cambridge, Mass.

Wilson, E. O. 1975. Slavery in ants. *Sci. Amer.* 232(6): 32–36.

IV

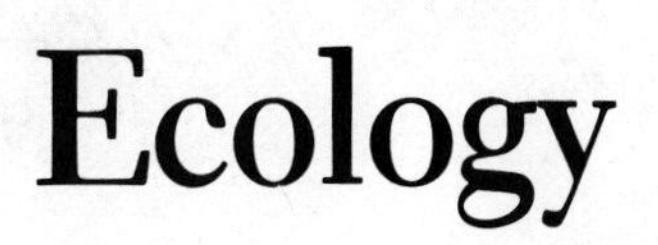
Ecology

*In the foregoing chapters, we have examined the structural and functional organization of insects. To this point, the emphasis has been process-oriented and concentrated on the structural and behavioral adaptations that have evolved among diverse insect groups. Repeated reference has been made to those features of the environment which have acted as* ***selective forces.*** *Clearly, no organism lives in a vacuum, so it is unnatural to consider either behavior or functional anatomy without considering environmental cues and constraints as well. In this sense, almost all of the chapters in this book are tied together by a thread of ecology. The main difference between what has gone before and what will be considered from this point on lies in the focus. In the previous chapters, we examined single organisms or species as influenced by past environments (evolution) as well as by current conditions (physiology and behavior), so the focus was akin to* ***autecology.*** *In the chapters that follow, the approach will be* ***synecological*** *in that it will focus on the interaction of populations of different species of insects with each other and with the plants, animals (including man), and physical factors that comprise their environment.*

*As is the case with most biological phenomena, it is virtually impossible to organize the treatment of entomology in such a way as to consider all of the information where it seems relevant and yet avoid repetition. For example, many aspects of defense are largely behavioral, so this subject was considered in the section on behavior. Cryptic coloration and mimicry are forms of protection against being eaten, and they too have a behavioral component. On the other hand, defensive coloration develops as the result of the coevolution of a group of species that interact as predators (or parasites) and prey (or hosts). The current state of that coevolution has a significant bearing on the population fluctuations of the interacting species.*

*From a practical standpoint, a synecological focus requires some basic understanding of population fluctuations. We need to examine both the factors that influence the direction and magnitude of numerical changes and the impact that different population levels have on the environment of the species in question. Insects are so numerous and function in so many different ways in virtually every terrestrial and fresh-water community that man has become increasingly aware of the need to understand the changes in their numbers. More and more natural communities are coming under man's management for his benefit, and many have been replaced by contrived communities designed to maximize the production of specific commodities. As a result, man must be able to exercise some control over the populations of harmful species. Furthermore, there is a growing need to manage beneficial species properly.*

*Although it seems inappropriate to some to consider insect populations without first discussing the various natural roles they play in different ecosystems, it is just as inappropriate to others to consider insect-plant and predator-prey relations without some knowledge of the other factors that affect their numbers. In this section, therefore, we will begin with a discussion of insect populations in general and then proceed to a consideration of their importance and management.*

# Insect populations and their environment

# 17

Each species is usually subdivided over its natural range into more or less well-defined groups called populations. Sometimes populations of the same species are totally isolated from one another geographically and never exchange genetic material; eventually such populations may become distinct species. More often a species has populations which may be separated at some times and contiguous at others. Consequently, there is either a continuous or periodic mixing of the species gene pool.

Within a particular geographic area a species displays a characteristic abundance; some species are common, whereas others are rare. All species tend to exhibit some form of seasonal, numerical periodicity that is related to their reproductive cycles, dispersal, and other features of their life history (Figure 17-1). But there are many factors which influence the number of individuals that occur in a region at the beginning of each reproductive cycle. Consequently, populations tend to fluctuate from year to year. If these fluctuations are plotted over a sequence of years, they appear to oscillate around an average population level. However, the periodicity of population oscillations does not necessarily correspond to a generation of the species or one calendar year. Populations may increase for several years before a downward trend begins or vice versa. Some species populations display marked numerical changes, whereas others are characterized by fluctuations of rather low amplitude (Figure 17-2). Species which have populations that fluctuate widely are often considered to be more unstable than those which fluctuate very little. But this is a dangerous interpretation in the absence of information about the normal fluctuations in the availability of resources to which such species are adapted.

There are a multitude of factors that can permit a population to increase or cause it to decline. Not all of these factors are operative all of the time, and different ones seem to be important to the population fluctuations of different species. For most factors, it is the extreme conditions that have the most impact on a population. For example, considerable fluctuations in the middle range of temperature may not affect a population to any extent, whereas either very low or very high temperatures may cause substantial mortality. Each species has limits of tolerance for each environmental factor it must contend with; beyond these limits of tolerance, a factor influences the species in a way that usually has a negative impact on its population. However, each species evolves under a composite regime of interacting variables and its theoretical average population is approximately that which can be attained under a theoretical average set of conditions. Although this is clearly a hypothetical situation, it helps to illustrate that if all or most of the important environmental factors are within a species' limits of tolerance, the over-all conditions would be favorable and the population would probably increase. On the other hand, if one or more factors made the over-all environment unfavorable, the population would probably decline.

This may seem rather obvious, but the problem is that it is not a simple task to determine which of the many environmental factors has the most

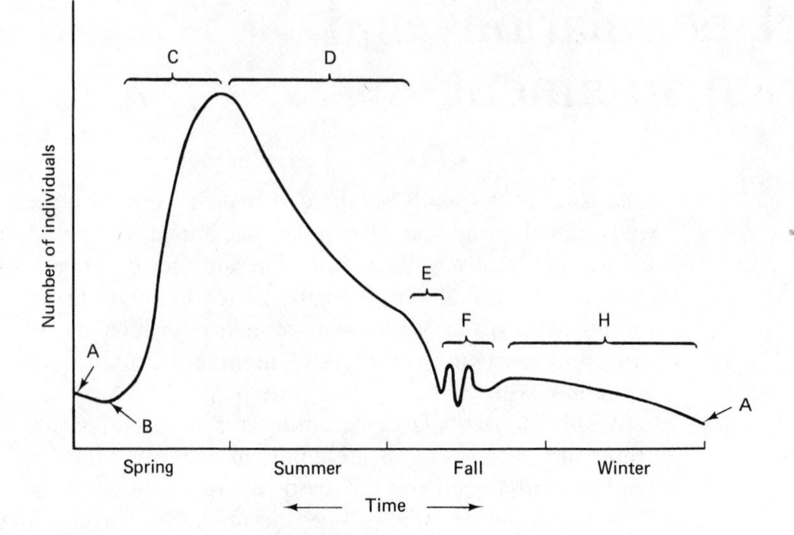


FIGURE 17-1. Graphical representation of population changes during a single generation of a hypothetical insect. The population cycle begins with a fixed number of viable females (A) only a portion of which become fertile so the population in effect declines up to the beginning of reproduction (B). The population then increases rapidly due to the recruitment of offspring during period (C). Subsequently the population declines during period (D) as adult and larval mortality exceed the addition of young. The population declines further during period (E) as a result of pupal mortality. The local population may then fluctuate during period (F) as a result of the migration (exodus) of young adults and an influx of adults from another locality. There may then be some mortality over the winter (H) before the population level at the beginning of the next generation (A′) is determined.

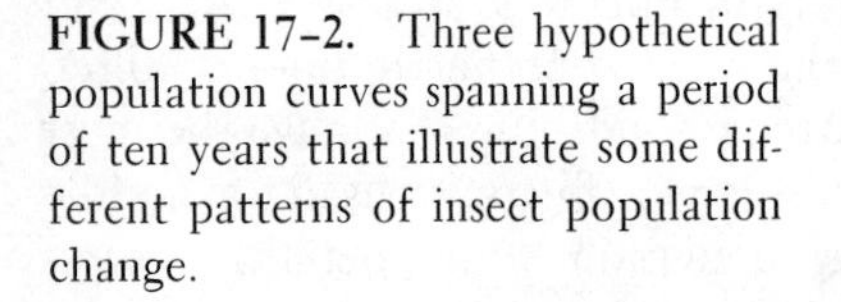

FIGURE 17-2. Three hypothetical population curves spanning a period of ten years that illustrate some different patterns of insect population change.

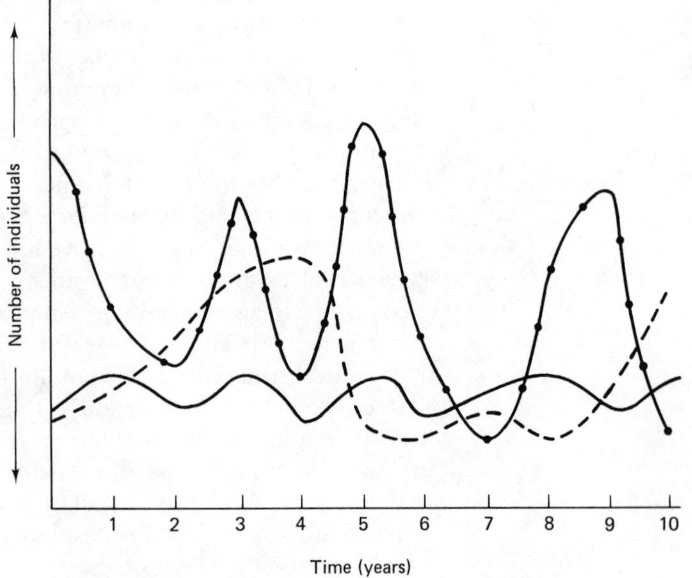

influence and whether one unfavorable factor can be compensated for by several favorable factors or at all. With insects, the situation is further complicated by the fact that different life stages have different environmental requirements. The situation may be even more difficult to interpret for those species which have evolved a high degree of immature-adult divergence.

The components of the environment that have been observed to have a significant influence on plant and animal populations have been variously grouped, but many entomologists subscribe to the following subdivisions proposed by Andrewartha and Birch (1954): (1) weather, (2) food, (3) a place in which to live (habitat), and (4) other animals and pathogens. In the following pages, we will examine the more important aspects of each of these components in terms of its population effects.

## *Weather*

Weather is the current combination of temperature, moisture, sky condition, wind, and so forth that can be recorded at a particular place. Climate, on the other hand, is the customary average pattern of weather for a specific locality. The weather that affects other living things may be quite different from what we perceive as the weather in that it may be substantially modified by the habitat in which each organism lives. An insect that lives several inches beneath the surface of the soil obviously does not experience the temperature and wind factors reported by a television weather broadcaster. It is possible, however, to obtain a correlation between the weather in an insect's habitat and the "standard" weather we experience. For many practical purposes, this synoptic approach is just as useful and certainly easier than taking an elaborate set of measurements in a specific insectan habitat (Wellington, 1957). The main elements of weather are temperature, moisture, wind, and light. These will be considered in terms of the major way each influences insects in general.

TEMPERATURE. Each species of insect has a range of temperature within which it can survive. This is called the **zone of effective temperature** or the **tolerable zone.** Within the tolerable zone, there are different optimal temperature ranges for a variety of vital functions. For example, a species' tolerable zone may lie between 5° and 45°C. However, most individuals may only crawl between 8° and 40°C or fly between 18° and 36°C, as illustrated in Figure 17-3. Different developmental stages may display a tolerable zone with different limits and different functional zones within it. Exposure to a

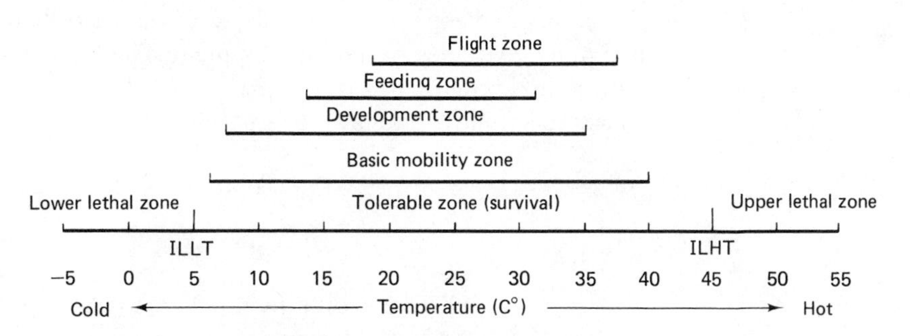


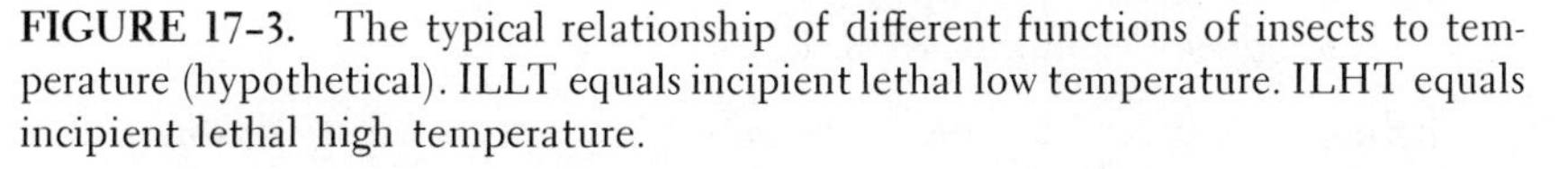

FIGURE 17-3. The typical relationship of different functions of insects to temperature (hypothetical). ILLT equals incipient lethal low temperature. ILHT equals incipient lethal high temperature.

temperature outside the tolerable zone will usually result in death if it persists for a long enough time; the further beyond the upper or lower limits, the shorter the duration of exposure that will cause death.

Obviously, the actual temperatures that limit the tolerable zone vary from species to species, but there are some finite limits that apply to all insects. No insect is known that can survive temperatures in excess of 63°C and most insects have an upper temperature tolerance between 40° and 50°C. The temperature at which an insect freezes depends upon how much of the body water is absorbed on colloidal particles (bound water) and how much is free water. Bound water may freeze as isolated crystals, surrounded by colloidal material that insulates it from other small drops of water and therefore delays tha wave of crystallization that causes death. The absolute minimum temperature tolerated by any insect is not known but is almost certainly below −30°C.

Insects generally seem to be separable into groups characterized by their response to low temperature. Some cannot survive for long if the temperature drops below the lower threshold for development. In other words, development must proceed or the individual dies. Other species can become dormant at low temperatures; activity and development cease but begin again as soon as the temperature exceeds the activity or developmental thresholds. A third group that usually overwinters in a particular stage adapts physiologically and can survive periods of extreme cold. Such insects undergo cold-hardiness development and remain dormant in a condition known as **diapause** until the danger of severe cold has passed. Other insects can adapt physiologically so that they can survive several months of hot dry weather in a dormant state called summer diapause.

Insects become acclimatized to the temperature under which they develop and live. Consequently, in the northern hemisphere a population of a species at the northern limits of its range may have a tolerable zone lower on the temperature scale than members of a southern population. Acclimatization to high temperature usually raises both the incipient lethal high and low temperatures, and vice versa. Such adjustment thereby improves the fitness of a species to different environments within its range.

In addition to establishing the physical limits of the range that a species can potentially occupy, temperature also regulates the rate of all other functions of cold-blooded animals. As metabolic processes are temperature-dependent, temperature exercises control over insect development, as noted in Figure 17-4. Although development will proceed throughout the entire developmental zone, it will usually occur more slowly toward both the lower and upper extremes. When development is plotted against temperature, the data points roughly describe a hyperbola, the exact form of which can be calculated from the following equation:

$$Y = \frac{K}{t - d}$$

in which

$Y$ = the time required to complete development
$t$ = the temperature
$K$ and $d$ = empirical constants

The curve described by this equation usually does not pass through all of

**FIGURE 17-4.** Graphical representation of the hypothetical effect of temperature on insect development. The descent of the curve indicates that development from egg to adult takes less time as the temperature increases.

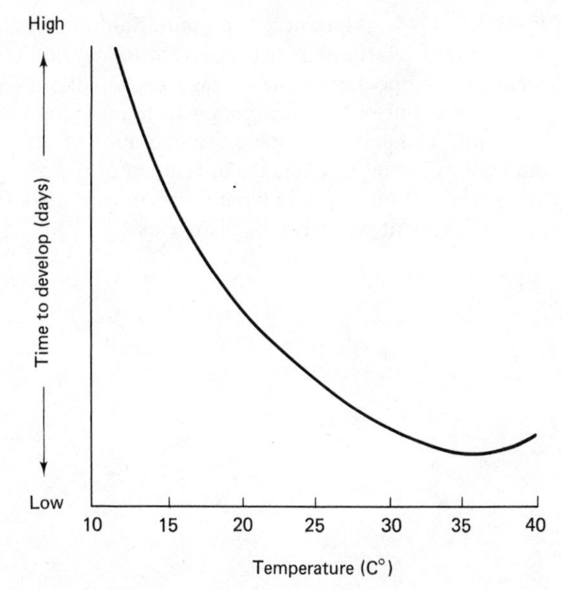


the observed data points and can be difficult to fit to raw data. A reciprocal of the preceding curve can be plotted from the linear equation:

$$\frac{1}{Y} = k + bt$$

in which $Y$ and $t$ are the same as in the previous equation, $b$ is the slope of the line, and $k$ is a new constant equal to $bd$. This equation describes that portion of development which occurs per unit of time at a given temperature. When $\frac{1}{Y}$ is zero, $d$ can be shown to equal $t$ and theoretically represents the lower temperature threshold for development. For general purposes, the developmental threshold is a reasonably good approximation, but the absolute threshold would normally be found to be lower if more data points were obtained at lower temperatures. Likewise, there is usually a decline in the developmental rate at higher temperatures (Figure 17-5).

These equations describe the pattern of development that results from the effect of a constant temperature on the organism's complex set of metabolic processes. This raises the question of what happens when insects are subjected to diurnal temperature fluctuations as so many are. Numerous studies have been conducted on the effect of constant versus fluctuating temperatures, but the results do not support a generality. Some workers have found that fluctuating temperatures either increase or decrease the rate of development compared to that which occurs at a constant mean equivalent temperature (for example, Huffaker, 1944; Matheson and Decker, 1965). Others have found fluctuating and equivalent constant temperatures to produce basically the same rate of development (for example, Howe, 1967). Insects live in a wide variety of habitats and must be acclimatized to the conditions that prevail there, so perhaps different responses to temperature should be expected. For example, species that live in the soil or fresh water are well buffered against diurnal temperature changes, and we might expect them to develop most rapidly under the constant temperature regimes that characterize their habitat. To these species regular fluctuations in temperature could

FIGURE 17-5. Graphical representation of the hypothetical relationship between rate of development and temperature. The average percent development per unit of time increases with temperature. The broken straight line drawn through most of the data points in the middle range of temperature gives an approximation of the low temperature threshold for development when extended to $x$ axis.

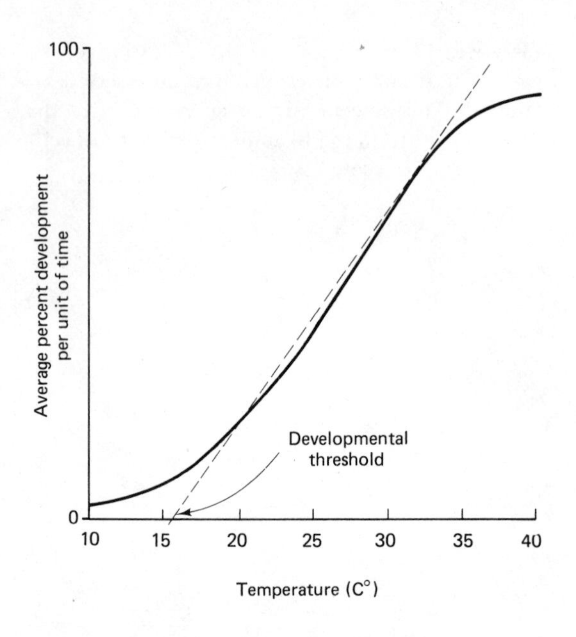


well be so unnatural as to represent a form of environmental stress that would impede normal development.

We must remember, however, that the temperature of an insect's body is not necessarily the same as that of the surrounding air. Insects do have some capacity for altering their body temperature and thereby affecting the rate of their metabolism. Dark insects absorb more solar radiation than light colored ones, and hairy insects cool more slowly than naked ones. Some insects enhance the warming or cooling of their bodies by orienting in such a way as to maximize or minimize the effect of solar radiation, much as lizards do. The heat generated by working muscles may also raise the body temperature; the thoracic temperature of flying insects always exceeds the ambient temperature. Similarly, insects can utilize evaporational cooling, especially if they are producing a surplus of water metabolically as may be the case during flight.

From a population point of view, temperature has a profound effect. Its regulation of metabolism and development determines the basic life history pattern of many species. Species that have only one generation a year are either directly limited by temperature or indirectly by the effects of temperature on other organisms they depend on for food or shelter. For those species which do not have their life history synchronized to the life history of another organism, the number of generations per year will be limited largely by temperature, and the number of generations per year can greatly influence the population late in the season as well as the population at the start of the next season.

If a careful study is made to determine the developmental zero and the rate of development under a series of constant and fluctuating temperatures, it is possible to estimate the total heat required to complete one generation. The units used are usually **degree-hours** or **degree-days;** each such unit is the difference between the established developmental zero and the actual temperature multiplied by time. For example, if the developmental zero of a species was found to be 5°C and it took 21 days for newly laid eggs to become adults at a constant temperature of 25°C, the heat requirement for one generation would be 21 (25 − 5) or 420 degree-days. From this value it is possible to predict in a general way how many generations are likely to occur during the season (i.e., between dormant periods if they exist).

Temperature also affects behavior, as reviewed in Chapter 11. Insects may remain totally inactive at both high and low temperatures or actively move along a temperature gradient until a preferred zone is encountered. From a population standpoint, the influence of temperature on dispersal, mating, and reproduction is of great importance. If conditions are adverse for dispersal, local populations neither increase because of influx nor decline because of exodus. If temperatures that are not suitable for mating prevail for several days, some adults may die without leaving offspring and others may become less fertile due to age.

MOISTURE. Insects occupy habitats that vary widely in their moisture characteristics from the maximum wetness of fresh-water ecosystems to the relative dryness of saltwater ponds and the almost absolute dryness of deserts and stored grains or flour. Water is essential to life and its conservation is extremely important to small terrestrial organisms, as was discussed at some length in earlier chapters.

Environmental moisture occurs as precipitation (rain, snow, hail, dew), water vapor in the air (**humidity**), surface water, and the water content of an organism's food (including metabolic water). When moisture is considered as a component of weather, it is usually recorded as precipitation or humidity. Humidity refers to the amount of water vapor in the air as influenced by temperature and atmospheric pressure. It is usually stated in terms of the amount of water vapor present as a percentage of the total water holding capacity of the existing air mass (**relative humidity**). When the temperature and pressure of the air change, the relative humidity changes even though the absolute amount of water does not. For example, as an air mass cools, its capacity to hold water vapor declines, so the relative humidity rises; when it cannot hold any more moisture than already present, we say it has reached saturation. If the air cools further, some of the water will condense and leave the air as precipitation. Dew is a common form of precipitation that occurs when air comes in contact with a cold surface and is cooled beyond its saturation point (**dew point**).

All forms of environmental moisture influence the water balance of terrestrial insects. The water content and requirements vary markedly from species to species and between life-history stages. Active stages generally contain more water than dormant stages, and more of the insect's body water is in a free state.

There are, of course, indirect effects of moisture on insect populations as a result of its influence on plant growth. Many insects' life histories are synchronized to the seasonal availability of resources, the temporal distribution of which are regulated by moisture. Under the climate of southern California, for example, many insect herbivores are dormant during the summer when there is not enough moisture to support plant growth, and if such conditions prevail longer than normal, an increase in herbivore mortality may occur. However, like temperature, moisture conditions vary greatly from insect habitat to insect habitat even within a rather limited area, but unlike temperature, moisture conditions can change very rapidly.

Precipitation can act as a direct cause of mortality. Insect eggs and small larvae can be permanently washed from their host plants by heavy rain. Rain may also saturate the soil and drown insects that are unable to escape. Many insects cease feeding during periods of precipitation and may seek refuges in which to pass a rainy period. A protracted period of rain, particularly when the temperature is suitable for development, may lengthen the time required to complete development or cause mortality by starvation.

The moisture content of an insect's habitat may determine directly whether or not an individual survives. Terrestrial species drown when totally emersed in water and most cannot survive complete dryness. It is much more difficult, however, to define a tolerable zone for moisture than for temperature. Relative humidity is particularly difficult to evaluate because it changes so markedly with time and space. Saturation may lead to indirect death by drowning, whereas under very dry conditions, death is usually the direct result of excessive water loss. As the moisture stress most commonly encountered by terrestrial insects is related to the problem of water conservation, **evaporation rate** may be a more meaningful measurement than relative humidity. Wellington (1949) provided support for this contention when he discovered that caterpillars of the spruce budworm congregated in the same zone of an evaporation gradient under different conditions of temperature and humidity. However, relative humidity has remained the popular measure of moisture, probably because it can be recorded with standard devices by a rather simple procedure.

Laboratory studies have demonstrated the effect of humidity on a variety of insect life-functions. But as is the case with temperature, the influence of moisture on a particular species depends on the physical characteristics of the habitat to which it is adapted. Andrewartha and Birch (1954) presented two sets of hypothetical curves summarizing the effects of moisture on a series of life functions that have a marked influence on insect populations. These curves are reproduced in Figure 17-6 as a substitute for a sampling of data from a number of unrelated species.

The humidity in an insect's habitat may have some indirect effects, as were noted for precipitation. In particular, relative humidity can substantially affect natural controls. Some parasites do not search for hosts or oviposit in them if the relative humidity is either low or high. The susceptability of insects to fungal, bacterial, and viral diseases also changes with environmental moisture. Moist conditions seem to facilitate the spread of some insect pathogens and may also affect their survival and virulence. Entomophagous fungi typically invade the soft tissues of insects by penetrating the cutical rather than by gaining entry through the digestive tract. Normally, the spores require warm, moist conditions to germinate and successfully invade their host. These environmental requirements have impeded the use of fungi in applied biological control (see Chapter 25). Humidity may also have a pronounced effect on interacting predator and prey populations. For example, some aphid populations thrive under moist conditions and benefit further from a decline in predation by lady beetles (Coccinellidae), which feed more actively when it is warmer and drier.

LIGHT. Unlike temperature and moisture, light is rarely, if ever, directly lethal under natural conditions. However, it exercises a great deal of control over processes directly related to survival. Light intensity greatly influences insect behavior, and the response of different species to light serves to partition the use of various resources. Many species are active during the

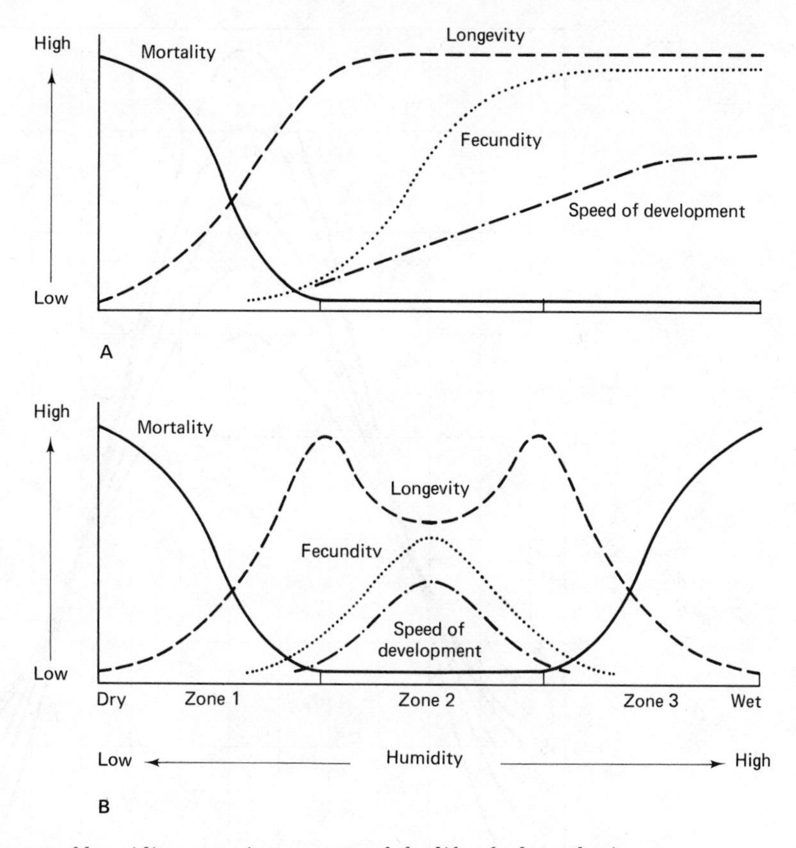


**FIGURE 17-6.** Influence of humidity on various aspects of the life of a hypothetical insect. **A.** An insect species not harmed by high humidity, **B.** An insect adversely affected by high humidity. Zone 1, lethal dryness; Zone 2, favorable moisture level; Zone 3, lethal wetness (species B). (Redrawn from Andrewartha and Birch, 1954.)

hours of full sunlight but remain quiescent at night when the nocturnal species assume their active routine. Other species, often called **crepuscular,** are active during the faint light that accompanies the dawn and dusk.

The most important characteristic of natural light is its periodicity. Because of the rotation of the earth on its tilted axis and its motion around the sun, the length of the light period for any date at any place on the globe can be predicted, as shown diagrammatically in Figure 17-7. This regularity provides insects and other organisms with an annual clock which can be used to accurately time events in their lives. The survival value of having growth and reproduction synchronized with the occurrence of favorable periods is obvious. The day-to-day duration of the light period provides a major environmental cue that is not subject to variation. Thus, many organisms have evolved responses to the pattern of light and darkness such that they can prepare physiologically for the onset of unfavorable conditions well in advance of their occurrence. For example, winter temperatures vary greatly from year to year. Insects that are not preadapted for overwintering may well be killed by an early cold snap. However, if they enter diapause in response to the photoperiod that safely predates unfavorable conditions, such as a period of lethal temperatures, their chance of survival is greatly enhanced.

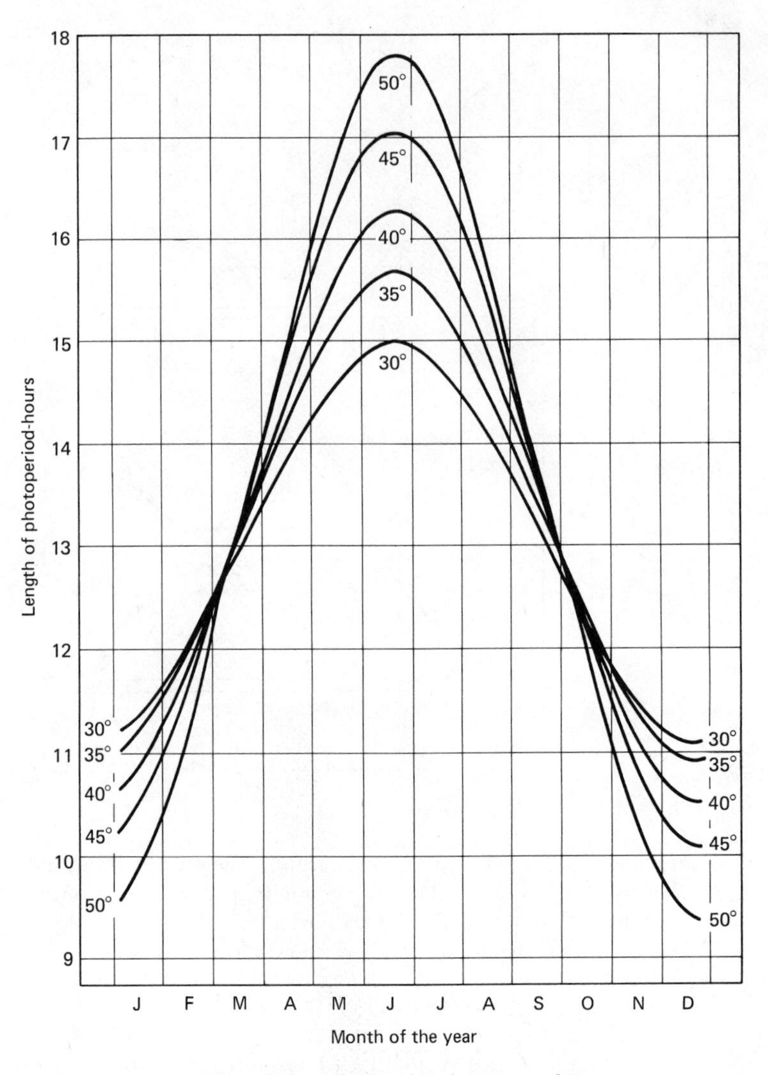


FIGURE 17-7. Diagrammatic representation of the relationship between photoperiod, date, and latitude.

WIND. The differential heating of the surface of the oceans and land masses affects the temperature of the overlying atmosphere such that adjacent air masses develop different temperature, moisture, and pressure characteristics and are thereby set in motion. This results in two basic types of air currents, the familiar and more or less horizontal currents we call wind and the vertically moving **convection currents.**

Air currents are of great importance to insect displacement and therefore affect population changes by influencing the numbers moving into or out of an area. Most insects will not take flight when the speed of the wind exceeds their normal flight speed, simply because a loss control over the direction of movement has been selected against in the course of evolution. Insect migration and dispersal and the adaptive behaviors that enable insects to exploit the wind as an external source of energy were discussed in considerable detail in Chapter 12.

***Food*** The wide range of foods that can be utilized by insects in general and the numerous mouth part adaptations that insects have evolved were discussed in some detail in Chapter 6. The food relationships that are displayed by a given species may be generalized or highly specific, depending upon the selective pressures that have molded the behavior of food location and acquisition. When the food is another living organism, the relationship that exists represents the current status of the coevolution of the interacting species. The coevolution of plants and herbivores, and predators and prey is discussed in Chapter 18. At a given time, all the factors that have an impact on a species' population, including survival, longevity, reproduction, rate of development, and dispersal, are affected by the quantity and/or quality of food available. Quality and quantity are not really separable characteristics as far as the food effects on a population are concerned. Food of high nutrient value cannot make up for an absolute shortage, and a vast quantity may not be a substitute for poor quality.

Insect growth and development are greatly affected by food quality. House fly larvae reared on manure were found to complete development in two to three weeks, whereas those reared at the same temperature on bananas required about four weeks (Folsom and Wardle, 1934). Furthermore, the complete development of fly larvae takes longer in dry manure than in moist manure. Many insects that consume foods such as wood, from which nutrients are chemically expensive to extract, tend to have protracted larval development. In most species, poorly nourished larvae become small, less fecund adults, and in some cases this can influence reproduction markedly. Several studies have demonstrated that flies of various species do not produce a normal complement of eggs when raised on a diet deficient in protein (Monroe and Lamb, 1968). Bee larvae fed entirely on royal jelly become reproductives called queens, whereas larvae fed on a combination of royal jelly, pollen, and nectar become sterile female workers. The workers are not poorly nourished, but the difference in diet retards the development of their reproductive organs.

The quantitative relationships between insects and their food supply can be grouped into three basic patterns: (1) Food is usually abundant, and the insect populations associated with it are usually limited by some other environmental component. This is the type of relationship that exists between most insect herbivores and their plant hosts. For these insects, food is of minimal importance as a limiting factor under normal conditions. (2) The food available is such that the insects which utilize it have the capacity to reproduce to the limit of their food supply. This may occur among predators and prey. In many such cases, the prey may not all be consumed, but they become so scarce that the predators are unable to expend the energy required to locate them. In such a case, food becomes an important limiting factor. (3) Food can be abundant but distributed in such a way as to be not readily accessible. This has been referred to as a **relative shortage of food.** For example, some insects, such as the tsetse fly that feed on the blood of ungulates preparatory to egg formation, disperse and oviposit between feedings. A low host population may reduce the number of offspring produced not because there is a shortage of blood but because of the way it is distributed.

Among the more important factors that affect the quantity of food available to individuals of a species are the developmental stage and number of individuals per unit of food, the number of species utilizing the same food resource, and whether or not feeding influences the food in some way. A large population of young maggots in a piece of meat may appear to have an abundance of food, but as the maggots increase in size, the food may limit the number able to complete development. When two or more species utilize the same food, the one which can develop most rapidly has a better chance to complete development before the food is depleted. The removal of plant products by sap feeders may actually prolong active photosynthesis to the benefit of other herbivores; conversely, leaf-feeders can reduce the level of photosynthesis and adversely affect the food relationship of sap-feeding species. Likewise, the annoyance caused by one species of the biting fly may stimulate avoidance behavior on the part of the host which could adversely affect the feeding of a second species.

Food shortages usually occur as a mosaic pattern over the total range of a population or species. Consequently, some groups of individuals may be more or less eliminated because of a shortage of food, but the population or species survives in areas of adequacy or abundance. Nevertheless, there is an over-all effect on the total population. In patches where insects are not able to survive, the food supply will be released from feeding pressure and gradually increase. These areas may then be repopulated by individuals from crowded areas. Thus, food greatly influences population movement as well as numbers.

In terms of insect populations, changes from generation to generation or from year to year are both influenced by food supply and superimposed over it. The consumption of food by one generation of many species does not necessarily affect the quantity or quality of food available to the next generation. For example, herbivores that consume the leaves of a tree one year do not necessarily reduce the amount of foliage available one year later. Likewise, scavengers and carrion feeders utilize foods that are only temporarily available, and whether or not they are consumed does not affect the food supply for the next generation. On the other hand, the food consumed by one generation of some species greatly affects that which will be available to their offspring. For example, a large population of parasites can substantially reduce a host population, thereby leaving the food for the next generation in short supply. The population of those species that have different food requirements as immatures than as adults is usually limited by whichever of the foods is in the shortest supply relative to need.

### *Other Organisms*

The chance that an organism will survive and leave offspring can be influenced greatly by the kind and the number of other organisms with which it is associated. In that insects as a group feed on virtually every species of plant and most species of animals, they are influenced by a great number of other organisms, which are best considered under the food component of environment. As a nonfood component, we are concerned with those other organisms that act either as various types of symbionts (including mutuals, commensals, parasites, predators, and pathogens) which enhance or impede survival or act as competitors for available resources. Consequently, the other organisms may be either of the same kind as the subject species or of a different kind.

The population effects of organisms of the same kind may result from too few individuals (underpopulation) or too many individuals (overpopulation or crowding). When a population is very low, the main problem is one of mate location. However, the chances of parasites and predators finding scattered individuals are also greatly reduced, so the probability of survival for individual hosts or prey may be improved at low population levels.

The problems of overpopulation are somewhat different, and some of them are rather poorly understood. If we think of overpopulation as simply a case of competition for some resource, we are really dealing with a shortage of the resource relative to the population exploiting it. Consequently, the limitation is one of too little food or too few places to live rather than too many individuals. There are, however, cases in which crowding affects the population dynamics of an insect before there is evidence of a resource shortage effect. These may be observed as either changes in behavior or changes in reproduction and survival. Some insects, such as locusts, are stimulated to migrate by the presence of numerous siblings. Other insects, when crowded, condition their environment in such a way as to reduce their own reproductive rate. For example, high populations of flour beetles, *Tribolium* spp, seem to contaminate the flour in which they live and as a result do not live as long or lay as many eggs as they do in unconditioned flour.

When two or more species share a resource, they must compete with each other for its use unless it is so abundant that it is not a limiting factor. Consequently, the populations of the interacting species are not as large as they would be if either was the sole user of the resource. When abundant resources are shared, the species involved may not be limited by the amount available but by their access to it. For example, several insect herbivores may feed on the foliage of the same species of plant with no danger of running short of food. However, they may greatly interfere with each other if they were to actively feed at the same place at the same time. Evolution normally works to ensure that species groups which use the same resources apportion the resource spatially, temporally, or both. One species may feed during the morning, the second during twilight, and the third at night; one may feed on the upper surface of the leaves, another on the lower surface, and so on.

Pathogens such as fungi, bacteria, and viruses present a somewhat special class of "other organisms." Although these microbes often extract nutrients from and develop at the expense of their insect hosts, we do not think of pathogens as having a food type of relationship with their host organisms as we do when we consider a predator and its prey or a parasite and its host. Yet, in reality, it is similar in that the pathogen population may increase at the expense of the insect population, which often declines. Insect pathogens are considered in more detail in Chapter 25 as an element of biological control.

### *Habitat*

An organism's habitat is simply the place in which it lives. The habitat of some species may be very general, such as the foliage matrix of a forest, but for others, it can be highly specific. Insects evolve habitat requirements in response to various selective pressures, just as they evolve food relationships and tolerances for physical factors. A species' habitat is an important environmental component because it influences the impact of other components such as weather and natural enemies on its chance to survive and reproduce. Many microhabitats exploited by insects are characterized by microclimates much less severe than the regional climate. Consequently, each species is adapted to the conditions of its particular habitat and tends to be found occupying the same kind of living place throughout its range, as any collector soon learns.

The extent of the habitat or the number of places suitable for certain functions such as resting, mating, or ovipositing can limit the size of the population that can survive in a given area. For example, prior to their management for pollination (see Chapter 19), populations of the leaf cutter bee *Megachile rotundata* were severely limited by the lack of natural nesting sites. For many insects, living places are abundant but vary greatly in quality, with the most suitable distributed rather unevenly over the landscape. When the environment is generally favorable, that is, the weather is clement, food is abundant, and the pressure from natural enemies is low, all the available living places may be suitable. But when the environment is in some way unfavorable, those places that are less adversely affected or favorably modify the adverse environmental component the most serve as refugia. Only the individuals in these refugia are able to survive, which means that the over-all population declines. Wellington (1965) provided one of the clearest examples of the importance of quality living places through his population studies on the western tent caterpillar. In those years characterized by favorable weather (generally warm and sunny), tents could be found throughout the study area, but in cool, cloudy years, the tents were almost totally restricted to those areas which received the most sun because of the effect of the local topography on cloud patterns (Figure 17-8) even though the larvae construct protective tents.

There are a number of reasons why an insect may be absent from a specific habitat. These include: (1) a lack of physical or biological requisites such as water or food, (2) the unsuitability of the weather components, (3) inacces-

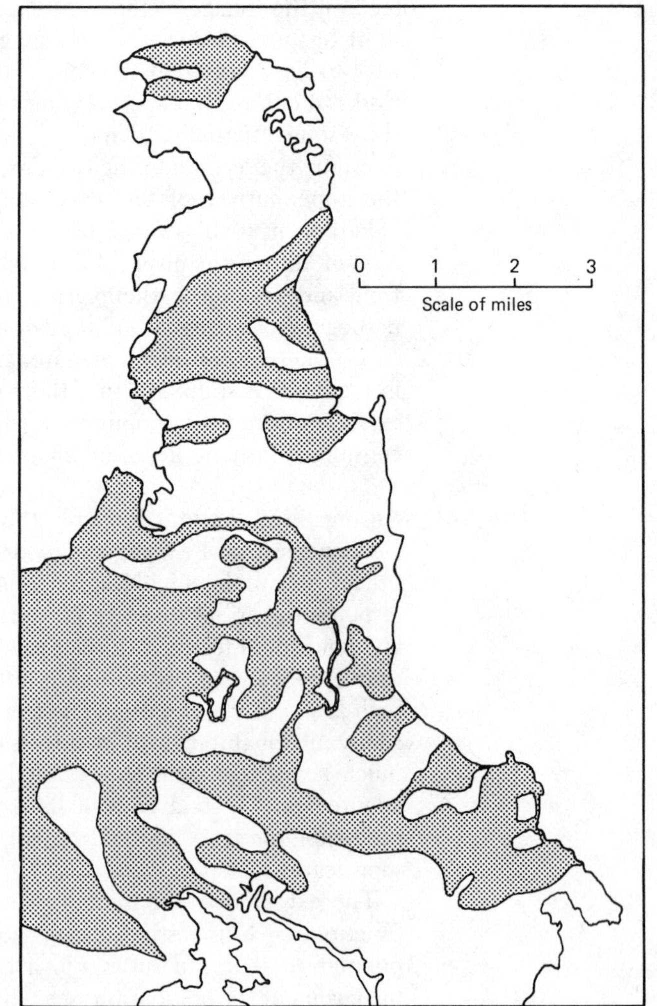


**FIGURE 17-8.** The distribution of sunny and cloudy habitats within the area of Wellington's study on the population behavior of the western tent caterpillar. Shaded areas indicate those parts of the study site frequently covered by diurnal clouds during March, April, and May when the tent caterpillar larvae are developing. (Reproduced with permission of W. G. Wellington and The Canadian Entomologist.)

sibility, as in the case of geographic isolation, and (4) unfavorable interactions with other species. Often we observe that insect species are absent from habitats that contain appropriate physical and biological requisites, have suitable weather, and are clearly accessible. This has led numerous biologists to conclude that a species can be excluded by the presence of another species which has basically the same requirements. The idea of **coexistence** or its antithesis, **competitive displacement**, known also as **Gause's Law**, implies that two species having the same ecological prerequisites for survival (**ecological homologues**) cannot live together in the same habitat indefinitely. Eventually, the one that is most effective in leaving fertile progeny will survive, and the other species will be eliminated. The reason for this is that since the two species are homologues, their numbers must be added together in terms of the holding capacity of the habitat.

The theory of competitive displacement has been surrounded by some controversy in spite of the fact that there are well-documented cases of one species becoming established in an area at the expense of an earlier occupant. For example, after its introduction into South America, the African variety of *Apis mellifera* spread rapidly and in some areas has completely displaced the domesticated honeybee. The biological control program directed against the California red scale produced a well-documented case of competitive displacement under applied circumstances. *Aphytis chrysomphali*, a parasite of the scale, was well established over much of the citrus-growing acreage of California. *A. lingnanensis*, imported in 1947–48, replaced the original species in all but some coastal pockets. *A. melinus* was introduced still later and soon displaced *A. lingnanensis* in the warmer interior areas (DeBach, 1974).

The composite survival value of an insect's environment is usually not the same over its entire range. In some areas, particularly those near the periphery of the range, conditions are suboptimal. In these marginal areas, a species will face a variety of difficulties and frequently leave a lower average number of offspring than in areas where the environment is optimal. The less favorable the conditions in an area are for a particular species, the more likely it is to be displaced there by another species. In this manner the total environment is divided among those species best suited to survive in each place.

## *Population Regulation*

A consideration of an organism's environment in terms of the foregoing basic components greatly simplifies the discussion of the way each component affects population density, but it tends to oversimplify the interpretation of natural situations. In actuality, the components interact with each other as well as with each subject species. For example, weather constantly exerts an influence on the food supply of insects, even though it may not be acting directly upon them in a limiting way. Birch and Andrewartha (1941) described one situation in which a species of Australian grasshopper was so abundant during moist years that widespread predation by birds had little effect on its population. However, during drought years, most of the grasshoppers were unable to complete their development because the grass on which they fed dried up too soon. Those that survived did so in moist refugia where the grass remained green, but these survivors were heavily predated by the birds that concentrated in these areas to feed. In a situation such as this, one is faced with the question as to whether an ensuing sharp decline in the grasshopper population was due to the weather, a shortage of food, or the avian predators.

In spite of these rather obvious interactions, population ecologists, many of whom were or are entomologists, have argued at some length about how animal populations are regulated. Some contend they are regulated primarily by factors such as the weather that act independently of population density (**density-independent factors**), while others argue that factors such as competition or predators that kill a variable number in accordance with population density (**density-dependent factors**) are most important. At times it has seemed that the argument was more important to some workers than the discovery of the truth about population regulation. But gradually there emerged some compromise theories that provide a more unified and realistic interpretation of population changes in nature.

One of the unifying approaches that has gained considerable popularity is the **Life-System Concept** of Clark, Geier, Hughs, and Morris (1967). They define a life system as . . . "that part of an ecosystem which determines the existence, abundance, and evolution of a particular population." This could be restated as the population plus those aspects of its environment that influence its size. The components of a typical life system and the basic form of the interactions that occur are presented in Figure 17-9.

The interaction of a subject population and its effective environment set in motion a variety of observable events, which Clark and others distinguish as being either primary or secondary. Primary events are those with basic demographic characteristics such as birth, death, and the movement of individuals. Secondary events are those which influence the magnitude, duration, or frequency of primary events, as would be anticipated from our earlier discussion. As examples, weather may directly influence the survival of individuals, or the size or quality of the food supply may influence whether survivors remain in an area or leave. These observable events involve ecologi-

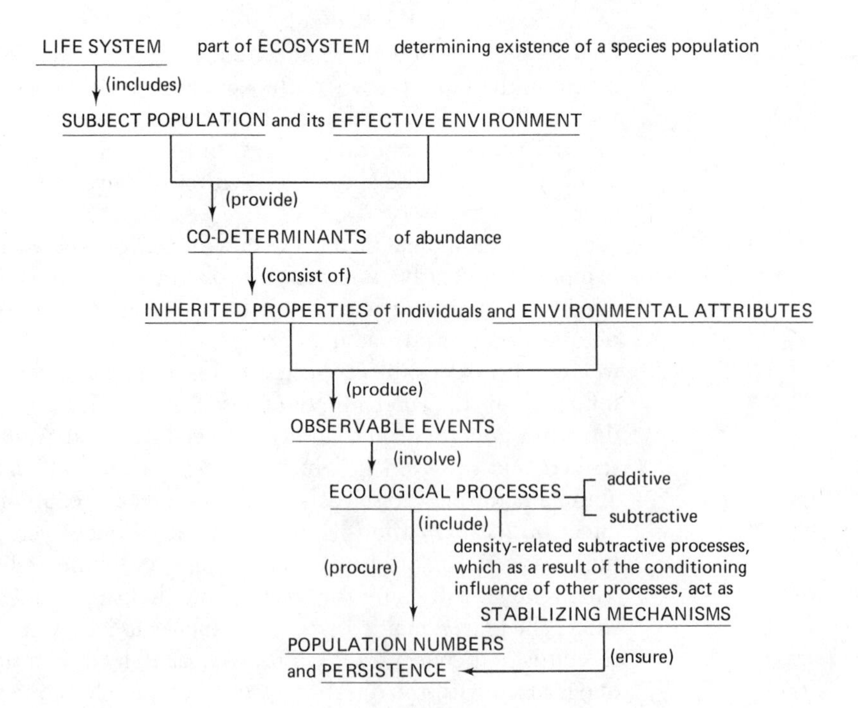


**FIGURE 17-9.** The components of a life system. (Redrawn from Clark and others, 1967.)

cal processes which acting alone would cause the population to either increase (additive) or decrease (subtractive).

Additive ecological processes are those which promote reproduction and influx. They include the provision of resources, the enhancement of mating opportunities and the production of viable offspring, the immigration of individuals, and the suppression of destructive forces. Subtractive processes are those which increase mortality, limit the production of offspring, and promote exodus; these processes include intraspecific competition, cannibalism, predation, parasitism, and indirect effects of a shortage of required resources.

The additive and subtractive ecological processes act in such a way that, barring a catastrophe, the species will survive, even though its population changes in relation to the relaxation or the strengthening of environmental constraints. Each species undergoes short-term numerical changes, which can be described by a simple mathematical expression based on broadly applicable generalizations. During a short interval of time, the numerical change in a population can be defined by the following equation:

$$N_t = N_o e^{(b-d)t} - E_t + I_t$$

in which

$t$ = the short interval of time
$N_t$ = the size of the population after time interval $t$
$N_o$ = the size of the population at the start of time interval $t$
$e$ = a constant equal to the base of Naperian logs
$b$ = the number of births during time interval $t$
$d$ = the number of deaths during time interval $t$
$E_t$ = the number of individuals that left during time interval $t$
$I_t$ = the number of individuals that entered during time interval $t$

Obviously then, even short-term changes in insect populations are greatly influenced by all of the environmental components we have discussed. The size of the population at the beginning of any time interval represents the history of the interaction of the codeterminants of abundance up to that point. The number of births from that time onward would be determined by the inherited reproductive potential and the normal sex ratio of the species as modified by the environment under which the population developed. The number of deaths that will occur is influenced by the age and the vitality of the population, the vigor of new offspring, the suitability of the weather, and the abundance of food, shelter, competitors, and enemies. Some of the major factors that lead to population increase because of influx or population decline from exodus are those which influence dispersal and migration, as discussed in Chapter 12.

Over long periods of time, populations fluctuate much less predictably than they do over a short term, and it is almost impossible to make any generalizations about which are the most important environmental factors involved. As mentioned at the beginning of this chapter, different species of insects display different long-term patterns of population change. Yet, it is over the long term that we are usually most concerned about insect population fluctuations. Regardless of whether a species of insect is beneficial or harmful, the benefit or damage it does is related to its population level. As we will see in the chapters that follow, man is constantly searching for ways to increase the size of populations of beneficial species and to suppress the populations of pests. Much research has been conducted in an effort to learn what causes insect populations to rise to pest status; the results have been varied, as one might expect.

When an investigator sets out to study a natural insect population, he can rarely count and keep track of every individual. As a result, he must rely on a sampling program designed in such a way that he is confident the small number of individuals and what happens to them is representative of the entire population. Obviously, the sampling method varies from species to species, depending on the problems posed by different types of habitats or the kinds of information that are being sought. Some sampling programs, especially those for sedentary stages like eggs, can be very simple, whereas others require a high level of sophistication. Population studies are normally conducted over several generations in an attempt to identify better those features of the environment which cause undesirable changes in numbers. Some mortality factors are easily identified, but they are not necessarily important in terms of population fluctuations. Factors that kill a more or less constant portion of a population do not induce numerical fluctuations, but those that kill a highly variable portion of the population often produce marked fluctuations. It is the latter type of mortality factor that must be identified, but this may be difficult if the absolute number of individuals killed is small.

One of the most widely used tools in population studies is called a **life-table.** A life-table is a systematic tabulation of the survival and causes of mortality that occur in a population throughout a period of time. The general scheme is to follow the population through a complete generation in order to obtain age or time specific information on the ecological processes that lead to mortality. Once again, the sampling plan that is used depends upon the objectives of the study and the basic biology of the subject species. The literature on population sampling techniques is extensive but has been reviewed by several authors (see Southwood, 1966; Harcourt, 1969).

According to Harcourt, the most commonly used headings in life-tables are as follows:

$x$ = the age interval (stage of development)
$l_x$ = number living at beginning of the age interval
$d_x$ = number that died within the age interval
$d_xF$ = the mortality factor responsible for $d_x$
$100qx$ = percentage mortality within the age interval
$S_x$ = survival level within the age interval

Additional columns can be included for specific purposes. A life-table based on hypothetical data and following the scheme recommended by Harcourt is presented in Table 17-1.

In this example, the sex ratio was considered as 1 : 1; if it departed in favor of males, it would be considered as mortality. The number of females is multiplied by two to maintain balance in the tabulated data. The mortality shown as due to fecundity reduction is obtained from the difference between the maximum eggs laid by females of the species and the average egg production observed. In this case, the maximum egg complement for the species was 120, whereas the observed average number of eggs laid was 72, or 40% lower. This has the effect of reducing to 240 the number of females that lay the maximum number of eggs. The number of eggs which begin the next

**TABLE 17-1** **Hypothetical Life-Table**

| $x$ | $lx$ | $dxF$ | $dx$ | $100qx$ | $Sx$ |
|---|---|---|---|---|---|
| Eggs ($N_1$) | 4,000 | Parasitism | 1,000 | 25 | 0.75 |
| Larvae | | | | | |
| Instar 1 | 3,000 | Weather | 1,500 | 50 | 0.50 |
| Instar 2 | 1,500 | Parasitism | 500 | 33 | 0.67 |
| Instar 3-5 | 1,000 | Predation | 500 | 50 | 0.50 |
| Pupae | 500 | Developmental | 100 | 20 | 0.80 |
| Adults | 400 | Sex (0.5) | — | — | 1.00 |
| Females × 2 | 400 | Fecundity Reduction | 160 | 40 | 0.60 |
| Normal Females × 2 | 240 | Adult Emigration | 180 | 75 | 0.25 |

$$\text{Expected eggs} = \frac{240 \times 120}{2} = 14{,}400$$

$$\text{Actual eggs laid} = \frac{60 \times 120}{2} = 3{,}600$$

$$\text{Trend Index (I)} = \frac{N_2}{N_1} = \frac{3{,}600}{4{,}000} = 0.90$$

generation ($N_2$) would be determined by the number of females left after emigration.

The trend index, $I$, gives an indication of whether the population is increasing or decreasing. When the index value is less than 1.0 as it is in this example, it indicates that the trend of the population is downward. Different workers base the trend index on different stages. The egg stage seems to be most widely used, but some workers insist it should be based on the reproductive stage. Economic entomologists usually determine the trend index from the stage that causes the maximum amount of damage in one generation to the same stage in the next generation. To really be useful, however, the data for life-tables spanning several generations or years must be combined.

As mentioned earlier, it is not necessarily the factor that causes the most mortality that produces fluctuations in a population. Consequently, it is necessary to identify all of the mortality factors operative throughout the life-cycle and to determine which one has the most impact on the population in the next generation; this mortality agent is called a **key factor.** The identification of key factors is handled by various kinds of data manipulation and analysis which need not be discussed here. The importance of the approach lies in the fact that once a key factor has been identified, it is possible to concentrate the investigative effort at critical periods when the factor or factors that will determine the population trend are both operative and measurable.

Our understanding of insect population dynamics has increased tremendously in recent years. Although we will not be able to accurately predict populations until we can make long-range predictions of environmental conditions, particularly weather, we can make trend projections with a reasonable level of confidence. But the real value of understanding population changes lies not in prediction but in management decision-making. Unfortunately, pest control decisions have been based largely on numerical data, and the predominant approach to control has been to induce an indiscriminate increase in mortality, often with somewhat limited effectiveness. This preoccupation with numbers needs to be changed. As Wellington (1977) so aptly stated, "For decades, a "population" has been viewed as a passive, monolithic lump of protoplasm, out of which chunks are carved by all the lethal agents assailing it. The comparative evolutionary or practical values of any agent have been judged primarily by the size of the chunk it removes. Neither in theory nor in practice has there been much interest in the qualities of the residual lump. . . ."

The individuals that make up any population vary according to their developmental history, the quality and quantity of their food supply, and a number of other factors. Most populations display a considerable degree of genetic variability as well. The range of differences between the individuals of a population serve as an effective buffer against changes in the environment, as so beautifully illustrated by examples of industrial melanism (see Chapter 4). In spite of the efforts of Wellington and a few others, population theorists have not adequately taken into account the importance of changes in population quality. Until this is done our management of insect populations, beneficial and pestiferous, will be imperfect to say the least.

# References Cited


Andrewartha, H. G. and L. C. Birch. 1954. The distribution and abundance of animals. University of Chicago Press, Chicago, Ill.

Birch, L. C. and H. G. Andrewartha. 1941. The influence of weather on grasshopper plagues in South Australia. *J. Dep. Agric. S. Aust.* 45: 95–100.

Clark, L. R., P. W. Geier, R. D. Hughes, and R. F. Morris. 1967. The ecology of insect populations in theory and practice. Methuen & Co., Ltd., London.

DeBach, Paul. 1974. Biological control by natural enemies. Cambridge University Press, London and New York.

Folsom, J. W. and R. A. Wardle. 1934. Entomology, with special reference to its ecological aspects. 4th ed. McGraw-Hill Book Company, New York, N. Y.

Harcourt, D. G. 1969. The development and use of life tables in the study of natural insect populations. *Ann. Rev. Entomol.* 14: 175–196.

Howe, R. W. 1967. Temperature effects on embryonic development in insects. *Ann. Rev. Entomol.* 12: 15–42.

Huffaker, C. 1944. The temperature relations of the immature stages of the malarial mosquito *Aedes aegypti* with a comparison of the developmental powers of constant and variable temperature in insect metabolism. *Ann. Entomol. Soc. Am.* 37: 1–27.

Matheson, J. W. and G. C. Decker. 1965. Development of the European corn borer at controlled constant and variable temperatures. *J. Econ. Entomol.* 58: 344–349.

Southwood, T. R. E. 1966. Ecological methods with particular reference to the study of insect populations. Methuen & Co., Ltd., London.

Wellington, W. G. 1949. The effects of temperature and moisture upon the behaviour of the spruce budworm *Choristoneura fumiferana* (Clem.) (Lepidoptera: Tortricidae). I. The relative importance of graded temperature and rates of evaporation in producing aggregations of larvae. *Sci. Agricul.* 29: 201–215.

WELLINGTON, W. G. 1957. The synoptic approach to studies of insects and climate. *Ann. Rev. Entomol.* 2: 143-162.
WELLINGTON, W. G. 1965. The use of cloud patterns to outline areas with different climates during population studies. *Can. Entomol.* 97: 617-631.
WELLINGTON, W. G. 1977. Returning the insect to insect ecology. Some consequences for pest management. *Environ. Entomol.* 6: 1-8.

# General Reference

PRICE, PETER W. 1975. Insect ecology. John Wiley & Sons, Inc., New York, N. Y.

# Major ecological roles of insects 18

An assemblage of interacting species that form the biotic components of an ecosystem can be viewed in terms of the food relationships that exist between them. An analysis of these relationships provides a picture of the feeding, or **trophic**, structure of the ecosystem and a basic understanding of the roles played by different organisms. The energy for all of the work done in an ecosystem comes initially from the sun. Solar energy drives the process of primary production, in which inorganic compounds are converted into the organic substance of green plants by photosynthesis. The plant tissue thus produced is fed upon by **primary consumers**, called herbivores, which are preyed upon by carnivores (**secondary consumers**) that in turn are fed upon or consumed by other carnivores (**tertiary consumers**). In communities comprised of relatively few species of plants and animals, the trophic structure may be quite obvious and simple. In communities characterized by greater species diversity, the trophic structure can be extremely complex; yet, it is still possible to determine the food or trophic level to which most species belong.

This natural grouping of species on trophic levels through which there is a unidirectional flow of energy forms what Elton (1927) called a **food chain.** In simple communities, energy does in fact seem to pass from link to link along a chain of life, but in complex communities, energy flows through a netlike arrangement of pathways, often referred to as a **food web.**

Most communities have two food chains of almost equal importance, the **herbivore food chain,** described previously, and the **detritus food chain.** Herbivore food chains are usually highly visible in that they typically consist of green plants, herbivores, and one or more levels of carnivores. A typical herbivore food chain would be:

> Willow trees → Caterpillars → Predaceous carabid beetles →
> Insect-eating birds → Hawks.

Insects are usually components of the second and third trophic levels of a herbivore food chain, as in the preceding example. However, there are numerous species of insect parasites and predators that would be part of the fourth or even higher levels. Although not as conspicuous as vertebrate herbivores and predators, insects are important components of most herbivore food chains.

The organic material that falls out of a herbivore food chain as litter, feces, and other exudates represents a substantial quantity of chemical energy. Although this energy is lost to the herbivore food chain, it is not lost to the ecosystem as a whole. It is used by a complex of organisms that comprise the detritus food chain. Detritus food chains are less easily characterized than herbivore food chains because it is difficult to assign the melange of worms, arthropods, bacteria, fungi, and other organisms to specific trophic levels. Nevertheless, myriads of soil-dwelling insects play an important role in utilizing the energy bound in the detritus outfall from the herbivore food chain and in hastening the recycling of nutrients back to the green plants. We will return to the role of insects in this recycling process a little later.

**FIGURE 18-1.** Diagram of a simplified food web in which omnivores increase the complexity of the routes of energy and nutrient flow.

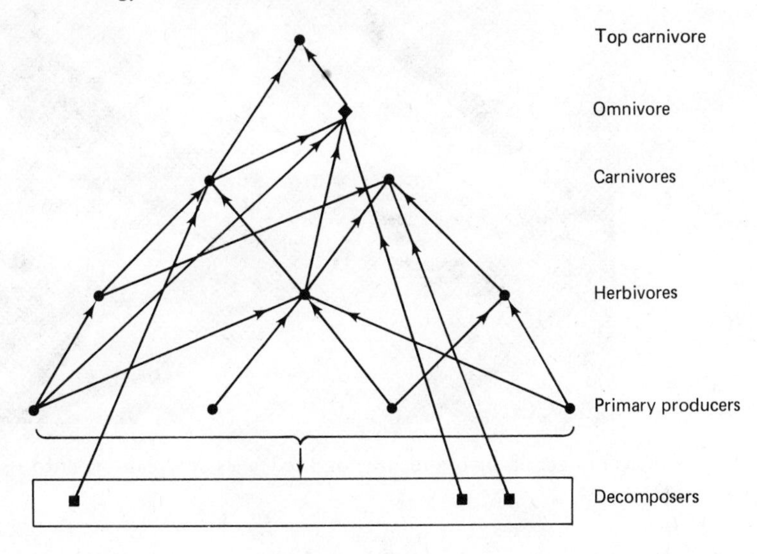


Many animals feed on several kinds of food, according to their nutritional requirements and the abundance of certain foods relative to the ease with which they can be obtained. Bears, for example, eat green grass, berries, insects, small rodents, fish, and a variety of other things, depending on availability. When they are concentrating on grass and berries, they are herbivores, but at other times they are primary or secondary carnivores and sometimes even scavengers. We call animals with this kind of feeding habit **omnivores**; numerous species of insects fit into this category. In many cases the omnivory of insects is regimented by their life history. This is particularly true of those groups that display complete metamorphosis and a trophic divergence of the immatures and adults. For example, many adults of hymenopterous parasites feed on pollen and nectar and are therefore specialized herbivores, whereas their larvae are carnivorous. Other insects not only occupy different trophic levels during different stages of development but as adults may display different dietary preferences which are related to sex. For example, the larvae of some biting flies are scavengers or herbivores, but the adult males feed on nectar, and the adult females, on blood.

The degree of omnivory displayed by the insects, particularly that which results from the trophic divergence of larvae and adults, has undoubtedly contributed greatly to their success. Complete metamorphosis in particular has provided a unique mechanism by which the competition for food among individuals of the same species is greatly lessened. All of the species on a particular trophic level are dependent on the trophic level immediately below for their survival. A dynamic balance, which is regulated by a system of feed-back controls, exists between trophic levels. Insects in general, and those which are components of several trophic levels in particular, play a major role in regulating the flow of energy and nutrients through ecosystems by providing a number of alternative trophic linkages, as illustrated in Figure 18-1.

## *Insects as Herbivores*

Approximately 50% of all insects are herbivorous (Figure 18-2) and very few, if any, species of plants are immune to their feeding. Insects feed on roots,

**FIGURE 18-2.** A typical insect herbivore feeding on the leaves of a plant. (Photograph courtesy USDA.)

stems, leaves, buds, flowers, fruits, and seeds; they feed both internally and on the surface. Plant-feeding (phytophagous) insects are common components of both terrestrial and fresh-water ecosystems. Usually there is some evidence of insect feeding on most plants by the end of a normal growing season, though often not severe. Occasionally, however, an insect herbivore escapes the forces that normally contain its numbers, and the plants on which it feeds suffer severe damage.

Obviously, plants do not grow, flower, and set seeds to provide food for insects or any other animal for that matter. Any plant species that did not exploit some method of defense against the feeding of herbivores would surely not survive for very long. Consequently, in the course of their evolution, plants have retained structural and chemical characteristics that deter would-be consumers. The consumers in turn have evolved mechanisms to overcome these defenses and thereby enhance their own survival. In other words, the feeding of an insect species on a plant species serves as a selective pressure, which leads to the persistence of those traits of the plant that tend to impede the destruction of structural parts or the consumption of its stored energy. These traits then act as a selective pressure that leads to the evolution of some method of counterattack by the insect. This is the process of **coevolution** that has been occurring between plants and insects for more than 300 million years. Over this span of time, a number of highly refined insect-plant relationships have developed. As Ehrlich and Raven (1964) pointed out, this may have been a major factor in the development of the organic diversity we see around us.

Plants tend to produce a wide variety of secondary metabolic products, many of which have been incorporated into the chemical defense against herbivores. The production of these chemicals is the result of enzymatic actions that are under genetic control. Thus, new **biotypes** that are biochemically different can result from single gene mutations. Initially these biotypes may be resistant to insect attack, and man has exploited this phenomenon in the selection and breeding of his crops (see Chapter 24). However, any herbivore species that is able to overcome a plant's defenses is at least initially freed from the competition of other herbivores that have not and is therefore at a considerable advantage.

Insects seem to have keyed to these odd chemical substances in plants in such a way that many have become specialists on single species or small groups of related species. Not only do these compounds limit the competition for food by reducing the number of herbivore species but they also serve as chemical cues which aid the insects that can tolerate or detoxify them in locating their host plants. For example, the family Solanaceae contains a number of plants, such as tomato, tobacco, potato, and nightshade, which contain large amounts of specific alkaloids that are powerful feeding deterents or poisonous to many phytophagous insects. However, they serve as attractants to the few insects that feed on these plants.

Some of the secondary metabolic compounds of plants have such a pronounced adverse effect on nonadapted insect species that man has extracted and purified them for use in insect control. Nicotine, pyrethrum, ryania, and rotenone are fairly well-known botanical insecticides used in insect pest control (see Chapter 23). On the other hand, alkaloids and other plant compounds, particularly terpenes, are responsible for the strong flavor or odor of plants that man has domesticated as crop species. When crop varieties are developed that are more palatable to us because they contain less of these chemicals, they are sometimes more palatable to insects as well. Growing these "defenseless" varieties can result in severe insect damage.

In addition to chemicals that act as feeding deterrents or as conventional metabolic poisons, plants have also produced a number of compounds that mimic the activity of insect hormones. Since the discovery of a juvenile hormone mimic called **juvabione**, present in the wood of the balsam fir *Abies balsamea* (Sláma and Williams, 1966), numerous plant components that are structurally and actively similar to the juvenile and molting hormones of insects have been isolated. Some plants, such as ferns, which have a high concentration of these compounds, are almost devoid of insect herbivores. All of these defensive chemicals are so widespread in the plant kingdom that their occurrence as a chance phenomenon must be ruled out. They are clearly the result of the coevolution of plants and the herbivores that eat them.

Not all insects have been able to evolve biochemical mechanisms that enable them to overcome plant defenses. Instead, they seem to have evolved life history patterns that allow them to avoid the chemicals involved. Larvae of the winter moth, for example, hatch early in the spring and complete their feeding and development on young oak leaves before they accumulate a large quantity of tannin, which adversely affects the growth and development of the caterpillars (Feeny, 1970).

Defensive chemicals and structures are produced at some expense to the plant, so their occurrence must provide the plant with some net benefit. Plants growing in harsh environments may be unable to expend the energy required for protection throughout the entire year or growing season, in which case the defensive chemicals may be produced only when they are particularly beneficial. Some shrubs adapted to arid areas seem to be completely free from insect attack during the period of new growth and the setting of seeds, but later in the season when water and nutrients are in short supply, insects attack them freely.

Some plants are aided by specialized insect herbivores and produce desirable chemicals to encourage the insects that are important to their survival.

The most obvious examples involve the production of perfumes and nectar attractive to pollinators, as we will see in more detail a little later. A lesser-known example is provided by the mutualistic relationship between certain tropical plants of the genus *Acacia* and ants. Small colonies of ants nest in the *Acacias'* large hollow thorns and feed on the sweet secretion produced by nodules at the ends of the leaves. The ants in turn provide protection by warding off herbivores. Plants from which the ants were experimentally removed displayed substantial herbivore damage (Janzen, 1966, 1967).

Some distasteful plant products are simply eaten by tolerant insect herbivores. These chemicals may then be regurgitated or expelled by the insect for its own protection against predators or parasites. The caterpillars of the anise swallowtail butterfly accumulate anise oil in a pronotal pouch, which is everted when the larvae are irritated by an attacker. The larvae of a chrysomelid beetle that feeds on datura leaves pile their own fecal material, which is laden with a distasteful alkaloid from the plant host, on their backs for protection (see Chapter 14).

The pressure that herbivores exert on plants, although often difficult to observe, clearly leads to a gradual change in the genetic composition of many plant populations. In spite of the defensive capabilities that evolve, most populations of plants are subject to some degree of feeding by phytophagous insects. We must assume, therefore, that in addition to defensive mechanisms, plants have evolved growth characteristics that permit them to maintain some average level of productivity in the presence of herbivores to which they are susceptible. While we do not normally think in terms of outbreaks of plants, the overproduction of biomass by a plant species released from the regulatory influence of a herbivore can result in substantial community disruption. There are numerous examples of exotic plants, such as the prickly-pear cactus, which have grown and spread abnormally in new environments where their herbivores are lacking. When the appropriate herbivores are introduced into the new area, there is usually a pronounced reduction in the abundance of the host plant. It seems quite reasonable, therefore, to think of insect herbivores as regulators of plant populations.

Irregularities in the relationship between insects and plants often lead to the conclusion that nature's strategy and man's strategy are not always compatible. When insect herbivores fail to exert control over the abundance of an unwanted plant, man may decide to intervene with some form of weed control. On the other hand, when phytophagous species increase in abundance to a level at which they reduce the productivity of a desired plant, man may attempt to control the herbivore.

There is no doubt that insects can and do cause serious damage to both wild and cultivated vegetation. When wild plants with no economic value are attacked, the feeding is largely ignored. But when insect feeding causes a decline in the yield of a commodity that man had hoped to harvest, the insect is classified as a pest. The subject of insects as plant pests is reviewed in Chapter 20, so it will not be dwelt upon here. However, it is necessary to realize that the impact of a herbivore on a species of plant depends on the type of damage it causes. Phytophagous insects that feed on leaves can cause 30 to 40% defoliation without producing a serious impact on the community, and even almost total defoliation of many perennials will not kill the plant if such feeding does not occur two years in a row. On the other hand, if a relatively small percentage of the flowers or seeds are consumed by insects, there may be a marked impact on the plant population.

Plant juice feeders are different from leaf feeders in that they do not destroy photosynthetic tissue to the same extent. Consequently, insects such as aphids can extract more energy per unit area of vegetation without destroying their food source than grazers. Way and Cammell (1970) suggested that these insects actually increase the total amount of carbon fixed by their host plants by removing nutrients which, if allowed to accumulate, would cause a decline in the rate of photosynthesis. However, some plant juice feeders actually destroy green tissue cells and cause a yellowing condition known as chlorosis, which is often accompanied by a decline in photosynthesis. Photosynthesis is often impaired additionally by sooty molds and dust that grow on or adhere to the honeydew excreted by sap feeders (see Figure 20-6).

Under normal or endemic conditions, the grazing of insect herbivores can increase the flow of energy and nutrients through an ecosystem (abiotic flux). According to Mattson and Addy (1975), grazing increases light penetration through the foliage canopy, reduces competition between plants, alters the plant species' composition of communities, increases the rate of nutrient leaching from the foliage, and increases the rate of litter fall and its accumulation. Grazing also stimulates the redistribution of nutrients from woody components where turnover is slow, to leaves and flowers where turnover is rapid, and stimulates the activity of decomposers. Unfortunately, most workers have been concerned with the detrimental effects of insect herbivores on plants and consequently have generally ignored the foregoing beneficial effects.

The few studies that have examined the possible beneficial aspects of herbivore-plant coactions have produced some highly interesting results. For example, when moderate defoliation occurs, there may be a substantial rise in the nitrogen, phosphorus, and potassium content of the litter as a result of increased production of material containing insect excrement and cadavers. The release of these nutrients may result in increased growth of both the overstory and understory vegetation. As Turnbull pointed out (1969), insect herbivores prevent the self-defeating perpetuation of mature coniferous forests. Such forests are characterized by a thick, closed canopy, which shuts out the sunlight so effectively that important understory plants and even seedlings of the dominant conifers cannot survive. Outbreaks of defoliators and bark beetles act to open up these stands, thereby allowing sunlight and rain to reach the soil surface and stimulate plant succession to begin anew. Thus, the accumulation of humic acids and resins in the soil and the nutrients locked up in the woody biomass of the dominant trees are eventually recycled. The net result is a healthier, more diverse, and more productive community.

### *Insects as Pollinators*

Insects that feed on pollen and nectar are specialized herbivores that have a mutualistic, rather than parasitic, relationship to their plant hosts (Figure 18-3). As with the more typical herbivores just discussed, pollinators and flowering plants have been coevolving for millions of years. The fossil record indicates that the first flowers appeared when the only potentially efficient pollinators available were insects, and the blossoms were correspondingly small. Later, there was a substantial increase in the size of the flowers, a development which roughly coincided with the rise of birds (Smart and Hughs, 1973). The flowering plants proper (Angiosperms) arose after the first appearance of birds, but the diversification of these new plants was more

**FIGURE 18-3.** A bumble bee collecting nectar from a daisy showing pollen grains adhering to the forelegs and body hairs. (Photograph courtesy of William G. Wellington.)

rapid and extensive than one might expect if birds served as the principal pollinators. Furthermore, many of the most primitive angiosperms are today pollinated by insects; this, in conjunction with the explosive **adaptive radiation** that apparently occurred among the Hymenoptera and Lepidoptera at about the same time that it did among the flowering plants, suggests that insects were the major pollinators. Birds probably contributed to the radiation of the angiosperms more as agents of seed dispersal than as pollinators (see Regal, 1977).

Insects that pollinate plants do not provide their services without cost. To obtain the benefits of outcrossing through pollen transport, the plants have to expend energy in the production of colorful flowers, fragrances, and nectar to attract and reward the pollinators. Part of the energy expended in this way comes from savings derived from the fact that pollen production could be reduced when plants changed from wind pollination to animal pollination. However, this was a major change that set in motion a long sequence of evolutionary events that improved the efficiency of sexual reproduction in plants.

As with the coevolution of plants and traditional herbivores, the plays and counterplays that have led to the highly specialized relationships between some pollinators and their plants form a fascinating subject. Obviously, if a number of different species of flowering plants occur in the same area, they must compete for the services of pollinators. Under such conditions, any plant species which through a mutation became more attractive to pollinators would receive the benefits of more visits and a higher level of outcrossing. That plant species would then be more successful. However, there is a limit to how far different species can proceed with adaptations that attract a larger portion of a group of generalized pollinators, because there is a limit to how much energy can effectively be allocated to colored petals, nectar, and fragrances.

Nectar production is one factor in the pollinator-plant relationship that must evolve a rather delicate balance. There must be a reward to the pollinator that bears a satisfactory relationship between its energy and general nutritional requirements and the energy expenditure required to obtain them. If a plant evolved the capacity to produce a copious supply of nectar, visiting pollinators might be able to fulfill their needs at a single flower and consequently not transfer pollen effectively to other flowers. On the other hand, a small supply of highly desirable nectar could result in the pollinators visiting numerous flowers and thereby increasing the level of outcrossing. An interesting interrelationship is illustrated by the short-tongued bumblebees that "rob" the nectar from red clover flowers by piercing the side of the corolla and not making contact with the pollen-bearing parts. This behavior may actually contribute to an increase in seed yield because long-tongued bumblebees that later enter the flowers directly must then visit more flowers to fulfill their requirements for nectar (Heinrich and Raven, 1972).

Until recently, nectar was considered to contain little more than a variety of simple sugars that serve as a source of energy for insects and birds that frequent flowers. The insects were thought to have to obtain amino acids, the building blocks of proteins, somewhere else. Recently Baker and Baker (1973) discovered significant concentrations of amino acids in the nectar of a large number of plant species. The occurrence of some basic nutritional requirements and a major source of energy in the same flowers would tend to concentrate the foraging activity of insects and thereby improve their effectiveness as pollinators. The Bakers found that flowers from which both nectar and pollen can be easily obtained have low concentrations of amino acids in their nectar. On the other hand, flowers which have become specialized for pollination by lepidopterans, which do not ingest pollen or have alternate sources of protein, produce nectar high in amino acids.

When a number of flowering plants are pollinated by generalists, they often display differences in blooming periodicity or nectar production. This is more than likely an evolutionary response to competition for pollinators. Consequently, some plants open their flowers under different light and/or temperature conditions. Likewise, different species have evolved different periods of seasonal flowering so that they are not all competing for the services of pollinators at the same time. From the point of view of the insects, this extends the availability of pollen and nectar to generalized species and may well have played a major role in the determination of which species of pollinators are univoltine and which are multivoltine. It must also have contributed substantially to the protracted foraging and brood-rearing behavior displayed by the subsocial and social bees.

Some flowers have evolved toward a restriction of the number of species that pollinate them. In some cases, this has been achieved by balancing the nectar supply against the size and energy requirements of certain pollinators and thereby selecting against others. In many cases, however, the flowers have changed structurally so that only certain insects can obtain the nectar. The elongation of the corolla and the depth of the nectaries within the flower have restricted some plants to visits by insects such as the butterflies and moths with greatly elongated mouth parts. Flower shape, color, and fragrance may also serve as features that limit the variety of pollinators. Insects see red as black, whereas birds see it as red; bird-pollinated plants are often shades of red and lack perfume. Bee-pollinated plants frequently reflect light in the blue to violet portion of the spectrum, which is favored by bees, and are often fragrant as well.

Nevertheless, none of these plant strategies would have proved to be successful means of assuring pollination if the insects had not been able to benefit from related adaptations in their own evolution. As with any form of generalist, generalized pollen and nectar gatherers are subject to a great deal of competition. Competition provides the selective pressure that at least under some circumstances leads to specialization. Any mutation that provides one species with an advantage over rival species in terms of nectar or pollen acquisition is likely to persist. Consequently, insects and plants have been able to evolve together to the point at which they are totally dependent on each other. This is well illustrated by the orchid family.

Orchids frequently grow as widely separated individual plants rather than in groups, but outcrossing is assured by precise pollen transfer systems involving insects. Some orchids offer nectar rewards that attract pollinators, but usually only one or two species with the right form and behavior actually transfer the pollen that is presented in little wads called **pollinia.** The pollinia are usually sticky and attach to the pollinating insect in such a position that they are only removed when the insect enters another flower with an appropriate removal structure. For example, when a hovering clearwing moth (Aegeriidae) that pollinates the purple-fringed orchid pauses before a blossom, uncoils its siphon, and thrusts it into the flower, a stalked pollinium sticks to the siphon near the moth's head (Figure 18-4*A*). As the moth withdraws its siphon, the pollinium remains attached to it, but the pull flexes the pollinium stalk so that the wad of pollen points forward. When the moth inserts its proboscis in another purple-fringed orchid, the pollinium makes contact with the sticky female stigma and remains there (Figure 18-4*B*). Some orchids do not produce nectar but attract insects by deception. This involves a variety of mechanisms, including mimicry of nectar-bearing flowers, mimicry of female insects to attract pollinating males, and mimicry of the host or prey species of parasites and predators.

Although some of our staple crops, such as corn, wheat, and other cereals, are pollinated by the wind, most fruits and vegetables are pollinated by insects. A wide variety of wild insects play a vital role in man's food production systems. In some crops, however, man has found that yields can be greatly increased through the management of honeybees and the domestication of various species of wild bees for pollination purposes (see Chapter 19).

The following are examples of important crops that depend almost exclusively on insects for fruit production: apples, pears, stone fruits, most berries, citrus, melons, peas, beans, tomatoes, cotton, and figs. The Smyrna fig is a classic example of the dependence of a crop plant on an insect for its production. The superior flavor of the Smyrna fig over other varieties was found to be the result of pollination of the flowers by a tiny wasp. A fig is an oval receptacle with a large number of flowers lining its inner surface. The only access to the flowers is through a small opening at the distal end of the fig. When the flowers are not fertilized, seeds do not form within the fig and the fleshy receptacle does not become sweet and flavorful. The Smyrna fig is always female so must be fertilized by pollen from an inedible variety (usually wild caprifig). A tiny chalcid wasp enters the wild figs to lay eggs within the fruit where the larvae develop. When the daughter fig wasps leave, they

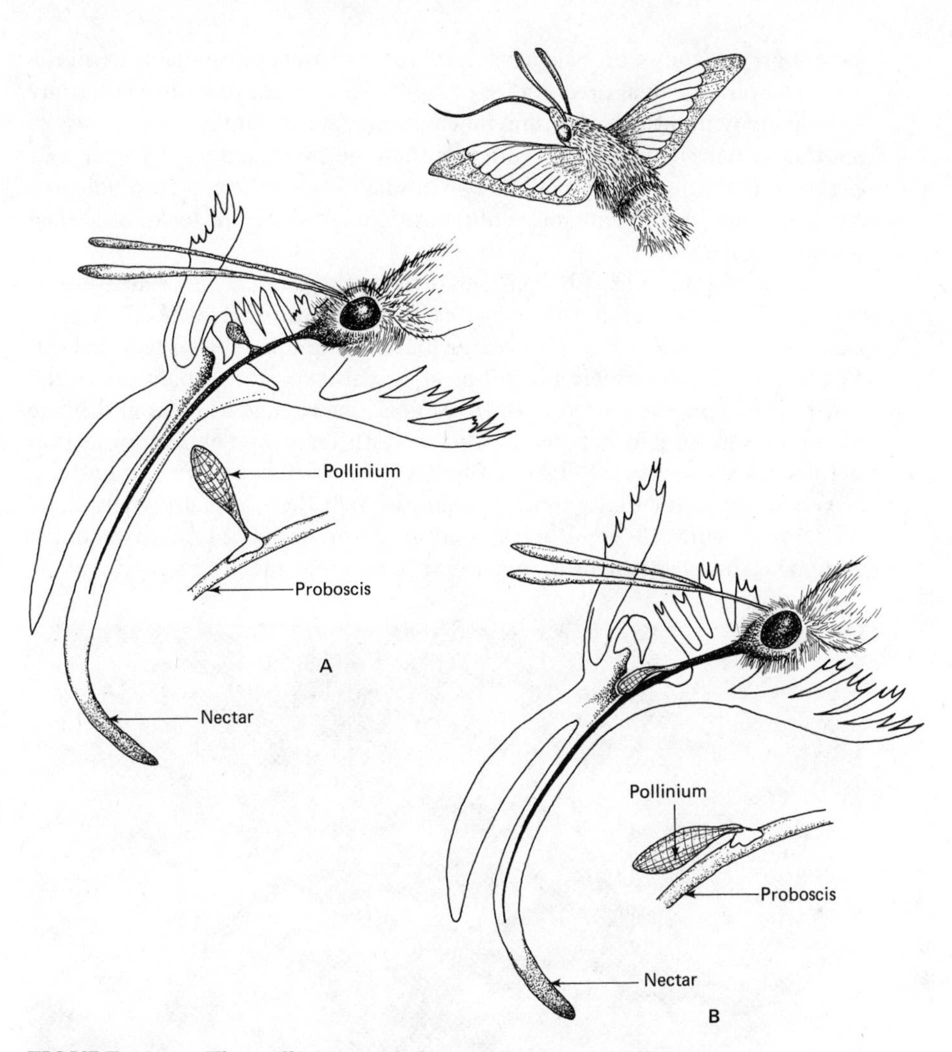


**FIGURE 18-4.** The pollination of the purple-fringed orchid by a clear-winged moth. **A.** The pollinium adheres to the moth's proboscis when it enters the blossom, **B.** The pollinium comes in contact with the sticky female stigma of the next blossom visited. (Adapted from Dowden, 1975.)

become covered with pollen on their way to freedom. In their search for oviposition sites, the wasps enter Smyrna figs as well as the wild figs. For some structural reason, the wasps do not lay eggs in the Smyrna figs but their entry does result in pollination. Consequently, in Smyrna fig groves, there is always an interplanting of caprifigs to provide an adequate supply of breeding sites for the fig wasp.

Insect pollination is a highly complex and fascinating subject that involves a variety of fundamental evolutionary and ecological principles. The subject has not been done justice here, but students who wish to probe deeper are urged to examine either Faegri and van der Pijl (1971) or Proctor and Yeo (1972).

## *Insects as Predators and Parasites*

In accordance with the sequence of trophic levels that make up typical herbivore food chains of the type described earlier in this chapter, animals that eat plants may in turn be eaten by other animals, and they in their turn by still others, and so on. We often refer to the carnivorous organisms on the higher trophic levels as predators or parasites. In a sense this is rather arbitrary in that any symbiotic relationship in which one species gains at the expense of another is parasitism regardless of whether the interacting species are two plants, a plant and an animal, or two animals. Nevertheless, entomologists tend to think of different forms of parasitic relationship in terms of animal characteristics.

True parasites are usually considered to be those organisms which live at the expense of but do not normally cause the death of their hosts. A good example among the insects is a sucking louse (Anoplura) on a vertebrate host. When a mosquito extracts a blood meal, it is also acting as a parasite. In the case of the louse, the parasitic relationship is long lasting, whereas in the case of the mosquito, it is ephemeral. But in both cases it is obligatory in that neither insect could reproduce without the blood taken from the host.

When one animal kills and consumes several others, it is usually called a predator (Figure 18-5). Whereas parasites can obtain more than ample sustenance from a single host, predators usually consume several prey. Among

**FIGURE 18-5.** The lady beetle shown here feeding on an aphid is a common predator familiar to many home gardeners. (Photograph by M. Badgley.)

the insects, there are many typical parasites and typical predators, but there are a large number of species, often called **parasitoids**, which seem to form an intermediate category. Parasitoids feed on the nonvital tissues of their hosts for a prolonged period, much like a typical parasite, but toward the end of their development, they consume some vital tissues and the host is eventually killed. One host is almost always enough, and frequently hosts that are large relative to the parasitoid will provide the nutrition for several to many individuals.

Parasitic relationships have probably evolved gradually from rather casual and accidental feeding encounters and then developed in different directions to quite different levels of specialization as a result of the coevolution of the interacting species. Although it is convenient to subdivide the basic relationship into categories, we must realize that evolution usually produces a continuum of morphological and behavioral adaptations that contains numerous examples not easily categorized. For example, the triungulins (motile first instar larvae) of meloid beetles actively hunt for spider egg masses in the manner of a predator. When an egg mass is found, the triungulin makes its way inside where it molts to a legless larva, which feeds on the surrounding media of spider eggs much like the larva of a typical endoparasite. Although the egg mass corresponds to the single host utilized by a parasitoid, a number of spider individuals (eggs) are eventually killed.

The opportunities for highly specialized parasitic relationships between insects and other animals are increased by the fact that metamorphosis permits considerable divergence between the immature and the adult stages. In some groups like the lady beetles, the larvae and adults are active, free-living predators that actually compete with each other for prey. In other groups like the blood-sucking flies and fleas, the larvae are scavengers or herbivores, and the adults are parasites. In the typical insect parasitoids, the larvae are parasitic, and the adults are usually pollen and nectar feeders if they feed at all. When only the juveniles are parasitic, as is the most common occurrence among insects, it is called **protelean parasitism.**

Regardless of the specific details, parasitism is clearly an exploitation strategy of great importance. Parasites and predators are largely secondary consumers and as such play a major role in the flow of energy through communities. Just as in the case of plants and herbivores, we can observe a wide variety of strategies that have evolved for obtaining food on the one hand and avoiding being eaten on the other. Like plants, insects, too, have evolved defensive mechanisms to reduce the depletion of their numbers (see Chapter 13); these defensive mechanisms must in turn be overcome by the attackers. Coevolution therefore progresses from generalized to specialized relationships. Insect predators and parasites have thereby divided up the available food resources, and in addition to extracting blood from most, if not all, terrestrial vertebrates, they also consume worms, slugs, snails, mites, spiders, fish, and larval amphibians. However, the majority of parasitic and predaceous insects eat other insects and thereby play a major role in the natural regulation of insect populations. The biological control of insect pests (see Chapter 25) often involves the conservation of naturally occurring parasites and predators or their augmentation through the introduction of other such species.

The interaction between animals on different trophic levels must evolve to a fairly finely tuned state in order to avoid the extinction of species from the

FIGURE 18-6. Population fluctuations of interacting prey and predator species generated by the Lotka-Volterra equations.

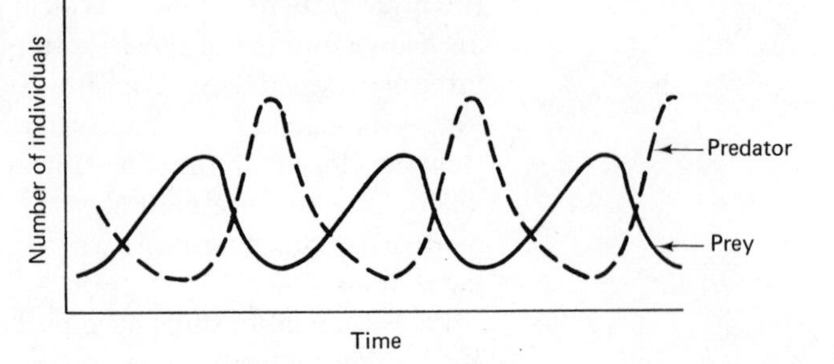


overexploitation of resources (too high a population) or a negative rate of increase (too low a population). Predation and parasitism are highly visible phenomena in most ecosystems, and the numerical relationships that develop between the eaters and the eaten have long been a source of fascination to population ecologists.

Although one animal eating another seems to be a rather simple relationship, when it is projected to the level of the two species populations, it becomes highly complex. Various workers have examined the puzzle of predator-prey population intereactions, but there has been considerable difficulty in developing a general description that meets everyone's approval. Some of the early attempts, while not accurate in their prediction of what happened in nature, were invaluable as a stimulus to further investigation. For example, in the 1920s, Lotka (1923) and Volterra (1931), working independently, showed that the trends in density of populations of a predator and its prey change direction in a systematic way (what Volterra called "the law of periodic cycle"). However, because of the assumptions they made concerning uniformity in the age of all individuals in the populations and instantaneous population reactions, their model predicted population oscillations that did not change in magnitude, as illustrated in Figure 18-6. Nicholson and Bailey (1935) proposed a new model based on populations of an insect parasite and its host. They, too, made a number of unrealistic assumptions and developed an equation that predicted population oscillations of increasing amplitude, as illustrated in Figure 18-7. In nature, periodic increases in the amplitude of predator-prey populations are usually followed by a decline in the magnitude of the fluctuations as the two populations damp each other and return to a more stable condition. Such was the prediction of some more recent models such as that of Leslie and Gower (1970) (Figure 18-8).

FIGURE 18-7. Population fluctuations of interacting host and parasite species predicted by the Nicholson-Bailey equation.

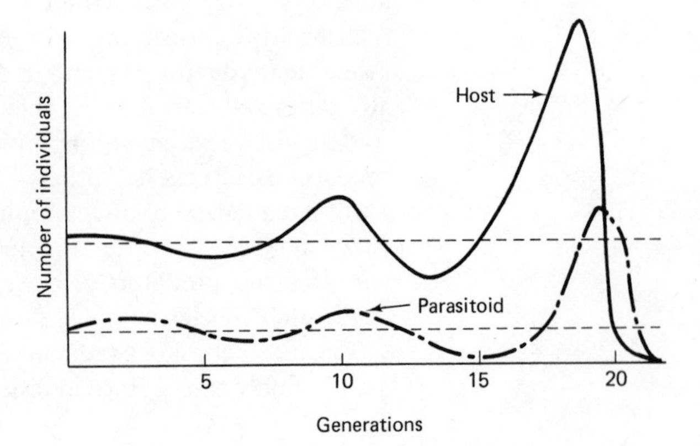

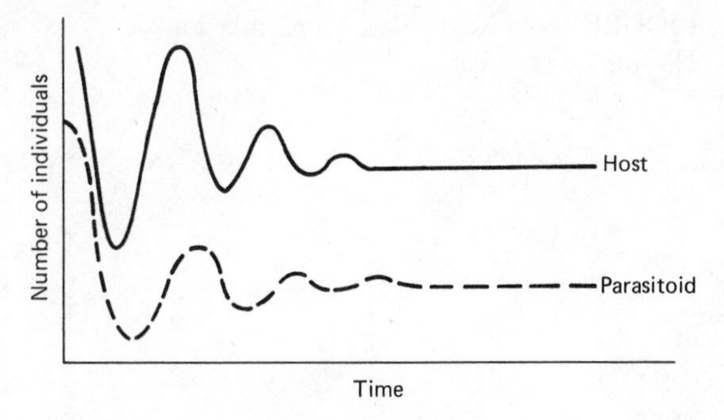


**FIGURE 18-8.** The damping population fluctuations predicted by the Leslie-Grower model when the rate of change of the parasite is dependent upon density.

The early population theorists did not have either the sophisticated computers or mathematics that are available now, so it is understandable that they made assumptions as a means of reducing the number of variables with which they had to deal. Among the factors that were not considered by the early workers are the effect of predator hunger on response to prey, the time required to handle individual prey items, the inhibition of predation by the prey, interactions between predator individuals, nonrandom versus random search behavior, uneven distributions of prey, and the presence of other prey species.

Since the earlier attempts to elucidate the way that populations of predators and prey interact, numerous workers have produced more refined models that included additional components of the predation process. A major contribution was made by a Canadian entomologist named C. S. Holling, who championed the idea of analyzing complex processes such as predator-prey interactions by breaking them into components. He identified which components are universally a part of such processes and which are present only in some situations. He then set out to examine each component experimentally in order to construct a complete model of the process in a step-by-step manner (Holling, 1959, 1964, 1966).

Holling has a gift for devising remarkably simple experiments to elucidate complex questions. In one example, a blindfolded assistant (the predator) searched a tabletop for sandpaper discs of uniform size (the prey). The data obtained enabled Holling to generate an equation, now widely known as "the disc equation," which describes the curvilinear relationship between prey density and prey captured per unit of time (Figure 18-9). The curve, which has been shown to describe what occurs in a number of diverse natural situations, indicates that the rate of prey capture is slow at first, accelerates quickly, and then levels off. Holling pointed out that at low prey densities most of the predator's time is spent searching but capture efficiency increases as the number of available prey increases. However, a point is reached at which more prey are available than can be handled in the time available. Thus, handling time had to be included in the predator model.

Since hunting by a predator is usually initiated by hunger and terminated by satiation, the threshold for pursuit of prey must change after a prey is consumed. Similarly, the search for hosts by insect parasites is affected by the number of eggs ready to be laid. These and other components of predation and parasitism can be subjected to careful experimental examination. As each component is analyzed, a new mathematical expression can be added to the model of the over-all process. Through this component analysis approach, we are gradually gaining a better and better understanding of the role of insects as predators and parasites in the regulation of pest populations.

FIGURE 18-9. Graphical representation of Holling's disc equation.

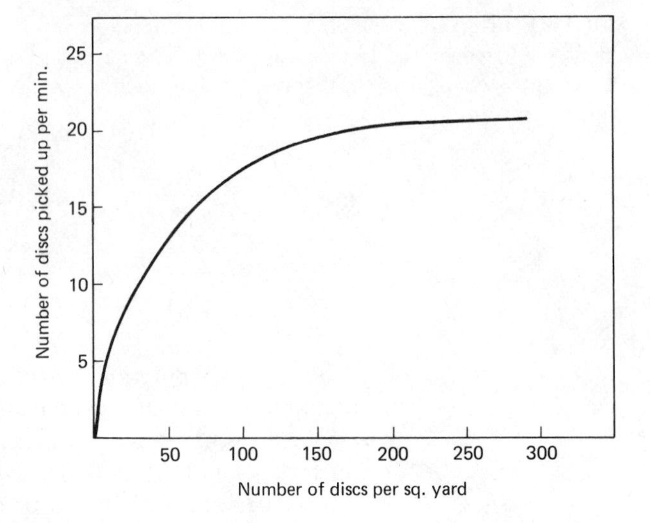


***Insects as Scavengers***

When plants and animals die they become part of the ecosystem's accumulation of organic material called detritus. The material that accumulates varies in form from the trunks of large trees and the carcasses of large animals to small pieces of plant litter and animal waste. Regardless of the form, these materials contain a large quantity of energy in the form of chemical bonds. In the detritus food chain there is not the discrete step-by-step transfer of energy from trophic level to trophic level that typifies herbivore food chains. The roles of decomposer organisms tend to overlap and there is more of a continuous flow of energy through them. The larger organisms ingest particles of organic matter, break it down in order to extract the energy to fulfill their metabolic requirements, and then excrete the unused portion in a slightly more simple form. This process continues until all of the available energy is utilized and all that remains are soluble nutrients, some of which can be utilized by plants.

Many insects feed on dead plant and animal tissue (Figure 18-10) and thereby contribute to its decomposition and to the recycling of its nutrient constituents. The fraction of the plant biomass that is recycled by the total assemblage of scavengers varies considerably, depending on the nature of the ecosystem. In forest systems, where much of the biomass is tied up in the large woody stems and branches of trees after years and years of accumulation, as much as 90% of the material is broken down by saprophages. In ecosystems dominated by shorter-lived plants, a relatively large portion of the biomass is consumed by herbivores, so much less finds its way into the detritus food chain. Consequently, the turnover rate of nutrients is slower in forests than in grasslands, and if it were not for insects it would be even slower in both systems.

The ultimate breakdown of plant and animal matter is accomplished largely by micro-organisms that attack the surface of the material rather than ingest and break it down internally. Consequently, the larger the surface area to volume ratio, the more rapidly decomposition can occur. Insects contribute in a major way to increasing the surface to volume ratio. For example, wood-boring insects tunnel deeply into logs, thereby permitting the penetration of other saprophages such as fungi. In some cases, such as the ambrosia beetles (Scolytidae), the boring insects actually introduce wood-attacking fungi as they tunnel. Many other insects simply reduce large particles into smaller ones and increase the surface to volume ratio in that way.

The total efficiency of a detritus food chain is also influenced by the temperature, moisture, and oxygen content of the organic medium. Decomposition progresses more rapidly under warm, moist, well-aerated conditions. If oxygen is not available, anaerobic decomposition occurs, but slowly. When adequate oxygen is present, the decomposer organisms can respire rapidly and the detritus is quickly broken down. Numerous soil-inhabiting insects facilitate the oxygenation of the soil by churning through it much as earthworms do. In fact, in tropical and arid environments, subterranean termites and ants take over the ecological role of aerating and mixing the soil that is performed by worms in the moist temperate zone. In combination with insect herbivores, insect scavengers contribute immeasurably to the abiotic flux of all terrestrial ecosystems.

**FIGURE 18-10.** Insect scavengers; soldier fly larvae (Stratiomyidae) in decaying plant matter.

***Insects as Food***

As well as being eaten by their own kind, insects are an important source of food for fish, amphibians, reptiles, birds, and small mammals. Most of the insects consumed by other animals are herbivores, but insect predators, parasites, and scavengers also become dietary items of many other groups. Insects, therefore, play a major role in the energy flow of a community. They also tend to increase the length of herbivore food chains and increase the cross links of community food webs. For example, insect scavengers extract energy and nutrients from the detritus food chain and return it immediately to the herbivore food chain when eaten by an insectivore.

Under normal circumstances, insect herbivores consume a smaller portion of the plant biomass of a community than do vertebrate herbivores. However, even though a large portion of the plant food eaten by an insect passes out as feces, more of the food consumed is converted to animal tissue than it is by larger herbivores. Some insects convert as much as 40% of the food they assimilate into body tissue. This is largely because they are cold-blooded and less energy is utilized for maintenance metabolism. Consequently, insects actually make more of the energy consumed available to carnivores than do warm-blooded herbivores. In addition, insects are so abundant that they assume great importance as converters of plant to animal biomass.

Most land birds become largely insectivorous during the raising of their young regardless of their normal food habits. Insects provide a diet rich in protein and lipids so necessary to the rapid rate of growth of young birds. It seems to be no coincidence that the nesting period of land birds usually coincides with the peak periods of insect emergence and rapid population increase.

Insects play the same fundamental roles in aquatic communities as they do in terrestrial ones, but it is in a variety of fresh-water ecosystems that insects stand out as an important source of food (see Cummins, 1973). Insect grazers and filter feeders such as mayfly nymphs, caddisfly larvae, stonefly larvae, and various midge larvae serve as the main converters of plant material into animal protoplasm. Along with a variety of scavenging and predaceous species, they comprise a major part of the staple diet of fishes, particularly in shallow lakes, ponds, and streams. In North America, mayflies are said to be the single most important dietary component of trout, which, like many fishes, feed not only on the aquatic larvae of insects but also on the adults as they emerge.

The importance of insects in the diet of many fresh-water fish is exemplified by the long-established art of making simulations of insects out of feathers as lures for sport fish (Figure 18-11). This practice apparently dates back at least to the second or third century when Macedonian fishermen used artificial flies to catch river fish for food rather than for recreation. Numerous articles and books have since been written on the subject, and some such as Izaak Walton's *The compleat angler* (1653) are known to every serious fly fisherman. The art of catching fish with feather flies does not end with the selection of one which resembles some insect that is abundant at the time. The skilled angler knows much about the aquatic biology of his favorite haunts and is thus able to bring the fish and his fly together by casting to the appropriate place at the appropriate time.

While on the subject of aquatic biology, we should recognize that many fresh-water environments are more fragile than most terrestrial ones. When man disturbs a terrestrial ecosystem, the impact of the disturbance is fre-

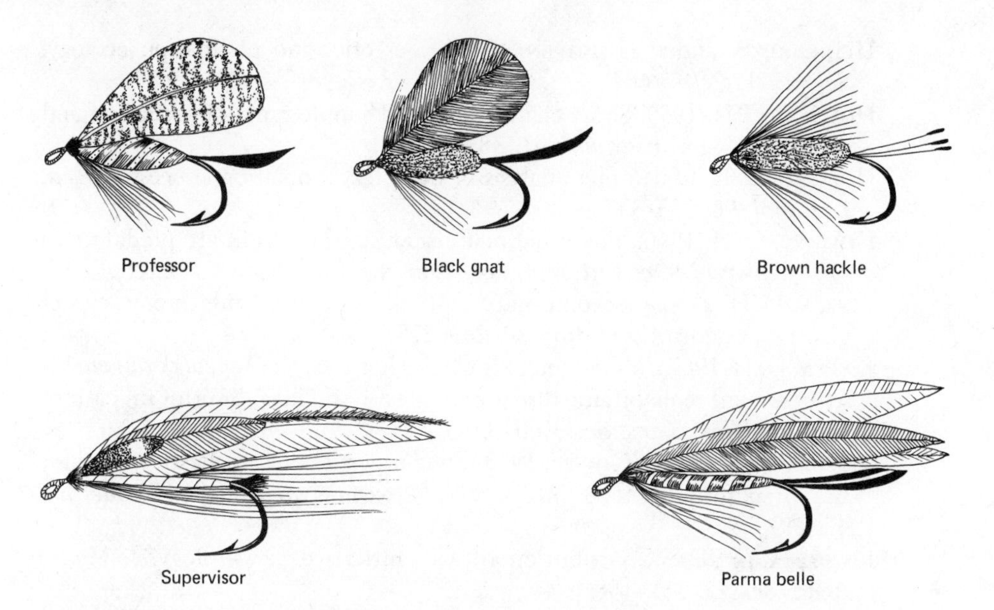

FIGURE 18-11. A selection of fishing flies.

quently rather visible. But we often disturb aquatic ecosystems more subtly and may not notice the damage we have done until it is too late. A wide range of chemical wastes have been entering our fresh-water systems for many years, but we still do not have full information on the effects that varying concentrations of thousands of different compounds have on even test species. It is fully within our capability to monitor changes in water chemistry, but since these changes occur as a result of irregular discharges of waste products, the data may not mean much in terms of estimating long-term impacts. However, any good aquatic entomologist knows which insects require what conditions. Certain insects are found only in clear, well-oxygenated water, whereas others do best where the organic content is high and dissolved oxygen is low. As a result, aquatic biologists are studying changes in the insect fauna that occur as bodies of water become polluted. Insects which are adapted to average conditions rather than to momentary fluctuations serve as useful biological indicators of the conditions that prevail over a longer period of time.

# References Cited

Baker, H. G. and I. Baker. 1973. Amino-acids in nectar and their evolutionary significance. *Nature* 241: 543–545.

Cummins, K. W. 1973. Trophic relations of aquatic insects. *Ann. Rev. Entomol.* 18: 183–206.

Ehrlich, P. R. and P. H. Raven. 1964. Butterflies and plants: A study in coevolution. *Evolution* 18: 586–608.

Elton, C. 1927. Animal ecology. Macmillan Publishing Co., Inc., New York, N. Y.

Feeny, P. P. 1970. Seasonal changes in oak leaf tannins and nutrients as a cause of spring feeding by winter moth caterpillars. *Ecology* 51: 565–581.

HEINRICH, B. and P. H. RAVEN. 1972. Energetics and pollination ecology. *Science* 176: 597–602.

HOLLING, C. H. 1959. Some characteristics of simple types of predation and parasitism. *Can. Entomol.* 91: 385–398.

HOLLING, C. H. 1964. The analysis of complex population processes. *Can. Entomol.* 96: 335–347.

HOLLING, C. H. 1966. The functional response of invertebrate predators to prey density. *Mem. Entomol. Soc. Can.* 48: 1–86.

JANZEN, D. H. 1966. Coevolution of mutualism between ants and acacias in Central America. *Evolution* 20: 249–275.

JANZEN, D. H. 1967. Interaction of the bull's horn acacia (*Acacia cornigera* L.) with an ant cohabitant (Pseudomyrmex ferruginea Smith) in eastern Mexico. *Univ. Kans. Sci. Bull.* 47: 315–558.

LESLIE, P. H. and J. C. GOWER. 1960. The properties of a stochastic model for the predator-prey type interaction between two species. *Biometrika* 47: 219–234.

LOTKA, A. J. 1923. Contribution to a quantitative parasitology. *J. Wash. Acad. Sci.* 13: 152–158.

MATTSON, W. J. and N. D. ADDY. 1975. Phytophagous insects as regulators of forest primary production. *Science* 190: 515–522.

NICHOLSON, A. J. and V. A. BAILEY. 1935. The balance of animal populations. Part I. *Proc. Zool. Soc. London.* pp. 551–598.

REGAL, P. J. 1977. Ecology and the evolution of flowering plant dominance. *Science* 196: 622–629.

SLÁMA, K. and C. M. WILLIAMS. 1966. The juvenile hormone. V. The sensitivity of the bug, *Pyrrhocaris apterus,* to a hormonally active factor in American paper-pulp. *Biol. Bull.* 130: 235–246.

SMART, J. and N. F. HUGHES. 1973. The insect and plant: progressive palaeoecological integration. Pages 143–155 *in* H. F. van Emden, ed. Insect plant relationships. *Roy. Entomol. Soc. London.* Symp. 6.

VOLTERRA, V. 1931. Variations and fluctuations of the number of individuals in animal species living together. Pages 409–448 *in* R. N. Chapman, ed. *Animal Ecology.*

WALTON, IZAAK. 1653. *The compleat angler or the contemplative man's recreation. Being a discourse of fish and fishing, not unworthy the perusal of most anglers.* Maxey, London.

WAY, M. J. and M. CAMMELL. 1970. Aggregation behaviour in relation to food utilization by aphids. Pages 229–246 *in* A. Watson, ed. Animal populations in relation to their food resources. *Symp. Brit. Ecol. Soc.* 10.

# Insects as a manageable resource 19

The discussion presented in Chapter 18 dealt with the roles of insects in natural ecosystems. If it were possible to totally eliminate insects from the ecosystems of which they are a part, the new systems would be characterized by shorter food chains, fewer links in the weblike pattern of energy flow, greater instability, and slower turnover rates in the abiotic flux. Insects have, therefore, become extremely important components of terrestrial and freshwater ecosystems and, under natural conditions, are highly beneficial. Insects can be considered as nonbeneficial or harmful only when we superimpose man's objectives over the basic strategy of nature. Man, as the only animal with sufficient intelligence (though not always the wisdom and motivation) to modify his environment and have an awareness of death, attempts to manipulate nature for his own material gain and personal well-being. In the process of these manipulations, man often has adopted strategies which conflict with the strategy of nature, and under such circumstances, some organisms become recognized as pests. The reasons that insects become pests and the ways in which they interfere with man are discussed in more detail in the next few chapters.

If all insects are really beneficial components of natural ecosystems, then it seems inappropriate to have a chapter entitled "beneficial insects." I feel it is more realistic to follow the view of Stephen (1973), who suggested that some species of insects can be considered as renewable or flow resources. This concept is based on the fact that successive generations of a desired species become available for use at predictable intervals of time. Furthermore, it is possible to manage or manipulate the population levels of such species in order to obtain the desired amount of a beneficial product or service.

Dr. Stephen points out that insects which can be or have been manipulated as flow resources can be separated into two categories that have long been recognized by economists. These categories are (1) **Primary resources,** those species that interact with other components of a system in either a complementary or competitive way to yield a desired product. For example, silkworms interact with mulberry to produce silk. (2) **Intermediate resources,** those species that interact with other components of a system, usually in a competitive way, to produce a benefit or derivative value rather than a product. For example, a herbivore released for purposes of biologically controlling a weed provides a benefit but does not yield a product.

Relatively few species of insects are managed as primary resources, but those that are yield highly desirable products. In recent years the expansion of biological control has added numerous insect species to the category of intermediate resources. The honeybee has long been managed as both a primary resource, mainly as a producer of honey and wax, and as an intermediate resource for pollination. The possibilities of managing other species to yield both a product and a service are attracting more and more attention, and these promise to produce some rather imaginative developments in the years ahead. In the following pages, we will examine some of the insects that have long been managed as renewable resources, some that have been managed only in recent years, and some that have potential for management in the future.

**Silk Moth** The fine strands that are spun into threads and woven into the luxuriously soft fabric we call silk consist of a secretion from the salivary glands of a large caterpillar, often called the silkworm. The product is called *sericum* in Latin which is the root from which the term **sericulture** (the production of silk) was derived. The beginning of sericulture is lost in history but is believed to have predated the birth of Christ by 2,500 years or more. The art of silk production was developed in China and remained a well-guarded secret until the year 195 A.D., when silk moth eggs and the rearing techniques were obtained by the Japanese (Yokoyama, 1973). About 555 A.D. two monks smuggled some eggs of the silk moth to Constantinople and thereby introduced the industry to Europe.

Silkworms are the larvae or caterpillars of the moth *Bombyx mori*, which has been a creature of domestication for so long that it no longer survives in the wild and is the sole living species belonging to the family Bombycidae. After three to four weeks of feeding on mulberry leaves, the fully grown larvae produce a proteinaceous salivary secretion as a continuous delicate strand that becomes remarkably strong, though pliable, when it comes in contact with the air. The silken thread, which may attain a length of 1,000 feet and take three days to produce, is spun into a cocoon that protects the pupa during metamorphosis (Figure 19-1). When the moth has completed its development, it secretes an alkaline material that weakens the silk at the head end of the cocoon to facilitate the adult's escape.

If the moths were allowed to emerge, the ruptured cocoon would be

FIGURE 19-1. Cocoon of the silkmoth *Bombyx mori*, cut away to show pupa within.

worthless because the silken thread would be broken into numerous short, useless pieces. Consequently, the pupae are killed by heat or fumigation. The intact cocoons are then soaked in warm water to soften the sticky material that binds the thread together so that it can be unwound by a highly skilled worker. The threads from several cocoons are spun together to form thicker strands of raw silk. The raw silk is then boiled, treated chemically, and washed in preparation for weaving into the most beautiful of fabrics.

It takes more than a ton of mulberry leaves to feed the 25,000 larvae needed to produce one pound of silk. Most of the world's silk is produced in China, Japan, Italy, Spain, and France. According to the *FAO Production Yearbook*, the total world silk production in 1972 was 45,360 metric tons, which was down somewhat from the peak production years prior to World War II. Nevertheless, even at current levels of production, more than 2,000 billion silkworms must be reared to the pupal stage on more than 90,000,000 tons of mulberry leaves each year.

Many insects produce silk, and numerous other lepidopterans pass the pupal stage in a silken cocoon. Several species of moths, particularly members of the family Saturniidae, have been investigated as alternate sources of silk. The beautiful Cynthia moth, *Samia cynthia*, was introduced into the United States along with its host plant, "tree of heaven," but *Samia* cocoons were found to be difficult to unwind, and the silk is of poor quality. The gypsy moth, *Portheria dispar*, now a serious forest pest in the eastern states, was also imported in the hope it could be used for commercial silk production. In recent years, man-made fibers have largely replaced silk for most common uses, but natural silk will almost certainly remain a highly desirable cloth for those who can afford it.

### Lac Scale

The material from which shellac is made is produced by a scale insect (Lacciferidae:Homoptera), which is quite closely related to a serious agricultural pest, the San Jose scale (Diaspididae:Homoptera). The lac scale, *Laccifer lacca*, is a native of India and Burma where it infests native plants related to the fig (*Ficus*). Lac comes from the Sanskrit word which means 100,000 and refers to the large number of individuals that can completely encrust the branches of their host plant. These heavy infestations result because scale insects are sedentary for most of their lives and several generations often live side by side. The soft bodies of lac scales are protected by the secretion of a resinous material that hardens into a shield over their backs. When the density of scales is very high, the entire branch to which they are attached becomes coated with the resinous substances; such branches are gathered as **stick lac.**

Stick lac is ground up by hand to free granules of the lac. These granules are then crushed and boiled in water to remove pigments and other impurities. The lac floats to the surface of the water, from which it is scooped up and placed in the sun to dry and bleach. The dried lac is then placed in long, slender burlap bags that are stretched in front of a charcoal fire. As the lac is heated, two men twist the ends of the bag to squeeze out the melted lac, which drops to the floor. Before the lac hardens again, it is gathered by other workers who stretch it much as one pulls toffee. The lac eventually hardens and is broken into small, thin flakes, which are shipped to manufacturers. The flaked lac is dissolved in methyl hydrate to produce a liquid coating that has many uses. Lac is also used in sealing wax, special inks, and plasticlike compounds. It requires approximately 150,000 lac scales to produce a pound of lac. Like silk, lac and shellac are being replaced by man-made compounds such as plastics and synthetic lacquers, but there are still craftsmen who prefer the natural product.

Other scale insects produce resinous coverings that can be extracted, but the material is not of as high quality as lac, and none of them has been used for commercial production. The Indians of the southwestern United States used the secretion of a local lac insect, *Tachardiella larreae*, to mend pottery and to waterproof baskets.

### *Cochineal Scale*

Another scale insect, *Dactylopius coccus*, which lives on a prickly-pearlike cactus, *Nopalea cochinellifera*, produces a beautiful carmine pigment. The insect is a native of Mexico and was raised by the Aztecs as a source of red pigment for paints and fabric dyes long before the arrival of Cortez. Because scarlet was considered a sign of wealth and nobility in Europe, the Spaniards became interested in the production of the dye for shipment to their homeland.

The mature female cochineal scales are covered by white fluffy tufts of wax while they feed on their cactus host, but when ruptured, the nearly black females exude a deep red-purple fluid. The Mexicans exercise great care in their cultivation of these insects, taking them into their homes during the winter and infecting new cactus pads each spring. When the insects are fully grown, they are brushed into containers and killed with hot water. The dead insects are then dried in the sun and cleansed of impurities. About 70,000 insects are needed to produce a pound of dye; a man working all day can collect enough insects to make about two ounces.

Because of the potential value of a high-quality red dye, the Spaniards attempted to increase production by forcing the native people of Mexico into slavery to cultivate the insect. They also sent the cactus and the insect to Spain to establish the industry there. Through the process of selection, the Spaniards actually developed a more highly prized product they called *Sangre de la tierra*.

Later, cochineal became widely used as a permanent red dye, a cosmetic pigment, as food coloring, and even as a pain suppressant. However, it was found to be carcinogenic and was banned from use in foodstuffs. The cochineal industry was struck another blow when aniline dyes were developed, but cochineal production continues and has recently been revived in the Canary Islands.

A related scale insect, *Dactylopius tomentosus*, occurs on *Opuntia* cactus in the southwestern United States (Figure 19-2). This scale also produces a deep red-purple dye, which was probably used by the local Indians for the decoration of fabrics, pottery, and baskets.

### *Honeybee*

The familiar honeybee was until quite recently the only insect managed as both a primary resource and an intermediate resource. Many species of bees gather nectar and convert it into honey, but only *Apis mellifera* lends itself to domestication and management for continuous honey production. *A. mellifera* is a native of the Old World that evolved long before the rise of man. Cave paintings dating back to 7,000 B.C. are believed to depict how primitive man collected honey from wild hives, using methods that were probably developed when man was no more than a hunter and gatherer. Wild bee

FIGURE 19-2. *Dactylopius tomentosus*, a relative of the cochineal scale, on *Opuntia* cactus in southern California. (Photograph courtesy of D. Mahr.)

colonies are still hunted as a source of honey, but most of the honey consumed now is produced according to highly refined management methods. The exploration and colonization of remote parts of the world in the sixteenth century led to the spread of the honeybee, which now occupies all but the polar regions of the globe.

The historical development of beekeeping, or **apriculture** as it is often called, is a fascinating subject which unfortunately cannot be treated in much detail in the space available here. However, a number of good reviews, such as that by Townsend (1973), are available to the interested student. According to Townsend, beekeeping proper most likely began with the realization that it was better to safeguard the future of colonies rather than to destroy them in the process of gathering the honey. To accomplish this, various substitutes for natural nesting sites were brought into use. The substitutes depended upon the materials that were locally available and varied from hollowed sections of tree trunk to pottery vessels, woven baskets, and coiled rope skeps (Figure 19-3). Although these early hives provided the bees with protection and made them easier to handle, the bees usually had to be killed in order to harvest the honey and wax. Clearly, what was needed was a hive that would facilitate the removal of the honey without injuring the bees. The major breakthrough came in 1851 with Langstroth's discovery that there is a rather precise space of five sixteenths of an inch that honeybees will pass through without instinctively connecting two adjacent surfaces with bridges of wax. The concept of the "bee space" opened the way for the development of the removable-frame hive, which permitted both the close examination of the bees and the nondestructive removal of the honey. There was a brief period during which numerous hive designs were proposed, but a combination of trial, error, and research eventually led to the type of hive in widespread use today (Figure 19-4).

As discussed in Chapter 17, many flowering plants produce a small drop of nectar that attracts pollinating insects. The bees as a group collect the nectar along with pollen as food for their young. *A. mellifera* has evolved a pattern of behavior that differs from most species in that some of the nectar collected is cured into honey and stored in waxcombs for use during periods when the weather restricts the availability of nectar. Thus, the common honeybee can survive from year to year as complete colonies. This allows for ongoing management by man and, of course, provides a crop of honey that can be harvested. Good management can result in crops in excess of 200 pounds of honey per hive in addition to the stores left for use by the bees during the winter. Some beekeepers remove most of the honey and then feed the bees on sugar syrup during the off-season.

The total quantity of nectar produced by plants is immeasurable but is presented in such small quantities that man could not utilize it without the help of insects. A worker bee visits from a few to many blossoms in order to fill its crop with nectar before returning to the hive and it has been estimated that more than 50,000 trips, often in excess of a mile, are required to produce a single pound of honey.

The nectar collected from flowers is carried in the crop of the bee, where it

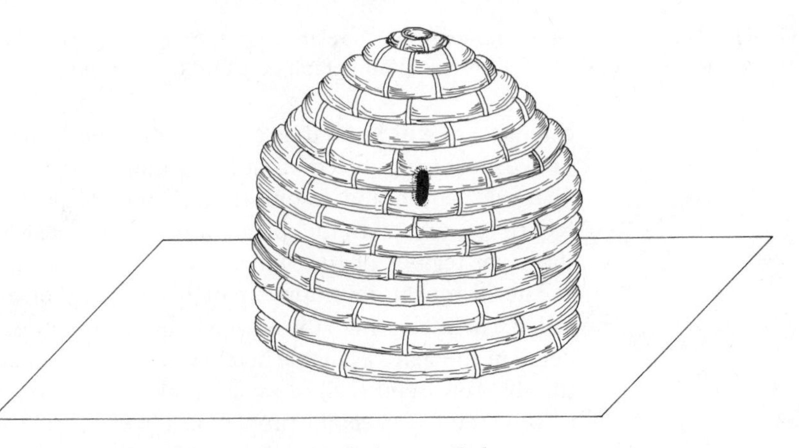

**FIGURE 19-3.** An old type of beehive, called a skep, made from coiled straw or rope.

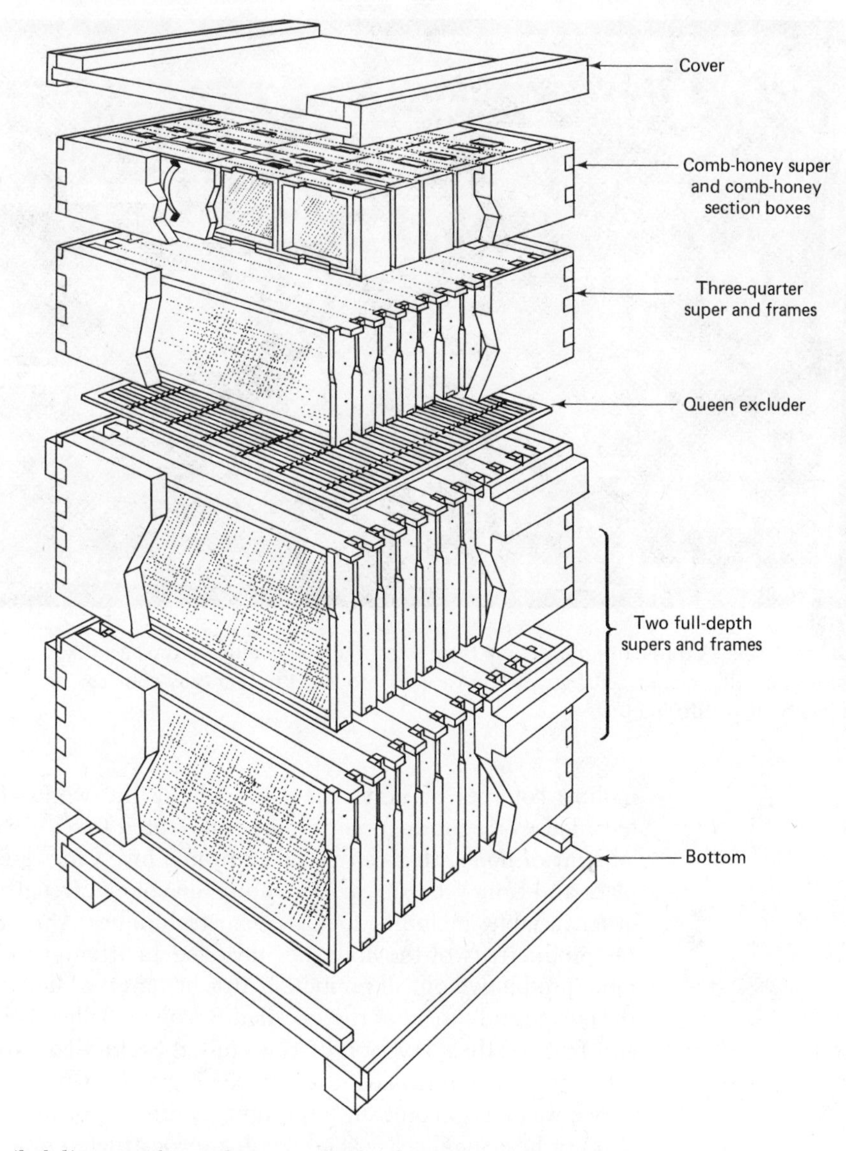


**FIGURE 19-4.** Detailed diagram of a modern, wooden beehive commonly used by American bee keepers. The basic hive consists of a base or bottom board, one or more full supers with frames, and a cover. The other components are optional. (Reproduced with permission of the University of California.)

is mixed with the bee's saliva and undergoes some changes as a result of the action of enzymes. When the worker bee returns to the hive, the nectar and saliva mixture is regurgitated through the mouth into the hexagonal wax cells that make up the honeycomb (Figure 19-5). The cells remain open for a period of time during which much of the water is removed from the nectar through evaporation; this is enhanced by worker bees, which create a flow of air over the combs by beating their wings. If the excess water were not removed, the nectar would ferment and not be usable by the bees later on. When groups of cells are filled with cured honey, they are sealed with a thin capping of wax.

According to Crane (1963), up to five million beekeepers manage 40 to 45

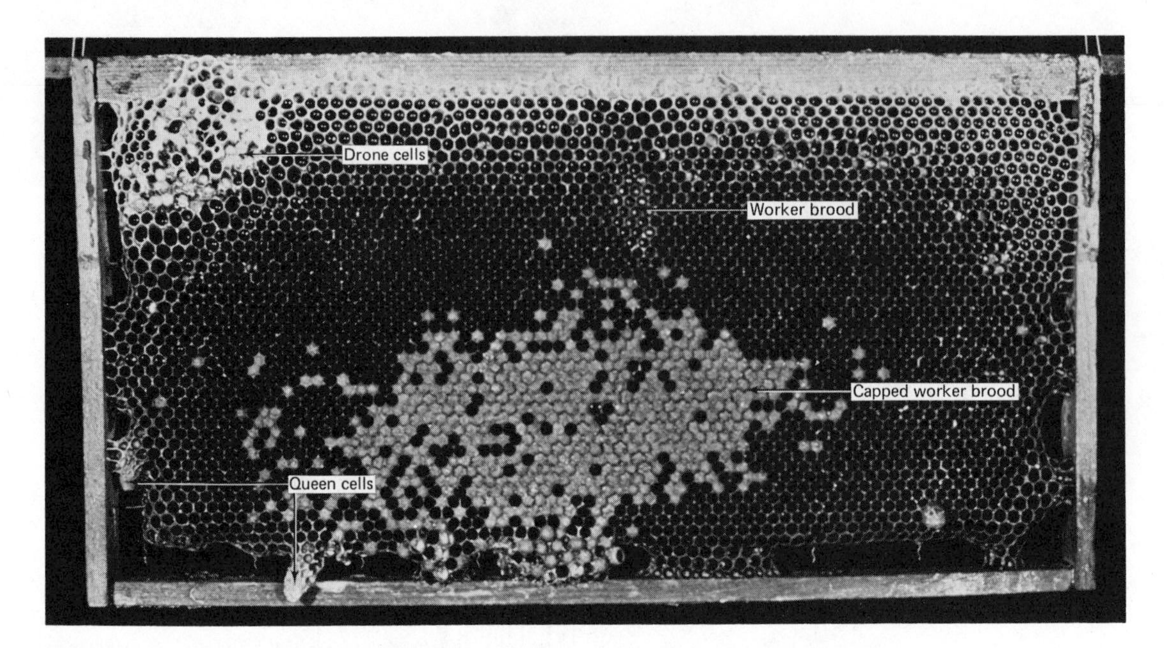


**FIGURE 19-5.** A frame of honey comb from a commercial beehive showing worker cells, drone cells, queen cells, and some capped brood. (Photographs courtesy of University of California.)

million colonies of honeybees throughout the world. In 1972 the total recorded world production of honey amounted to 850,569 metric tons. The amount of honey that can be taken from a hive varies greatly from place to place and from year to year, depending on colony strength, the weather, and other variables including management techniques. A recent plan to increase the productivity of the honeybee involved an attempt to hybridize it with a more productive but ill-tempered African variety. These experiments led to the escape and spread of the so-called Brazilian "killer" bee in South America and fears of their invasion of the United States. To date the bee is not a problem here and it may never be (see Figure 20-1 for present distribution).

Beeswax has become an important by-product of honey production. The ranks of hexagonal cells called comb are constructed of wax, secreted by the hypodermal glands of the workers and extruded between their abdominal sternites as small scales. The scales of wax are then molded into cells with the blunt mandibles. Cappings and surplus combs are melted down and sold for use in candles, furniture polish, car wax, cosmetics, and numerous other products. Hives managed by American beekeepers produce roughly six million pounds of surplus wax each year, but this is probably only about half of the nation's consumption.

In recent years pollen and royal jelly have become additional products harvested from honeybees. Some areas produce far more pollen than the bees need to feed their young. In these areas hives can be fitted with pollen traps, designed to knock the balls of pollen from the bees' pollen baskets on the hind legs as they enter the hive. The pollen is then collected every two to three days and placed in cold storage. The pollen is sold commercially to other beekeepers and health food outlets. Royal jelly is the special food that bees feed to their young larvae and is the sole diet of larvae destined to become queens. Royal jelly is used in high-priced cosmetics and in treating certain skin disorders. With careful management, about one pound of royal jelly can be produced from 200 colonies.

These products plus queens and packages of worker bees form the basis for the management of bees as a primary renewable resource. But the managing of honeybee colonies principally as an intermediate resource, with crop pollination the provided service, has become a big business. Adequate pollination can substantially increase the yield of fruit, nut, vegetable, and seed crops, but when they are grown in extensive monocultures, there are just not enough native pollinators to visit all of the blossoms. Consequently, many beekeepers have contracts with farmers to pollinate their crops. The beekeepers lease their colonies for a specific amount of time during the crop's blooming period. Since the colonies may only be required for a short time, they are usually kept smaller than those used for honey production, and the beekeeper is well organized to maximize his mobility.

The number of colonies used per acre for pollination depends on the kind of crop, the weather, colony strength, and the competition provided by other flowering plants. For most field and orchard crops, two colonies per acre is more or less standard, but in some seed crops such as alfalfa, three to five hives per acre may be necessary for maximum yield. Since bees will work the flowers closest to their hive, better pollination is achieved when the hives are distributed throughout the crop rather than grouped in one place. The contract between the beekeeper and the grower usually specifies the minimum colony strength, number of hives per acre, hive distribution, movement times, rental fee, and other pertinent information. In some crops such as almonds the bees produce an undesirable honey, so pollination rental fees are usually higher.

### *Other Bees*

Although no other species of insect lends itself to the commercial production of honey or wax, there are other bees which are as good or better pollinators than *Apis mellifera.* Recently two such bees have been brought under management, and because the bees themselves can be sold as a product, they share the distinction of being both a primary and intermediate resource. These species are the alkali bee, *Nomia melanderi* (Halictidae) and the leaf cutter bee, *Megachile rotundata* (Megachilidae), both of which are gregarious and have a distinct preference for foraging on alfalfa. The alfalfa flower has a tripping mechanism which *A. mellifera* learns to avoid by extracting the nectar with their long proboscis while penetrating the flowers from the side and thereby not serving as efficient pollinators.

The native soil-nesting alkali bee was the first wild species to be managed for alfalfa pollination (Stephen, 1960). The female bee constructs a more or less vertical burrow in soil of appropriate texture and moisture content. A varying number of larvae are raised in each burrow on a diet predominated by alfalfa pollen. Following a study of the physical conditions of natural nesting sites, Stephen was able to develop artificial "bee beds" from soil of suitable texture, moisture retention capacity, and adequate surface salts to reduce evaporation. These beds were constructed adjacent to alfalfa fields and innoculated with bee-containing cores removed from natural nesting sites. Whereas the bees construct up to 40 holes per square foot under natural conditions, they can construct up to 230 holes per square foot in the artificial beds because of the more uniformly packed soil (Figure 19-6). Furthermore, the beds can be constructed to encourage the bees to form nest holes of the most efficient depth.

At high population densities, the number of larval cells per nest hole

A

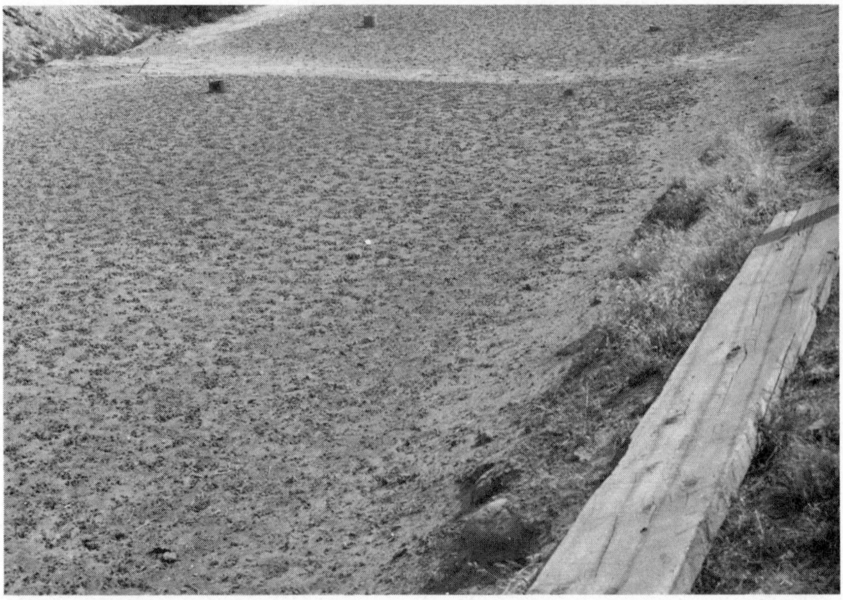

B

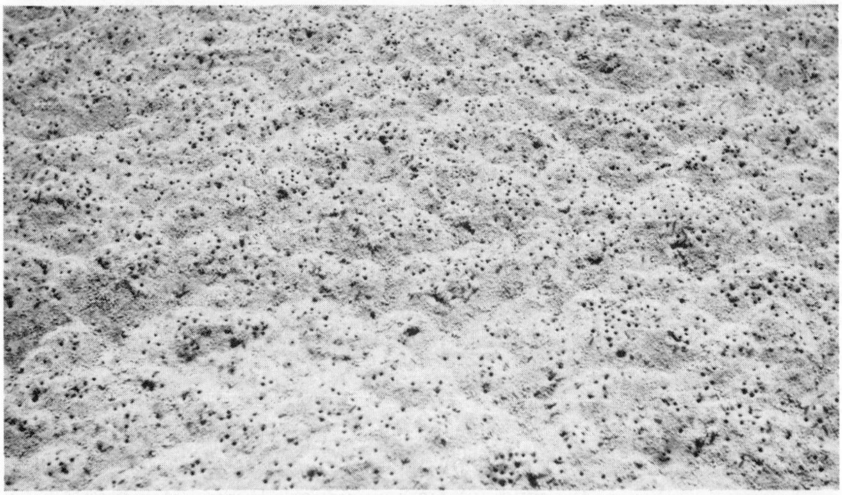

C

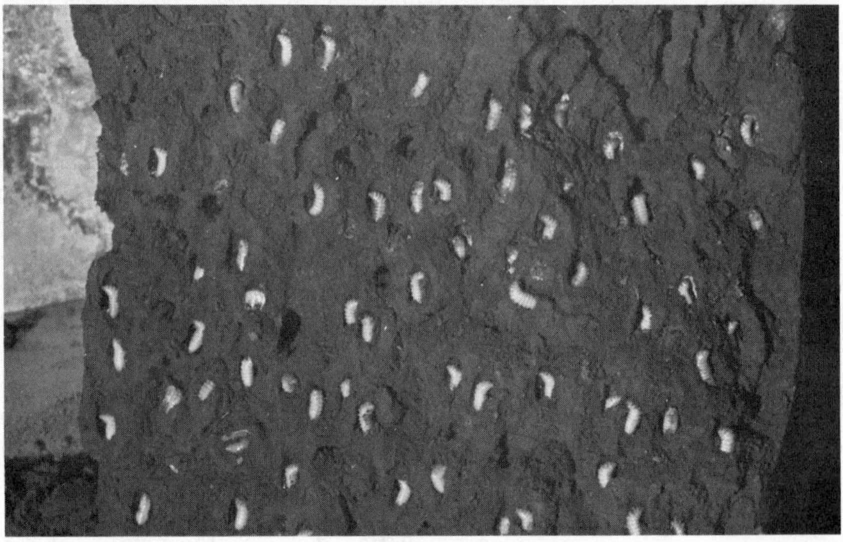

FIGURE 19-6. [OPPOSITE] A man-made bee bed showing a large number of entrance holes to the burrows of the alkali bee *Nomia melanderi*. A. View from a distance, B. Close-up of the surface, C. Section through core of soil showing larvae in the burrows. (Photographs courtesy of William Stephen.)

declines markedly, but populations in most artificial beds tend to stabilize at 70 to 100 holes per square foot. Another benefit of the artificial beds is a reduction in brood parasitism by bee flies, which are kept away by the high level of bee flight activity above the nest site (Stephen, 1973). The consequences of these factors are a substantial increase in the bee population working the alfalfa and a large increase in seed production resulting from the improved pollination.

Unlike the alkali bee, the introduced leaf cutter bee nests above ground in pre-existing holes of appropriate size. Although widespread throughout some of the alfalfa seed production areas, the leaf cutter bee populations tended to be kept low by an inadequate supply of suitable nesting sites. Stephen (1962) found that the leaf cutter bee would nest in drilled boards and in boxes of small drinking straws. These kinds of contrived nesting sites have several advantages over the artificial alkali bee beds in that they are less expensive to construct and can be moved from place to place to coincide with the blooming period of different alfalfa fields.

The leaf cutter bees overwinter as prepupae in a series of cells separated by discs of leaf material (Figure 19-7). These prepupal cells can be held under

FIGURE 19-7. A section of an artificial nesting site for the leaf-cutter bee *Megachile rotundata*, showing larval cells constructed of leaf material placed in the channels of a piece of corrugated cardboard. (Photograph courtesy of William Stephen.)

refrigeration and protected from attack by natural enemies until they are placed outdoors in the spring. After emergence and mating, the young females proceed to construct and provision cells in available nest holes. Over a life span of five to seven weeks, each female constructs a number of larval cells (usually less than 30), each of which is provisioned with 7 to 12 loads of alfalfa pollen (Stephen, 1973). Unlike the alkali bee that has only one generation per year, the leaf cutter bee has two or more generations per year and the population undergoes considerable growth during the season. According to Stephen, small alfalfa growers manage populations of one half to three million bees whereas large growers develop populations of 30 million or more (Figure 19-8).

The pollination service provided by managed populations of alkali and leaf cutter bees has increased alfalfa seed production from an average of 150 to 200 pounds per acre to 1,500 to 1,800 pounds per acre while the price of the seed has remained stable. In addition to this valuable service, there is a demand for the bees to populate other alfalfa-growing areas. The sale of soil cores containing the alkali bee and of both extruded cells and tunnel series of the leaf cutter bee has led to both species being managed as a dual resource much like the honeybee.

### *Miscellaneous Insect Products and Services*

Other than pollination, the most important service provided by managed insect populations is clearly the biological control of insect and weed pests. However, the subject will not be discussed here because natural predator-prey and parasite-host relations were considered in Chapter 17 and applied biological control is the subject of Chapter 25.

Several insects have been found to be useful for medical purposes. However, with few exceptions, the insects that provide the desired product or service have not been reared or managed. They are simply collected from their natural habitats as required. One of the more bizarre examples comes from Latin America where native Indians use the mandibles of soldier army ants to close wounds. The vicious soldier ant is held such that when it bites the patient one mandible enters on each side of the cut clamping it closed. The ant is then beheaded so that the head and jaws remain as a suture.

The drug cantharidin is derived from the integument of certain blister beetles (Meloidae). One European species, *Lytta vesicatoria*, commonly known as Spanish fly, has been collected as the principal source of cantharidin. This drug is a strong mucous membrane irritant with a pronounced effect on the urogenital tract. As a result, it has been used as an aphrodisiac of both humans and domestic animals, particularly cattle. Although still used in cattle breeding, it has been withdrawn from human medicine because it can be fatal.

Numerous observations made over hundreds of years have indicated that unattended wounded soldiers fared better than their treated comrades. The reason was that the wounds became infested by the larvae of certain blow flies. These maggots devoured the necrotic tissue and thereby cleansed the wound. Much later, it was found that a nitrogenous waste, **allantoin,** excreted by the maggots, acted to inhibit infection. For a time, allantoin was produced commercially but has largely been replaced by modern antibiotics.

Honeybees and other aculeate (stinging) hymenoptera are reared or collected as a source of venom. The venoms are purified and used to develop antivenom serums. Bee venom was once widely recommended as a treatment

FIGURE 19-8. Mobile field domicile for the leaf-cutter bee *Megachile rotundata*. Units of this size contain 360,000 straws and 200,000 nesting females. Lower photograph shows a close-up of boxes of straws. (Photographs courtesy of William Stephen.)

for rheumatism, and some people still believe in its effectiveness. Another extract made from the bodies of bees was once widely used as a treatment for a variety of complaints including hives, dropsy, and scarlet fever.

Numerous species of insects have been reared for purposes of scientific study. Insects are generally easy to handle and many can be raised on artificial diets. This alleviates the problem of reliance on natural food resources that may be available only seasonally. Under laboratory conditions, many insects can be reared without interruption, making it possible to have an adequate supply of all life stages available at all times. The generally short life cycle makes insects desirable for breeding experiments and genetic studies. Several insects have contributed immeasurably to our understanding of basic biological principles. The small fruit fly *Drosophila* has been the principal subject of studies in animal genetics, and flour beetles of the genus *Triboluim* have been used extensively in population, development, and ecological projects. Studies of ants and honeybees have contributed greatly to our understanding of behavior, and other common species such as the house fly *Musca domestica* have led to major advances in nutrition, toxicology, and endocrinology.

Another important service to science performed by insects came to my attention for the first time on a recent trip to Wyoming. A paleontologist friend routinely examines harvester ant mounds when he surveys an area for fossil-bearing deposits. Harvester ants construct piles of particulate debris collected from the area around their nest site. If one of these piles contains fossils such as those shown in Figure 19-9, the paleontologist knows that a more careful search of the adjacent area is warranted.

**FIGURE 19-9.** A collection of small fossil teeth obtained from the nest mound of a harvester ant in Wyoming. (Photograph courtesy of Jason Lillegraven.)

**Insects as Food** Many kinds of insects have been mass-reared as fish bait and pet food, and insects have also been collected during periods of great abundance as food for domestic animals; various species of swarming locusts are periodically fed to pigs and poultry or are spread on the soil as fertilizer. In many parts of the world, insects such as butterflies, termites, and beetles are considered a luxury food and much effort is applied to their collection, but for many human beings less fortunate than ourselves, insects become an important source of protein and fat in an otherwise restricted diet. Apart from the fact that insects probably exceed the biomass of all other land animals combined and that of man by more than ten times, their bodies contain exceptionally large amounts of high-quality protein and lipids. Analyses of house fly pupae by Calvert and others (1969) and Teotia and Miller (1974) showed that they contain slightly more than 60% protein and between 10% and 15% fat. The same research groups found the pupae to contain 17 and 16 amino acids, respectively, including a good supply of all those essential in the diet of humans. Calvert and others found the pattern of fatty acids present to resemble that of some fish oils, and Teotia and Miller found the fly pupae to be a good source of important minerals. Analyses of termites and locusts have indicated that they too form a highly nutritious source of food. About 80% of the dry weight of termites consists of roughly equal portions of protein and fat, and their caloric value exceeds 500 cal/100 g (Tihon, 1946). The various species of locusts that have been analyzed show that they range from approximately 50 to 75% protein and from 5 to 18% fat. In my own research, I have found the fat content of a small bark beetle to vary widely according to the conditions of development, but fat contents in excess of 30% of dry weight were common among both larvae and adults (Atkins, 1975). Insects are also rich in certain vitamins; bee pupae, for example, are reported to be very high in vitamins A and D.

On the basis of their general abundance and great nutritional value, we should not be surprised to find insects as a component of the human diet in many parts of the world. Indeed, it is rather surprising to entomologists that insects are not more widely reared for food in a world with an unevenly distributed and barely adequate supply of protein.

Europeans and their descendants are the only major group of people who do not seem to enjoy a tasty meal of insects. The prejudice is rather difficult to understand in that the Greeks and the Romans, both of whom influenced European culture so greatly, consumed a variety of insects and even reared some as specialty foods. Yet, in spite of our insistence that insects are not fit for human consumption unless prepared as a novelty item coated in chocolate, we actually consume more insects than most people realize. It is virtually impossible to grow crops under a system of mechanized agriculture and to process the produce by automation without some contamination of food by insects or insect parts. As a result, the FDA has set upper limits for the quantity of insect material that can legally contaminate foods. Such items as catsup and cereals are particularly difficult to prepare without insect contamination. However, there is really nothing to fear; higher levels of insect contamination might increase consumer resistance but would not create a health hazard nearly as great as the pesticide residues that would result from the level of crop pest control necessary to avoid it. Is it not strange that consumers would rather eat food contaminated with toxic chemicals than with nutritious insects?

The following quotation from Vincent Holt's little book, *Why Not Eat*

*Insects*, published in 1885, seems rather appropriate: "Cheese-mites, the grubs of a small fly, are freely eaten by many persons, whom I have often heard say 'they are only cheese.' There is certainly some ground for this assertion, as these grubs live entirely upon cheese; but what would one of these epicures say if I served up to him a cabbage boiled with its own grubs? Yet my argument that 'they are only cabbage' would be fully as good as his."

Insects belonging to almost all orders are eaten by people around the world with some rather interesting local preferences or, perhaps, specialties. Probably the grasshoppers and their relatives are among the most widely eaten. Ancient Greek and Roman authors reported that locusts were collected, dried, and ground into a flour used in baking some form of cake or bread. American Indians ate roasted locusts, and it may be that seeing them do it prompted the settlers on the western plains to also eat locusts during the plagues of the late 1800s. Ants also seem to be widely consumed by aboriginal peoples in many parts of the world.

Termites are probably the second most commonly consumed group of insects, particularly in the tropics where they abound. The giant physogastric queens are considered a delicacy by a number of African tribes. Because of their seasonal swarming behavior, winged reproductives are easily captured in large numbers at certain times. The wings are removed and the bodies are fried lightly in oil. Termites prepared in this way can be purchased in many African village markets. They are also rendered to produce an oil used for cooking.

Hemipterans and beetles are also fairly widely used as human food. Beetle larvae, particularly large white grubs (Scarabaeidae), are rather substantial morsels; I have eaten the prepupae of a bark beetle (Scolytidae) and found them to resemble the flavor of the white meat of walnuts. Roman epicureans apparently raised certain aquatic beetles for the table. Large aquatic bugs are consumed in various parts of the world. The Indians of southern California and Mexico apparently collected vast quantities of back swimmer eggs (Notonectidae). They would submerge sheets of matting in the shallow water around the margin of ponds where the bugs were depositing their rather large eggs. The matting was raised to collect the eggs, which were dried, sacked, and used or sold at market as a flour. The southern Californian Indians also collected the highly conspicuous yellow oak scale, *Cerococcus sp.*, which they used as a chewing gum.

Finally, mention must be made of the Lepidoptera, the larger species of which provide rather large, tasty, and nutritious morsels of food. The nectar-feeding females, laden with eggs, are rich in protein and fat and have a sweet flavor as well; the late instar larvae are also very high in protein and lipids. It is not surprising that they are widely eaten; yet, if we encounter an ear of corn infested by a corn earworm, we destroy the highly nutritious larva and eat the less nutritious corn. The Aztecs apparently did not share our prejudice in that they prized the caterpillar-infested ears. Indians in California and Nevada collected large numbers of the pine-feeding larvae of the Pandora moth. Fires were lit beneath infested trees, and the smoke would cause the two-inch-long caterpillars to drop. They were then dried or smoked for later use. Probably the caterpillar known to most Americans as a food item is the larva that feeds on agave and ends up in the bottom of bottles of some brands of tequilla; the same larvae are fried and sold in gourmet shops as "fusanos."

Although the foregoing examples are but a few of the many that could be presented, they provide ample evidence that among groups of people who are not hampered by prejudice, insects are a significant, desired, and sometimes necessary dietary component. The management of insects as a renewable resource for food purposes may well be the next major development in entomology. More and more research papers are published each year that deal with the usefulness of insects as food or as the source from which dietary supplements can be prepared. A number of species could be reared as a primary resource to yield a marketable food commodity. Some of these species could be managed as both a primary and an intermediate resource with insect flesh as the product and the recycling of organic waste as the service.

The honeybee, which is already managed as a dual resource, looks like a good candidate to become valued as a food item as well as for all its other merits. In many areas where the winters are cold and long, apiculturists exterminate their colonies in the fall and begin again in the spring with packages of bees imported from warmer areas. If this procedure were not followed, the colonies would have to be fed or left with a large quantity of the honey gathered during the previous summer, and many would still not survive the winter. Hocking and Matsumura (1960) reported that more than 130 tons of bee larvae and pupae are destroyed each year in the prairie provinces of Canada alone. The authors found that bee brood contains about 18% protein, so in that region alone about 13.5 tons of quality protein goes to waste each year.

The honeybee is ideally suited to management as a food source for a variety of reasons. First, the fact that honey is widely liked removes some of the distasteful stigma that surrounds most insects. Furthermore, pollen and royal jelly have become increasingly popular, particularly among health food advocates. Also, bees have a high level of brood production during the times that nectar and pollen are available; they closely regulate conditions within the hive, making brood development predictable and easy to manage. Finally, there is far more nectar and pollen available than is ever exploited. Because it is available in such small quantities, it can be used by man only if the bees act as an intermediary in its harvest. Vast areas exist where the nectar flow would not support an apiary managed for honey production, but hives in such areas could produce a large amount of brood without requiring much care.

Insects generally are equivalent to chickens in the efficiency with which they convert their food to body weight. This conversion rate, which can run to more than 30%, averages twice that of feeder-lot cattle and four times that of pastured sheep. Among the insects that have high levels of food conversion efficiency and are easy to rear are the common wax moth, *Galaria mellonella* (Figure 19-10), the flour beetles, *Tribolium* spp., and a number of flies, including the house fly and the screwworm fly. The screwworm was mass-reared for the sterile male control program (see Chapter 24) on an artificial diet with a large component of animal blood. At the peak of the project, in the vicinity of 190 million flies were reared each week. This amounted to more than 13 tons of larvae containing about eight tons of protein and two tons of lipid.

An exciting aspect of managing insects for food stems from the fact that many of them could be reared on materials that are not only of little value but actually pose problems. For example, very low-value range land is hardly

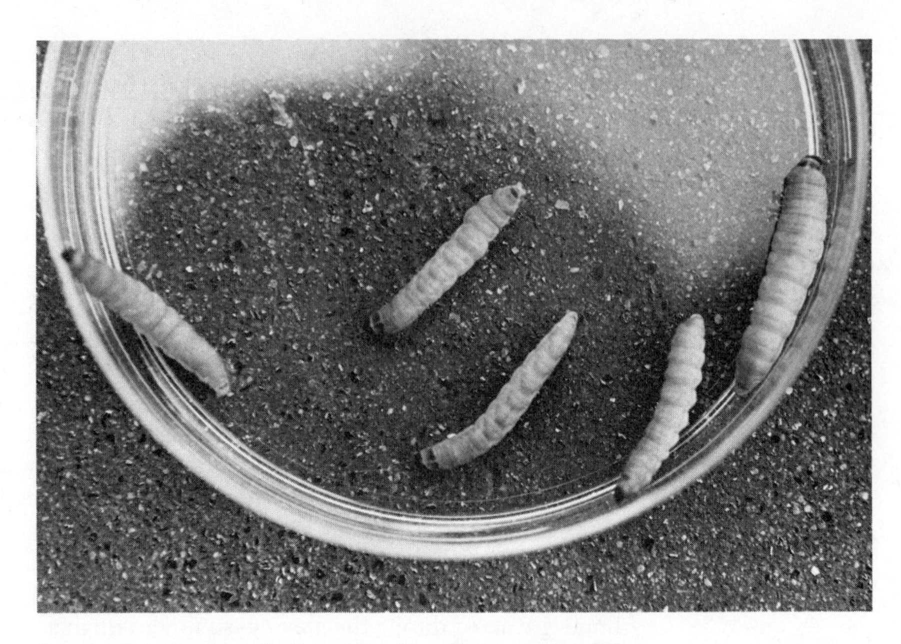

**FIGURE 19-10.** Large larvae of the wax moth Galaria mellonella in a petri dish 75 mm in diameter.

worth managing for cattle or sheep production, but it might produce a fairly good crop of grasshoppers. Even more realistic projects involving the mass rearing of insects on wastes could be undertaken. Poor-quality grain can be fed either to cattle which convert it to flesh with about 10 to 12% efficiency or to *Triboluim* which will convert better than 30%. Termites, which, as we have already seen, are one of the favored groups for food purposes, can be mass-reared on newspapers and forest wastes. Various species of flies could be mass-reared on the thousands of millions of tons of human and animal waste produced each year. The solid waste generated by cattle feed lots in the United States has become a disposal problem that could be alleviated by converting it into fly protein.

We may be a long way from educating the general public to incorporate insects into their everyday diet, but we really need to move in this direction by first using insects as intermediaries in the recycling of animal waste back to poultry and livestock and by then developing insect protein supplements for human foods. Several investigators have conducted trials in which insects have been fed to poultry and pigs in place of other foods. The work of Teotia and Miller (1973, 1974) showed that chicks which were fed fly pupae as their sole source of protein for seven weeks gained only slightly less weight than those fed a regular grain-based diet. The weight gain of chicks fed on fly pupae during the first four weeks of their life was similar to that of chicks fed on a regular diet of soybean oil meal.

As mentioned earlier, the biomass of insects probably exceeds that of all other land animals. We are literally surrounded by insect protein and lipids that are of high quality. Much more research will be needed to find the methods to efficiently manage and harvest this relatively untapped resource, but as the world becomes more heavily populated and the supply of high-quality protein grows shorter, man may be forced to turn to the insects as a solution to the food crisis.

# References Cited


Atkins, M. D. 1975. On factors affecting the size, fat content and behavior of a scolytid. *Z. ang. Entomol.* 78: 209-218.

Calvert, C. C., R. D. Martin, and N. O. Morgan. 1969. House fly pupae as food for poultry. *J. Econ. Entomol.* 62: 938-939.

Crane, E. 1963. The world's beekeeping—past and present. Pages 1-18 *in* R. A. Grant, ed. The hive and the honeybee. Dadant and Sons, Inc., Hamilton, Ill.

Hocking, B. and F. Matsumura. 1960. Bee brood as food. *Bee World* 41: 113-120.

Holt, V. M. 1885. Why not eat insects. E. W. Classey, Ltd., Middlesex, England.

Stephen, W. P. 1960. Artificial bee beds for the propagation of the alkali bee, *Nomia melanderi. J. Econ. Entomol.* 53: 1025-1030.

Stephen, W. P. 1962. Propagation of the leaf cutter bee for alfalfa seed production. *Oregon State Univ. Agric. Exp. Stn. Bull.* No. 586.

Stephen, W. P. 1973. Insects as natural resources and tools of management. Pages 31-44 *in* P. W. Geier, L. R. Clark, D. J. Anderson, and H. A. Nix, editors. *Ecol. Soc. Aust.* (*Memoirs 1*): *Canberra.*

Teotia, J. S. and B. F. Miller. 1973. Fly pupae as a dietary ingredient for starting chicks. *Poultry Sci.* 52: 1830-1835.

Teotia, J. S. and B. F. Miller. 1974. Nutritive content of house fly pupae and manure residue. *Br. Poult. Sci.* 15: 177-182.

Tihon, L. 1946. A propos des termites au point de vue alimentaire. *Bull. Agric. Congo Belge.* 37: 865-868.

Yokoyama, T. 1973. The history of sericultural science in relation to industry. Pages 267-284 *in* R. F. Smith, T. E. Mittler, and C. N. Smith, editors. History of entomology. Annual Reviews Inc. Palo Alto, Ca.

# Insects as pests—a human perspective 20

On the basis of what has been said in foregoing chapters, it should be clear that in playing their natural ecological roles insects are almost certainly going to come into contact with man. This contact can only increase as we strive to meet the needs of our growing population by expanding the exploitation of natural ecosystems and developing new agroecosystems. For the most part, insects are highly beneficial, and if man were still simply a part of natural history, insects would do him little harm as a species. Even those that have become adapted as human ectoparasites and thereby transmit some serious diseases would probably be beneficial in the sense that they would regulate our population and remove the less fit individuals. Therefore, insects cannot be considered as pests in an ecosystem in which man and the insects play natural roles, since all of the organisms in such systems interact as part of a complex scheme that evolves over a period of time toward a state of dynamic equilibrium.

An organism becomes a pest only when we deem it to be so. And we make such a judgment solely on the basis of a conflict that develops between ourselves and the normal activities of the pest organism. As our population has increased, we have attempted to fulfill our needs by extracting more and more materials from our environment and by maximizing the productivity of agriculture and forestry. At the same time, we have applied considerable effort to extending our longevity and decreasing deaths caused by disease, the results of which are to increase our population still further and to create additional demands for food, fiber, and other resources. As a result we often carry the designation of insects as pests to an extreme, as in situations where they simply cause a little inconvenience or discomfort. In fact, we have become so pest conscious that we tend to think of some organisms only as pests, and the philosophy of "once a pest, always a pest" is far too commonplace. From an ecological point of view, this is undesirable because it leads far too frequently to pest control programs that are more disruptive than beneficial. We should realize that situations change and only consider organisms to be pests when and where they cause a significant amount of damage or, in our best judgment, will do so if we do not intervene. This, of course, requires some understanding of the relationship between numbers and damage as well as of the factors that cause populations to increase to the point at which significant harm will result.

Many pest situations are rather natural phenomena. For example, large numbers of mosquitoes and biting flies emerge more or less simultaneously in areas where the climate synchronizes their development, as it does throughout the subarctic. When the bulk of the adult population is present at one time and many individuals are competing for their needs, the discomfort they cause can be quite severe. Similarly, the ecologically simple and relatively unstable northern forests periodically experience large-scale outbreaks of defoliating insects. The tree mortality that results is part of the natural management of such systems, but it is not always in the best interest of man, who may have plans to harvest the fiber at some future time. Many other similar examples could be given, but the point I wish to make is that numerous insect pest problems have their origin in natural functional relationships but cause concern because of the special emphasis that we place on our personal needs and comfort.

The crop plants and animals that form the main components of modern agroecosystems had their origin in the natural ecosystems from which man originally gathered his food and fiber. In their ancestral form, these species interacted with other components of their environment according to the coadaptations that arose during their evolution. When first cultivated or domesticated, the selected plants and animals were scarcely removed from the conditions of their established range and many of the natural interactions between them and other species were perpetuated. In the 10,000-year history of agriculture, many domesticated species have been improved through selection and natural hybridization, but for the most part the changes were gradual until the more recent efforts of modern plant and animal breeders. Prior to these developments, many symbionts were probably able to adapt quite readily as agriculture slowly changed.

Once an organized form of agriculture permitted man to establish centers of civilization, he began to explore and expand into those new areas that provided the requisites for a higher-quality life. As he moved about, he initially relied heavily on the crop plants and the domesticated animals with which he was familiar, carrying them with him wherever he went. Once established in a new area, he added native species to his list of desirable cultivars. By the Renaissance, the transport of domesticated species from region to region was commonplace and continued at a high level throughout the colonial period. At the same time, a number of insects associated with crop plants and domestic animals were transported along with their hosts. A good example is the white cabbage butterfly, *Pieris rapae*, which was first introduced to North America from Europe about 1860. *P. rapae* is now a cosmopolitan pest of crop plants belonging to the cabbage family (crucifers).

Many times crop plants were introduced to new areas without the introduction of some of their more important natural enemies, but serious pest problems have also arisen because of introductions that came after the establishment of a crop. The Hessian fly, *Mayetiola destructor*, frequently a serious pest of wheat, was not introduced to America along with wheat but was brought in later in the straw used as animal bedding by the Hessian troops who came to fight in the Revolutionary War. The cereal leaf beetle, *Oulema melanopus*, is another pest of wheat in America that developed following a more recent introduction (ca. 1958), and there are many similar examples involving other crops.

Pest problems, particularly in forestry and agriculture, are population problems. Often there is something about a new environment that allows a species to increase in numbers much more rapidly once it has become established there than it did in its place of origin (Elton, 1958). This has happened on numerous occasions following the introduction of species associated with related crop plants or forest trees to other countries. Frequently the introduced species are not serious pests in their place of origin because of various environmental constraints, including weather and natural enemies. But in the new environment they are freed from such constraints and become pests. For example, the European corn borer, *Ostrinia nubilalis*, was imported to the United States in broom corn, probably from Hungary where it was not a serious pest. Over its native range the population is held down by a complex of native parasites and predators and because under the climate regime of eastern Europe it can complete only one generation per year. Following its introduction into the northeastern United States, the corn borer spread quite rapidly in the absence of natural enemies and became established throughout the corn-growing areas of this country. After a period of about 20 years, a second generation of corn borer began to appear in some of the warm areas and a bivoltine biotype soon became widespread. Now it is the characteristic of two generations per year that forms the basis of the European corn borer problem in the western hemisphere.

Similarly, the Balsam Wooly Aphid, *Adelges piceae,* became a pest of North American forests. Rarely a pest in the coniferous forests of Europe, this adelgid was introduced (probably on nursery stock) into the more moderate coastal climates of the northeast (ca. 1900), Oregon (ca. 1928), and British Columbia (ca. 1958). Following each introduction, the insect spread rapidly through the local fir (*Abies*) forests and is now a serious pest, particularly in areas where the climate permits several generations per year combined with a high level of survival during the mild winter months.

Far too often, species have been purposely introduced to new areas in the belief that they will provide either tangible or intangible benefits. This has been particularly true for birds, fish, and plants, but there are some rather notable cases among the insects as well. The gypsy moth, *Porthetria dispar,* was imported to Massachusetts from Europe in 1869 by a misguided naturalist named Leopold Trouvelot who hoped to crossbreed the hardy gypsy with the silkworm moth as the basis of a new textile industry. The gypsy moth escaped to become one of the most dreaded pests in the eastern United States (see Graham, 1972). Because of its hardiness and the fact that the larvae can feed on a wide variety of host trees, ranging from pine to holly, the gypsy moth is a threat to forest and ornamental trees throughout the country. Vast sums of money are being spent in an attempt to contain the pest within the presently infested areas.

The African bee, *Apis mellifera adansonii,* is more aggressive, produces stronger colonies, and gathers more honey than its Italian and Caucasian counterparts, *Apis mellifera ligustica* and *A. m. caucasica,* respectively. As a result, the African bee was imported to South America by a group of Brazilian scientists who were interested in bee genetics and in the development of a hybrid that would be a better honey producer. Unfortunately, the African bee escaped and has become established over a wide area of South America (Michener, 1975). Because of its aggressiveness, the African subspecies and hybrids have replaced the Italian and Caucasian subspecies in some areas, and are creating a problem because of their vicious nature and the fact that their venom is about twice as toxic as that of *A. m. ligustica* or *caucasica.* The rapid spread of the so called Brazilian bee in South America (Figure 20-1) is thought by some to pose a threat to apiculture in the United States.

Many pests have their origins as components of natural communities in areas to which crops or domestic animals are introduced. Weeds, for example, are simply plants growing where they are not wanted, so almost any native plant with the capacity to invade a disturbed site can become a weed pest of agriculture. The same is true of ectoparasites of native animals when they readily attack introduced domestic breeds; the screwworm *Cochliomyia hominivorax* in the southern United States is a typical example, having transferred from native deer and rodents to cattle.

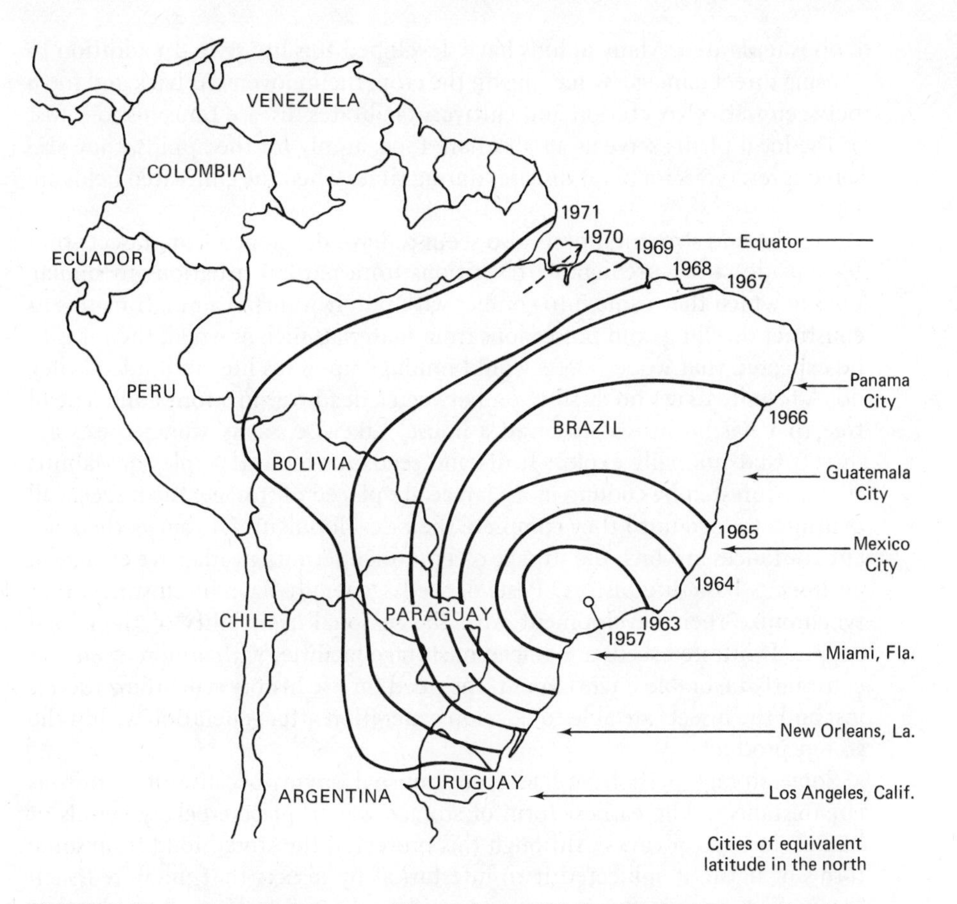


**FIGURE 20-1.** Spread of *Apis mellifera adansonii* following its introduction into South America near Rio Claro, São Paulo, Brazil in 1957. (Reproduced, with permission, from *Annual Review of Entomology*, Volume 20. Copyright © 1975 by Annual Reviews, Inc. All rights reserved.)

Quite often insects that feed on the foliage of native plants switch over to related introduced crop plants and respond to the increased food supply with a large increase in population. The alfalfa butterfly, *Colias eurytheme*, a California native that fed naturally on scattered wild legumes, was once quite a rare species, but the widespread cultivation of alfalfa throughout the state has made the butterfly a common sight in agricultural areas where it often attains pest status. Likewise, the Colorado potato beetle, *Leptinotarsa decemlineata*, was a little-known native of the eastern slopes of the Rocky Mountains where it fed on buffalo bur, a member of the family Solanaceae. The pioneer settlers introduced another solanaceous plant, the potato, along the trail westward, and the beetle soon spread eastward moving from potato patch to potato patch, often totally destroying the settlers' plantings. *L. decemlineata* remains a major pest of potatoes in the United States and also attacks a variety of additional solanaceous crops, including tobacco, peppers, potatoes, and eggplant.

Although numerous native species that become pests of introduced crops have a limited range of hosts, as is true of the Colorado potato beetle, they may be rather general feeders that increase in numbers as a result of the favorable conditions provided by a crop. Once the crop is harvested, such insects move back to native plants, which serve as a holdover site until the crop is replanted. Many aphids have developed this life style. In addition to causing direct damage by feeding on the crop, their movement back and forth between native vegetation and cultivars facilitates disease transmission. Just as the local plants serve as an alternate food supply for the aphids, they also serve as reservoirs for plant diseases during times when the cultivated fields are bare.

Storage and structural pests also seem to have developed from insects that have made rather predictable transitions from natural situations to similar ones in which they came into conflict with purely human values. If man is to construct dwellings and possessions from materials such as wood, then it is to be expected that wood borers would impinge upon his life, and indeed they do. A termite draws no distinction between a dead tree in a forest and a dead tree that has been used to build a house. Likewise, many storage pests are insects that normally exploit fruits and seeds for food. If apples or walnuts that are infested by codling moth larvae are placed in storage, the insects will continue to feed until they complete their development. Of course, there are circumstances in which the storage of a product permits an adaptive change in an insect's basic life history. Pests of seeds generally have life histories that synchronize their development with the seasonal availability of their food supply. If infested seeds are placed in storage facilities with a more even and constantly favorable environment, the need for life history synchronization is lost and the insects are able to develop generation after generation within the stored product.

Some storage pests have had more unusual origins because of fortuitous circumstances. The earliest form of storage was to place crockery vessels or baskets in trees or caves. Although this protected the stored food from some forms of harm, it subjected it to infestation by insects that inhabited such places. For example, insects such as psocids and a variety of small beetles that fed on organic detritus associated with bird nests or animal lairs probably gained access to man's caches of food and were probably carried from place to place in them. Many of these insects have since been spread by world trade and are recognized as cosmopolitan species of general concern.

There are a few insects the mere presence of which causes some alarm. In most cases these are species with a history of causing economic damage in their home range, and, as a result, regulations are enforced to keep them from spreading to new areas. If a single specimen is encountered within a restricted zone, an eradication program is usually proposed, as was the case following the recent discovery of both the Japanese beetle and Oriental fruit fly in southern California. However, most pest problems are problems of high population. A real pest problem develops only when the population of an injurious species reaches or exceeds the level at which we cannot tolerate the damage, discomfort, or health hazard it will cause.

### *Sources of Pest Problems*

The populations of all animals, with the exception of man, tend to fluctuate up and down around some average level we refer to as the animal's population equilibrium. Some species display population oscillations of greater amplitude than others, and reference is repeatedly made to populations as being either unstable (highly variable) or stable. Communities in which the populations of some species oscillate widely are likewise labelled as unstable systems. Many ecologists have drawn attention to the fact that community stability seems to be related to diversity, and instability to a lack of diversity, but others see no causal relationship between them. For a review of stability and diversity in agroecosystems, see van Emden and Williams (1974). I do not wish to enter the controversy, but there is ample evidence that the management practices employed in agriculture and forestry, which are basically designed to increase productivity by stalling succession, often cause reduced diversity and at the same time stimulate the development of pest problems.

Periodic animal population outbreaks, such as those of forest defoliators, seem to occur in relatively simple ecosystems, such as boreal forests, more frequently than they do in complex systems, such as the topical rain forests. The natural role of some defoliators in the northern forest is to disturb the simple climax community and to induce diversity by opening up areas in the stand to succession.

Many agricultural crops closely resemble the highly productive early stages of succession and are also relatively simple in terms of age, spatial, and species diversity. The insect pests of these crops, like those of the boreal forests, seem to have evolved a high degree of vagility and a high inherent rate of increase such that their populations oscillate widely. In northern forests they increase in numbers rapidly during periods of favorable climate. In agroecosystems they respond to other factors such as the quality and quantity of food resources. The cultivation of a single uniform crop over an extensive area (**monoculture**) provides a basis for a population increase among many potential pests. As Marchal (1908) stated, "Man in planting over a vast extent of country certain plants to the exclusion of others, offers to the insects which live at the expense of these plants conditions eminently favorable to their excessive multiplication." Pimentel (1961) demonstrated the wisdom of Marchal's statement by showing experimentally that several insect species attained outbreak status in single-species plantings of crucifers but did not do so in mixed-species plantings.

Although Pimentel's study indicated that increased species diversity was correlated with reduced herbivore outbreaks, it did not indicate clearly the nature of the regulatory process. Complexity can only contribute to stability if there are more or less direct interrelations between the species. Certainly the cultivation of a single species of plant increases the ease with which herbivores that feed upon it can locate food, and the availability of a large quantity of food decreases competition and density related mortality. The addition of different plant species therefore, could reduce the population of a herbivore restricted to the original plant by disrupting the continuity and abundance of its food supply. However, the increased plant species diversity need not affect the herbivore in any other way. If new herbivores that entered the system to feed on the added species of plants could serve as alternate hosts for enemies of the pest herbivore these natural enemies might be maintained at a higher population and have a stabilizing effect on the pest. In contrast, Watt (1965) contends that an increase in the diversity of parasites and predators of a pest may lead to stability on their trophic level. But this may impede their ability to respond to a rapid increase in the pest population and thereby contribute to overall instability.

Most of the highly productive agricultural acreage in the world is under monoculture, and, generally speaking, the intensification of cropping practices or the cultivation of a single crop with a fairly long developmental period provides an ideal environment for the development of pest problems. This can be an even more risky procedure in areas where the climate is favorable to the continuous cultivation of the same crop. In many regions with favorable year-round temperatures, multiple cropping has been impossible because of the occurrence of a dry season, but this has been overcome by irrigation, and pest problems have resulted. In India, recent leafhopper outbreaks in rice have been attributed to irrigation (Pradhan, 1971), and in Ghana, a second crop of yams is often seriously damaged by increased populations of yam beetles (Nye and Greenland, 1960).

Nevertheless, we cannot simply label monocultures uniformly as a source of pest problems. In some cases they appear to have advantages over mixed culture, especially when other crops provide alternate hosts for potentially damaging species. For example, the southern masked chafer (a scarab beetle) is a more serious pest of cotton and of soybeans when the two crops are planted adjacent to wheat, because the larvae feed on the roots of the wheat, whereas the adults feed on the leaves of the cotton and soybeans.

The life history of the crop under cultivation is also an important consideration. A mature stand of alfalfa grown as a perennial may be a highly diverse ecosystem, as is apparent to anyone who has walked through such a field. On the other hand, a crop such as radish grown as a root crop, may progress from a barren seed bed to harvest in a matter of weeks, scarcely sufficient time for a local pest population to increase, unless as a result of influx.

Obviously, each situation is different, and we cannot make a blanket statement that increasing species, age, or spatial diversity will reduce pest problems. As pointed out by van Emden and Williams (1974), adding additional species of herbivores may serve to damp the increase of individual herbivore populations, but we would probably lose just as much or more of the crop biomass; such a procedure could only be justified, therefore, if the addition of a nondisease vector reduced the population of a vector.

There are, however, a number of features of the way in which crops and forests are managed that enhance the build-up of plant-feeding insects. For example, the damage done to residual trees plus the slash left behind when a forest is harvested may provide an abundance of suitable breeding sites for bark beetles, causing their population to increase to outbreak proportions the following year. The practice of crop fertilization, designed to increase productivity, may also cause an increase in pest populations. Often the crop plants become a more nutritious and attractive source of food. To cite a few examples, leafhopper populations were highest on rice fertilized with heavy applications of nitrogen (Nene, 1971), and mite populations increase on a variety of fertilized crops (LeRoux, 1954; Cannon and Connel, 1965). The same is true for aphids and a variety of other insects. On the other hand, there are reports that fertilizer applications create less favorable conditions for some phytophagous pests, and the more prolific growth of fertilized crops may increase the level of insect damage that can be tolerated.

Many plants have evolved mechanisms that prevent or discourage the feeding of herbivores or provide resistance to disease. These mechanisms range from the bearing of stout spines that act as feeding deterrents to large browsers, to leaf pubescence or sticky exudates that inhibit insect attack, to complex chemical substances that act as poisons, repellents, or hormone mimics. But sometimes the characteristics that make a plant undesirable to its natural enemies also make them undesirable as agricultural crops. Consequently, these characteristics have often been selected out in the process of crop plant development. This is particularly true of certain chemicals such as terpenes and alkaloids that affect the flavor of the crop. In other cases, resistance factors that do not affect flavor are accidentally lost in breeding programs designed to improve other characteristics (van Emden, 1966).

In modern agriculture, one of the more exciting developments has been and still is the development of high-yielding, and insect- and disease-resistant varieties. In the latter case, it may mean breeding back into a crop characteristics that were previously bred out intentionally or inadvertently. For example, wheat varieties with pubescent leaves appear to be less attractive to the cereal leaf beetle, but pubescence is no longer a characteristic of commercial wheat varieties (Haynes, 1972). There are many examples of pest problems that have been lessened by the introduction of resistant varieties. However, it must be realized that the introduction of new high-yield varieties in place of traditional varieties carries with it the danger of pests developing from species that were of minor importance in more standard crops.

The high degree of genetic diversity that characterizes the wide variety of crop plants grown in an area provides some degree of protection against any one of a number of potential local pests becoming injurious. For example, Smith (1971) reported that more than 600 varieties of Bulu rice were grown throughout Indonesia, whereas the new high-yielding varieties that now dominate rice culture in the Far East were all developed from a single cross, which narrowed the germ plasm of the crop considerably. As a result, new pest problems involving leafhoppers and tungro virus have developed. Although it is necessary that we be aware of such problems, it would be unfair not to mention that the plant breeders who develop new crop plants are well aware of the potential dangers and are usually ready to respond with new breeding programs.

Irrigation permits sequential cropping in areas with suitable temperatures for most of the year and can thereby create pest problems, as mentioned earlier. In addition, irrigation produces more succulent plants that are more attractive for both feeding and reproduction and may facilitate the spread of plant diseases. Rivnay (1964) lists a number of instances of increased insect attack attributable to irrigation, particularly in cotton and peanuts. On the other hand, insect populations and damage may be reduced by irrigation; in fact, flooding is a practical way to reduce populations of some soil insects.

Many important crops are annuals planted in well-prepared seedbeds. In such a situation, many pest problems must begin with the invasion of the crop from less disturbed areas elsewhere. Interestingly, the tendency for insects to disperse and their postdispersal rate of increase are related to the stability of their habitat. Those associated with unstable vegetation (such as annual crops) have a greater tendency to disperse and a greater potential for increase than those associated with stable vegetation (MacArthur, 1960; Southwood, 1962). Many species associated with temporary habitats disperse between generations as a characteristic of their life history and may also disperse readily when their habitat is disturbed by practices of cultivation. These possibilities can be important in crops grown along a transect through areas ranging from early to late planting dates, since the cropping practices in the early crop may generate potential invaders of the late crop. If the prevailing wind tends to blow along the transect in the direction of the later plantings, serious pest problems can result.

Population increases within an agroecosystem as the result of invasion are also influenced by the distance between the crop and the source of the invaders, as well as by the over-all attractiveness of the crop to the invading species. Wind may play a significant role in determining the movement of potential pests between crop areas, but it may not always be the prevailing winds that cause problems. For example, Cook (1967) showed that the beet leafhopper in the San Juaquin Valley of California often spreads against the prevailing wind because of behavior patterns that make the leafhoppers susceptible to dispersal on local winds related to the valley's topography.

Actively dispersing insects are able to respond to a variety of stimuli, including crop color, form, and odor or the microclimatic gradients associated with the cultivated land. This can lead to preferential settling in vegetation with desirable characteristics. Many agroecosystems contrast with the surrounding vegetation in a way which can increase the attractiveness of the crop to dispersing insects. Passively dispersed organisms also enhance resident populations and thereby contribute to the creation of pest problems. The number of wind-blown individuals that impinge upon a crop will, of course, depend on its location relative to the source of invaders and on the size of the target formed by the crop.

In the previous paragraphs, we have really been enumerating the ways in which agricultural and forestry practices create an environment more favorable for growth of populations of insects that have the potential for becoming pests. Once we determine that a pest problem exists, we normally respond with a control tactic designed to make the environment less favorable to the pest and thereby reduce its numbers. But direct pest control, such as the application of a pesticide, while making the environment unfavorable for a short time, may in fact make it more favorable in the long run. Furthermore, we must be aware of the fact that there may be some reaction elsewhere in the system or in a neighboring system. For example, Blair (1964) suggested that the elimination of the screwworm in the southeastern United States may permit an increase in deer and jack rabbit populations, thereby increasing the competition for range vegetation between these species and cattle. Similarly, the use of systemic insecticides against livestock pests could contaminate their dung and destroy the insects that aid in its decomposition (Anderson, 1966). These examples are rather subtle, however, compared to the impact of applications of nonspecific pesticides to crops.

Although pesticide treatment of a crop ecosystem frequently produces a spectacular decline in the pest population, the results may be rather temporary. If the destruction of the target pest is accompanied by the destruction of other phytophagous species plus most of the target's natural enemies, there may be a subsequent rapid pest population increase called **pest resurgence.** This kind of population rebound can result from a variety of factors. In a simple case, the surviving pests greatly outnumber the surviving natural enemies and also have a higher fecundity, so increase very rapidly (Figure 20-2). The rapid resurgence of the pest may also occur because the survivors are more vigorous and have a higher reproductive potential than many of those that did not survive, combined with a marked reduction in both interspecific and intraspecific competition.

In most agroecosystems there are numerous phytophagous species, most of which do not cause any significant damage because they do not normally occur in sufficient numbers. Nevertheless, they have the potential to become pests. A pesticide treatment may release such a population from the regulatory factors that normally keep it in check, thereby causing a new (**secondary**) pest outbreak to develop (Figure 20-3). A situation of this type developed

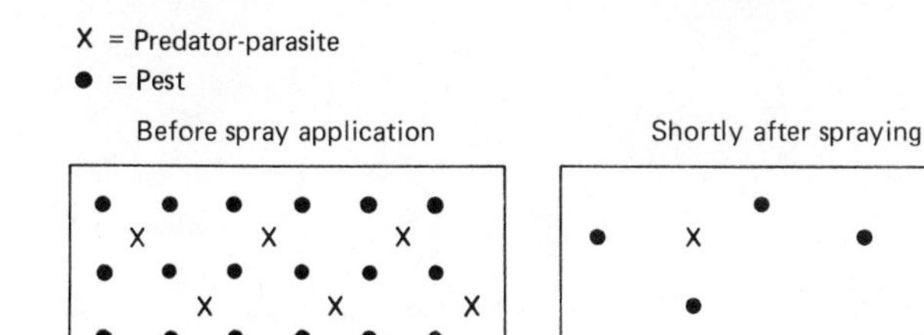
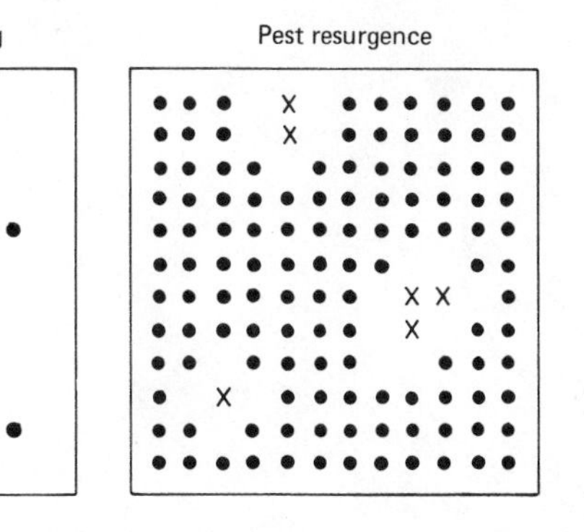


FIGURE 20-2. Diagrammatic representation of pest resurgence. The three squares represent an agricultural plot immediately before, immediately after, and some time after treatment with an insecticide to control the pest (solid dots). The immediate result is significant reduction in the pest accompanied by an even greater reduction in the population of natural enemies (x's). Because of the unfavorable ratio between the pest and its natural enemies and the tendency for pest individuals to invade the crop, there is a rapid resurgence of the pest. (Reproduced with permission, from Smith and van den Bosch, in *Pest Control,* W. Kilgore and R. Doutt eds., 1967, Academic Press, Inc.)

early in the use of DDT as an agricultural insecticide against apple pests. The DDT killed a wide range of phytophagous insects as well as a variety of parasites and predators but was not lethal to spider mites. In the absence of the competition and natural enemies that normally regulated their populations, the spider mites often increased greatly in numbers and became pests of the orchards in their own right.

Finally, the favorable nature of most crop environments, combined with pest resurgence, secondary pest outbreaks, and more susceptible host plants, have too often resulted in the need to repeat the use of some pesticides. In a number of cases, this new selective pressure led rapidly to the expression of various forms of pesticide resistance by a large variety of insects. Once resistant to chemical suppression, an insect becomes in a sense a new pest with an increased potential for causing economic damage.

## *Major Types of Insect Pests*

Approximately half of all of the insects known are plant feeders. Of the other half, most are entomophagous (insect-eaters), a large number are scavengers, and a relatively small number feed as external and internal symbionts of animals other than insects, particularly vertebrates, including man and his domesticated animals. Exactly how many insects can actually be labelled as pests is difficult to determine because some are eliminated and others are added as we change our approach to the management of insect habitats and to pest control. However, if we consider only those species that cause significant problems and therefore require regular control procedures to reduce the damage they cause, the list would contain only about 3,500 species. In the United States, probably only about 600 species of insects can be considered as serious pests with any regularity.

Inasmuch as about 50% of all insects are herbivores, it is not surprising that

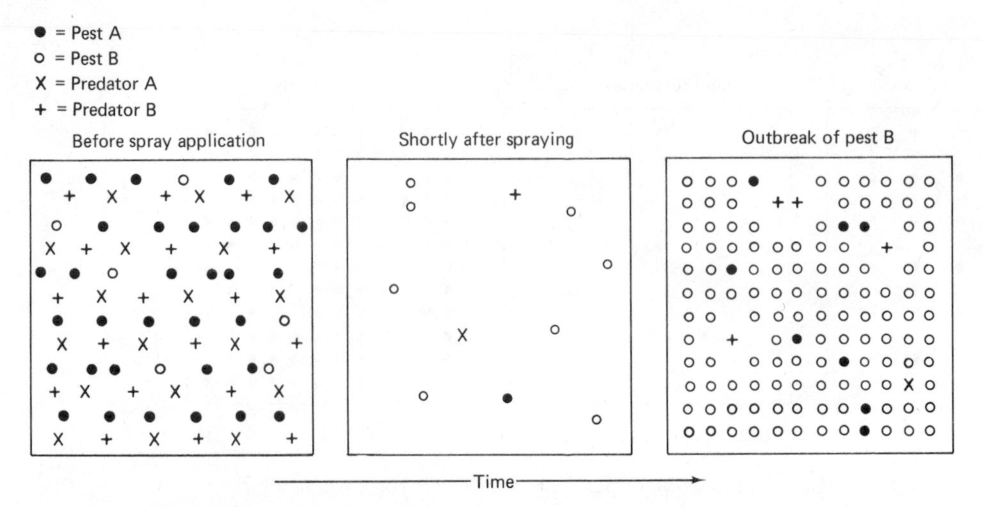


FIGURE 20-3. Diagrammatic representation of a secondary pest outbreak. The three squares represent an agricultural plot immediately before, immediately after, and some time after treatment with an insecticide to control a pest (solid dots). The insecticide effectively reduces the pest population but reduces the population of natural enemies (x's and +'s) as well. A noninjurious population of a potential pest (o's) is not affected by the insecticide. Released from predation and competition, the second species increases to a damaging level of abundance. (Reproduced with permission from Smith and van den Bosch, in *Pest Control,* W. Kilgore and R. Doutt eds., 1967, Academic Press, Inc.)

most insect pests are injurious to plants. In addition to causing damage by directly injuring, destroying, and consuming plant tissue, insects are also involved in the dissemination of plant diseases, consideration of which will be left to Chapter 21. Insects also attain pest status by annoying or injuring man and his domesticated poultry and animals and, to a lesser extent, wild life. Annoyance to humans is sometimes difficult to evaluate, but in the case of poultry and livestock, it can often be measured by a decline in egg or milk production or weight gain. Feeding by internal and external parasites may lead to death, either directly or by predisposing an animal to other mortality factors. Most insect pests that are important from the viewpoint of human health are disseminators of disease; they will be considered in Chapter 22. Finally, there are insects that depreciate or destroy our possessions, including stored food, furniture, books, clothing, and many of our structures, ranging from telephone poles and railroad ties to houses and other buildings.

In terms of evaluating insect-caused losses in agriculture and forestry, it is often useful to distinguish between direct damage and indirect damage. Direct damage means a commodity that is of some value is directly destroyed or downgraded. Stored products that are eaten or spoiled by insect feeding or contamination, the infestation of a piece of fruit or a seed, a tree that is killed, feeding damage to a piece of lumber, and the death of a calf from a parasite attack are all examples of direct damage. A decline in the yield of a crop caused by insect injury to nonusable parts of the plant, reduced incremental growth of forest trees resulting from defoliation, and a decline in milk production as a result of parasite-caused irritation are examples of indirect damage.

### *Insect Damage to Plants*

The most conspicuous damage to plants results from insects chewing directly on external parts, including the leaves and roots, to acquire food. The results of such damage may be quite spectacular on an individual plant (Figure 20-4) or over a large area (Figure 20-5). Other chewing insects may feed within stems, fruits, seeds, and even between the surfaces of the leaves. In the latter case, the damage may be quite noticeable, but most internal injury is often difficult to detect until serious damage has been done.

The second major form of plant damage is caused by insects with piercing-sucking mouth parts. Although not as conspicuous as damage caused by chewing, it can be no less significant because of the withdrawal of sap, the creation of puncture wounds, and the cell mortality that may be followed by scar tissue formation. Severely damaged foliage may lose much of its photosynthetic capacity, and photosynthesis may be reduced by mold that grows in the sticky exudate of plant juice-feeding insects (Figure 20-6).

**FIGURE 20-4.** A corn plant severely damaged by the feeding of larvae of the European corn borer. (Photograph courtesy USDA.)

FIGURE 20-5. [OPPOSITE] An autumn-like scene photographed during midsummer in Connecticut following an outbreak of the gypsy moth. Note that both deciduous and coniferous trees have been almost completely defoliated. (Photograph courtesy USDA.)

Other forms of plant damage result from the injection of toxins by insects at the time of feeding and from oviposition wounds. The toxins may either cause the plant to die or result in deformation and galling. Quite a variety of insects create wounds in the process of laying eggs and these may lead to breakage or invasion by pathogens.

All of the activities mentioned can cause both direct and indirect damage to crops, forests, and ornamentals. Most direct damage in agriculture is caused by insects that feed or oviposit on fruiting structures. A few well-known examples include the feeding of codling moth larvae in English walnuts and apples (Figure 20-7), related caterpillars such as Oriental fruit moth and cherry fruit worm, and a number of insects, including lygus bugs, cotton boll

FIGURE 20-6. The discoloration of a citrus leaf caused by sooty mold growing in the honey dew produced by the woolly white fly. A clean leaf is shown for contrast. A heavy coating of the mold can substantially reduce photosynthesis.

**FIGURE 20-7.** An apple cut open to show the damage caused by a codling moth larva. This pest causes direct damage because each infested fruit is unmarketable. (Photograph courtesy USDA.)

worm, (Figure 20-8), pink boll worm, and boll weevil that damage the squares and bolls of cotton. The direct destruction of the leaves of lettuce and cruciferous crops by caterpillars such as the imported cabbage worm and cabbage looper and damage to root crops by wireworms can also cause serious losses. Bark beetles are the most important direct pests of forests. Some of the more important direct pests of American agriculture are listed in Table 20-1.

Sometimes a direct loss results from insect activities that affect the appearance of a product rather than cause a reduction in yield. Probably the best example is to be found in the citrus industry where fruit marked by mites or scale insects is downgraded (Figure 20-9) even though the edible parts are unaffected—truly a foible of our society. The attack of timber by ambrosia beetles is a significant cause of damage that results in the degrading of lumber, often more because of appearance than a decline in structural quality.

Generally, most indirect damage results from insect attack on vegetative parts (leaves, stems, roots) which causes a decline in the production of the reproductive parts or storage parts (fruits, seeds, tubers) that comprise the crop. Most cereal pests are of this type, including a variety of aphids that are juice feeders, the cereal leaf beetle and grasshoppers that chew on the leaves, wireworms, and cutworms that attack the roots, wheat stem maggot that bores in the stalk, and the wheat stem sawfly that injures the stalk with oviposition punctures.

The European corn borer has become one of the most serious indirect agricultural pests in the United States, not only because of its attacks on corn

FIGURE 20-8. The cotton boll worm, another direct pest, emerging from a damaged boll in which it developed. (Photograph courtesy USDA.)

**TABLE 20-1** **Representative Important Direct Pests of Agricultural Crops (Arranged alphabetically by family)**

| Common Name | Order | Family | Scientific Name | Crop Damaged |
|---|---|---|---|---|
| Boll weevil (I) | Coleoptera | Curculionidae | *Anthonomous grandis* Boheman | Bolls of cotton |
| Pink bollworm (I) | Lepidoptera | Gelechiidae | *Pectinophera gossypiella* (Saunders) | Bolls of cotton |
| Cabbage looper (N) | Lepidoptera | Geometridae | *Trichoplusia ni* (Hübner) | Lettuce and cuciferous crops |
| Lygus bug (N) | Hemiptera | Miridae | *Lygus hesperus* Knight | Legumes and cotton |
| Corn earworm (I)<br>Tomato fruitworm<br>Cotton bollworm | Lepidoptera | Noctuidae | *Heliothis zea* (Boddie) | Corn and vegetable crops<br>Tomatoes and tobacco<br>Bolls of cotton |
| Codling moth (I) | Lepidoptera | Olethreutidae | *Laspeyresia pomonella* (Linnaeus) | Apples and English walnuts |
| Oriental fruit moth (I) | Lepidoptera | Olethreutidae | *Grapholitha molesta* (Busck) | Stone fruits |
| Imported cabbage worm (I) | Lepidoptera | Pieridae | *Pieris rapae* (Linnaeus) | Cuciferous crops and vegetables |
| Japanese beetle (I) | Coleoptera | Scarabaeidae | *Popillia japonica* Newman | Tree fruits, truck crops, and turf |
| Cherry fruit fly (I) | Diptera | Tephritidae | *Rhagoletis cingulata* (Loew) | Sweet cherries |
| Mediterranean fruit fly (I) | Diptera | Tephritidae | *Ceratitis capitata* (Wiedemann) | All fruits including citrus |
| Oriental fruit fly (I) | Diptera | Tephritidae | *Dacus dorsalis* Hendel | Peach, apricot, plum, apple, pear, quince |

I = Introduced species N = Native species

FIGURE 20-9. Cosmetic damage to a valencia orange caused by California red scale. (Photograph by M. Badgley.)

(Figure 20-10) but also because it feeds on more than 200 other varieties of plants, including soybeans, sorghums, cereals, potatoes, and beans. Another important pest, the Colorado potato beetle, also reduces yield indirectly by devouring the crown of the potato plant.

The order Homoptera contains a variety of important indirect pests, which seriously affect the yield of many crops, particularly tree fruits. Probably the most important pest in this group is the San José scale, which was introduced into California about 1870 (Figure 20-11). The San Jose scale does not cause much direct damage to the fruit but causes more serious permanent damage by killing twigs and branches. It is particularly destructive to apples, pears, peaches, and sweet cherries. Other members of this group include a number of other scales, aphids (Figure 20-12), white flies (Figure 20-13), and mealy bugs. Some of the more important indirect pests of American agriculture are listed in Table 20-2.

In forestry, as in agriculture, there is a wide variety of insects that act as indirect pests, but the defoliators are the most important. Most are lepidopterous larvae, including the infamous spruce budworm and gypsy moth, the hemlock looper and other geometrids, the Douglas-fir tussock moth, and the eastern tent caterpillar. The other main group of forest defoliators are sawfly larvae (Hymenoptera), among which those that attack pines or larch are the most serious pests. Some of the most important forest pests of North America are listed in Table 20-3.

**FIGURE 20-10.** A corn field devastated by the European corn borer. (Photograph courtesy USDA.)

**FIGURE 20-11.** The San José scale on developing apples. (Photograph courtesy USDA.)

FIGURE 20-12. Damage to the developing head of a cabbage caused by aphids.

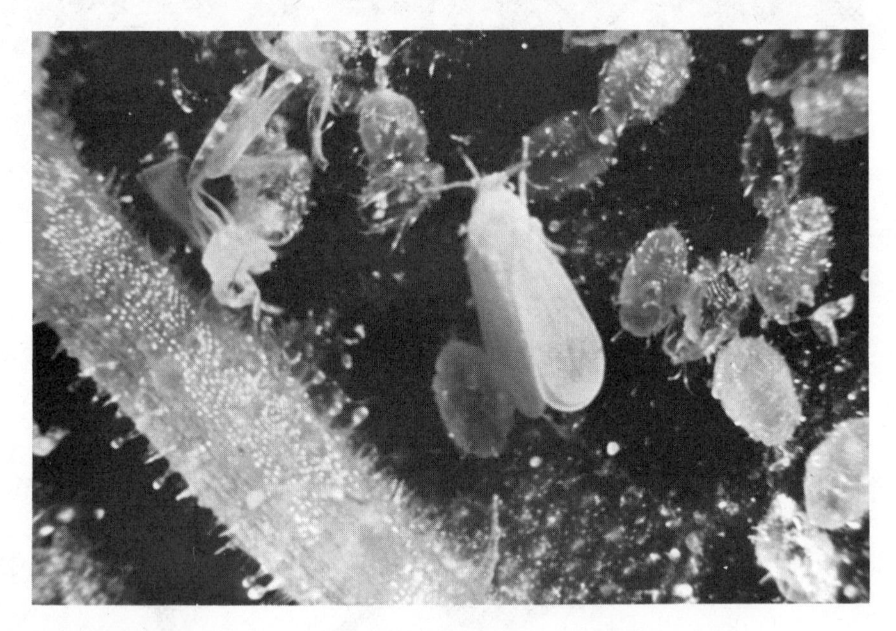

FIGURE 20-13. An infestation of greenhouse whitefly on the leaves of a squash plant. (Photograph by M. Badgley.)

### *Insect Injury to Livestock and Poultry*

Although heavy infestations of parasites cause some livestock mortality, deaths are rather limited and generally restricted to young individuals. Most insect-caused losses are indirect, resulting from poor performance because of annoyance or a weakened condition. Just as with insects that attack plants, insects that attack livestock and poultry may be either quite host specific or attack a variety of domesticated species. Members of the order Diptera are the most general pests of livestock; these include the stable fly, horse flies, black flies, the horn fly, and the face fly. Livestock are sensitive to insect bites

**TABLE 20-2** **Representative Important Indirect Pests of Agricultural Crops (Arranged alphabetically by family)**

| Common Name | Order | Family | Scientific Name | Crop Damaged |
|---|---|---|---|---|
| Migratory grasshopper (N) | Orthoptera | Acrididae | *Melanoplus sanguinipes* (Fabricius) | Cereals and corn |
| Green bug (I) | Homoptera[1] | Aphididae | *Schizaphis graminum* (Rondani) | Cereals, particularly wheat |
| Wheat stem sawfly (N) | Diptera | Cephidae | *Cephus cinctus* Norton | Wheat and other cereals |
| Cereal leaf beetle (I) | Coleoptera | Chrysomelidae | *Oulema melanopus* (Linnaeus) | Wheat and other cereals |
| Colorado potato beetle (N) | Coleoptera | Chrysomelidae | *Leptinotarsa decemlineata* (Say) | Potato, tomato, pepper, tobacco |
| Mexican bean beetle (N) | Coleoptera | Coccinellidae | *Epilachna varivestis* Mulsant | Beans |
| White-fringed beetle (N) | Coleoptera | Curculionidae | *Graphognathus* spp. | Corn, peanuts, potato, cotton |
| San Jose scale (I) | Homoptera[1] | Diaspidae | *Quadraspidiotus perniciosus* (Comstock) | Apple, pear, peach, cherry |
| Chinch bug (N) | Hemiptera | Lygacidae | *Blissus leucopterus* (Say) | Wheat, corn, sorghums |
| Army worm (N) | Lepidoptera | Noctuidae | *Pseudaletia unipuncta* (Haworth) | Cereals |
| Fall army worm (N) | Lepidoptera | Noctuidae | *Spodoptera frugiperda* (Smith) | Cotton, corn, peanuts |
| European corn borer (I) | Lepidoptera | Pyralidae | *Ostrinia nubilalis* (Hübner) | Corn and field crops |
| Tobacco horn worm (N) | Lepidoptera | Sphingidae | *Manduca sexta* (Linnaeus) | Tobacco and tomato |

[1] Suborder Homoptera
I = Introduced species  N = Native species

and the feeding of parasitic larvae. The annoyance and irritation caused by biting flies buzzing about and feeding can cause livestock to become restless, to curtail feeding, and even to stampede. Such activities result in weakness and weight loss rather than gain.

Black flies can be particularly irritating, and in northern areas, where the more or less simultaneous emergence of the flies results in outbreaks almost every year, they have curtailed economic development, particularly the establishment of a cattle industry. In northern Saskatchewan, *Simulium arcticum* results in cattle mortality. An outbreak in 1946 is claimed to have killed 600 animals. *Simulium venustum,* which attacks man as well as domestic animals, can reduce milk production in northern areas by as much as 50% (Fredeen, 1956).

Horn or heel flies, which get their name from the fact that they cluster around the base of the horns and low on the hind legs where they escape the defensive actions of the host, may average as many as 5,000 individuals per animal. When the flies are controlled, cattle show increased gains of up to 55 pounds per animal, and a calf will gain an extra 75 pounds.

The native screwworm (Figure 20-14), although now largely under control, has been a serious livestock pest. In this species, it is the feeding of the maggots in healthy tissue adjacent to minor wounds that causes the damage. During a severe outbreak in 1935, 1.2 million infected animals and 180,000 deaths were recorded. Prior to effective control, the screwworm caused an estimated $100 to $140 million in damage each year to the livestock industry in the southern states.

The most important pests of poultry are biting lice of the Order Mallophaga. Of nine species that attack chickens in this country, the chicken body louse, chicken head louse, and shaft louse are the most important. The

**TABLE 20-3** **Representative Important Forest Pests of North America (Arranged Alphabetically by Family)**

| Common Name | Order | Family | Scientific Name | Principal Hosts | Type of Damage |
|---|---|---|---|---|---|
| Balsam woolly aphid (I) | Homoptera | Chermidae | *Adelges piceae* (Ratz.) | True firs | Galling and shoot damage |
| White pine weevel (N) | Coleoptera | Curculionidae | *Pissodes strobi* (Peck) | Pines | Kill terminal shoots |
| European spruce sawfly (I) | Hymenoptera | Diprionidae | *Diprion hercyniae* (Htg.) | Spruce | Defoliation |
| Hemlock looper (N) | Lepidoptera | Geometridae | *Lambdina fiscellaria* Guen. | Spruce, Hemlock, Fir | Defoliation |
| Douglas-fir tussock moth (N) | Lepidoptera | Lymantriidae | *Hemerocampa pseudotsugata* McD. | Douglas-fir | Defoliation |
| Gypsy moth (I) | Lepidoptera | Lymantriidae | *Porthetria dispar* (L.) | Variety of trees and shrubs | Defoliation |
| Red pine scale (I) | Homoptera | Margarodidae | *Matsucoccus resinosae* Rich. | Red pine | Removal of sap |
| Cone moth (n) | Lepidoptera | Olethreutidae | *Barbara colfaxiana* Kearf. | Douglas-fir | Destroys cones and seeds |
| European pine shoot moth (I) | Lepidoptera | Olethreutidae | *Rhyacionia buoliana* (Schiff.) | Pines | Damage to young shoots |
| Bark beetles (N) | Coleoptera | Scolytidae | *Dendroctonus* spp | Pines, Spruce, Douglas-fir | Killing by girdling |
| Ambrosia bettle (N) | Coleoptera | Scolytidae | *Trypodendron lineatum* Lec. | Douglas-fir, Hemlock | Perforate lumber |
| European elm bark beetle (I) | Coleoptera | Scolytidae | *Scolytus multistriatus* (Marsh) | Elm | Transmit Dutch elm disease |
| Larch sawfly (*) | Hymenoptera | Tenthridinidae | *Pristiphora erichsoni* (Hartig.) | Larch | Defoliation |
| Spruce budworm (N) | Lepidoptera | Tortricidae | *Choristoneura fumiferana* (Clem.) | Spruce, Balsam fir | Defoliation |

I = Introduced Species N = Native species * = Origin uncertain

feeding of heavy infestations of these ectoparasites causes severe irritation which can result in failure of the birds to gain weight, reduced egg production, and some mortality among young birds.

### *Other Forms of Insect Damage*

As mentioned earlier, a wide variety of insects feed on and despoil stored products, and if precautions are not taken, the losses sustained can equal or exceed those sustained by crops prior to harvest. Various authors have claimed that in areas where little or no pest control is practiced insects can destroy 50% of a crop before harvest and the other 50% after harvest. In the United States where storage facilities are good and control methods are readily available, storage losses attributed to insect damage amount to $300 to $600 million annually. According to one estimate, 130 million people could have lived for a year on the grain consumed and spoiled by insect pests throughout the world in 1968 (Wilbur, 1971). The most important storage pests are listed alphabetically by family in Table 20-4.

Many insects have found the environment in man's habitations and institutional structures to be highly favorable; their expansion into cooler

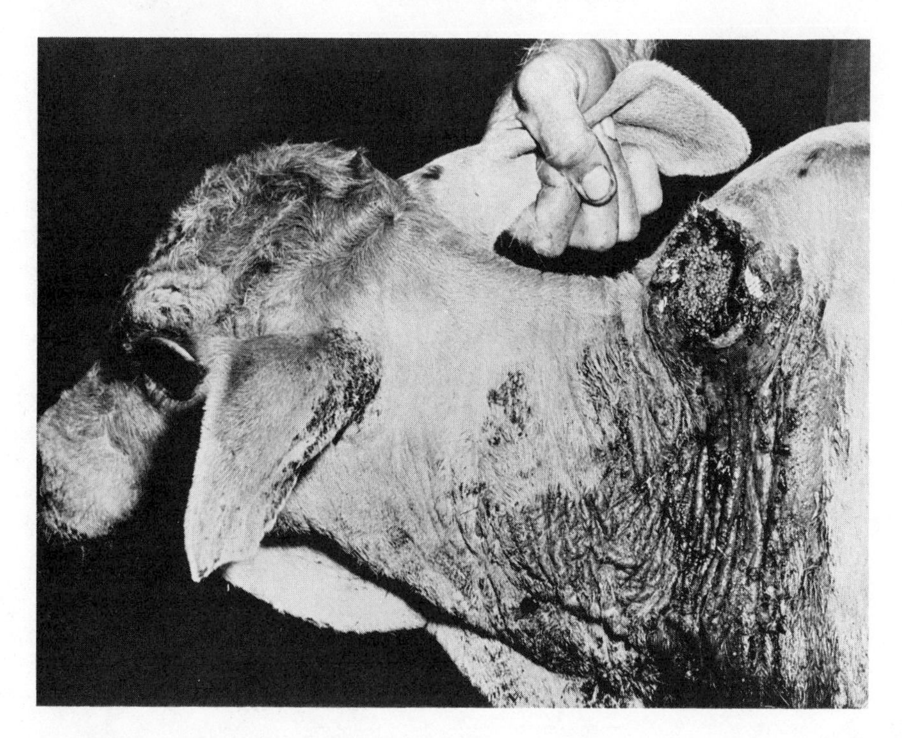

**FIGURE 20-14.** A typical example of a screw worm injury. Note the larvae in the wound in this Brahman calf and the adult flies attracted to the liquid exudate. (Photograph courtesy USDA.)

**TABLE 20-4** **Important Insect Pests of Stored Products (Arranged alphabetically by family). From Wilbur (1971)**

| Common Name | Scientific Name | Family | Order |
|---|---|---|---|
| Lesser grain borer | *Rhyzopertha dominica* (F.) | Bostrichidae | Coleoptera |
| Sawtoothed grain beetle | *Oryzaephilus surinamensis* (L.) | Cucujidae | Coleoptera |
| Merchant grain beetle | *Oryzaephilus mercator* (Fauvel) | Cucujidae | Coleoptera |
| Flat grain beetle | *Cryptolestes pusillus* (Schönherr) | Cucujidae | Coleoptera |
| Rusty grain beetle | *Cryptolestes ferrugineus* (Stephens) | Cucujidae | Coleoptera |
| Rice weevil | *Sitophilus oryzae* (L.) | Curculionidae | Coleoptera |
| Maize weevil | *Sitophilus zeamaize* Motschulsky | Curculionidae | Coleoptera |
| Granery weevil | *Sitophilus granarius* (L.) | Curculionidae | Coleoptera |
| Trogoderma complex (including khapra beetle) | *Trogoderma* spp. | Dermestidae | Coleoptera |
| Angoumois grain moth | *Sitotroga cerealella* (Olivier) | Gelechiidae | Lepidoptera |
| Cadelle | *Tenebroides mauntanicus* (L.) | Ostomidae | Coleoptera |
| Indian meal moth | *Plodia interpunctella* (Hübner) | Pyralidae | Lepidoptera |
| Mediterranean flour moth | *Anagasta kuehniella* (Zeller) | Pyralidae | Lepidoptera |
| Confused flour beetle | *Tribolium confusum* duVal | Tenebrionidae | Coleoptera |
| Red flour beetle | *Tribolium castaneum* (Herbst) | Tenebrionidae | Coleoptera |

climatic zones and their survival there over the cold winter months have become dependent upon man providing suitable living conditions. The larvae of clothes moths and carpet beetles feed on many items such as clothing, carpets, and upholstery that contain natural fibers. Numerous common household pests such as cockroaches and silverfish are opportunistic feeders on any starchy materials. In addition to consuming food products, they also eat starchy materials used in book bindings, stationery (Figure 20-15), and even wallpaper paste.

Carpenter ants, anobiid beetles, and termites are all damaging to wooden structures. Termites are major structural pests in the southern part of the United States and many tropical countries where they frequently necessitate extensive repairs as a result of the damage they do to structural timbers, flooring, and even wooden fences (Figure 20-16). In southern California, the law allows the purchaser of a home to request that the vendor provide a current termite inspection report that shows the dwelling to be free of termites. Frequently, wood used to construct furniture and other items is infested prior to manufacture, and the insects within continue to feed and damage such items for many years. I had one lady bring me a wooden-heeled sandal that she had purchased in Korea three years previously; the sandal had recently collapsed as a result of the feeding of a single larva of a cerambycid beetle.

**FIGURE 20-15.** Damage to an envelope caused by silverfish. (Photograph courtesy USDA.)

### *The Determination of Pest Status*

In the previous paragraphs we have briefly reviewed the various origins of potentially injurious insects and some of the features of crop and forest management that favor the increase in insect populations. However, as indicated earlier, insects do not warrant control on the basis of a history of having caused damage or simply because an effective pesticide is available.

In forestry and agriculture the objective of pest control is to prevent losses and thereby increase the profitability of the crop. In public health there are

**FIGURE 20-16.** Heavy structural damage to a house caused by termites. (Photograph courtesy USDA.)

other considerations that cannot be expressed in monetary terms, so pest control decisions in that sector are made on the basis of more intangible needs, and costs do not weigh as heavily.

Generally, the accounting used in agricultural pest control is primarily on the basis of the control costs and the value of the potential benefits in the year of the control program. The long-term effects of control procedures are rarely considered, regardless of whether they are favorable or unfavorable. In reality, any expenditures that have been made by governments to clean up pesticide pollution must be considered as a form of subsidy.

Frequently, pest species occur in a crop at a population level that does not result in damage sufficient to cause a loss of income. Except for a few rather special cases, the mere presence of a potential pest is not ample reason to engage in control. Arbitrary decisions to use pesticides as a prophylactic treatment have led to unnecessary residue and resistance problems and are no longer acceptable, even though it is not always easy to convince a grower who can see insects in his fields that control is not required. Control should only be undertaken, therefore, if the pest population reaches a density at which some economic loss will result if suppressive action is not taken. This population level is called the **economic threshold** and identifies a crucial element in any control decision process. The term **economic injury level** is often used synonymously with economic threshold but should be reserved to describe any level of pest population or damage of economic significance. Obviously these values require the establishment of some relationship between the density of the pest and the damage in terms of reduced yield that is likely to occur, and it is important that the population density be determined by an appropriate sampling procedure.

The relationship between pest density and damage can vary considerably and is seldom rectilinear. At low densities some pests act as thinning agents and can actually cause an increase in yield by reducing the competition between plants. Conversely, high pest populations experience intraspecific competition which may result in a lower per individual impact on the crop. Between these extremes there may be a more direct relationship between the pest population and reduced yield, but even then may vary from one situation to another. For example, the impact of a given density of a pest may vary depending on the stage of crop development; an insect population that can severely damage a crop in the early stages of development may not be significant in a mature crop close to harvest. Similarly, the economic threshold may need to be adjusted to take into account certain events that occurred earlier in the crop's development. Two different pests may be present at density levels that would be tolerable if they acted independently of one another, but the damage done by one pest early in the development of the crop may mean that the damage caused by a later attack by the second pest could not be tolerated.

Sometimes a large population develops in a crop or invades it beyond a critical stage of development. For instance, a large population of alfalfa caterpillars in a crop about to be mowed may be insignificant. Likewise, the invasion of a cotton crop by a potentially damaging population of pink boll worm or boll weevil may not result in a significant reduction in yield if the squares or bolls are developed beyond the stage that is preferred by the insects.

The situation may become even more complex if the potential problem involves insect transmission of a disease. A large vector population could be tolerated if the crop was a disease-resistant variety or if the pathogen reservoir was small or inaccessible. Conversely, a susceptible crop might suffer severe damage from a much smaller vector population if the insects were actively feeding and moving back and forth between the crop and an accessible pathogen reservoir. Consequently, in insect vector-disease situations, the establishment of an economic threshold requires information concerning the location and size of the pathogen reservoir as well as information about the vector population.

Whereas it might be possible to determine on the basis of experience that a given pest population attacking a particular crop at a particular time will cause a reduction in yield, it is often more difficult to determine what the reduced yield will mean in lost income. The latter is what usually influences pest control decisions. This may seem paradoxical in a world that is growing short of food, but it is nevertheless a factor in the crop production economics of intensive agriculture. There must be a favorable cost-benefit ratio for all phases of production. The costs of insect control include the price of the insecticide plus all aspects of its application and must be offset by at least an equivalent gain in income; to be realistic, the benefits from pest control should substantially exceed the costs. This means that the economic threshold must be closely related to the planned disposition of the crop. If a crop is grown for processing as in the case of apples to be used for cider or tomatoes to be used for paste, more damage of certain kinds can be tolerated, and the reduced overall value of the crop limits the amount that can be spent on pest control. Sometimes the market conditions change according to supply and demand, and it is unfortunate when money is spent and pesticides are used to protect a crop for which there is no market. These situations do develop and may be difficult to foresee when a control decision has to be made. For example, in 1970 only about 76% of the 793,000 ton peach crop in California was purchased by canners. An estimated 79,400 tons of peaches were shaken from the trees prior to harvest (Perelman and Shea, 1972). In 1976, millions of dollars worth of California's fruit and vegetable crops, previously treated with insecticides, rotted because of a strike in the processing industry.

As pointed out by Woods (1974), the use of pesticides is subject to the law of diminishing returns. The initial pest control input may provide less extra yield than subsequent units of input, but a point is reached at which further control expenditures result in little or no increase in yield as the pest population is suppressed below the economic threshold. The concept of total pest control that once prevailed does not make economic sense, even though some farmers adhere to the outdated clean field concept. Ideally, growers should be encouraged to curtail pest control at the point where the expected benefits of increased yield drop below the cost of additional control.

Additional considerations relate to whether the pest directly attacks the portion of the crop to be used or some other part, the damage to which causes a reduction in the yield of the desired part. In the first case, the greater the proportion of pests killed, the greater the proportion of saleable yield. In the second case, a lower level of control may result in no significant decrease in yield. For example, Wheatley and Coaker (1970) found that the percentage increase in yield of cauliflower rose only slightly as the dosage of pesticide directed against cabbage root fly was increased from 3 mg per plant to 50 mg per plant. Yet, the root damage was decreased from about 95% to 5%. Apparently, once established, the cauliflower plants were able to tolerate considerable root damage and only one low-level pesticide application was necessary just after transplanting.

At what point in the development of a stand of trees a potential pest problem occurs greatly affects the economics of pest control in forestry. The same philosophy may apply in some long-term agricultural crops, especially in the years prior to the formation of a harvestable crop. The annual increase in the value of a forest depends upon incremental growth and is very low on a per acre basis. There is therefore no annual income to offset the management costs of which pest control is a part. In the early stages of growth, protection against some pests such as shoot moths and weevils may be necessary to assure a growth form that will lead to marketable trees later on, but some tree mortality may be beneficial as a form of natural thinning. During intermediate years when growth is maximal, any losses in annual growth caused by defoliation, for example, may not be of sufficient ultimate worth to warrant a cash expenditure for insect control. However, when the forest is in a harvestable condition, it represents the value accumulated over a long time plus the accumulated management costs. At this stage, tree mortality represents a significant loss, so the degree of damage that can be tolerated is low and a fairly high-cost pest control program would be justified.

In many cases economic thresholds have not been established, and many of those established in the past were set too low just to be on the safe side. When economic thresholds are set too low, they result in unnecessary control programs and unnecessary production costs for growers. However, the establishment of ideal economic thresholds is not a simple matter, since, as the foregoing discussion indicates, it is often necessary to evaluate a number of variables, including the ultimate disposition of the crop, the stage of crop development, crop variety, climatic conditions, prior investment, and market conditions.

Various workers have struggled with this problem and have derived models of varying complexity in an attempt to formulate an approach that would optimize the relationship between yield and pest control costs. A rather simple example presented by Smith and van den Bosch (1967) illustrates that what may appear to be rather severe insect damage is not necessarily economic damage. Cotton has a limited capacity to set bolls, and the squares (buds) in excess of that number drop whether or not they are punctured by lygus bugs or attacked by other square-feeding insects. Consequently, the injury the insects cause to the excess buds does not result in a reduction in yield. But when the insect damage reduces the boll load below the plant's carrying capacity, some crop loss results. A small reduction in the number of bolls below carrying capacity may not be economically significant, but control could be required to prevent additional damage (Figure 20-17).

Ordish (1952) developed a tabulation of cost-potential benefit ratios for a series of crops that took into account a variety of situations. For example, he considered three types of apple orchards with different yield characteristics and determined the cost-benefit ratio for three levels of pest control (low, partial, and full) employed against severe, medium, and light attack by fungus disease and arthropods.

Headley (1972) developed a more sophisticated approach, aimed at establishing economic thresholds that permit optimization of the net return of a control procedure. The following summary of Headley's approach incorpo-

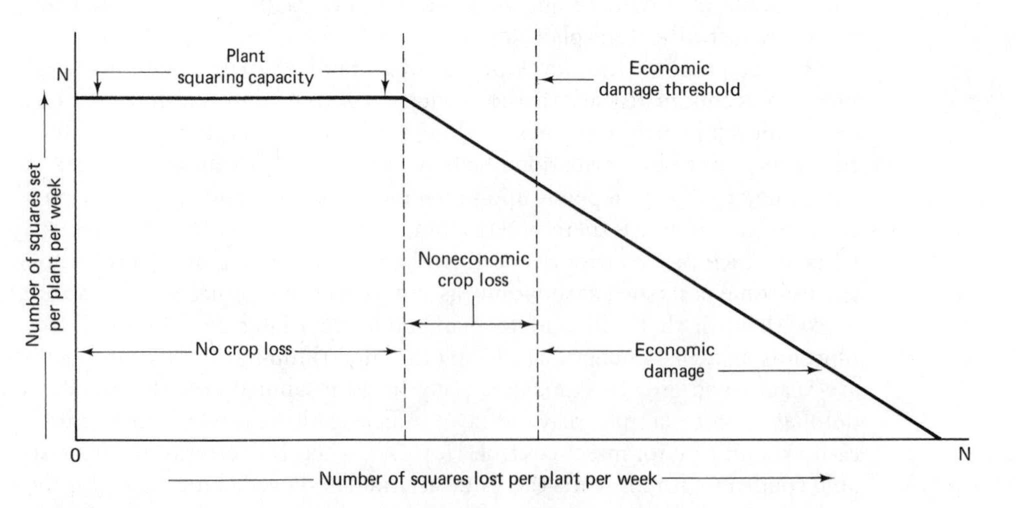


FIGURE 20-17. The effect of plant capacity and insect injury on the set of cotton bolls. (Reproduced with permission from Smith and van den Bosch, in *Pest Control*, W. Kilgore and R. Doutt eds., 1967, Academic Press, Inc.)

rates the modifications presented by Woods (1974) but does not alter the essence of the original concept.

At any time, the grower of a particular crop operates according to a production formula that relates the input of investment in land, labor, and capital to the output. Given a fixed amount of land and labor, applying the capital in the best possible way will give rise to a production curve similar to curve *A* in Figure 20-18. Assuming no pest attack, the most profitable relationship between cost (input) and crop value (output) occurs at the intersect of $O_A$ and $C_A$. Beyond this point, additional output is subject to the law of diminishing returns as additional units of capital input exceed the value of additional units of output.

If pest damage was to occur after the full input of capital was committed and no pest control was undertaken, a new production curve (*B* in Figure 20-18) would be generated. Since the costs remain the same but the output is reduced to $O'_B$, the grower would lose the difference between $O_A$ and $O'_B$. Point $O'_B$ lies on the new production curve beyond the point of profit maximization, represented by the intersect of $O_B$ and $C_B$. If the grower had anticipated the pest attack, he should have reduced his capital input from $C_A$ to $C_B$ in order to make a profit of $(O_B - C_B)$, which is larger than $(O'_B - C_A)$. The input of capital equal to $C_A$ without engaging in pest control actually increases the loss, since the increased output between $O_B$ and $O'_B$ costs more than it is worth.

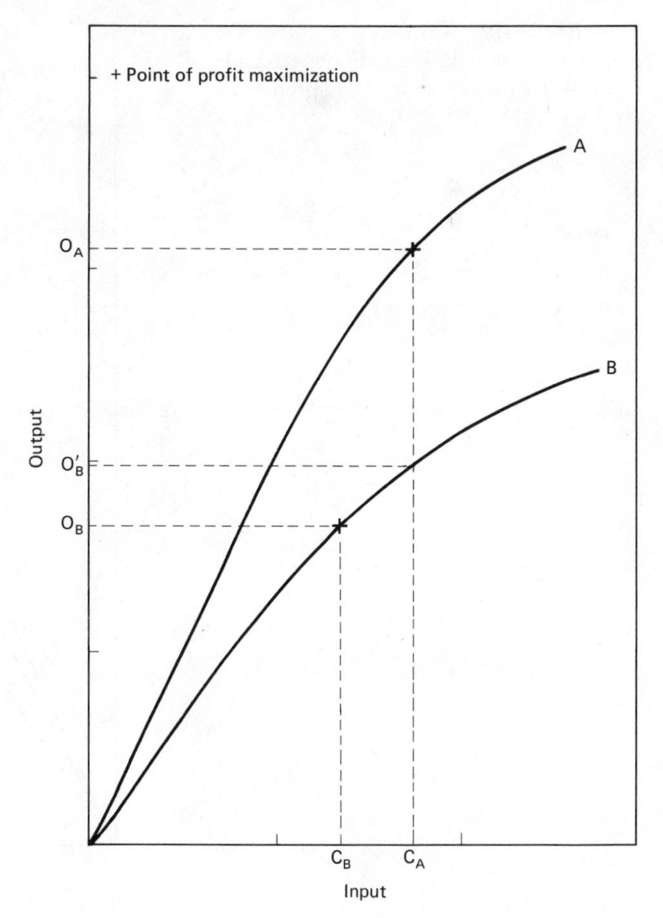


**FIGURE 20–18.** Cost-output curves for **A** an uninfested crop and **B** an infested crop in the absence of pest control. (Adapted from Headley, 1972.)

If the pest damage could be reduced partially by pest control, a third curve is generated that lies between the other two (Figure 20-19), and the point of maximum profit, excluding the cost of pest control, is indicated by the intersect of $O_C$ and $C_C$. The decision whether or not to engage in pest control can then be made on the basis of the control costs relative to increased profit. With pest control, the maximum profit the grower could obtain would be $(O_C - C_C)$ – (cost of pest control), and the extra profit he would gain above the maximum attainable in the absence of pest control would be $(O_C - C_C) - (O_B - C_B)$ – (cost of pest control). If the cost of pest control would be greater than the difference between $(O_C - C_C)$ and $(O_B - C_B)$, then it would not be worthwhile. On the other hand, if the grower could shift from production curve *B* to production curve *C* and thereby increase his profit, the control action could be justified.

Headley's approach introduces the concept of economic efficiency or the maximization of the difference between total benefits and total costs as a means of avoiding the overestimation of pest losses and a corresponding overinvestment in pest control. This may well prove to be unimportant in developing countries where an increase in the production of food is more important than the cost of each unit. Even in this country, we do not have the biological or economic data necessary to put Headley's method into operation. Nevertheless, the ideas indicate that a more sophisticated approach to the concept of the economic threshold is possible and could help to

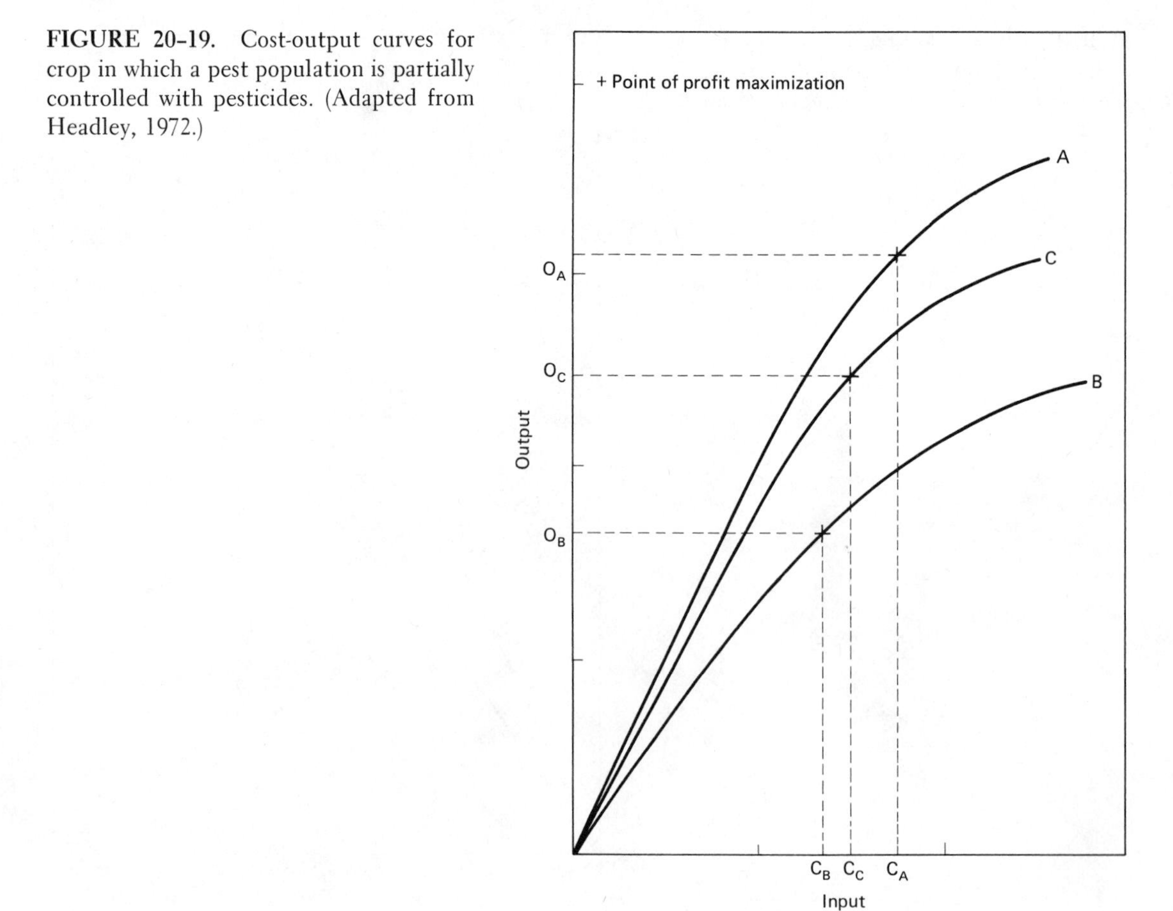


**FIGURE 20-19.** Cost-output curves for crop in which a pest population is partially controlled with pesticides. (Adapted from Headley, 1972.)

reduce the frequency with which pesticides are used unnecessarily. In the meantime, we should apply the best of the more simple methods developed by Ordish and others and strive to make economic thresholds realistic. The subject has been reviewed thoroughly by Stern (1973).

### *Estimates of Pest Damage*

There is no question that in the absence of pest control insect-caused damage can affect the yield of some crops substantially, but it is also true that some crops in certain areas suffer insignificant damage and no expenditure for pest control is necessary. The effects of indirect pests are certainly the most difficult to evaluate, since some low levels of feeding can actually increase yield whereas a high population can reduce a crop to zero. In the case of direct pests, every unit of produce rendered unsaleable represents a loss, and in the absence of control almost every unit may be infected. I have seen unattended apple orchards, for example, where almost every apple contained codling moth larvae. Likewise in forestry, bark beetles can kill almost every tree in a stand over a considerable area.

Nobody really knows the monetary value of insect-caused damage, even when only considered in terms of material losses. If we attempt to add some measure of the misery and suffering caused by medically important insects, the problem is insoluble, so this aspect will be considered separately in Chapter 22. Some good estimates of crop losses have been made in the United States and other more advanced countries, but it is difficult to arrive at global estimates because of the variability in yields per acre and the amount spent on pest control, which in a sense is part of the loss when measured in terms of money. Even when we confine our estimate to the economy of one nation, losses expressed in monetary terms can be misleading because of changes in prices and because the pest control industry, while adding to the cost of crop production, also contributes to the gross national product.

An expression of losses in terms of absolute yield or a percentage of yield is not an accurate portrayal since it does not permit any inclusion of value losses that result from a decline in the quality of the produce, which often results when adequate controls are not initiated. Furthermore, if losses are expressed as a reduction in yield from what would have been obtained in the absence of pests, the losses are based on an unknown quantity.

Any attempt to determine losses during a single season is, of course, complicated by a variety of nonpest factors, such as weather, that make year-to-year comparisons difficult. On the other hand, if pest losses are based on an average for several seasons, they would have to take into account yield increases throughout the period that have accrued because of improved crop varieties, more effective pest control, and increases in the use of fertilizer and irrigation.

Ordish (1952) attempted to overcome some of these problems with the introduction of a unit he called the untaken acre. This is a unit that represents the extra land that must be devoted to a crop to make up for the yield lost to pests. However, on an international scale it is difficult to apply such a system because of the vast differences in the productivity of different agricultural systems.

One of the best attempts to estimate the agricultural losses attributable to pests was made by Cramer (1967). He surveyed the available data for the entire world and, using the prices paid to the farmers in each country, came up with an estimate of losses amounting to between 70 and 90 thousand million United States dollars per year. He estimated that this amounts to about 55% of actual production or 35% of potential production. The latter figure could be broken down to 13.8% attributable to insects, 11.6% attributable to diseases, and 9.5% attributable to weeds.

Cramer's figures, however, represent losses incurred up to harvest. Additional losses usually occur during transport and storage. On a world-wide basis, these losses may be equivalent to preharvest losses. A friend who was in India when a large amount of grain arrived from the United States in an effort to relieve famine claims that half of the shipment was eaten by rats before it reached the people. In the Congo, insects in one year caused in-storage weight losses of 50% to sorghum, 20% to beans, and 15% to ground-nuts (Woods, 1974). Furthermore, it must be recognized that weight was not a perfect measure of the loss, since part of what was left was insect biomass, waste material, and badly damaged commodity.

In the days of the American pioneers there was a popular rhyme that instructed the farmer how to plant his kernels of corn:

One for the maggot,
One for the crow,
One for the cutworm,
And one to grow.

In spite of all of the technology that has been applied since, insects are still consuming about 10% of the crops grown in the United States at a cost of about 5.5 billion dollars per year.

## References Cited


ANDERSON, J. R. 1966. Recent developments in the control of some arthropods of public health and veterinary importance—muscoid flies. *Bull. Entomol. Soc. Am.* 12: 342–348.

BLAIR, W. F. 1964. An ecological context for the effects of chemical stress on population structure. *Bull. Entomol. Soc. Am.* 10: 225–228.

CANNON, W. N. and W. A. CONNELL. 1965. Populations of *Tetranychus atlanticus* McG. (Acarina: Tetranychidae) on soybean supplied with various levels of nitrogen, phosphorus and potassium. *Entomol. Exp. Appl.* 8: 153–161.

COOK, W. C. 1967. Life history, host plants and migrations of the beet leafhopper in the western United States. *Tech. Bull. U. S. Dept. Agric. No. 1365.*

CRAMER, H. H. 1967. Plant protection and world crop production. *Pflanzenschutz-Nachrichten,* Bayer. 20: 1–24.

ELTON, C. S. 1958. The ecology of invasions by animals and plants. Methuen & Co., Ltd., London.

FREDEEN, F. 1956. Black flies (Diptera: Simuliidae) of the agricultural areas of Manitoba, Saskatchewan and Alberta. *Proc. Int. Cong. Entomol.* 10(3): 819–823.

GRAHAM, F., JR. 1972. The war against the dreaded gypsies. *Audubon.* 74(2): 45–51.

HAYNES, D. L. 1973. Population management of the cereal leaf beetle. Pages

232–240 *in* P. W. Geier, L. R. Clark, D. J. Anderson, and H. A. Nix, editors. Insects: Studies in population management. *Ecol. Soc. Aust.* (*Memoirs 1*): *Canberra.*

HEADLEY, J. C. 1972. Economics of agricultural pest control. *Ann. Rev. Entomol.* 17: 273–286.

LEROUX, E. J. 1954. Effects of various levels of nitrogen, phosphorus, and potassium in nutrient solution on the fecundity of the two spotted spider mite *T. bimaculatus* reared on cucumber. *Can. J. Agr. Sci.* 34: 145–151.

MACARTHUR, R. 1960. On the relative abundance of species. *Amer. Natur.* 94: 25–34.

MARCHAL, P. 1908. The utilization of auxiliary entomophagous insects in the struggle against insects injurious to agriculture. *Pop. Sci.* 72: 352–419.

MICHENER, C. D. 1975. The Brazilian bee problem. *Ann. Rev. Entomol.* 20: 399–416.

NENE, Y. L. 1971. Plant protection: A key to maintaining present gains in food production. Pages 365–377 *in* K. L. Turk, ed. Some issues emerging from recent breakthroughs in food production. State College of Agriculture, Ithaca, N. Y.

NYE, P. H. and D. J. GREENLAND. 1960. The soil under shifting cultivation. Farnham Royal, Commonwealth Agricultural Bureaux.

ORDISH, G. 1952. Untaken harvest. Constable & Company, Ltd., London.

PERELMAN, M. and K. P. SHEA. 1972. The big farm. *Environment* 14: 10–15.

PRADHAN, S. 1971. Revolution in pest control. *Pesticides* 5(8): 11–17.

PIMENTAL, D. 1961. Species diversity and insect outbreaks. *Annals Entomol. Soc. Am.* 54: 76–86.

RIVNAY, E. 1964. The influence of man on insect ecology in arid zones. *Ann. Rev. Entomol.* 9: 41–42.

SMITH, R. F. 1971. The impact of the Green Revolution on plant production in tropical and sub-tropical areas. Founders Memorial Lecture, 1970, presented to the Entomol. Soc. Am., Los Angeles, Ca., Nov. 29, 1971. 19 p. (mimeographed).

SMITH, R. F. and R. VAN DEN BOSCH. 1967. Integrated control. Pages 295–340 *in* W. W. Kilgore and R. L. Doutt, editors. Pest Control: Biological, physical and selected chemical methods. Academic Press, Inc., New York, N. Y.

SOUTHWOOD, T. R. E. 1962. Migration of terrestrial arthropods in relation to habitat. *Biol. Rev.* 37: 171–214.

STERN, V. M. 1973. Economic thresholds. *Ann. Rev. Entomol.* 18: 259–280.

VAN EMDEN, H. F. 1966. Plant insect relationships and pest control. *Wld. Rev. Pest Control* 5: 115–123.

VAN EMDEN, H. G. and G. C. WILLIAMS. 1974. Insect stability and diversity in agro-ecosystems. *Ann. Rev. Entomol.* 19: 455–475.

WATT, K. E. F. 1965. Community stability and the strategy of biological control. *Can. Entomol.* 97: 887–895.

WHEATLEY, G. A. and T. H. COAKER. 1970. Pest control objectives in relation to changing practices in agricultural crop production. *In* Technological economics of crop protection and pest control. Society of Chemical Industry, Monograph No. 36.

WILBUR, D. A. 1971. Stored grain insects. Pages 495–522 *in* R. E. Pfadt, ed. Fundamentals of applied entomology. Macmillan Publishing Co., Inc., New York, N. Y.

# Insects as vectors of plant pathogens

# 21

In the foregoing chapter we examined the kinds of damage that insects do to crops and forests. We will now examine the importance of insects as agents of dissemination and inoculation of micro-organisms that cause plant diseases. The subject of insects as vectors of plant diseases is exceedingly broad because it involves the interaction of insects, pathogenic organisms, one or more species of host plant, and the effects of the physical environment on each of them. This aspect of entomology, probably more than any other, speaks to the need for a broad, basic education in the biological sciences before any consideration of specialization is entertained.

Insects act as vectors of hundreds of species of plant pathogens, including **bacteria, fungi, viruses,** and **mycoplasmas.** Of the pathogens mentioned, viruses are the most important and insects are the main means by which they are introduced into agricultural crops. Several good reviews of insect-plant-virus relationships are available (Ossiannilsson, 1966; Maramorosch, 1969; Watson and Plumb, 1972), and a broader coverage of insects and pathogens of all types is presented by Carter (1973).

The associations of plant pathogens with insects range from casual, facultative relationships to highly specific, obligatory relationships that pose some rather interesting questions regarding their evolution. Obviously, insects that feed or oviposit on plants and move about in the process could easily carry plant parasites on their bodies and therefore serve as accidental disseminators of disease-causing agents. However, it is much less obvious as to how a plant parasite would evolve a symbiotic relationship in which it multiplies within the insect as some plant viruses do. The advantages to the pathogen are clear. The insect can serve as a place in which to pass periods when the plant host is absent or provide a means of dispersal and then inoculation into the proper host. The fact that the pathogen can pass time outside of its plant host may reduce the over-all impact of the pathogen, thereby reducing selective pressures that could lead to the development of host resistance. But what does the insect gain to counter what it gives up to the micro-organism it harbors? The answer to this question seems uncertain. Some workers have found that insects are better able to survive and reproduce on diseased hosts. This is certainly the case with bark beetles that infect their host trees with blue-stain fungi, which block the conductive tissue and render the host more suitable. But studies on aphids and leaf hoppers have produced conflicting evidence. Several species of aphids were found to prefer sugar beets infected with mosaic virus and to reproduce more rapidly on diseased plants (Baker, 1960). On the other hand, some leaf hoppers leave fewer offspring when they feed on infected plants (Watson and Sinha, 1959). Other workers have found evidence that the plant pathogens reproduce in their vectors at the latter's expense. For example, Jensen (1959) showed that a strain of peach-yellows virus caused the early death of its leaf hopper vector; other workers have discovered cytological evidence of pathogenicity of viruses to their insect vectors but no clear outward evidence of a detrimental nature.

Inconsistencies of this type are common in biology, and students should not be uneasy about what appears to be a lack of clearly defined situations.

Evolution is a long, on-going process and what we witness at any time is the state to which it has progressed at that point. In a sense, it is like viewing a single frame of an unfinished motion picture. At a particular time, each pair or group of interacting species is at a different stage of its evolutionary experiment, which will either lead to a more specialized interrelationship or its termination.

### *Insects and Plant-pathogenic Bacteria*

Many bacteria cause diseases of plants but relatively few are entirely dependent on insects for their dissemination and inoculation. Plant-pathogenic bacteria cannot penetrate plant tissue unless they enter through natural openings such as stomata or through wounds. Furthermore, bacterial cells do not survive well when exposed to the elements, so they normally pass the period between their host's growing seasons in the soil. However, bacterial cells are often sticky and readily adhere to the surface of insects, which then carry them to feeding or oviposition wounds through which they can invade a new host. This is a very generalized insect-pathogen relationship but one which involves the transmission of some important plant diseases.

The apple maggot, which is one of the most common pests of orchards, enhances the distribution of bacterial rot of apples. Oviposition punctures are frequently the site of initial infection, but the disease also spreads away from the feeding tunnels of larvae, indicating that the latter disseminate the pathogen within the fruit. Additional examples of bacterial plant pathogens that are spread mechanically by insects include ring rot of potatoes by grasshoppers, bacterial disease of sugar cane by white flies (Aleyrodidae), and fire blight of orchard trees by aphids, codling moth, flies, ants, bees, and other insects. However, in the case of fire blight, there is some evidence to suggest the pathogen has evolved a more than casual relationship with some vector species.

There are several cases in which plant-pathogenic bacteria are ingested by insects, retained internally over the winter months, and then disseminated by them in the spring. Probably the best-known example of this type of relationship involves bacterial wilt of members of the cucumber family. This disease is apparently dependent upon two different insect vectors for overwintering, dissemination, and inoculation. A small number of overwintered striped cucumber beetles and twelve-spotted cucumber beetles (Chrysomelidae) are infected with bacteria ingested the year before and, consequently, infect the young plants they feed upon. Other cases in which the bacteria are harbored over the winter in the gut of insects and then spread by them the following spring include bacterial wilt of corn by flea beetles (Chrysomelidae), potato blackleg by seed corn maggot (Anthomyiidae), and olive-knot disease by the olive fly (Tryptidae).

### *Insects and Plant-pathogenic Fungi*

The spores of fungi are quite resistant to the harsh conditions of an unmodified environment so tend to survive fairly well when hosts are unavailable. They are also readily dispersed by wind, water, and a variety of animals. Consequently, few fungal diseases depend on insects for their transmission, so most insect-plant fungus relationships are rather casual. Carter (1973) lists the following seven types of fungus infection of plants in which insects have some involvement:

1. Contamination of the insect as a result of feeding on fungal masses.
2. Infection of plants through traumatic injury.

3. Fungal development on insect exudates.
4. Plant infection during pollination.
5. Plant infection through feeding and oviposition wounds.
6. Plant infection through feeding punctures.
7. Plant infection through symbiotic association between insect and fungus.

Ergot disease of cereals, particularly rye, is probably the most important example of a few diseases spread by insects that feed on fungal masses. Infected flowers exude a sticky material that contains fungal spores. A variety of insects, including flies, thrips, and beetles are attracted to the sweet, amino acid-rich exudate as a source of food. In the process of insect feeding, the fungal spores adhere to the body and are taken into the gut. The primary infection of cereal crops results mainly from wind dissemination of the fungus spores, but insects are the principal agents of secondary infection or spread within the crop.

Plant infection by pathogenic fungi through traumatic injury, fungal growth in insect exudates, and infection incident to pollination are all of rather minor importance. However, endosepsis of figs is rather interesting in that it results from the rather peculiar pollination of this fruit that was discussed in Chapter 18. The female fig wasps pick up fungus spores from infected male caprifigs in which they developed. The spores are then transmitted to edible smyrna figs when the wasps enter them in search of a place to oviposit. The fungus develops inside the fig, eventually converting the pulp to a wet, sticky mass.

Quite a wide variety of fungal diseases result from spores entering plants through gross feeding wounds or oviposition punctures caused by insects. This type of disease-insect relationship encompasses several important crop pest problems. Black mold and yellow mold of corn often begins their infection at the locations where the corn earworm, *Heliothis zea*, has entered the developing ear. The fungus grows well in the insect's excrement, and spores are readily carried on the bodies of earworm larvae. Similarly, boll rot of cotton results from two fungi that gain entry to the developing bolls through feeding and oviposition wounds made by a variety of insects, including cotton bollworm and lygus bugs (Bagga and Lester, 1968). A variety of fungus-caused fruit spoilage diseases, including brown rot of stone fruits and apples, are initiated by the feeding and egg laying of insects.

The infection of plants with fungal pathogens through insect feeding punctures is similar to the examples given, except the insects are sap feeders and the site of initial infection tends to be more specific. One such disease is yeast spot of soybean, transmitted by stinkbugs (Pentatomidae). Unsuccessful attempts to infect plants through artificial punctures suggest there is a specialized relationship between the pathogen and the bugs that transmit it (Daugherty, 1967).

Of greatest interest are those fungal diseases of plants caused by pathogens that have a symbiotic association with an insect vector. Two important diseases illustrate the extremes of the symbiotic associations that exist. The disease known as perennial canker of apple is caused by a fungus carried by the woolly apple aphid (Aphididae). The aphid causes minor galling where it feeds on the stem just below the soil surface. The galls rupture during the winter, creating entry courts through which the fungus can enter its host. The disease is expressed as fruit rot and branch canker. The cankers become sealed by callus tissue that develops during the summer following infection. The aphids show a preference for the new callus tissue as a feeding site, thereby creating a new wound through which the fungus can re-enter. Austwick (1958) considers this to represent a mutualistic relationship between the aphid and the fungus. The insects create entry courts for the fungus, which in turn stimulates the production of new aphid feeding sites.

The causal agent of Dutch elm disease and its bark beetle vectors share a more specialized relationship. Both the most important bark beetle and the fungus were introduced to America from Europe, apparently at different times. The smaller European elm bark beetle, *Scolytus multistriatus* (Scolytidae), had been known in the United States for a number of years when the disease was discovered in 1930. Both organisms spread widely, but the bark beetle has expanded its range more rapidly than the fungus. The natural vector is known to occur in all of the contiguous states except Florida, Montana, and Arizona, whereas the disease was absent from these states plus Washington, Oregon, California, Nevada, Utah, New Mexico, and Mississippi until it was found in central California in 1975. Now that the pathogen is widespread, it is also transmitted by the native elm bark beetle *Hylurgopinus rufipes*, but the imported beetle out-competes the native and thereby serves as the primary vector.

The spores of the Dutch elm disease fungus *Ceratostomella ulmi* are carried on the surface of the beetles. The fungus then grows within the insect's galleries (Figure 21-1). When young adults emerge, sticky, newly formed spores adhere to them and are thereby carried to another host. In the case of *S. multistriatus*, the young beetles disperse to healthy elms where they feed on twigs and in the crotches of small branches before locating sickly trees in which to establish their brood. The fungus is thereby inoculated into feeding wounds as well as gallery sites. In brood trees, the fungus spores spread through the functional conducting vessels, which are later blocked by the fungus. Consequently, both the spread and the inoculation of the fungus is dependent upon its insect vector. The bark beetles probably derive some benefit from the fact that the host elms are weakened by the fungus. Other bark beetles, particularly those that attack healthy pines, carry the spores of blue-stain fungi and benefit greatly from the fungal blockage of their host's vascular tissue.

### *Insect and Plant-pathogenic Viruses and Mycoplasmas*

Plant viruses and viruslike mycoplasmas cause some of the most destructive diseases of beneficial plants. Most of them are transmissible by insects, particularly those with piercing-sucking mouth parts, and a number have evolved highly specialized relationships with their vectors. A few viruses are transmitted by chewing insects such as grasshoppers and beetles, but about 90% are transmitted by homopterans, mainly aphids (Aphididae) and leafhoppers (Cicadellidae).

Plant viruses can be divided into two groups on the basis of their association with insect vectors. The majority belong to a group having a rather casual relationship with insects in that they are carried on their surface; these are alternatively called **stylet-borne** or **nonpersistent** viruses. The second group consists of viruses that are carried internally and are referred to as **circulative** or **persistent.** This group is sometimes divided in **propagative** and **nonpropagative**, depending on whether or not they multiply within their vector. The

**FIGURE 21-1.** Galleries of *Scolytus multistriatus*, the principal vector of Dutch elm disease. Note the staining of the wood adjacent to the galleries caused by the fungus (arrow). (Photograph courtesy USDA.)

important difference between the major groups is that stylet-borne viruses are lost soon after they are picked up, whereas circulative viruses are retained by their vector for long periods. Consequently, some insects are able to infect only the first plant or two they feed on after picking up a virus from an infected host. Others cannot infect new plants until some time has lapsed but then might remain infective for the remainder of their lives.

Aphids mainly serve as vectors of stylet-borne viruses. In terms of the number of viruses transmitted, most of which cause mosaic symptoms (Figure 21-2), and the number of insect species involved, aphids are the most impor-

FIGURE 21-2. Symptoms of common bean mosaic virus. Mosaic disease viruses are most commonly transmitted by aphids. (Photograph courtesy USDA.)

tant plant disease vectors. Many of the viruses transmitted by aphids occur in high concentrations in the host plant epidermis so are readily picked up when the aphids probe the tissue with their mouth parts. Some aphids, like *Myzus persicae*, transmit many kinds of stylet-borne viruses, whereas others are quite specific as to the viruses they transmit. Inasmuch as this group of viruses does not enter the aphid, no satisfactory explanation has been found for the varying degrees of aphid-virus specificity.

Aphids serve as vectors for a few persistent viruses in a manner analogous to that found among most viruses borne by leafhoppers. Such viruses are not picked up by probing but are acquired only when the aphid engages in a period of sustained feeding. The aphids cannot transmit these viruses immediately because they pass through the vector's body to either the salivary glands or ovaries (Figure 21-3). A few viruses have been found to multiply within the body of their aphid vector, and at least one is passed to some offspring through the egg (transovarially). The potato leaf roll virus multiplies within the green peach aphid and is passed on transovarially (Stegwee and Ponson, 1958; Miyamoto and Miyamoto, 1966). Both the lettuce necrotic yellows and the sow thistle yellow vein viruses multiply in an aphid vector (O'Loughlin and Chambers, 1967; Richardson and Sylvester, 1968).

Almost all viruses transmitted by the leafhopper are characterized by persistence within their vector, and in a number of cases they are propagative and passed transovarially. It is rather paradoxical that perhaps the only case of a leafhopper-transmitted, nonpersistent virus is tungro disease of rice, vectored by *Nephotettix impicticeps*, which caused a severe setback to the so-called "Green Revolution." Most leafhopper-transmitted viruses cause

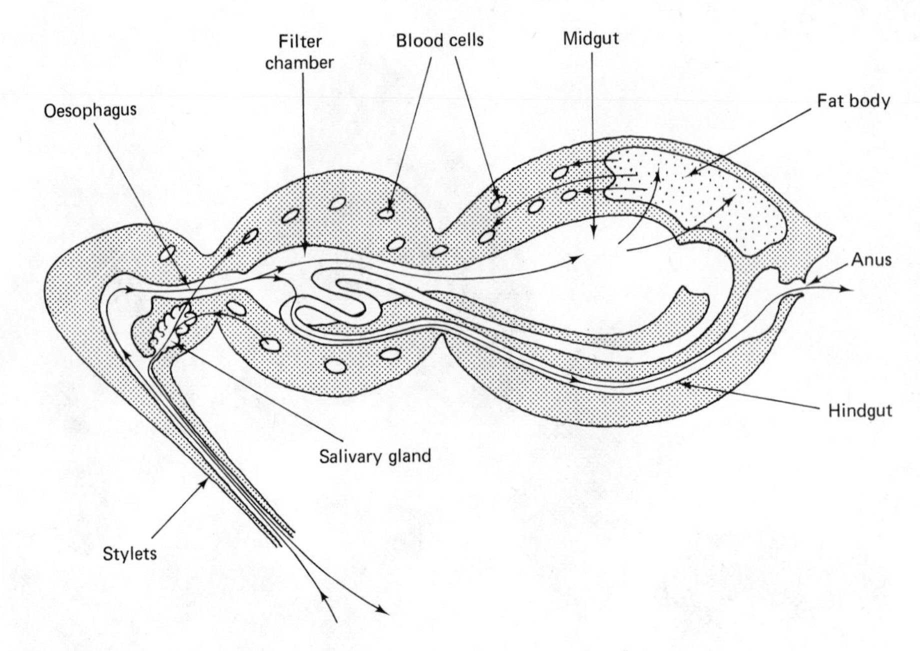


**FIGURE 21-3.** Diagrammatic section through an aphid showing the route taken by a circulative or persistent virus before it can infect another host after being picked up from an infected host plant.

symptoms referred to generally as yellows (Figure 21-4). Some examples of important plant virus diseases and the type of relationship they have with their insect vectors are presented in Table 21-1.

## Ecological Aspects

In the pest situations discussed in Chapter 20, the population aspects, though often complex, mainly involved the pest species. For the most part, the populations of crop host plants or host animals could be reasonably well defined. Decisions about pest control could therefore be made mainly on the basis of the numbers of pests and the anticipated damage they would cause relative to that which could be economically tolerated. When one is concerned with a plant disease, the causal agent of which is transmitted by one or more insects, the situation becomes more complex. The host population is less predictable, because it may contain individuals with a range of susceptibility or be a complex of wild and cultivated species. The size and location of the **pathogen reservoir** (population) may be difficult to evaluate. The size of the vector population, although important, might be less important than its mobility and feeding behavior. The impact of an insect-vectored plant pathogen on a crop thus involves three or more interacting species, all of which may be affected differently by the prevailing physical environment.

Because plant-pathogenic bacteria and fungi are disseminated by a variety of agents and only a few are dependent on insects for their spread and inoculation, the following discussion will apply mainly to plant viruses and mycoplasmas. Our understanding of the ecological aspects of plant disease transmission by insects is based on the results of an extensive number of

**FIGURE 21-4.** Typical symptoms displayed by an onion plant infected with aster yellows virus. Yellows disease viruses are most commonly transmitted by leaf hoppers. (Photograph courtesy USDA.)

**TABLE 21-1** **Some Well-known Virus Diseases of Crop Plants and the Type of Relationship Shared with Their Insect Vectors**

| Virus Disease | Principal Crop | Vector Group | Transmission | Propagative | Transovarial |
|---|---|---|---|---|---|
| Tomato spotted wilt | Tomato | Thrips | Stylet-borne | no | no |
| Swollen shoot | Cacao | Mealybugs | Stylet-borne | no | no |
| Cassava mosaic | Cassava | White flies | Stylet-borne | no | no |
| Tobacco mosaic | Tobacco, tomato | Aphids | Stylet-borne | no | no |
| Lettuce necrotic yellows | Lettuce | Aphids | Circulative | yes | no |
| Potato leaf roll | Potato | Aphids | Circulative | yes | yes |
| Tungro disease | Rice | Leafhoppers | Stylet-borne | no | no |
| Beet curly top | Sugar beet, tomato | Leafhoppers | Circulative | no | no |
| Aster yellows | Aster, celery | Leafhoppers | Circulative | yes | no |
| Rice stunt | Rice | Leafhoppers | Circulative | yes | yes |

research projects. However, few, if any, diseases have been studied as thoroughly as would be desirable, and many have received only passing attention. Consequently, generalities are risky at best, and the following discussion is intended only to present some feeling for the complexity of the subject.

Assuming that the host plant species has a known degree of susceptibility, one of the most important considerations involves the movement of the vector from one host plant to another. Regardless of whether this movement is from plant to plant within a field or from field to field or plant species to plant species, it is influenced by a wide range of interacting biotic and physical factors. The movement of insects can range from trivial flights that seem to have no directionality to long-distance movements over a well-navigated course. Some of the controversy concerning the terms applied to different types of movement were discussed in Chapter 12, as were some of the underlying causes of such movements. They will not be treated in detail again here.

All levels of movement are important. Movement from plant to plant within the crop increases the number of infected plants (secondary infection) and increases the size of the pathogen reservoir available to species feeding prior to long-distance movements or to transient vectors. Long-distance movements out of established breeding grounds are important to the seasonal spread of diseases and to their introduction into newly established crops (primary infection). Sometimes these seasonal patterns of movement are quite predictable as with the vector of beet curly top in the San Joaquin Valley, described in Chapter 12. Numerous vectors overwinter or develop early spring populations in the south and then move northward, infecting later planted crops as they go.

As stated earlier, insect movements often involve a combination of active and passive forces. In those situations in which an adaptive strategy has evolved that allows a vector to exploit passive dispersal, the outcome may be quite predictable. However, if the passive dispersal is largely a matter of accident, it would be necessary to await the outcome.

The fact that moving vector populations are of great importance to the spread and establishment of plant diseases creates a variety of assessment problems. Many different sampling methods are used to estimate "resident" populations and the number of vectors moving in or out of the crop. Since the basic reason for vector surveys is to provide information necessary for control decisions, standardization in both population assessment and reporting is highly desirable.

The physical environment has a pronounced effect on the movement of vectors. Different conditions have different effects on different types of insects, but some generalizations that apply to groups of species are possible. For example, Thomas and Vevai (1940) suggested that aphids usually fly when the temperature is above 70°F, the relative humidity is below 80%, the wind velocity is below 5 mph, and the diurnal temperature fluctuation is in excess of 12°F. This is, of course, an oversimplification in that it establishes no upper limit for temperature nor lower limit for humidity, both of which are important. Wind tends to be a major factor, because few small insects fly voluntarily at higher wind speeds. Local movements of aphids and white flies are most common during warm, calm periods, but strong, turbulent winds and convectional currents can carry small insects aloft and transport them long distances. Cool, cloudy, or rainy periods can act as a strong deterrent to vector movement and may substantially slow disease development.

Movement between plants of the same host species is affected by biological factors as well. High population levels can have a disturbing effect that results in more frequent short flights. As indicated earlier, host plant condition may also influence whether a vector stays or leaves. Different vectors display different preferences for infected or uninfected plants. Some aphids such as the green peach aphid show a preference for sugar beets infected with different viruses, depending on its stage of development (Macias and Mink, 1969).

The movement of vectors between different plant species that serve as hosts for both the insect and the virus is common. The regular cropping practices of modern agriculture enhance such movements. When crop plants are removed, weeds growing in field margins and pockets or strips of wild vegetation serve as reservoirs for viruses. The same plants that serve as hosts for the pathogens serve as alternate hosts for the insect vectors that leave the crops during harvest. This is one of the reasons for weed control adjacent to crop land, but there are both pros and cons to clean field margins, as discussed in Chapters 24 and 26.

The basic life history of many vectors, particularly aphids, imposes an obligatory change in the host plant at regular intervals. The life history of an aphid, presented in Figure 9-13, shows the typical pattern in which overwintered eggs hatch on the primary host. After several generations, migrants are produced that move to summer hosts, which often include weeds. After several generations on summer hosts, fall migrants that return to the primary host develop. This pattern of movements often creates a regular flow of virus pathogens from crop plants to wild species.

Apart from the sizes of the interacting vector population and virus reservoir, the normal and climatically induced movement of the vector, and the distribution of susceptible plant hosts, there are other factors that complicate the pattern of the spread of plant diseases by insects. The physical environment may have a pronounced effect on the length of the vector's life span as well as its behavior, but, once again, generalities seem inadvisable. In the case of nonpersistent viruses, conditions such as calm combined with warmth and midrange humidity may substantially increase dispersive behavior that is often characterized by an alternation of flight, landing, probing, and taking flight again. Since the time span between probes is shortened, the virus is spread more rapidly. On the other hand, if the time between probes is long, the virus might not remain viable on the stylets, thereby reducing the likelihood of infection of new plant hosts.

Persistent viruses are characterized by a **latent period.** This is the period between the time that the virus is picked up from an infected plant and the time when it is capable of infecting another plant. In the case of those persistent viruses which multiply in their insect vector, the latent period is usually protracted. Because insects are cold-blooded, the body temperature of the vector, which influences the length of the latent period of the virus, is affected by the air temperature. Consequently, the age of the vector population when it comes in contact with a virus reservoir, and the weather conditions thereafter can have a great impact on the pattern of disease transmission. If the population is composed largely of older adults, or if the environmental conditions are such that the latent period is lengthened while the vector's life span is shortened, there may be little or no disease transmission.

Clearly then, the control considerations in the area of insect-vectored plant pathogens can be quite complex. If insects are not obligate vectors of a pathogen, disease reduction can be achieved best by directing the control against the pathogen. The more intimately the pathogen and vector become related, the more necessary it becomes to treat the host plant species, the disease agent, and the vector as a system. Gaining an understanding of these complex systems that involve three or more dynamic populations is a difficult process, and making the evaluations necessary for control decisions can be quite a problem. But from a positive point of view, the complexity makes additional approaches to pest control possible. Furthermore, varied approaches to weed control, pathogen control, and insect vector control can be combined with crop resistance to form a management program that would be less likely to generate pest resistance, because no single selective pressure would be acting on the genetic make-up of the pests.

## References Cited


Austwick, P. K. C. 1958. Insects and the spread of fungal disease. Biological aspects of the transmission of disease. Oliver and Boyd, London.

Bagga, H. S. and M. L. Laster. 1968. Relation of insects to the initiation and development of boll rot of cotton. *J. Econ. Entomol.* 61: 1141.

Baker, P. F. 1960. Aphid behaviour on healthy and on yellows infected sugar beet plants. *Ann. Appl. Biol.* 48: 384–391.

Carter, W. 1973. Insects in relation to plant disease. 2nd ed. John Wiley & Sons, Inc., New York, N. Y.

Daugherty, D. M. 1967. Pentatomidae as vectors of yeast-spot disease of soybeans. *J. Econ. Entomol.* 60: 147–152.

Jensen, D. D. 1959. A plant virus lethal to its insect vector. *Virology* 8: 164–175.

Macias, M. and G. I. Mink. 1969. Preference of green peach aphids for virus infected sugar beet leaves. *J. Econ. Entomol.* 62: 28.

Maramorosch, K., ed. 1969. Viruses, vectors, and vegetation. Interscience Publishers, New York, N. Y.

Miyamoto, S. and Y. Miyamoto. 1966. Notes on the transmission of potato leaf-roll virus. Sci. Rept. Hyogo. Univ. Agric. Ser. *Plant Protection* 7: 51–66.

O'Loughlin, G. T. and T. C. Chambers. 1967. The systemic infection of an aphid by a plant virus. *Virology* 33: 262–271.

Ossiannilsson, F. 1966. Insects in the epidemiology of plant viruses. *Ann. Rev. Entomol.* 11: 213–232.

Richardson, J. and E. S. Sylvester. 1965. Aphid honeydew as inoculum for the injection of pea aphids (*Macrosiphon, Acyrthosiphon*) *pisum* (Harris) with pea-enaction mosaic virus. *Virology* 25: 472–475.

Stegwee, D. and M. B. Ponson. 1958. Multiplication of potato leaf-roll virus in the aphid *Myzus persicae* (Sulz.) *Entomol. Exp. Appl.* 1: 291–300.

Thomas, I. and E. J. Vevai. 1940. Aphis migration. An analysis of the results of five seasons' trapping in north Wales. *Ann. Appl. Biol.* 27: 393–405.

Watson, M. A. and R. T. Plumb. 1972. Transmission of plant pathogenic viruses by aphids. *Ann. Rev. Entomol.* 17: 425–452.

Watson, M. A. and R. C. Sinha. 1959. Studies on the transmission of European wheat striate mosaic virus by *Delphocodes pellucida* Fabr. *Virology* 8: 130–163.

# Insects and human health 22

Many of the serious diseases of humans are caused by infectious organisms transmitted by insects. The relationship of pathogens, their insect vectors, and man undoubtedly dates to the beginning of our existence. The relationships are natural ones arising from the exploitation by insects of vertebrates as a source of food. Blood-feeding bugs, lice, flies, and fleas evolved long before the advent of man. Many probably later adopted man as an alternate host, fortuitously, as a result of the close association of man with the animals that previously served as primary hosts. For example, the sucking lice (Anoplura) probably were occupants of bird nests and animal lairs in the caves that man also adopted for shelter. The transfer of insects from one type of warm-blooded host to another would not seem to be impeded by any major obstacles.

The awareness of disease-causing organisms and insects as pests of man is depicted widely in ancient art and literature. In the King James version of the Old Testament (Exodus 8:24) is the testimony, ". . . and there came a grievous swarm of flies into the house of Pharoah, and into his servants' houses, and into all the land of Egypt; the land corrupted by reason of the swarm of flies." By the sixteenth century a number of observers had expressed their belief that there was a relationship between certain arthropods and the incidence of human disease. But it was not until after the development of the microscope that man was able to observe the tiny organisms that infect his body. The first major discoveries that removed all doubt concerning the involvement of insects in the transmission of disease soon followed. In 1878, Patrick Manson reported his observations of the development of the nematode *Wuchereria bancrofti* in the bodies of mosquitoes. He and other workers later proved the mosquito to be the intermediate host and vector of this filariasis-causing worm. Proof of the association and transmission of a sleeping sickness type of disease (nagana) by tsetse flies, malaria by anopheline mosquitoes, yellow fever by *Aedes aegypti*, and other diseases by other insects, as well as by ticks and mites, soon followed, and the field known as medical entomology was on its way to becoming an important area of study.

Clearly, the medical importance of insects is far too vast a subject to be treated in more than a general way in a book of this kind. However, some of the more important types of human injury will be mentioned along with some specific examples of insect-borne diseases that are of global concern. Students whose appetite is whetted by this chapter are encouraged to delve into one of several good medical entomology texts, such as those by James and Harwood (1969) or Smith (1973).

## The Evolution of Medically Important Relationships

Many insects have evolved a parasitic way of life. Parasitism is a form of intimate relationship between two species in which one species, the **parasite**, derives some benefit from another species, the **host**, at the latter's expense.

Sometimes the relationship is a casual or accidental one such as when a scavenging species feeds at the site of a wound; this is referred to as **facultative parasitism.** In other cases, the parasite is dependent on its host for completion of development and is said to display **obligatory parasitism.** Some obligatory parasites require only occasional and temporary contact with their host, as in the case of female mosquitoes that periodically require a blood meal and satisfy this need during the period of egg development from several host individuals. Others, like the chewing and sucking lice, exhibit a continuous relationship with a single host. Likewise, some are able to obtain their needs from a variety of host species, whereas others have evolved a specific relationship with a single species of host. This specialization is extreme in the human louse, which has become two subspecies that are behaviorally, though not morphologically, distinct. *Pediculus humanus capitatus* occupies hairy parts of the body and glues its eggs to the hair shafts. *Pediculus humanus humanus* attaches to the clothing and deposits the eggs along seams in areas that often make contact with the body. Most insect parasites of man, like the lice, feed from an external position and are called **ectoparasites.** However, there are some, particularly the larvae of certain flies, which pass part of their life cycle within the host's body; these are called **endoparasites.**

As mentioned previously, insects seem to have evolved a parasitic way of life from more simple life styles that involved more casual associations with vertebrates in their nests and lairs. The nest habitats provide these insects with a rather stable environment and a supply of food in the form of detritus, continuously replenished by the vertebrates with which they are associated. From scavenging on the organic debris in the nest area to actually feeding on surface debris on the host, and later the host's skin, secretions or blood would seem a natural course of events that could take place without any major morphological changes. In many cases, the parasites that evolved in this way have remained associated with their host's nest, either directly or by occupying adjacent hiding places when the host is active and might endanger the parasite. Some, however, became more or less permanently associated with the host and transfer to each generation of host offspring in the nest or lair. This new way of life undoubtedly provided an array of benefits to the insects and contributed to a furthering of these parasite-host associations. Such was probably the general course of evolution followed by the bedbugs, biting lice, sucking lice, and fleas as well as a few highly specialized beetles, flies, ticks, and mites. Some of these parasites, in turn, became associated with man as a result of his sharing dwellings such as caves with other animals. Usinger (1966) believes that the bedbugs, *Cimex* spp, associated with man, transferred to him from bats which cohabited his caves, and then remained with man when he began to build his own habitations.

The other main line of evolution of a parasitic way of life seems to have occurred among more or less free-living forms that encountered vertebrate hosts away from their nests or lairs. Some of these insects more than likely also began as predators, perhaps rather general ones, or as opportunistic scavengers. There are a number of examples of insect families that contain predaceous species that feed on insect hemolymph and parasitic species that feed on vertebrate blood. For example, flies of the families Ceratopogonidae and Rhagionidae display this kind of feeding divergence. Even species of hemipterans and thrips that normally feed on plant juices will probe the skin of man when they are generally stimulated to feed. Large populations of thrips in field crops such as young cotton can be particularly irritating to agricultural workers.

Numerous muscoid flies tend to feed on whatever liquid nutrients happen to be available and accessible. Inasmuch as their sponging mouth parts are adapted for utilizing surface liquids, the generalized feeders in this group have no apparatus suitable for penetrating host tissue to obtain blood. However, these insects will imbibe blood that oozes from wounds, including those made by biting flies. Although there is no direct evidence, it seems likely that nutrition played a major part in the evolution of some parasitic flies. The high protein content of blood obtained opportunistically may have resulted in more offspring being left by those flies which were able to obtain a blood meal instead of, say, liquified vertebrate excrement. If a mutation occurred, such as the development of penetrating projections on the tip of the labium (see Chapter 6), which created an opportunity to increase the blood component of the diet, the new biotype may have experienced an advantage expressed as an increase in its reproductive potential.

Protein seems to be a dietary requirement of adult insects prerequisite to egg production. In many insects, sufficient protein is acquired by the immature stages to satisfy the need for egg formation, but if protein can be obtained by the adult female as well, more eggs may be produced and the ovipositional period can be extended. Both of these conditions would seem to be advantageous and, consequently, might have led to a proliferation of blood-feeding behavior.

Apart from host transfers within his dwellings and from vertebrates killed for food, man has most clearly become a suitable host for many nonspecific blood-feeding insects, such as mosquitoes. In fact, the early loss of a dense covering of hair may well have made man a desirable host for blood-sucking insects. Other insects also became associated with man without feeding on him directly. His construction of dwelling places and the use of fire created a new environment for many species and enabled them to expand their range into areas of cooler climate. The dependence on man for survival in new areas created a relationship called synanthropy that is enjoyed by a variety of insects such as cockroaches, houseflies, silverfish, and larder pests. Some of these tend to move back and forth between areas where filth accumulates, thereby increasing the likelihood of attaining medical importance as simple mechanical vectors of pathogenic organisms. It is rather easy to demonstrate that insects which frequent unsanitary environments carry a wide variety of micro-organisms, ranging from bacteria to the eggs of round worms on the surface of their bodies. Yet, there are relatively few cases in which an outbreak of disease has definitely been attributed to the presence of a large population of facultative vectors. Prior to the advent of the automobile and indoor plumbing, large accumulations of horse manure combined with the prevalence of outdoor privies resulted in large fly populations in most communities. Yet healthy human populations were not seriously affected by what would seem to be a potentially hazardous situation. There are cases in which medical entomologists have been almost certain that flies have been implicated in outbreaks of dysentary, but there is never absolute proof. Generally, a healthy person in a clean environment has little to fear from having a fly walk across his plate of food.

Most vertebrate pathogens transmitted by arthropods generally do not harm their arthropod vectors, but there are exceptions. According to Mussgay

(1964), typhus-causing rickettsiae shorten the adult life of their louse vectors, plague bacilli are pathogenic to bedbugs, and tularemia bacilli can cause the death of human lice. On the other hand, vertebrate viruses appear to have no effect on the insects that vector them; insects are, however, killed by their own viruses. The lack of pathogenicity of **arboviruses** (arthropod-transmitted viruses) to their vectors suggests an ancient association.

Some protozoans that infect vertebrates also seem to have had a long association with their vectors. In the malaria-causing sporozoans, the microorganisms have developed complex developmental cycles in their vectors, and the relationship between the sporozoans and their insect hosts has become highly specific. Manwell (1955) believes that malaria parasites originated from intestinal parasites of reptiles that were passed on to birds in the course of evolution. When the blood-sucking flies evolved, these parasites were picked up by the insects and became adapted to them as obligate vectors. He believes the mammals are more recent hosts, as indicated by less frequent infection and more severe pathogenicity.

Insect-transmitted filarial worms were probably also originally vertebrate, rather than arthropod, parasites. Anderson (1957) believes that a typical gut nematode became established in other regions of the body such as the eyes, skin, or peripheral blood vessels, from which the larvae could be picked up by bloodfeeding insects and transported to new hosts.

These diverse beginnings have led to a variety of associations, which form the bases for the many ways that insects have become a detriment to human health. For convenience, we can describe medically important insects as having either a direct or an indirect effect. Direct effects would include annoying behavior, stings, bites, tissue invasion, mechanical injury, allergies, and fear or delusion. The indirect effects, which are by far the most important, result from insects acting as facultative or obligatory vectors of vertebrate pathogens.

## Direct Effects of Medical Importance

### *Annoyance and Loss of Blood*

A wide variety of insects cause annoyance and discomfort simply as a result of their presence. Some people are disturbed by only a few individuals, but almost everyone becomes angered by the presence of large numbers of bothersome insects. Those which are particularly aggravating tend to crawl on our person, as is characteristic of some ants, or to fly about in a disturbing fashion as do many kinds of flies. The combination can be quite distracting. Whereas many annoying insects actually do no physical harm, the mere presence of a single individual that can bite or sting often sets up a defensive mood that can spoil an otherwise pleasant experience.

Numerous insects, a variety of flies in particular, feed on blood entirely or periodically as a prerequisite to egg maturation. In some instances, man makes a perfectly good source of food and is bitten readily. Anticoagulants or enzymes that cause a local irritation may be released during feeding, and the bite may well be painful, as is that of the horse fly (Tabinidae). Blood loss caused by insect feeding is not usually as significant in man as it often is in poultry and domestic animals. However, in northern regions, swarms of biting flies at certain times of the year make it necessary for workers outdoors to wear gloves and veils. There is also reasonable certainty that individuals lost in bad fly areas have been weakened and perhaps killed by the loss of blood and the **anaphylactic shock** that can result from numerous insect bites.

### *Envenomization*

The injection of venom or poison by insects is a common occurrence. The effect of insect venoms on humans is highly variable, ranging from mild, transitory irritation to death. The number of recorded insect-caused deaths are few, but according to Parrish (1963), stings by Hymenoptera caused half of the 460 deaths that occurred over a ten-year period in the United States as a result of venomous animals.

Insect venoms are injected into the body of man by way of a bite (penetration of the skin by mouth parts), a sting (penetration by the specialized ovipositor possessed by some Hymenoptera), or an abrasion as when the skin is broken by spines or spurs that have associated poison glands (e.g., some tiger moths: Artiidae). Although it does not represent envenomization in the strict sense, some insects produce a poison reaction on contact or by an active projection of a poisonous spray.

The most common bites are produced by blood-sucking insects which have mouth parts adapted for penetrating the skin prior to the removal of the blood. Except in the few species like horse flies that break the skin in a rather vicious fashion, the mechanical insertion of the mouth parts is not that painful. Interestingly, insects that have painful bites do not usually act as vectors of pathogens. Apparently, to become a successful vector of a pathogen, an insect must be able to prolong its feeding period on the host; this is better served by a relatively painless bite. The pain and subsequent irritation usually result from the injection of saliva into the puncture. Saliva contains enzymes and, sometimes, anticoagulants that produce a localized reaction. Most of us are all too familiar with the bites of mosquitoes, a few other small blood-sucking flies, and, occasionally, fleas. However, most of us are unaware how extensive the list of biting insects really is. Next to the flies, the Hemipterans are probably the most important group of biting insects, among which a few larger types such as the toe biters (Belostomatidae), assassin bugs (Reduviidae), and backswimmers (Notonectidae) can be quite painful. Most people who live in an urban setting have less likely experienced bites by these insects, and, under normal conditions, encounters with lice and bedbugs have declined.

In terms of both frequency and severity, stings are probably more important than bites. Among the insects, only the order Hymenoptera contains species that possess a specialized stinging apparatus with an associated venom gland (see Chapter 9). The so-called aculeate hymenopterans belong to seven superfamilies that include the cuckoo wasps, velvet ants, wasps, hornets, yellow jackets, and bees. Some species are more aggressive than others, but most stings are used for defense, rather than offense. In the social species, multiple stings can result from the release of an alarm pheromone with the venom that incites other individuals to attack. A few species of ants seem to sting in an offensive manner associated with their quest for food. The imported fire ant *Solenopsis saevissima* (Figure 22-1) has become a serious pest in agricultural areas, driving workers from the fields with its extremely painful stings.

A number of insects have integumentary secretions with irritating properties. Probably best known are the beetles of the family Meloidae, which are commonly called blister beetles because contact with them may result in

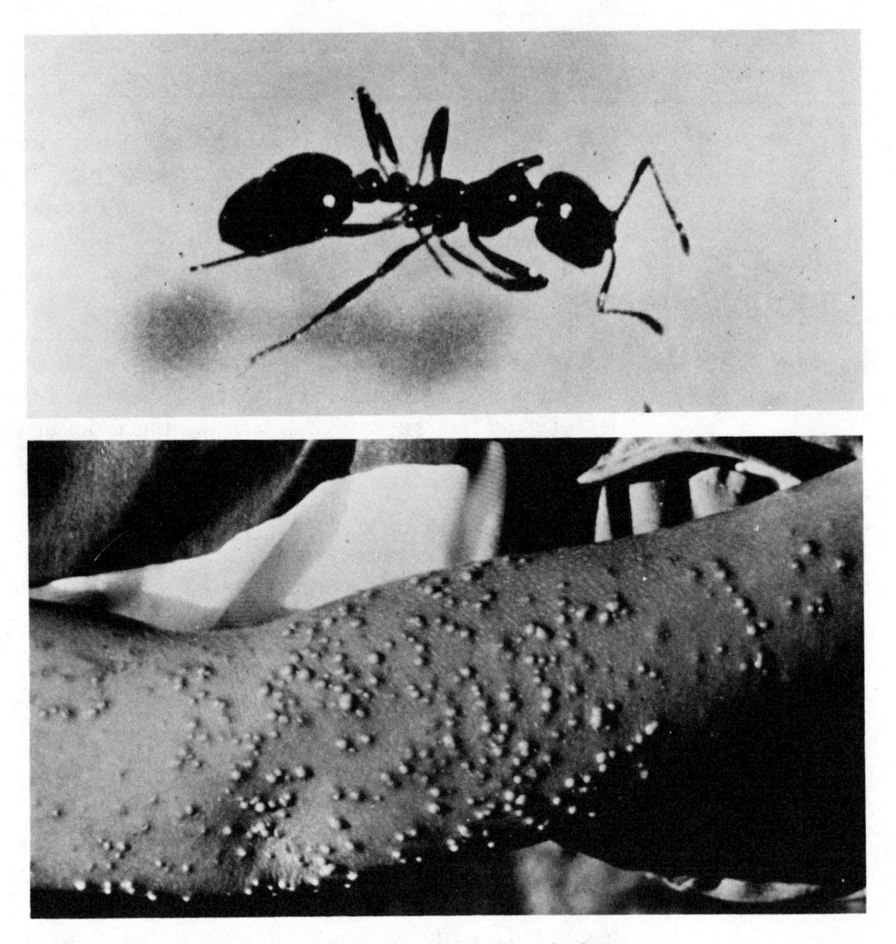

**FIGURE 22-1.** The imported fire ant *Solenopsis saevissima* and the arm of a boy showing a number of fire ant bites. (Photograph courtesy USDA.)

blistering of the skin. One such beetle, *Lytta vesicatoria*, commonly referred to as the Spanish fly, has been used as a source of an irritating substance called **cantharidin** which has some medical uses. At one time, a powder derived from ground up parts of *L. vesicatoria* was used as a human aphrodisiac. However, such practice is not recommended since the cantharidin is such a powerful mucous membrane irritant that its use can be extremely dangerous (see Chapter 19).

Quite a number of caterpillars possess urticating hairs. In some species, fragments of the hairs (setae) break off and cause mechanical irritation just like a fine bark sliver. However, these hairs are often hollow and filled with a toxin supplied by a poison gland cell in the epidermis. These hairs are clearly defensive but under some circumstances cause severe irritation accidentally. Larvae of the family Saturnidae, especially those of the genus *Hemileuca*, are particularly irritating. I have seen people who are not normally sensitive to foreign substances break out in an ugly rash when their skin has come in casual contact with the larvae of *Hemileuca electra* (Figure 22-2), an inhabitant of scrub vegetation in the southwest United States. The irritation caused by urticating hairs is not necessarily confined to contact with living larvae. A number of forest workers suffered severe irritation of the eyes and

FIGURE 22-2. A *Hemileuca electra* caterpillar showing the clumps of simple and branched urticating hairs.

nasal passages during the salvage of timber following a recent outbreak of the Douglas fir tussock moth (Lymantriidae) in Washington and Oregon. Apparently, felling the defoliated trees stirred up the urticating hairs of cast larval skins and those attached to the abandoned cocoons. The fine hairs drifting in the air were then inhaled by the workmen.

***Allergy***

An allergy is a condition of hypersensitivity to any of a wide number of materials in the environment. An allergic response to insect proteins is a common occurrence and is sometimes hard to separate from a more normal response when venomous compounds are involved. Since an allergy is just a degree of response, what can be described as an allergic response can occur as a result of almost any form of contact with insects or insect parts. Chronic forms of allergy include severe itching, eczema, or the symptoms of hay fever; deaths that occur are usually the result of anaphylaxis. However, the range of reaction is quite wide. For example, although everyone finds a bee or wasp sting to be painful, different individuals may have different side reactions. Some suffer little more than some temporary itching, some remain swollen around the sting for several days, while others become gravely ill and even die.

***Tissue Invasion***

The larvae of insects belonging to several orders are known to invade vertebrate tissue, including that of man. Most important in this regard are maggots, the larvae of Diptera, in which case the condition is called **myiasis.** Various tissues may be invaded, and a variety of terms are applied to describe localized infections, such as dermal myiasis (skin) or gastric myiasis (digestive tract). As with other forms of parasitism, myiasis may be casual or accidental (facultative) or a developmental requirement of the parasite (obligatory). Scott (1964) has reviewed human myiasis in North America.

There are no insects that produce myiasis for which man is the primary host, but there are numerous species which are obligate parasites of other vertebrates and which under some circumstances attain considerable medical importance. A good example is provided by the screwworm *Cochliomyia hominivorax*, which is a native obligate parasite of a variety of vertebrates in the southern United States and which became a serious pest of cattle. Like most screwworms, *C. hominivorax* lays eggs in or near wounds and the larvae proceed to invade the adjacent living tissue. During outbreaks, this insect has also infected humans and is known to cause **nasopharyngeal myiasis** as well as **dermal myiasis.**

Most cases of human myiasis are facultative but can be severe nevertheless. Fly larvae have been known to invade the intestine by way of the anus and also to enter the urogenital tract. This form of parasitism is usually confined to situations in which human beings are living under particularly filthy conditions. Species of flies, the larvae of which usually develop in carrion or dead flesh, are sometimes attracted to festered wounds for oviposition, and the larvae subsequently cause cutaneous myiasis. Some of the wound-infesting flies restrict their feeding to necrotic tissue and actually perform a beneficial service by cleaning out the dead tissue that might otherwise become putrified (see Chapter 19). Students interested in additional details about myiasis are referred to Zumpt (1965).

***Psychological Disturbances***

Some people have a genuine, though often unjustified, fear of insects that bears no relationship to whether or not the particular species is injurious. This irrational and lasting fear of insects is called **entomophobia.** This fear of real insects, however, is completely different from hallucinations of infestation by imaginary "bugs," which is more correctly called **delusory parasitosis.** Persons who are so affected by insects should be treated with compassion in that they may require professional psychological help in order to overcome their affliction. Delusory parasitosis, in particular, requires a great deal of patience on the part of professional entomologists because it is extremely difficult to persuade a sufferer that nonliving particles collected from their bodies are not "bugs."

# Insects as Vectors of Human Pathogens

The primary vectors of human pathogens are those arthropods in which the pathogenic micro-organisms pass part of their life cycle; they may undergo development (nematodes) or multiplication (rickettsiae, viruses, protozoa, bacteria). Although many arthropods associated with filth carry a wide range of pathogens on their body surface, their importance is well below that of the obligatory vectors. The groups to which medically important vectors belong include the bugs (Hemiptera), sucking lice (Anoplura), fleas (Siphonaptera), and flies (Diptera) plus the ticks and mites (Acarina) which will not be considered. By far the most important group are the flies, among which the mosquitoes transmit some of the most injurious human pathogens.

***Important Mosquito-borne Pathogens***

Numerous species of mosquitoes (Culicidae) are vectors of pathogenic organisms, but clearly the most important are a relatively small number of species belonging to the Anopheline genus *Anopheles* and the Culicine genera *Aedes* and *Culex*. These genera are quite easily separated in all stages of the life cycle (Figure 22-3). These important vectors display a great diversification in terms of their ecology, especially in their preference for breeding sites. As a group, the mosquitoes are able to reproduce in a variety of aquatic habitats, ranging from lakes, ponds, and rivers to rain barrels, tree holes, and the water held in the leaves of plants. The water in which the larvae develop can range

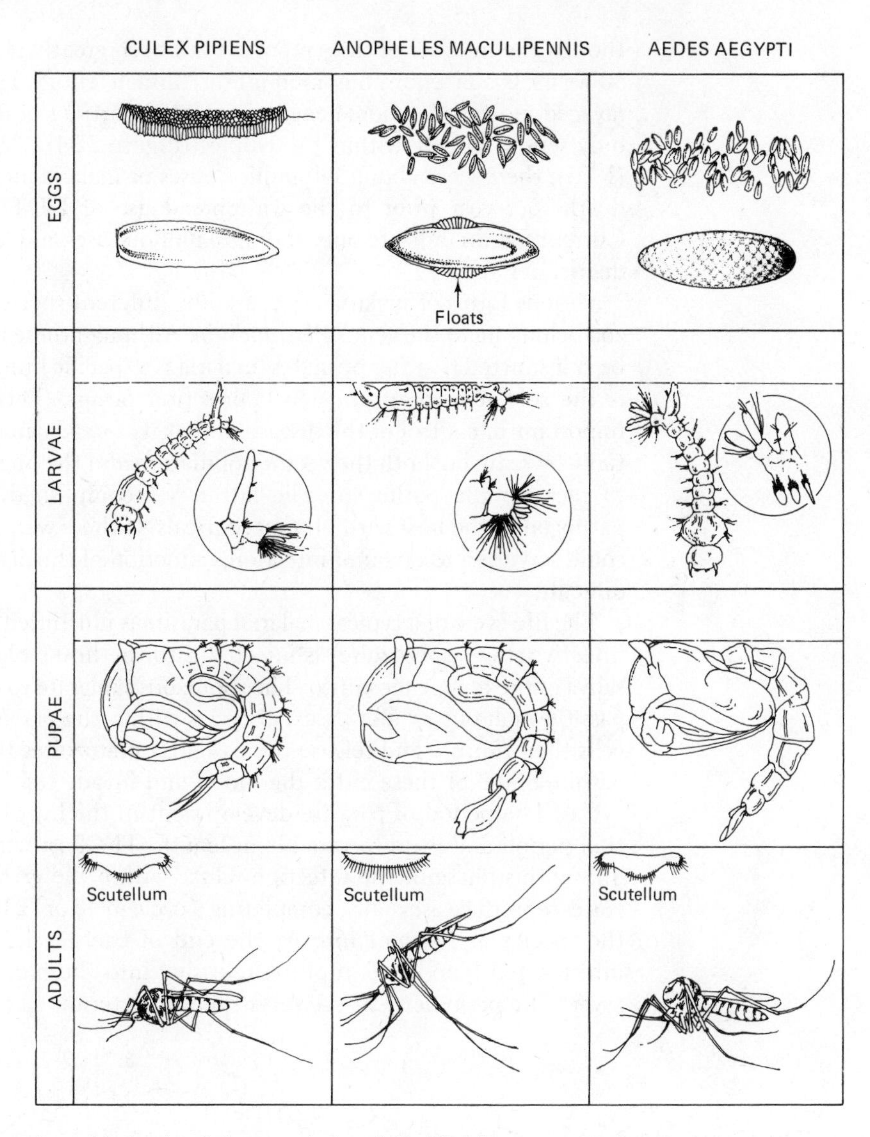


**FIGURE 22-3.** Anatomic and behavioral characteristics of the life history stages of mosquitoes that can be used to identify the three most medically important genera.

through fresh, polluted, and brackish. As is the case with most insects, the rate of development of mosquitoes is highly dependent on temperature. Arctic species that develop in pools of melted snow take a month or so to develop and complete only one generation per year; those that inhabit warm tropical areas develop in less than a week and have a number of generations during the period that water is available. Almost all kinds of vertebrates serve as hosts for the blood-feeding female mosquito. Generally, the range of hosts of a particular species is somewhat limited, and those species important in the transmission of human pathogens tend to have some preference for man.

The main diseases transmitted by mosquitoes are **malaria, yellow fever, filariasis,** and **encephalitides.** Malaria is caused by a protozoan, filariasis by filarial worms, and yellow fever and the encephalitides by viruses often called arboviruses—a contraction of arthropod-borne viruses.

Malaria is probably the most important arthropod-borne disease, even though the number of cases reported has been greatly reduced over the past 30 years by an enormous amount of human effort. In the past, malaria ravaged most of the more heavily populated portions of the world but now is only serious largely within the tropics (Figure 22-4). According to Busvine (1975), there were about 300 million cases of malaria and 3 million malarial deaths per year prior to the widespread use of DDT for vector control. Current figures indicate only about 50 million cases and less than one million deaths per year.

Various forms of malaria are caused by different species of parasitic protozoa belonging to the genus *Plasmodium.* Although these micro-organisms can be transmitted to other primates, malaria is a specific human disease and man is the sole reservoir from which infection occurs. This has proven to be important in control of the disease in that its spread can be reduced substantially by reducing both the vector population and the opportunity for vectors to pick up the pathogen. The latter is accomplished by controlling the pathogen in the host with pharmaceuticals. If there were also wild hosts that could serve as a reservoir of infection, reduction of the disease would be more difficult.

The life cycle of a typical malarial parasite is illustrated in Figure 22-5. The infective stage (**sporozoite**) is injected into the host's bloodstream with the saliva of the female mosquito. The sporozoites migrate to the liver where they pass through one or more asexual cycles within the parenchyma cells. These cells then rupture and release the daughter **merozoites** that have developed within. Some of these enter the blood and invade red blood cells (erythrocytes). The period of parasite development in the liver is called the incubation period and is quite variable in length. The host witnesses no ill-effects during this phase of the infection. Once within the erythrocytes, the merozoites reproduce asexually, completing a cycle in 48 or 72 hours, depending on the species of *Plasmodium.* At the end of each cycle, a large number of infected red blood cells rupture, releasing into the blood toxins that cause fever. The periodicity of the fever spells is determined by the length of the

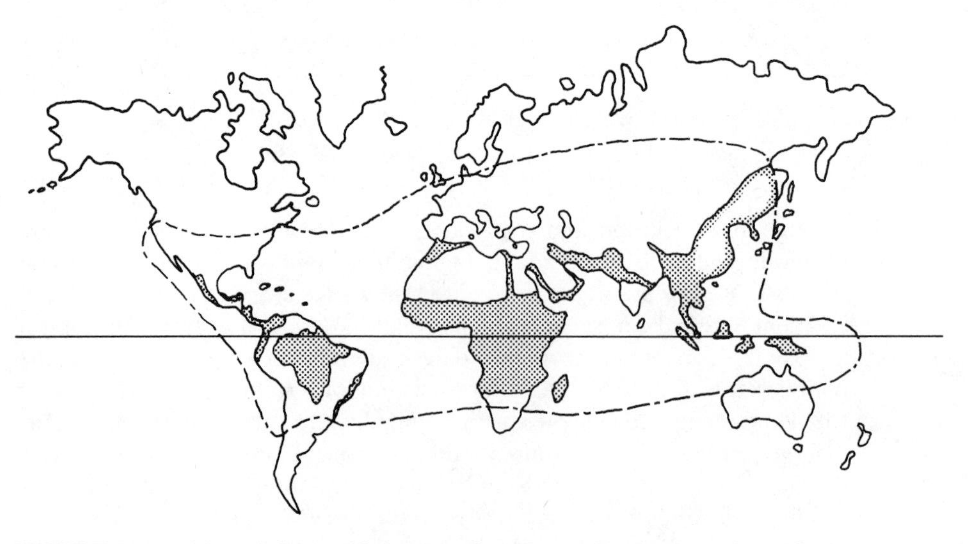

**FIGURE 22-4.** Map of the world showing the present distribution of significant numbers of cases of malaria (shaded) and estimated extent of its distribution in the past (outlined). (Redrawn from Buovine, 1975.)

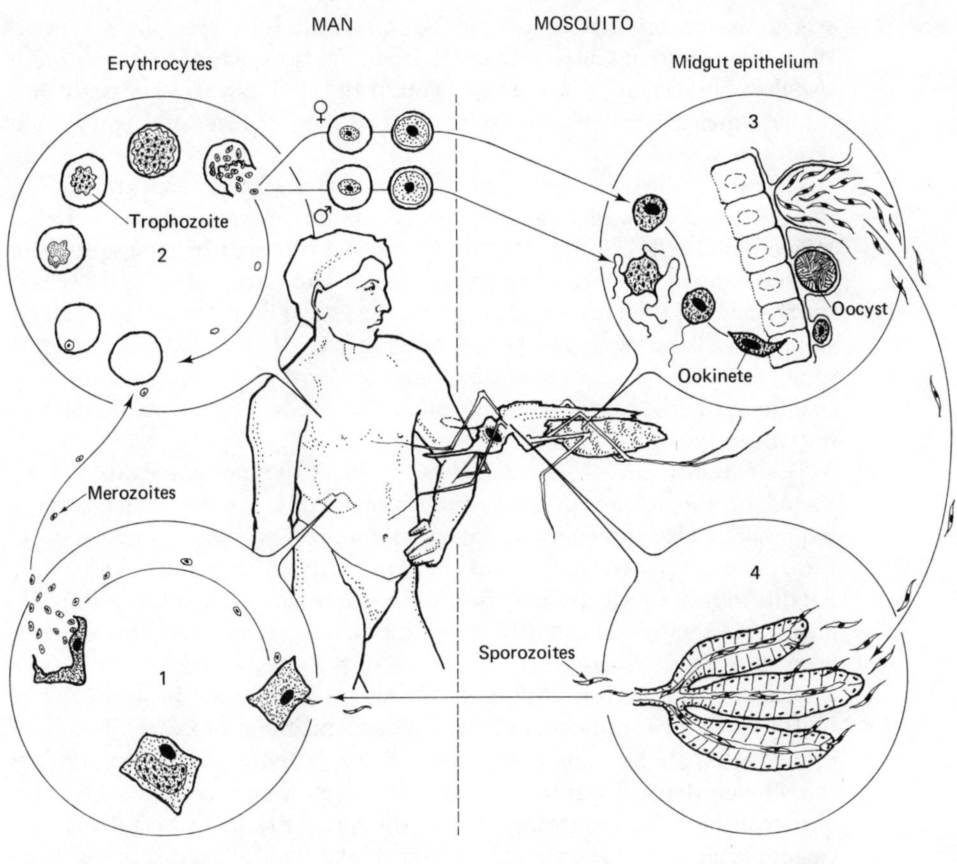


FIGURE 22-5. Life cycle of human malaria. (Adapted from James and Harwood, 1969.)

CYCLE IN MAN

**Sporozoites** are introduced into man from the salivary glands of a feeding mosquito (arrow from 4 to 1). Sporozoites enter liver parenchyma cells where they undergo primary exerythrocytic schizogony.

**Merozoites** are formed and invade (arrow from 1 to 2) red blood cells where the developing stage is called a **trophozoite.** The trophozoite grows, destroys its red blood cell to release additional merozoites which enter more red blood cells to continue the cycle. **Micro** (male) and **macro** (female) **gametocytes** are also periodically released from red blood cells and, when ingested by an *Anopheles* mosquito (arrows from 2 to 3), the microgametocyte divides to form motile **microgametes.**

CYCLE IN MOSQUITO

A microgamete enters a macrogamete to form a zygote that becomes a motile **ookinete.** Ookinetes pass through the mosquito midgut epithelium to form **oocysts** that grow, burst, and release sporozoites into the hemocoel (arrow from 3 to 4).

Sporozoites are active, pass freely throughout the body cavity, and concentrate in the salivary glands where they can be introduced into the blood of a vertebrate the next time the mosquito feeds.

synchronized reproductive cycles in the blood cells. If the cycle takes 48 hours, the result is **tertian malaria,** characterized by a three-day cycle consisting of two days of fever and a day of remission. If the cycle takes 72 hours, the host displays **quartan malaria,** comprised of two days of fever and two days of remission.

The cells of the parasite that are released when the erythrocytes rupture are the precursors of sexual gametes, which in turn invade other blood cells. When blood cells infected with this stage are swallowed by a mosquito, the gametes complete development. Female cells are fertilized by male cells in the stomach of the mosquito and form a new stage that burrows through the stomach wall and forms a cyst on its outer surface. Within the cyst numerous spindle-shaped sporozoites develop, burst out, and move through the hemocoel to invade the salivary glands, from which they are subsequently injected into a new host.

The parasites do not seem to adversely affect the mosquitoes that vector them; yet, this is a phase in the life history of the parasite that is of great importance. Development within the mosquito is governed more or less by the ambient temperature. In temperate climates, the development of the parasite within the vector may be too slow to permit completion within the latter's life span. This could be the main factor that has limited malaria to areas of warmer climate.

Mosquitoes transmit more than 150 viruses that cause diseases of vertebrates. As with the protozoans that cause malaria, these viruses multiply both in the vertebrate host and in the insect vector. The most important of these cause **hemorrhagic fevers** such as yellow fever, or what are collectively called **encephalitides.** In the former group, the virus attacks the capillaries and causes bleeding of the respiratory and urogenital tracts, accompanied by fever. The encephalitis viruses invade the central nervous system, producing pain in the joints and the back or, in severe cases, some degree of paralysis.

Yellow fever is still a serious problem in Central America and the tropical regions of South America and Africa; it does not seem to have been a problem in the Asian tropics. The disease occurs in two forms, **urban yellow fever** transmitted by *Aedes aegypti* and **jungle** or **sylvan yellow fever** transmitted by mosquitoes belonging to several genera. Mayan writings indicate that jungle yellow fever may have been a problem in the New World in pre-Columbian times, but the major outbreaks of urban yellow fever did not occur until after *A. aegypti* was introduced from Africa. *A. aegypti* is closely associated with human habitation, often breeding in small containers such as rain barrels. While this behavior increased virus transmission it also tended to simplify vector control, which, in combination with the development of vaccines that can be easily administered in populated areas, virtually eliminated outbreaks of the disease in urban environments.

There are a large variety of encephalitides of which western equine encephalitis, eastern equine encephalitis, St. Louis encephalitis, and Venezuelan encephalitis are the most important in the United States. None of these compare with malaria or yellow fever as serious problems, but they cause considerable concern nevertheless. These diseases are caused by viruses that normally infect birds but are transmitted to man and horses by culicine mosquitoes, which feed on both avian and mammalian hosts. Humans and horses can become infected, but there is no further transmission beyond these hosts. In 1972, an outbreak of Venezuelan equine encephalitis, transmitted by *Culex tarsalis*, led to the widespread vaccination of horses in the southwestern United States. In 1975, the United States experienced the worst epidemic of St. Louis encephalitis in more than a decade. The disease, which causes inflammation of the brain, fever, and sometimes death, was particularly prevalent in Mississippi where more than 125 cases and 32 deaths were reported as a consequence of abnormally high populations of the vector, *Culex pipiens*.

Of all the filarial nematodes transmitted by mosquitoes, *Wuchereria bancrofti*, which is restricted to man and extremely prevalent throughout the tropics, is particularly important. The adults are typical small nematodes which live in the lymphatic and connective tissues of man. The fertilized females produce hundreds of minute larvae called **microfilariae** which periodically enter the peripheral circulatory system. Development does not proceed until the larval worms are taken up by a feeding mosquito. Those microfilariae that successfully invade the mosquito hemocoel from the stomach undergo further development in the thoracic muscles and then accumulate in the proboscis. When an infected mosquito feeds, the nematodes leave the proboscis and enter their human host through the feeding puncture. Once in the blood of man, they proceed to the lymphatic tissue where they complete the cycle.

The early effects of filarial infection may go unnoticed, although there may be transient headaches and pains of the type that could be attributed to any number of causes. However, when the worms become mature, they release toxic materials, and if present in large numbers, the nematodes can cause obstruction of the lymph nodes. If this occurs, there is progressive swelling, usually of the legs and scrotum, until the grossly distorted condition known as elephantiasis occurs (Figure 22-6). Infection rarely results in death, and individuals who exhibit severe swelling are but a small fraction of the estimated 250 million people in the tropics and subtropics who are infected by these parasites (World Health Organization, 1974). The fact that the disease is not generally fatal, combined with the mild nature of the symptoms in the early stages of infection, has greatly slowed progress in eradication of the disease. However, like malaria, it can greatly reduce the efficiency and productivity of the populations of developing nations in the tropics, and in the near future more serious attempts to reduce the disease will become necessary.

### *Other Major Fly-borne Pathogens of Man*

There are numerous other diseases of man caused by pathogens that are vectored by members of the order Diptera. Of these, three diseases are of considerable importance; namely, trypanosomiasis transmitted by tsetse flies (Muscidae), leishmaniasis transmitted by sandflies (Psychodidae), and onchocerciasis transmitted by black flies (Simuliidae).

**Trypanosomiasis** provides an excellent example of the impact that insects can have on the development of an area. Tsetse flies (*Glossina* spp) are found throughout much of tropical Africa, and over much of their range they serve as vectors of the trypanosomes that cause sleeping sickness in man and nagana in horses and cattle. Millions of Africans have died of sleeping sickness, but of perhaps equal or greater importance is the fact that the human populations across the range of the disease have been unable to keep domestic cattle or draught animals for tilling the soil. Even after years of effort to reduce the problem, trypanosomiasis continues to impede the development and self-

FIGURE 22-6. Symptoms of the filarial disease encephalitis. (Photograph courtesy of Center for Disease Control, Atlanta, Georgia 30333.)

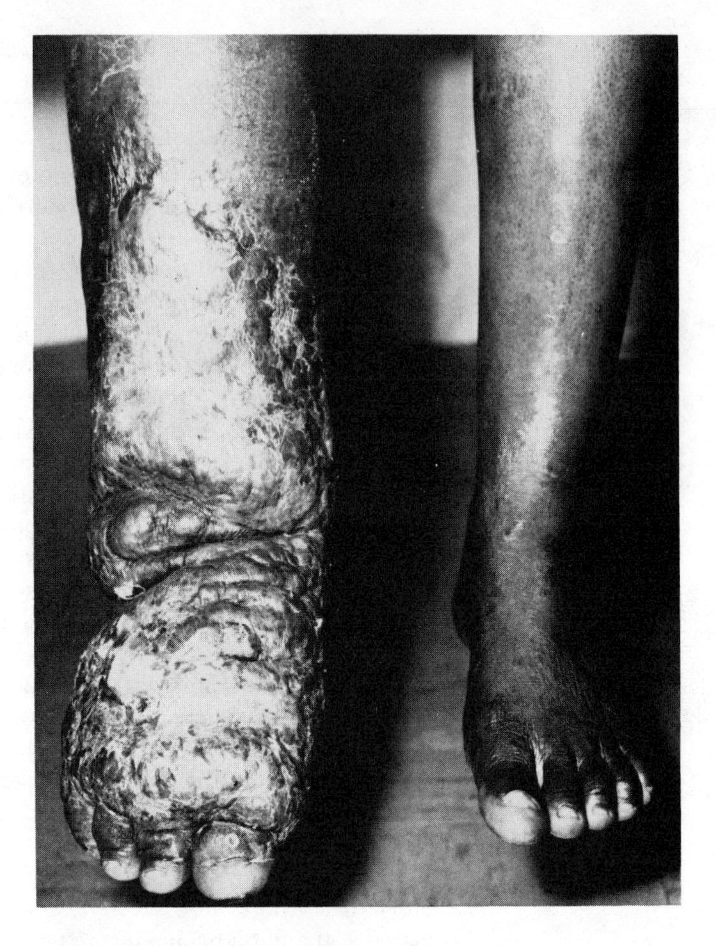

sufficiency of the nations of tropical Africa. African sleeping sickness is the last major insect-transmitted disease for which we have no control.

The disease-causing trypanosomes (Figure 22-7) live as blood parasites in vertebrates. In wild native game animals, particularly ungulates, they are generally harmless, but in hosts like man and domestic cattle, they cause debilitation and death. Trypanosomes do not reproduce sexually but they do undergo a cycle of development in the tsetse fly, which once infected can transmit the parasites for the remainder of its life.

There are two varieties of sleeping sickness caused by different species of the parasite. Rhodesian sleeping sickness develops rapidly and death can occur quite rapidly. The main reservoir of the parasite is in wild game, but some are transmitted from human to human. Gambian sleeping sickness, on the other hand, is characterized by a fairly long period of debility that accompanies the multiplication of the trypanosomes in the blood, followed by a severe phase in which the parasites invade the central nervous system, causing listlessness, emaciation, and eventually death. In this form, man constitutes the main reservoir and is an effective one owing to the protracted initial period of parasite proliferation in the bloodstream.

**Leishmaniasis** is a highly variable disease caused by the flagellate protozoan *Leishmania*, vectored by sandflies of the genus *Phlebotomus*. There is a visceral form in which the parasite attacks the spleen and liver, which become enlarged, and a cutaneous form which causes open sores to develop. The visceral form is most prevalent in the Old World where the parasite exists as three epidemiological varieties. In India, man is reported to be the sole reservoir of the parasite (Adler and Theodor, 1957), whereas in the Sudan, rodents are the reservoir; around the Mediterranean, dogs are frequently infected, and the disease is more prevalent among children.

The cutaneous form of leishmaniasis is common in warmer parts of the Old World where it is known as Oriental sore or Delhi boil and also in the tropical and subtropical Americas where it is vaiously named. One form of the disease is mucocutaneous, particularly in the mouth and nasal passages where it causes horrible disfigurement (Figure 22-8), as a result of destroying even the underlying cartilage and bone. In the forms that do not spread to the mucosa, both moist and dry ruptures of the skin occur.

In vertebrates, the parasite exists as an ovoid cell that lacks a flagellum, but once within the stomach of the sandfly, it is transformed to a true flagellate. The parasite then multiplies in the insect and later reinfects a vertebrate by way of infective fly feces or by regurgitation when the vector takes another blood meal.

**Onchocerciasis** is caused by the filarial nematode, *Onchocerca volvulus,* transmitted by *Simulium* spp. The disease is endemic to West and Central Africa but also occurs in Central and South America, perhaps as a result of introductions of the parasite with African slaves. Man appears to be the sole reservoir of this parasite. Microfilariae are taken up from the superficial layers of the skin by feeding black flies. Some of the young nematodes break through the fly's stomach wall, undergo development in the thoracic muscles, and then migrate to the proboscis. The parasite reinvades man when the vector takes another blood meal. In man, the mature female worms produce large numbers of microfilariae, which circulate in the skin where they can cause nodules and itching. If they penetrate the eye, the victim may become blind; hence the name river blindness is often applied to the disease. In West

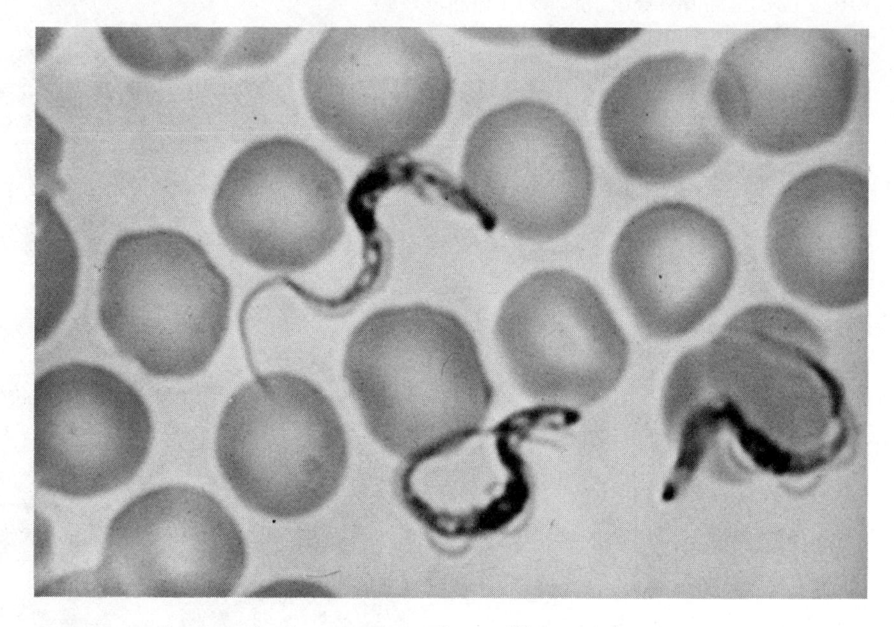

**FIGURE 22-7.** Trypanosomes in the blood of a vertebrate host. (Photograph courtesy of Neil Marshall.)

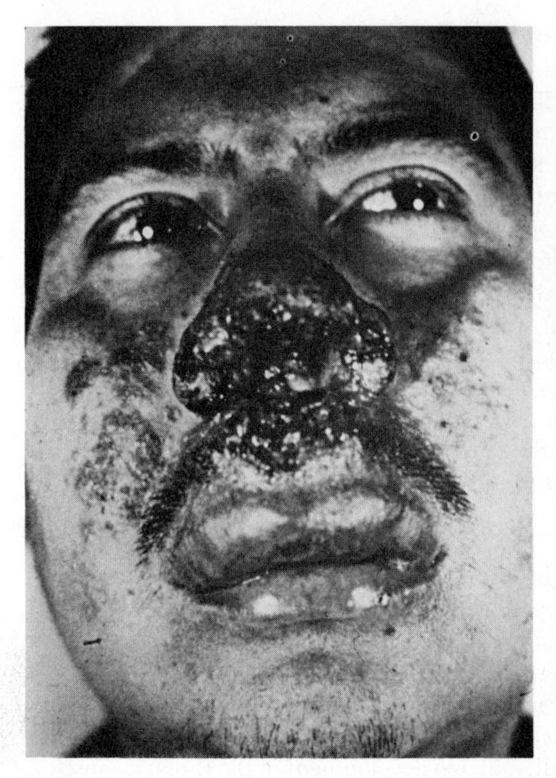

FIGURE 22-8. Mucocutaneous leishmaniasis. (*Top.*) Facial disfigurement. (*Middle.*) Skull of a pre-Columbian Indian from coastal Peru showing damage to the facial bones caused by leishmaniasis. (*Bottom.*) Pre-Columbian Indian olla ornamented with a face disfigured probably by leishmamiasis. (Middle and bottom photographs courtesy of Neil Marshall, origin of top photograph unknown.)

Africa, more than a million people, of which an estimated 60,000 are totally blind, have the disease (Pratt and Darsie, 1975). Like mosquito-borne filariasis, this disease also impairs the productivity of the native population.

### *Pathogens Transmitted by Other Insects*

Although the Diptera are the most important insect vectors of human pathogens, they are different from the disease-bearing bugs, lice, and fleas in that the immature stages live in a completely different habitat from the human beings on which the adults feed. The lice and fleas are wingless and remain on or very near their host. The disease-bearing bugs (Reduviidae) are winged but rarely fly and usually remain near their hosts.

Several genera, among which *Triatoma* (Figure 22-9) is probably the most important, transmit **Chagas' disease,** also known as American trypanosomiasis. The causal organism, *Trypanosoma cruzi,* is quite closely related to the flagellate that causes sleeping sickness. Chagas' disease occurs throughout the New World from the southern United States to Argentina, although, according to Pratt and Darsie (1975), no human cases have been recorded in the United States since 1955. However, in Central and South America more than 10 million people are believed to be infected. The disease tends to be most prevalent among the poor in rural areas and is frequently chronic. In such cases the most apparent symptoms are facial swelling and swelling around the site of infection, but the parasites also invade endothelial tissues; death usually results from damage to the heart muscles. Children suffer from an acute form of the disease which results in fever, facial swelling, and anemia.

Unlike the diseases considered thus far, the causal organism of Chagas' disease is not transmitted through the vector's mouth parts. The infectious stage of the trypanosome resides in the bug's hind-gut and passes out with the feces. Since the bugs almost always defecate while taking a blood meal, the parasites are left on the skin of the host near a site of irritation. When the

**FIGURE 22-9.** *Triatoma,* the vector of Chagas' disease. (Photograph courtesy of Neil Marshall.)

bite is scratched the trypanosomes are picked up on the fingers and subsequently transferred to the eyes, nose, mouth, or open sores, which provide suitable sites for invasion.

The vector tends to be a persistent occupant of human habitations (Figure 22-10). They conceal themselves in crevices from which they emerge every few days to search for a blood meal. If they feed on an infected host, they pick up the trypanosome with the blood; the parasites then complete their development in the bug's digestive tract. The trypanosomes can be transmitted from human to human, but birds, wild animals, and the bugs themselves form additional reservoirs from which the vector can acquire the parasite. In southern California *Triatoma* is often found in the houses of wood rats.

History is punctuated with epidemics of **bubonic plague**, a bacterial disease transmitted by fleas. Plague is mainly a disease of wild rodents and is transmitted among the members of a population by fleas. Human plague epidemics usually develop when wild rodents mix with rats on the edges of towns. The disease usually spreads rapidly among the highly susceptible domestic rats, and fleas that will feed on both rats and man tend to bite man more frequently as their rat hosts are killed by the disease.

Usually the plague bacilli are suppressed in man and only produce swellings called "buboes" from which the name of the disease is derived. In some susceptible individuals the bacteria multiply rapidly and cause death in a short time. Man does not usually form a reservoir from which fleas can transmit the disease to other individuals, but when the lungs are infected, there may be some spread between people by way of contaminated fluids expelled by coughing and sneezing.

FIGURE 22-10. A rural house in Venezuela with a metal roof instead of the more common thatched roof. The use of metal roofing eliminates one of the favored hiding places of *Triatoma*. (Photograph courtesy of Boyd Collier.)

Plague is still a potentially serious problem, and government agencies conduct plague surveillance in a number of states. In 1975, a campground in the Rocky Mountain National Park was closed temporarily following the collection of plague-infected fleas from a white-footed deer mouse. In the summer of 1976, several of California's mountain parks were closed because of the presence of infected rodents, and posters were set out to caution the public (Figure 22-11).

Finally, human lice serve as the vectors of several diseases, but these are no longer of great concern because of the relative ease with which the host

Wild rodents (Chipmunks and Ground Squirrels) at several locations in this region have been found infected with plague. This disease is occasionally transmitted to humans. Treatment is effective if started in time. Notify your physician if you should become ill within one week following your visit here.

Plague is transmitted by the bite of an infected flea, or by direct contact with an infected rodent. To prevent exposure to this and other rodent diseases, do the following:

1. AVOID ALL CONTACT WITH SQUIRRELS, CHIPMUNKS OR OTHER RODENTS AND THEIR FLEAS. PROTECT DOMESTIC PETS WITH FLEA COLLARS OR FLEA POWDER. DO NOT FEED RODENTS.

2. AVOID CAMPING, SITTING, OR SLEEPING ON THE GROUND CLOSE TO RODENT BURROWS.

3. REPORT ANY SICK OR DEAD ANIMALS TO PARK, FOREST, CAMPGROUND, OR LOCAL HEALTH AUTHORITIES. DO NOT TOUCH DEAD RODENTS.

El Dorado County Health Department
County Government Center
South Lake Tahoe, California 95705
Phone 916 541-7133

California Department of Health
Vector and Waste Management Section
714 P Street
Sacramento, California 95814
Phone 916 445-4408

Placerville Phone 916 626-2411

**FIGURE 22-11.** Plague warning poster. (Courtesy California Department of Health.)

specific vectors can be controlled. Over the course of history, epidemic typhus, caused by bacterialike organisms called rickettsiae, may have had as great an impact on civilization as the plague. Relapsing fever, caused by a parasitic spirochaete, is also largely a disease of the past.

The events that have brought man and insects into close association seem to be reasonably clear. But how some insects became obligate vectors of human pathogens is a question yet to be answered to everyone's satisfaction. Most of the evidence suggests that most of the arthropod-transmitted micro-organisms that cause diseases of humans were first associated with the vector species. In general, long-term associations between hosts and pathogens tend to reduce pathogenicity. For example, the rather severe effect of tsetse fly-borne trypanosomes in man compared to their lack of pathogenicity in wild game hosts probably results from the fact that since the contact between man and the pathogen is relatively recent, there has been insufficient time for them to fully adapt to each other.

### *Epidemiological Aspects and Control*

Insects of medical importance must be evaluated differently from those which become pests of forests, crops, livestock, or commodities. In these latter cases, it may be possible to determine quite satisfactorily the population levels that will cause significant economic loss and to make control recommendations based on a system of cost-benefit accounting (see Chapter 20). It is much more difficult to determine the worth of human comfort and impossible to place a value on even a single human life. The control of medically important insects therefore becomes a matter of morality rather than economics. Nevertheless, the suppression of insect-borne diseases often becomes economically necessary. The debilitation caused by diseases over which we can exercise direct control has greatly influenced the rate of development of a number of nations in the tropics and subtropics. More than ever before it is important that these countries engage in development activities that will enable them to become relatively self-sufficient. They will be unable to achieve this goal unless the people are freed from the misery and reduced efficiency that so often accompany insect-borne disease.

As in the case of the insect transmission of plant pathogens discussed in the previous chapter, it is not possible to predict roughly how much damage will result from a known population of insects alone. There is a complex of factors that determines whether a vector population will cause a disease epidemic; how these factors interact depends on the nature of the relationship that has evolved between the vector, the pathogen, and the host species. For a disease problem to exist there must be a sufficiently large number of vectors that have access to a reservoir of viable parasites and then the capability of transmitting the parasites to susceptible vertebrate hosts. When the reservoir is an animal other than man, its population dynamics may be just as important as that of the vector. The life span of the reservoir and the vector is also important. Long-lived vertebrates such as human beings provide a persistent reservoir. The life span of the vector determines the number of feedings and the opportunity to pick up and transmit the parasite. Furthermore, the life span of the vector determines the time available for development and, perhaps, proliferation of the parasite within the vector. The higher temperatures in the tropics may permit more parasites to complete their development in cold-blooded vectors; this may account for the higher incidence of insect-borne diseases in warmer climates. Interestingly, most of the insect-borne diseases that have ravaged civilizations in the temperate zone, like the plague, for example, have vectors that either live on or with warm-blooded hosts and are consequently less affected by cool, ambient temperatures.

The nature of the relationship between the host, parasite, and vector species can greatly influence the success with which a disease can be brought under control and the methodology employed. In the case of malaria and urban yellow fever for which man is almost the sole noninsect host, it is theoretically possible to almost eliminate the disease in urban areas by reducing the vector population and by killing the parasite in man pharmaceutically. Once the parasite is not available to the vector, the disease can be kept in check without engaging in widespread vector control. This seemed to be an entirely practical approach against yellow fever until the jungle monkey reservoir was discovered. Now the disease is periodically introduced into urban areas from these jungle reservoirs, and it is unlikely that the disease will ever be completely eliminated. On the other hand, it would be entirely possible to eradicate malaria from developed countries with mass medication combined with vector control.

Encephalitis control is quite different, because man and horses are secondary vertebrate hosts, whereas birds form the major parasite reservoir. The incidence of encephalitis cases seems to be influenced largely by the size of the vector population. Since it would not be practical to attempt to eliminate the viruses or the birds that harbor them, the only practical attack on these diseases is to attempt to reduce the populations of the mosquito vectors.

Plague is different again in that a disease threat occurs when a rodent reservoir develops such that the flea vector can transmit the disease from the rodents to humans. Control and eradication of plague involves surveillance and separation of the rodent reservoir from human populations. Rat control through sanitation, good construction, and poisoning is the most sound approach to disease reduction.

# References Cited


ADLER, S. and O. THEODOR. 1957. Transmission of disease agents by phlebotomine sand flies. *Ann. Rev. Entomol.* 2: 203–226.

ANDERSON, R. C. 1957. The life cycles of dipetalonematid nematodes (Filarioidea, Dipetalonematidae): the problem of their evolution. *J. Helminth.* 31: 203–244.

BUSVINE, J. R. 1975. Arthropod vectors of disease. Edward Arnold & Co., London.

JAMES, M. T. and R. F. HARWOOD. 1969. Herms' Medical Entomology. 6th ed. Macmillan Publishing Co., Inc., New York, N. Y.

MANWELL, R. D. 1955. Some evolutionary possibilities in the history of malarial parasites. *Indian J. Malar.* 9: 247–253.

PARRISH, H. M. 1963. Analysis of 460 fatalities from venomous animals in the United States. *Amer. J. Med. Sci.* 245: 129–141.

PRATT, H. D. and R. F. DARSIE, JR. 1975. Highlights in medical entomology in 1974. *Bull. Entomol. Soc. Amer.* 21: 173–176.

SCOTT, H. G. 1964. Human myiasis in North America (1952–1962 inclusive). *Fla. Entomol.* 47: 255–261.

Smith, K. G. V., ed. 1973. Insects and other arthropods of medical importance. British Museum (Natural History), London.
Usinger, R. L. 1966. Monograph of Cimicidae (Hemiptera-Heteroptera). Thomas Say Foundation (Entomol. Soc. Amer.) Vol. 7.
World Health Organization. 1974. WHO expert committee on filariasis: Third report. *WHO Tech. Rep. Ser. No. 542.*
Zumpt, F. 1965. Myiasis in man and animals in the Old World. Thornton Butterworth, Ltd., London.

# Chemical control—the good and bad of it

# 23

The historical beginning of chemical pest control is unknown, but there are references to insecticides in the writings of the Greeks, Romans, and Chinese, dating to about 1,200 years before the birth of Christ. Fumigation to kill insect pests by burning toxic plants also dates to about 1,200 B.C. in China. Homer suggested the use of sulfur as a means of averting pest problems, and by the first century A.D. both the Greeks and the Chinese were aware of the insecticidal properties of some metals. For example, Shen Nung Pen Tshao Ching (*Pharmacopoeia of the Heavenly Husbandman*), believed to have been written between 100 and 200 A.D., mentioned the use of mercury and arsenic for control of the body louse and the use of a Liliaceous plant for killing insect pests of man. Pig oil, lime, sulfur, copper, cinnabar (HgS), copper alum, and iron sulfate were all used for pest control in China before the tenth century (Konishi and Ito, 1973).

In spite of its early beginning, the era of chemical pest control belongs to the twentieth century. The modern use of insecticides dates to 1867, when Paris green was found to be effective against outbreaks of the Colorado potato beetle *Leptinotarsa decemlineata.* About the same time, powdered sulphur, lime sulphur, and a mixture of copper sulphate and lime called Bordeaux mixture gained widespread acceptance in England and France as fungicides, especially useful for controlling diseases of grapes. These successes, particularly that of Paris green, stimulated the experimentation with these and other compounds against a variety of plant pests. In the early years of the twentieth century, botanical insecticides and fluorine coumpounds were found to be effective.

The obvious value of chemical pesticides to crop protection quickly led to the production and the sale of a variety of ineffective remedies for a number of pest problems. This fraudulent activity led to the passage of the Federal Insecticide Act of 1910. The act set standards for commercially sold insecticides and fungicides, prohibited the inclusion of phytotoxic materials in the formulations, and forbade the printing of false claims on pesticide labels.

Within a few years, the use of pesticides in American agriculture had become so commonplace that concerns about residues in agricultural produce, particularly arsenic and lead on apples and pears, began to develop. As a result, the Federal Food, Drug, and Cosmetic Act of 1938 was passed, establishing levels of tolerance for insecticide residues in agricultural products. The passage of this act was indeed fortunate timing, inasmuch as it established the basic regulations to protect the public against poisonous residues in crops before the spectacular increase in pesticide use that was to follow the discovery of the insecticidal properties of DDT in 1939.

The original Federal Food, Drug, and Cosmetic Act was improved in 1954 with the addition of the Miller Amendment and again in 1960 with the addition of the Food Additives Amendment. The purpose of the law as it exists today is to maintain the purity and safety of the public food and drug supply through the establishment of tolerances for poisonous ingredients, food additives, and contaminants (including insect parts and excrement) in foods and for pesticide residues in raw agricultural produce. Established pesticide tolerances set the maximum amount of the various chemicals that can remain in a food crop, including meat, fat, milk, and eggs. The task of setting these tolerances and conducting the field surveillance necessary to enforce them originally fell to the Food and Drug Administration (FDA), but in 1970 the setting of tolerances was transferred to the Environmental Protection Agency.

In 1947 the Federal Insecticide, Fungicide, and Rodenticide Act replaced the Federal Insecticide Act of 1910. This new law was expanded to cover other agricultural chemicals, including rodenticides, herbicides, defoliants, and disinfectants, and it expanded labeling regulations. It also instituted a requirement that any pesticide marketed across state lines or used on produce entering interstate commerce be registered with the United States Department of Agriculture (USDA). Before a product is registered, the manufacturer must provide scientific evidence that the product is effective for the intended purposes stated on the label and will not be injurious if used as directed.

The purpose of the 1947 act was to protect both the user of pesticides and the public from false claims and misuse. The label of every pesticide must show the name, brand, or trademark under which it is sold, the name and address of the manufacturer, the net contents, an ingredients statement, and a warning statement if necessary to prevent injury. Pesticide labels must also include directions for use which, if followed, will give the desired results and allow for compliance with established residue tolerances.

The Federal Insecticide, Fungicide, and Rodenticide Act of 1947 was substantially amended in 1972. The amendment strengthens and expands the authorities provided by the old law in the following ways:

1. Extends federal registration and regulation to all pesticides, including those distributed for use within a single state.
2. Requires proper application of pesticides to ensure greater protection of man and the environment.
3. Prohibits use in a manner inconsistent with labeling instructions.
4. Authorizes classification of pesticides into "general use" and "restricted use" categories.
5. Specified that federal standards for certification of applications of restricted use pesticides be established by October 21, 1974 as a basis for state programs for training and certifying pesticide applicators.
6. Requires registration of plants manufacturing pesticides for interstate commerce and export.

# Some Characteristics of Pesticides

All pesticides have a number of characteristics in common, some of which are desirable and some undesirable. Ideally, they should be highly effective against pests without being injurious to beneficial or desired organisms. Unfortunately, most pesticides do not completely comply with this ideal. Many are broad-spectrum poisons and therefore injure beneficial species. Some are toxic to crop plants, kill beneficial insects and micro-organisms, taint the flavor of foods, or stain the desired product. As many of these undesirable features as possible are naturally avoided, and a high priority in agriculture is for chemicals with some degree of **specificity**. It has been difficult to obtain highly specific compounds, partially because specificity narrows the number of uses and reduces marketability. However, a growing interest in conserving beneficial species is increasing the demand for more specific chemicals, and manufacturers are finding it more profitable to produce them.

Because of the regulations governing residues, it is usually desirable that pesticides break down after they have been applied; however, they must be stable in storage and not be rendered inactive by other chemicals with which they are mixed to facilitate their application. Furthermore, it may be desirable that certain pesticides remain active in the environment for several weeks to reduce the need for critical timing and to increase the time they are effective against the target pest. This forms one of the bases for pesticide selection as a crop approaches harvest, and statements in the directions for use usually refer to the minimum time prior to harvest that each pesticide can be applied. The present trend in agriculture is to use pesticides, particularly insecticides, that have a low level of **persistence.**

The **mode of action** of pesticides, which is the way they act upon organisms to cause death, has proven difficult to ascertain. Although many are protoplasmic poisons (e.g., arsenic), nerve poisons (e.g., organochlorines and organophosphates), respiratory inhibitors (e.g., oils), or growth regulators (e.g., many organic herbicides), the precise ways in which they interfere with normal biochemical processes often are unknown.

Populations are made up of individuals which vary in a variety of ways, including their susceptibility to poisons, so it is necessary to have a standard measure to determine the **toxicity** of pesticides, both for target and representative nontarget organisms. This is derived from a statistical method called probit analysis, which provides a measure of the response of a population sample over a predetermined period of time to various doses of the pesticide.

If several geometrically increased doses are applied to a population sample, the cumulative percentage killed increases with increased dose level. When cumulative percent mortality for each dose is plotted against the logarithm of the dose, a sigmoid curve similar to that shown in Figure 23-1 results. However, such a curve based on real data is difficult to plot, so the normal procedure is to convert the cumulative percent mortality values to probit values, using standard statistical tables. For convenience in calculation, 5 is added to each standard deviation. Thus, probits for 0, 50, and 100% mortality are $-\infty$ 5.0, and $+\infty$, and if the data fit a normal distribution, probits 4.0 and 6.0 would represent 16 and 84% mortality, respectively.

When the data are plotted in this manner, it is possible to fit a straight line to the data points, as shown in Figure 23-2. Each probit thus represents a theoretical mortality that can be expected to result from the application of a particular dosage of pesticide. However, as would be expected from the shape of the sigmoid curve, the relationship between dosage and mortality is more accurate at 50% than at the extremes. Consequently, the standard measure of toxicity is the $LD_{50}$, or the log dosage that will kill 50% of the population. The $LD_{50}$s of various pesticides are widely quoted for both pests and nontarget organisms (see for example, Pimentel, 1971). Since this value often varies according to the route of entry (oral or dermal) and sex, these are usually specified. The dosage may be expressed either in terms of milligrams per individual or as milligrams per kilogram of body weight. The $LC_{50}$, a less useful inhalation-concentration value, is expressed as micrograms per liter. Mammalian toxicity is usually based on experimental data obtained from rats.

Another important aspect of the probit transformation of dosage-mortality

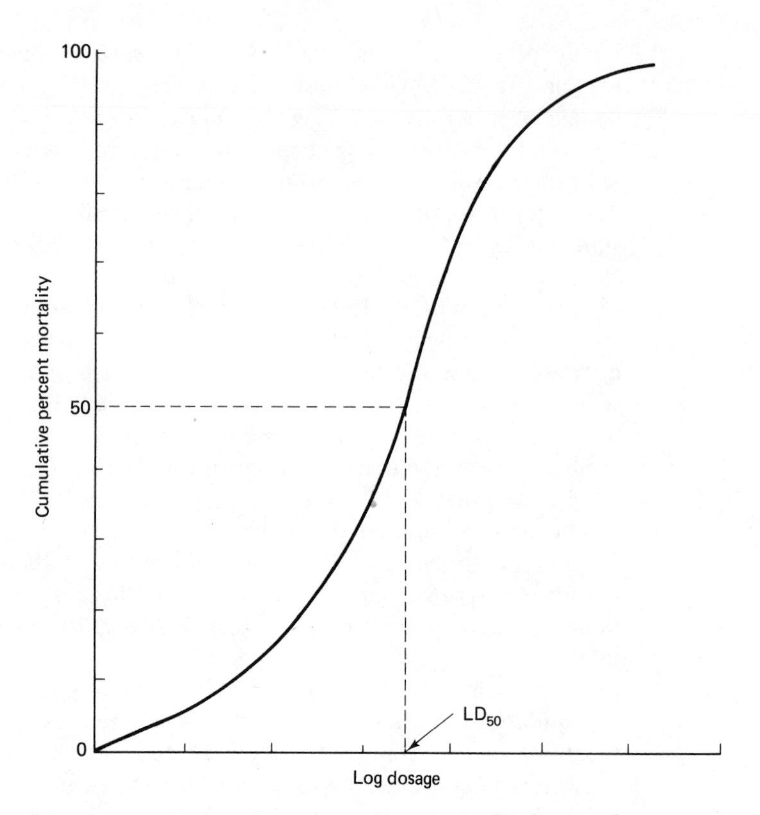


**FIGURE 23-1.** The cumulative mortality in relation to an increase in the dosage of a toxic substance.

data is that it permits the use of the slope of the line as a measure of the variance in the susceptibility of a population to the pesticide. The shallower the slope of the line, the greater the variation in the susceptibility to the toxicant of the individuals making up the population. Such a population would require a dose considerably higher than the $LD_{50}$ to effect a reasonable level of mortality. A steep slope indicates homogeneity in the susceptibility of a population and a more narrow range between nonlethal and lethal doses. Hoskins and Gordon (1956) indicated that changes in the slope of the line may indicate the development of resistance, particularly if the slope changed from steep to shallow and then back to steep accompanied by an increase in the $LD_{50}$ with each change.

It must be realized, of course, that the determination of dosage-mortality relationships requires good experimental design. Large samples, replicates, and controls must be used, and the controls must be handled the same as the treated samples in all respects except for the application of the pesticide being studied. Another important consideration is the possible synergistic action between the active pesticide ingredient and other chemicals used in its formulation. Consequently, different formulations must be used if the objective is to measure the toxicity of the pesticide by itself.

The fact that many pesticides are toxic to nontarget organisms makes it necessary that the labels carry relevant information. Even though there is not a direct relationship between toxicity and hazard because of differences in persistence, in the characteristics of particular formulations, and in the dosage

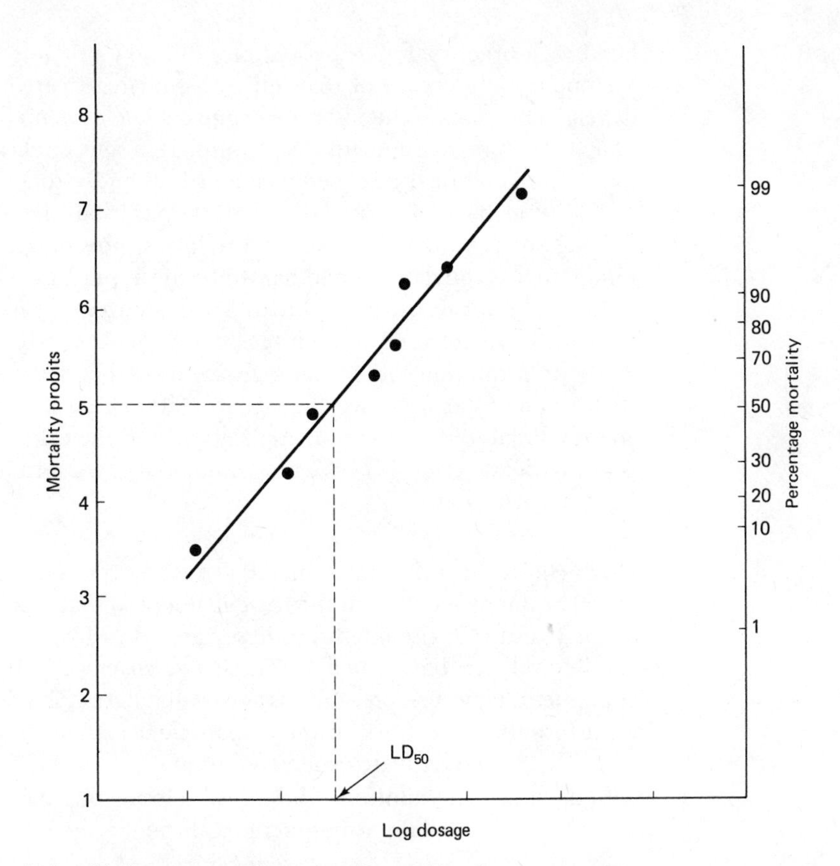


**FIGURE 23-2.** The relationship between percent mortality and log dosage when the mortality data is transformed to probits (see text for details).

commonly used, it is advisable to heed label notations that separate pesticides into four categories on the basis of their acute oral and dermal toxicities as follows:

1. Highly Toxic oral $LD_{50}$, 0–50 mg/kg; dermal $LD_{50}$, 0–200 mg/kg; inhalation $LC_{50}$ (lethal concentration), 0–2,000 ug/L. Label must bear the words DANGER and POISON and have skull and crossbones displayed.
2. Moderately Toxic: oral $LD_{50}$, 51–500 mg/kg; dermal $LD_{50}$, 201–2,000 mg/kg; inhalation $LC_{50}$, 2,001–20,000 ug/L. Label must bear a WARNING statement.
3. Slightly Toxic: oral $LD_{50}$, 501–5,000 mg/kg; dermal $LD_{50}$, 2,000–20,000 mg/kg. Label must bear a CAUTION statement and directions for avoiding hazards.
4. Nontoxic: oral $LD_{50}$, more than 5,000 mg/kg; dermal $LD_{50}$, more than 20,000 mg/kg. No special wording on label.

# Pesticide Formulation

The active poisonous ingredient of many pesticides is too toxic and too expensive to be used in a pure form and therefore must be mixed or formulated with other materials to facilitate dilution and an even distribution of the appropriate dosage of toxicant over the area treated. Although heavily dependent upon the chemists and engineers who develop the compounds and design the spray equipment, the formulation and application of pesticides involves a good deal of experience and trial and error.

The objectives of formulation and application are to provide optimal coverage of the treated surface at the lowest possible per unit cost, with a minimum of crop damage and hazard to the applicator. The least expensive, safest, and most readily available dilutent is water, but many of the modern toxic materials are not water soluble, so other means of dilution must be used. Aside from dilution, other materials are used that act as carriers, spreaders, stickers, emulsifiers, and wetting agents. Although it is inappropriate here to go into details of formulation, application, and the calculation of concentrations, a brief description of the most commonly used formulations is provided for familiarization.

***Dusts*** Dusts consist of a dry, inert material, such as ground walnut shell or talc, mixed or impregnated with the toxic material so that the final concentration of the toxicant in the mixture is less than 10%. Most dusts are fairly safe to handle and can be applied to give good coverage with relatively simple equipment. However, precautions against inhalation should be exercised, and if the formulation consists of a small particle size, there may be problems with drift. Dusts are relatively expensive because they are used as manufactured with no additional dilution. They are also losing popularity because they are a source of particulate atmospheric pollution.

***Wettable Powders*** Wettable powders consist of an inert, dry material plus the toxicant in a concentration of 25 to 75% and a dispersing or wetting agent. The dustlike material is then diluted further by suspension in water. The reduction in the amount of inert material makes these formulations more portable than dusts in terms of the quantity of active ingredient, and the use of water as a carrier reduces the over-all cost. The main drawback to wettable powders is that they must be agitated to prevent settling, so the application equipment used is more complex than that used in dusting.

***Granules*** Granules are much like dusts except that the particle size of the inert material is larger and the toxicant concentration is much lower. Since the entire bulk must be transported and there is no further dilution, this type of formulation is more inconvenient and costs more per unit of toxicant than other formulations. However, these materials are mainly used to treat the soil and can be safely broadcast by hand.

***Emulsifiable Concentrates*** Emulsifiable concentrates consist of about 15 to 75% toxicant (by weight) plus a solvent and an emulsifier. The concentrate is mixed with water to form an oil-in-water type of emulsion, which is probably the most versatile of the pesticide formulations. Sprays of this type make effective contact poisons that spread and stick well on the surface treated and make good contact with the waxy cuticle of plants and insects. The main problems with emulsifiable concentrates are that some of the solvents are phytotoxic, and the formulation will break up into its component parts if not agitated.

***Solutions*** Solutions are simply a toxicant dissolved in a solvent. They have good spreading characteristics and can be applied with simple equipment, because no agitation or mixing is required. Most synthetic, organic pesticide compounds are insoluble in water, so more expensive organic solvents must be used. This increases phytotoxicity and hazard to the applicator. However, the solvents can increase the effectiveness of the pesticide by enhancing penetration.

***Fumigants*** Fumigants are pesticides applied in a gaseous form and are normally used where the vapor can be confined as in greenhouses and storage facilities, although soil fumigation is a common practice. Most fumigants are held as a liquid under pressure and change to a gas when the pressure is released.

# Classification of Pesticides

There are several different ways that pesticides can be grouped. They may be classed according to the kind of pest they are used to control, in which case we designate those that kill insects as insecticides, mites as acaricides, weeds as herbicides, fungi as fungicides, and so on. They may also be classified according to their chemical composition. Broadly speaking, there are inorganics such as arsenic, copper, and mercury, and organics such as those derived from plants (botanicals) and those synthesized from organic materials such as coal, tar, and petroleum. The synthetic organic compounds are then subdivided further on the basis of their general chemical nature; DDT, for example, belongs to the organochlorines. Other rather broad classifications are based on the type of formulation (dust, spray, wettable powder, and so on), or the basic mode of action (stomach poison, contact poison, nerve poison, desiccant, and so on). Each system of classification has its own usefulness, but all require some degree of cross-referencing. For example, the organophosphate parathion may be formulated in a variety of ways and used as an insecticide, acaricide, or nematicide. Nevertheless, I will discuss the major pesticides grouped according to the type of pest to be controlled. The coverage presented here will be concerned mainly with insecticides, although fungicides and herbicides are mentioned, because insects, diseases, and weeds are often closely interrelated components of man-managed ecosystems. The discussion will, by necessity, be short, but a number of rather extensive treatments that include most of the practical details are readily available.

***Insecticides and Acaricides*** Insecticides and acaricides can conveniently be considered together because of the similarity in the ways that phytophagous insects and mites cause damage, either directly or as vectors of plant diseases. Both insects and mites are potentially exposed to applied chemicals at some time during their life history, and both feed either by chewing or sucking on plant or animal tissue. Consequently, they can be controlled with either stomach poisons, contact poisons, or fumigants, depending on their specific characteristics.

Stomach poisons are usually sprayed on the surface of the host on which the pest feeds. It is necessary therefore that the coverage be as complete as possible and, in the case of growing plants, the application be repeated to compensate for coverage dilution because of plant growth. Surface applications of stomach poisons are less effective against tunnelling species and those with piercing-sucking mouth parts, since they frequently do not ingest a lethal dose of the chemical. A number of stomach poisons, however, have some systemic qualities. That is, they can penetrate the host in some way and either the unmodified molecules or a toxic derivative is carried by the conducting system of the host to the feeding site of the pest. Some compounds of this type can be applied to seeds or to the soil prior to planting and provide protection for a short time following germination.

Contact poisons do not have to be ingested because of their ability to penetrate the targets' cuticle or gas exchange system once contact has been made. Contact can be achieved by the pest walking over a treated surface or by drops or dust particles impinging on the pest directly during application. Such compounds are sometimes effective against all stages of pest development, including the eggs.

Fumigants are compounds that enter a vapor state under normal crop conditions or are applied as a gas; they enter the pest through its breathing apparatus. Fumigants are particularly useful in confined spaces such as storage structures and greenhouses or for treatment of the soil. Those applied to the soil can be formulated to provide for a fairly slow rate of release from various types of carrier compounds, thereby spreading the treatment over a longer period of time.

INORGANIC COMPOUNDS. Inorganic compounds, a group which was more widely used in the past, include calcium arsenate, lead arsenate, cryolite, and sulfur, most of which act as protoplasic poisons following ingestion. They have been phased out of control programs mainly because of their unfavorable characteristics, which include high mammalian toxicity, persistence as residues, and the fact that several target species developed resistance.

BOTANICALS. The main insecticides and acaricides derived from plants are pyrethrum, nicotine, rotenone, and ryania, all of which are contact poisons. Only pyrethrum, obtained from a *Chrysanthemum*, is still used to any extent. It is an ephemeral compound, being oxidized rapidly in the presence of sunlight, and therefore is seldom used outdoors. Although it has low mammalian toxicity, it is an effective nerve poison with a rapid debilitation (knock-down) effect when used against insects. Both natural and synthetic pyrethroids are expensive to produce, so they tend to be used mainly in relatively high-cost household sprays.

The insecticidal qualities of nicotine have been known for a long time, but nicotine compounds have fallen into disfavor mainly because of their high mammalian toxicity. Nicotine acts mainly as a fumigant and its volatility is most commonly reduced by combining it into nicotine sulphate. It is low in its injurious effects on plants (phytotoxicity), is effective against insect larvae such as caterpillars, and has a low level of persistence in the environment.

Rotenone, formulated from the roots of *Derris*, is used to a limited extent to control both crop and livestock pests. Its main usefulness now is as a fish killer, applied to clear bodies of water containing unwanted species, prior to restocking. Ryania is a rather specific contact and stomach poison of caterpillars, which is useful in integrated control programs where broad spectrum pesticides might prove disruptive.

ORGANOCHLORINES. The organochlorines, among which DDT is by far the best known, were the first of the synthetic organic insecticides. Their high efficacy as both contact and stomach poisons and their inexpensiveness have made them the most widely used of all insecticides and acaricides.

Although synthesized first in 1874, DDT was not known to have insecticidal properties until it was tested along with many other compounds in the search for chemicals that would provide woolens with extended protection against clothes moths, and fighting men with protection against the vectors of insect-borne diseases. The use of DDT dust against the flea vectors of typhus in Naples in 1944 is credited with preventing an epidemic of disastrous proportions. After the war DDT was used widely, almost as a panacea, to control insect pests of man, his livestock, forests, crops, and gardens. But the initial effectiveness and popularity of DDT was, in a sense, its undoing in that overuse and misuse led to the development of a series of ecological problems that have resulted in its being banned in Sweden, Great Britain, Canada, and the United States.

The early effectiveness of DDT led to the search of other man-made compounds for pest control. The phenomenal proliferation of these substances since the mid-1940s is attributable in part to the wide variety of possibilities for making minor changes in their molecular structure. The DDT group provides an excellent example of four similar compounds with different attributes as pesticides (Figure 23-3). DDT is a persistent, broad-spectrum insecticide (but not useful against mites) with a high affinity for milk and the

**FIGURE 23-3.** Structural formulas of four closely related organochlorine pesticides with quite different uses (see text for details).

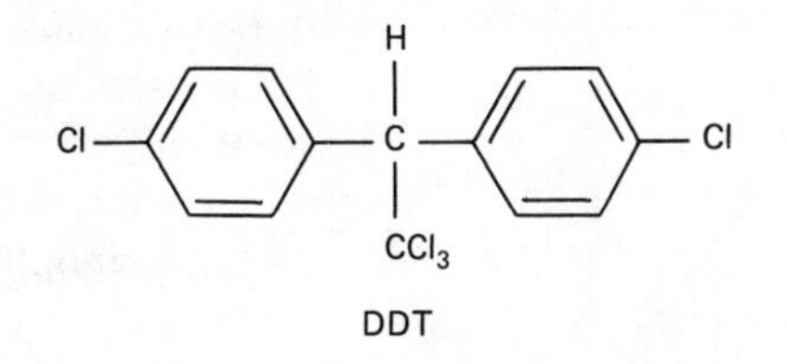


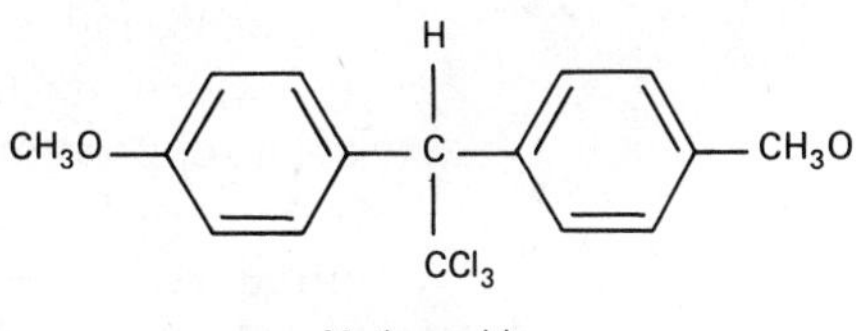


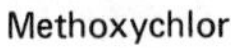


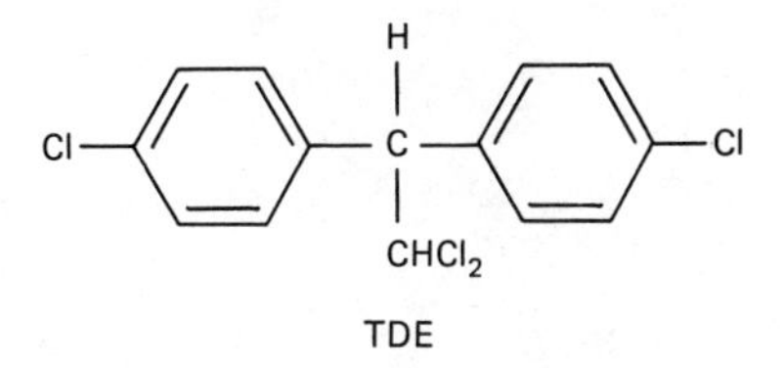


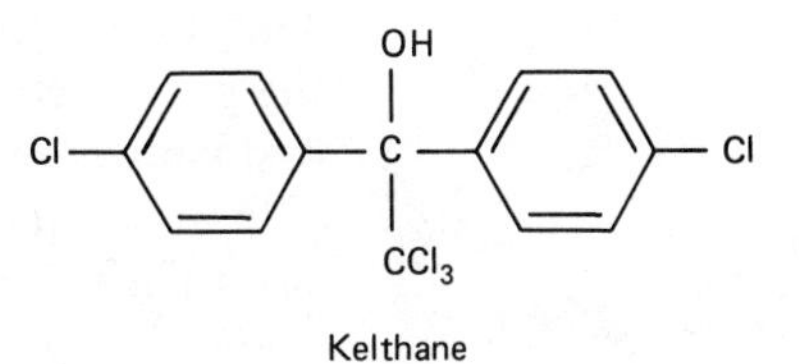

fatty tissues of vertebrates. Methoxychlor is also quite nonspecific as an insecticide but has the virtue of a low affinity for milk and fatty tissue. TDE (also known as DDD) is a rather narrow-spectrum insecticide that is particularly effective against caterpillars. Kelthane is a poorer insecticide than any of the other three, but it is a very effective miticide.

Other well-known organochlorines include benzene hexachloride (BHC) and the cyclodiene insecticides. Crude BHC tends to taint food products so is generally not used on crops unless in the form of the gamma isomer Lindane, which is relatively expensive. BHC is particularly useful against soil pests.

Probably the best-known cyclodienes are Aldrin and Dieldrin, both of which have been used extensively against grasshoppers and locusts, soil pests, and cotton pests but are now being replaced by organophosphates and carbamates as a result of residue and resistance problems. Other well-known representatives of this group include heptachlor, isodrin, endrin, toxaphene, and chlordane. Chlordane, although recently banned in the United States, is particularly effective against ants and is used to prevent ants from interfering with biological control programs that are directed against a variety of plant sap feeders, such as white flies and scale insects.

ORGANOPHOSPHATES. The organophosphates (OP's) form the largest and most versatile group of insecticides and acaricides. They can be used as contact poisons, stomach poisons, and fumigants. Several are excellent plant systemics, which provide a high level of protection against aphids, spider mites, and other plant juice-feeders. A few are also useful as animal systemics for the control of livestock parasites.

The organophosphates are related to nerve gases developed for military use, and they have a mode of action of irreversibly inhibiting cholinesterase and thereby disrupting nerve synapses. However, the range of mammalian toxicity of the OP compounds is great. For example, parathion is extremely toxic to mammals and is falling into disfavor, whereas malathion is so safe that it is sold widely for use by home gardeners.

The opportunities for synthesizing organophosphates are almost unlimited, and more than 100 commercially produced compounds have already been introduced. The widespread interest in OP compounds stems from the fact that they are far less persistent in the environment than the organochlorines, and so far there has been less resistance development among target species. Many OP's are also quite selective.

A few of the better known organophosphates and their uses are Diazinon for livestock pest control; Fonofos, a highly persistent soil insecticide; Demeton, Disulfoton, and Dimethoate, highly effective systemics especially useful against sucking pests and insect vectors of plant viruses.

CARBAMATES. The carbamates are largely contact insecticides which, like the organophosphates, appear to be cholinesterase inhibitors. Most have rather low mammalian toxicity and are readily degraded in the environment. They have some systemic action, but only a few are used in this way. Some of the newer compounds have proven to be quite selective. Probably the best-known and most widely used carbamates are methomyl and carbaryl. Although injurious to a variety of beneficial species, carbaryl has been useful against leaf-feeding caterpillars and is widely used in the control of cotton pests. Zectran® is one of the more effective systemic carbamates. In addition to its effectiveness against a wide range of insects, it is also toxic to mites, snails, and slugs.

OILS. Hydrocarbon oils have long been used in the control of both insects and weeds. As insecticides and acaricides, the less-refined products are referred to as "dormant oils" because they are injurious to actively growing vegetation and must be used during the plant's dormant period. The more highly refined "summer oils" can be used on verdant vegetation, although phytotoxicity is the major limitation in the use of all oil sprays. Oils are particularly effective against the eggs and dormant stages of mites and scale insects, because they act to interfere with the organisms' gas exchange and respiration.

Oil sprays are gaining in popularity because they are not toxic to mammals and therefore are not subject to residue restrictions. Furthermore, they are less injurious to beneficial parasites and predators, making them highly compatible with biological control methods. There are no known cases of resistance development after more than 80 years of use.

## *Fungicides*

There are basically two types of fungicides. One type is used to kill exposed fungus mycelia, whereas the other is used as a coating that prevents spore germination or the penetration of the plant by the mycelia after spore germination. In recent years, some promising compounds have been discovered with systemic qualities which may permit an attack against the mycelia within the plant host, which has heretofore not been possible. Fungicides are applied either as a foliar spray, as a seed dressing, or as a soil treatment.

INORGANIC FUNGICIDES. As indicated earlier, the use of sulphur as a pesticide has a long history, but it came into prominence as a means of controlling mildews on fruit trees and grapes during the early nineteenth century. Various formulations of sulphur are still used against fungus diseases of apples and pears. Sulphur is particularly useful against powdery mildews because they grow entirely on the surface of their host. The more finely divided the sulphur, the better are its fungicidal properties. Commercially produced lime sulphur, consisting mainly of calcium polysulphides that release elemental sulphur, is the form most commonly used.

Copper sulphate, like sulphur, has been used for a long time as a fungicide but gained popularity as a control agent for fungus diseases of grapes. Bordeaux mixture, developed in France in 1885 by Millardt, is still widely used. However, the lime component of the mixture tends to abrade and block the nozzles of spray equipment, so less problematical copper formulations have been developed, the most important being copper oxychloride. The main use of copper fungicides is directed against potato blights, downy mildew of grapes, *Phytophthera* diseases of cotton, rubber and cacao, and seed-borne diseases of grains (Ordish and Mitchell, 1967).

ORGANIC FUNGICIDES. The first organic fungicides were all derivatives of dethiocarbamic acid used in the production of synthetic rubber. The first to be produced commercially was thiram, which has been used widely for the treatment of seeds to prevent the development of blight and other fungal diseases. Although still used, thiram has been replaced by a series of metallic salts of dithiocarbamic acid such as zineb (zinc), maneb (manganese), and nabam (sodium). While effective against damping off, foot rot, and various blights, these earlier organic fungicides were not very effective against powdery mildews. More recently, several organic compounds have been developed that provide good protection against these diseases, and some organophosphates have proved useful against rice blast and several powdery mildews. Further research on organophosphates for disease control may lead to the discovery of additional fungicides that can be used systemically. Systemic fungicides are of particular usefulness against soil-borne pathogens such as *Verticillium* wilts that cause a variety of vascular diseases. Two fairly new materials, Vitavax® and Plantvax®, show considerable promise against cereal smuts and rust diseases.

***Herbicides*** As with the chemical control of arthropods and diseases, the chemical control of weeds has become a complex aspect of agriculture, important to both crop production and animal husbandry. As clearly stated by Woods (1974), ". . . weeds, unlike fungi and insect pests, are not directly dependent upon their victims, but merely share the same habitat." Furthermore, weeds are closely related to crop and pasture plants, often making it difficult to kill one and not the other if they share the same habitat at the same time. In addition to selecting the right chemical, it is, therefore, important that it is applied in such a way that the desirable plants are not injured. This can be accomplished by selecting the appropriate herbicide relative to the nature of the crop and the weeds to be controlled and by adjusting the type and time of treatment. Herbicides are applied either to the soil or to the foliage of the weeds. The application can be made before the crop is planted (preplanting), after planting but before the crop germinates (pre-emergence), or after the crop is actively growing (postemergence). Depending on the herbicide, the weeds may be killed by contact with either roots or the foliage, or by translocation after entry through the roots or the leaves. The destruction of weeds with a broad-spectrum contact herbicide prior to planting the crop is clearly the most straightforward method. Pre-emergence treatment with a contact herbicide is also straightforward but requires more critical timing. Postemergence treatments are more difficult because of the possible sensitivity of the crop.

# Trends in Pesticide Production and Use

Data on the world production and use of pesticides are fragmentary and often difficult to interpret. Even in the United States, complete and up-to-date statistics on pesticides are difficult to obtain, because some types of survey that provide important information are conducted irregularly or at rather long intervals.

In 1974, the last year for which data are available, United States manufacturers produced 1.4 billion pounds of synthetic organic pesticides with a value of over 1.9 billion dollars (USDA, 1974*a*). On the basis of earlier estimates, the United States accounts for about 40% of the total world production. World consumption therefore must be in the order of 3.5 billion pounds per year, costing in excess of 4.5 billion dollars. The production and sales data presented in Table 23-1 show a steady increase from 1969 to 1974, a period during which a great deal has been heard concerning the curtailment of pesticide use. Table 23-2 shows the breakdown of production by class of

**TABLE 23-1** **Synthetic Organic Pesticide Production and Sales in the United States from 1969 to 1974[1]**

| Year | Production | | Sales (Domestic and Export) | |
|---|---|---|---|---|
| | *1,000 Pounds* | *1,000 Dollars* | *1,000 Pounds* | *1,000 Dollars* |
| 1969 | 1,104,381 | 953,592 | 928,663 | 851,166 |
| 1970 | 1,034,075 | 1,058,389 | 880,914 | 870,314 |
| 1971 | 1,135,717 | 1,282,630 | 946,337 | 979,083 |
| 1972 | 1,157,698 | 1,344,832 | 1,021,565 | 1,091,708 |
| 1973 | 1,288,952 | 1,492,770 | 1,198,568 | 1,343,581 |
| 1974 | 1,417,158 | 1,984,794 | 1,365,214 | 1,815,433 |

[1] *Source:* "The Pesticide Review 1975," Agricultural Stabilization and Conservation Service, USDA.

pesticide for the same period. Table 23-3 presents the United States sales volume of pesticides by class for 1967, 1970, and 1973. These data show only small increases in the production and sale of insecticide and fungicides but a marked increase in the use of herbicides.

The last survey of the farm use of pesticides in the United States was conducted in 1971 (USDA, 1974*b*). During 1971, about 833 million pounds of pesticides were used for pest control in the United States. Of this amount, 494 million pounds were used by farmers; the remainder were used mainly by the forest industry, government agencies, and home-owners. Farmers used an additional 112 million pounds of sulfur and 222 million pounds of petroleum for pest control. The use of pesticides in households and gardens has continued to increase; the value of these products rose from $146 million in 1971 to $182 million in 1972.

Of the 565 million pounds of insecticides (including inorganics) produced in the United States in 1971, roughly 170 million pounds or about 30% were used by farmers. Of this amount, 154 million pounds were applied to crops, and 15 million pounds were applied to livestock. In addition, farmers used 74 million pounds of petroleum for insect control. Table 23-4 shows the quantities of major types of insecticide used by farmers for various purposes in 1971 and compares the 1971 totals with those for 1966. These data indicate a decline of about 22% in the use of organochlorines but increases of 77% and

**TABLE 23-2** **Production (1,000 pounds) of Pesticidal Chemicals[2] by Class in the United States from 1969 to 1974[1]**

| Year | Insecticides[2] | Herbicides | Fungicides | Total[3] |
|---|---|---|---|---|
| 1969 | 570,522 | 393,840 | 140,019 | 1,104,381 |
| 1970 | 490,132 | 404,241 | 139,702 | 1,034,075 |
| 1971 | 567,710 | 428,849 | 149,158 | 1,135,717 |
| 1972 | 563,575 | 451,618 | 142,505 | 1,157,698 |
| 1973 | 639,209 | 496,109 | 153,674 | 1,288,952 |
| 1974 | 650,209 | 604,288 | 162,661 | 1,417,158 |

[1] *Source:* "The Pesticide Review 1975," Agricultural Stabilization and Conservation Service, USDA.
[2] Includes small quantity of inorganic materials plus fumigants and rodenticides.
[3] Includes small amount of synthetic soil conditioners, rodenticides, and multipurpose fumigants.

TABLE 23-3 United States Sales Volume (1000 pounds) and Per Cent of Total Sales by Class for 1967, 1970 and 1973[1].

| Year | Insecticides[2] | | Herbicides | | Fungicides | |
|---|---|---|---|---|---|---|
| | *Production* | *Per cent* | *Production* | *Per cent* | *Production* | *Per cent* |
| 1967 | 503,796 | 44.8 | 439,965 | 38.3 | 177,886 | 15.9 |
| 1970 | 443,943 | 50.4 | 308,112 | 35.0 | 128,859 | 14.6 |
| 1973 | 605,315 | 50.5 | 446,849 | 37.3 | 146,404 | 12.2 |

[1] *Source:* "The Pesticide Review 1974," Agricultural Stabilization and Conservation Service, USDA.
[2] Includes a small quantity of synthetic soil conditioners, rodenticides, and multipurpose fumigants.

93% in the use of organophosphates and carbamates, respectively. Most striking, as well as encouraging, is the approximate 540% increase in the use of petroleum.

The trend away from organochlorines probably resulted largely from the development of resistance to these compounds by a number of pests against which they were once highly effective. Notably, the cotton bollworm and tobacco budworm became resistant to DDT, and the corn rootworm became resistant to aldrin. Furthermore, DDT had fallen into general disfavor as a result of environmental contamination and was eventually banned. The large increase in the use of petroleum, most of which was applied to citrus, apples, and other fruits, probably reflects the development of integrated control programs for these crops (see Chapter 26). Petroleum is less harmful to beneficial species than are most organic insecticides, and its use helps to avoid secondary pest problems.

In spite of the impression often created that this country is continuously saturated with pesticides, only about 5% of our agricultural land is treated with insecticides and about 12% with herbicides. If the land devoted to pasture and range is excluded, still only 12% of our crop acreage is treated with insecticides, 27% with herbicides, and 1.3% with fungicides (Pimentel, 1973). However, pesticide use in the United States is not evenly distributed by region or among crop types.

In 1971 cotton accounted for 73.3 million pounds, or 47% of all the insecticides used by American farmers (Table 23-5). Large quantities of insecticides were also used on corn and vegetables. For any specific crop, the need for pesticide treatment varies according to the geographic area (Figure

TABLE 23-4 Comparison of the Quantity (X 1,000 pounds) of Different Types of Insecticides Used by Farmers in 1966 and 1971, and Pattern of Use for 1971

| Type of Insecticide | 1966 Total | 1971 *Total* | 1971 *Crops* | 1971 *Livestock* | 1971 *Other* |
|---|---|---|---|---|---|
| Inorganics | 5,784 | 3,232 | 3,042 | 189 | 1 |
| Botanicals | 204 | 213 | 69 | 144 | |
| Organochlorines | 89,239 | 69,873 | 61,876 | 7,627 | 370 |
| Organophosphates | 39,966 | 70,706 | 65,031 | 5,369 | 306 |
| Carbamates | 12,933 | 25,412 | 24,166 | 1,194 | 52 |
| Other synthetics | 798 | 334 | 72 | 261 | 1 |
| Petroleum | 11,419 | 73,950 | 60,721 | 13,126 | 103 |
| Total (including petroleum) | 160,343 | 243,720 | 214,977 | 27,910 | 833 |

TABLE 23-5 Farm Use of Insecticides, by Crop, 1966 and 1971[1]

| Crop | 1966 Pounds of Active Ingredients[2] (Million pounds) | 1966 Percentage of Farm Insecticides Used (Percent) | 1971 Pounds of Active Ingredients[2] (Million pounds) | 1971 Percentage of Farm Insecticides Used (Percent) |
|---|---|---|---|---|
| Cotton | 64.9 | 47 | 73.3 | 47 |
| Corn | 23.6 | 17 | 25.5 | 17 |
| Other field crops[3] | 8.7 | 6 | 17.5 | 11 |
| Vegetables | 11.1 | 8 | 11.1 | 7 |
| Fruits (not including apples and citrus) | 6.6 | 5 | 6.3 | 4 |
| Soybeans | 3.2 | 2 | 5.6 | 4 |
| Apples | 8.5 | 6 | 4.8 | 3 |
| Tobacco | 3.8 | 3 | 4.0 | 3 |
| Citrus | 2.9 | 2 | 3.1 | 2 |
| Hay and pasture | 4.1 | 3 | 2.6 | 2 |
| Other | 0.2 | ([4]) | 0.5 | ([4]) |
| All crops | 137.6 | 100 | 154.3 | 100 |

[1]Does not include Alaska.
[2]Does not include petroleum.
[3]Includes wheat, sorghum, rice, peanuts, and sugarbeets.
[4]Less than 0.5 percent.

23-4) in which the crop is grown. The South traditionally accounts for a large share of the farm use of insecticides. In 1971, the Southeast absorbed 40.4 million pounds (26%), the Delta states, 32.3 million pounds (21%), and the Southern Plains, 18.5 million pounds (12%) (USDA 1974*a*). This uneven pattern of insecticide use clearly reflects the type of crops, such as cotton, that are grown mainly in southern areas.

According to Pimentel (1973), annual losses in American agriculture caused by pests have increased steadily throughout this century. The annual losses caused by insect damage for the 1951–1960 period were estimated at $3.8 billion, or about 13% of the total agricultural production. Pimentel calculated that in the absence of insecticide use the annual damage would increase to $4.8 billion, or 16.3%. Pimentel also estimated that the return on each dollar invested in pesticides of all types is about $2.82. This is somewhat lower than previous estimates, which have been as high as $5.00. However, even the best cost-benefit ratios for pesticides are below those calculated for other control methods.

## The Pesticide Controversy

After the use of DDT against lice averted a disastrous outbreak of typhus in Naples toward the end of World War II, many felt that a solution had been found for the world's insect pest problems. The high insect toxicity and persistence of DDT, combined with the ease and inexpensiveness of its use, soon led to the replacement of more conventional cultural control methods

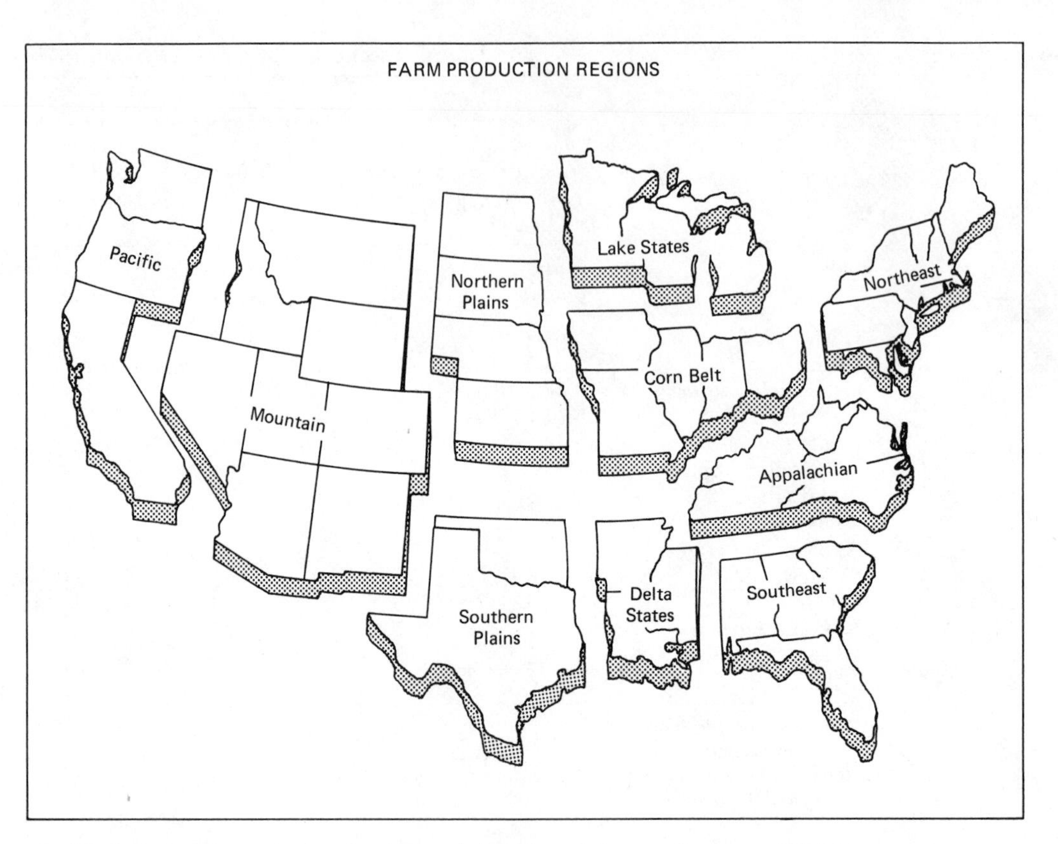


**FIGURE 23-4.** The agricultural regions of the United States. (Courtesy USDA.)

by the widespread use of synthetic organic chemicals. In spite of early warnings by prominent insect physiologists such as Sir Vincent Wigglesworth, pest-free crops became the ideal, and many agriculturalists felt a panacea had been found. It was only a short time, however, before the unheeded warnings were proved to be justified, as evidence appeared that DDT and related compounds had begun to lose their effectiveness. The insecticide industry responded with new materials, but these, too, lost their initial potency as time passed.

When the new organic pesticides first became popular, little was known about their mode of action or what happened to them after they were introduced into the environment. In fact, as pointed out by Moore (1967), it is regrettable that pesticides are often thought of as substances applied to a pest population rather than to the ecosystem of which the pests are a part. The first indication that all was not well came in 1946 when resistance to DDT was encountered in several populations of the housefly.

Resistance, as defined by the WHO Expert Committee on Insecticides (1957), is "the development of an ability in a strain of insects to tolerate doses of toxicants which would prove lethal to the majority of individuals in a normal population of the same species." Resistance develops as a result of natural selection, not as a result of the mutagenic action of the pesticides. Insects and mites that have displayed resistance development have done so because their populations contained genotypes that were preadapted for resistance. Normally they were rare, probably because in an insecticide-free environment they were less well suited for survival than their susceptible siblings. But when subjected to the presence of an insecticide, the resistant forms were better suited, so they survived and multiplied. When the selective pressure of the pesticide is continuously a component of the environment, the nonresistant forms are gradually eliminated and only the offspring of preadapted individuals survive. Clearly then, resistance results more from the misuse of pesticides than from the pesticides themselves, because it is misuse that increases the intensity of the selective pressure and allows the resistant strains to burgeon forth.

When resistance to DDT was encountered, other cyclodiens, organophosphates, and carbamates were used instead, but many pests demonstrated the ability to develop resistance to them as well. Pests that have developed what we commonly call multiple resistance have become increasingly difficult to control. Although some pests do not have the ability to develop multiple resistance, and it is possible that some populations will undergo reversal once the selective pressure of pesticides is removed from their environment, it appears as though the trend is against us. Over 200 species of arthropods are now resistant to one or more pesticides (Brown, 1968; FAO, 1969).

Resistance is not confined to insects and mites that have been treated with insecticides and acaricides. Bacteria are known to have developed resistance to antibiotics, and rabbits in Australia developed resistance to myxomitosis. Even chickweed *Stellaria media* has been known to become resistant to the herbicide 2,4-D by the third generation (Woods, 1974). Certainly resistance development is something we should keep in the back of our minds whenever a new approach to pest control is devised. It is folly to state arbitrarily that behavioral changes will not occur as a result of pheromone trapping (Atkins, 1968), that insects cannot become resistant to compounds that mimic their hormones, or that sterile male releases will not lead to the development of parthenogenetic strains. All of these are possibilities if we engage in the misuse of the control methods.

The recognition of other drawbacks to pesticides developed more slowly. They first came to public attention following the publication of Carson's *Silent Spring* in 1962. Although in the opinion of some to be more emotional than scientifically accurate, this book did serve as a stimulus for a closer examination of the environmental side effects of pesticides. The literature of the last few years is permeated with reports on the undesirable aspects of pesticides in the environment; some are exaggerations, but many are factual and are a matter for real concern (Rudd, 1964; Moore, 1967; Slater, 1967).

The problems of pest resurgence and secondary pest outbreaks that were discussed briefly in Chapter 22 have occurred as the result of pesticides killing beneficial species and disturbing the normal population regulatory processes. There are hazards to human health that arise during the application of poisons and from toxic residues in agricultural products. There are also cases where crop productivity has declined because of pesticide misuse, as was the case with cotton in the Canete Valley of Peru (Smith, 1969), and there is documented evidence that pesticides have entered food chains with severe consequences for animals on higher trophic levels.

However, this is not the place to review the details of the problems that have arisen from inadequate knowledge, an overzealous desire to increase food and fiber production, and efforts to stamp out misery-causing diseases. These problems have been discussed many times before. What is more important is to recognize that pesticides, properly used, are still the most effective means of reducing pest-induced losses. As Smith (1970) stated, "When pest populations approach economic levels, there is little other than pesticides we can use to avoid damage and which will have the desired immediate effect." Brown (1970) believes that pesticides will also be important, at least initially, in the protection of the new high-yielding varieties, some of which have led to the development of new pest problems.

There are, indeed, alternative approaches to pest control, as we will see in the following chapters. Unfortunately, one of the most serious consequences of relying so heavily on chemicals for pest control over two decades is that these alternative methods were not perfected sooner. To again quote from Smith (1970), ". . . it is irresponsible to advocate the total replacement of pesticides with sophisticated, but poorly tested alternative pest-management techniques. It is a disservice to society to discard the good crop protection currently available and to adopt in its place a glamorous new but untested methodology."

Clearly, we must adopt a balanced point of view, the aim of which is not to ban the use of pesticides but rather to prevent their misuse. This must involve not only an elimination of unnecessary prophylactic treatments but the development of methods that will reduce the quantity and frequency of use. It is always easy to look back and criticize what was done in the absence of more adequate information, but it is more difficult to learn from the mistakes and correct them in the future. We now realize that many of the problems that developed could have been avoided simply by applying no more pesticide than was necessary. Yet there are still growers who apply pesticides that are not needed, and the concept of weed-free farms that prevails in some sectors today is no less dangerous than the consequences of the insect-free crops of the 1950s.

## References Cited


Atkins, M. D. 1968. Scolytid pheromones—ready or not. *Can. Entomol.* 100: 1115–1117.

Brown, A. W. A. 1968. Insecticide resistance comes of age. *Bull. Ent. Soc. Amer.* 14: 3–9.

Brown, Lester. 1970. Seeds of change. Praeger Publishers, New York, N. Y.

Carson, Rachel. 1962. Silent spring. Houghton Mifflin Company, Boston.

Food and Agricultural Organization. 1969. Fifth session of the FAO working party of experts on resistance of pests to pesticides. Rome, Italy, September 1968.

Hoskins, W. M. and H. T. Gordon. 1956. Arthropod resistance to chemicals. *Ann. Rev. Entomol.* 1: 89–122.

Konishi, Masayasu and Yosiaki Ito. 1973. Early entomology in East Asia. Pages 1–20 *in* R. F. Smith, T. E. Mittler and C. N. Smith, (editors.) History of entomology, Annual Reviews, Inc., Palo Alto, Calif.

Moore, N. W. 1967. A synopsis of the pesticide problem. *Adv. Ecolog. Res.* 4: 75–130.

Ordish, G. and J. F. Mitchell. 1967. World fungicide usage. Pages 39–62 *in* D. C. Torgeson (ed.), An advanced treatise, Academic Press, Inc., New York.

Pimentel, David. 1971. Ecological effects of pesticides on nontarget species.

Executive Office of the President, Office of Science and Technology, Washington, D. C.

PIMENTEL, DAVID. 1973. Extent of pesticide use, food supply and pollution. *N.Y. Entomol. Soc.* 81: 13–33.

RUDD, R. L. 1965. Pesticides and the living landscape. Faber & Faber, Ltd., London.

SLATER, W. (Ed.) 1967. A discussion on pesticides: benefits and dangers. *Proc. Roy. Soc. London* (*B*). 167: 88–163.

SMITH, R. F. 1970. Pesticides: Their use and limitations in pest management. Pages 103–113 *in* R. L. Rabb and F. E. Guthrie, editors. Concepts of pest management. North Carolina State University, Raleigh, N. C.

U.S.D.A. 1974*a*. The Pesticide Review 1973. Agr. Stab. and Cons. Ser., 60 pp.

U.S.D.A. 1974*b*. Farmers' Use of Pesticides in 1971. Agr. Econ. Rep. No. 252. Econ. Res. Ser., 56 pp.

WOODS, ARTHUR. 1974. Pest control: A survey. John Wiley & Sons, Inc., New York, N. Y.

# Alternatives to conventional chemical control

All forms of agricultural pests can be suppressed by a variety of special chemical or nonchemical methods. Some of these methods are as old as agriculture itself and have been time-tested as practical and ecologically sound approaches to increasing crop yields. Others are new and in some cases have not yet advanced beyond the experimental or conceptual stage. Although the use of chemical pesticides has become the mainstay of pest control since the introduction of DDT, and will certainly remain important for some time to come, it is not the most ideal approach from an ecological point of view. Problems of pesticide pollution, pest resistance, and new pesticide-induced pest problems have rekindled the interest in alternative control tactics that can be used alone or in combination to provide a more ecologically sound approach to pest control.

Nonchemical pest control has been the subject of a number of rather lengthy books. It involves a diversity of methodologies that draw from and integrate a wide variety of scientific and technological disciplines. In addition, each different tactic has a number of ecological and economic implications. Consequently, the subject cannot be treated completely here. My objective is to survey the field in order to demonstrate that there is no basis for a total reliance on the use of pesticides. At the same time, I wish to illustrate that each alternative tactic has some limitations and that much exciting and imaginative work is left to be done. In the following survey, therefore, I will describe those tactics that have been demonstrated to be effective or to have potential either alone or in combination as solutions to a variety of pest problems.

The organization of reviews of pest control tactics seems to be based largely on personal preference. For example, biological control in the broadest sense includes host resistance and autocidal methods in addition to the manipulation of parasites, predators, and pathogens. Others choose to consider host resistance as a form of cultural control. However, each of these approaches is distinctly different and will be treated separately. This chapter will be concerned with cultural control, host resistance, legislative or regulatory control, special chemical methods, autocidal control, and some miscellaneous physical and mechanical methods. Some of these methods are useful against weeds and plant diseases as well as against insects. Mention is made of the usefulness of the control methods against these other pests because it has a bearing on their use in integrated pest management, discussed in Chapter 26.

## Cultural Control

The term cultural control as used here encompasses the manipulation of the crop (plant or animal) and its environment in such a way as to make conditions less favorable to pest species and thereby slow their rate of increase or suppress their injurious effects. The kinds of manipulation include only those that involve existing environmental factors, as opposed to those that involve the addition of a mortality-causing agent such as a pesticide or additional natural enemies. Consequently, the cultivation of resistant varieties of plants or animals will not be included here. The extent of environmental manipulation can range from rather commonly used procedures of cultivation such as tillage to more radical environmental changes such as the conversion of a natural community to a site for crop culture; this includes what some refer to as ecological control.

Cultural control follows two basic approaches and involves some of the oldest and simplest methods of pest control such as hand weeding or the removal of diseased plants (roguing). The two approaches can be distinguished as either the avoidance of pest increase and related damage or the destruction or eradication of an existing pest condition. In some instances, the techniques of cultural control have become required by law and form a basis of what is often referred to as **legislative control**, which will be considered briefly a little later.

One of the important aspects of cultural control is that many of the methods employed are primarily concerned with the growth of the crop. Most of these functions are obvious and will not be considered in detail.

### *Tillage*

Tillage can be defined as the disturbance of the soil as a means of preparing it for planting, for improving water penetration and conservation, or as a means of destroying soil-inhabiting pests by exposure or mechanical injury. The development of ways to till the soil was a major factor in the transition of civilization from food gathering to the establishment of permanent agriculture (Cole and Matthews, 1938) and has been a major factor in increasing crop production. It is a standard longstanding technique for weed control, accomplished both by hand and by specially designed machines. Autumn cultivation is extremely effective against some weeds such as chickweed and wild oats, but the timing of effective cultivation depends upon the germination cycle of the weeds that create a problem, so no generalization can be made. Because most weeds are pioneer species well adapted to invade severely disturbed sites, seedbed preparation is often followed by a rapid and substantial weed invasion. In some circumstances, such as the establishment of a perennial forage crop, a second tilling prior to planting but after the weeds have emerged can produce good results.

Tilling the soil at the appropriate time can be an effective way of destroying the soil-inhabiting stages of some insect pests or life stages in crop residues. Fall ploughing is an effective method of exposing grasshopper egg pods to desiccation and of killing the pupae of the grape berry moth. Deep ploughing following harvest to bury infected plant parts and stubble destroys the larvae of wheat-stem sawfly, corn earworm, and European corn borer.

However, tillage is not always advantageous and may actually aggravate some pest problems. Haynes (1973) stated that conventional tillage practices have no direct effect on cereal leaf beetle populations but destroy more than 97% of the parasite population. If the soil surface tends to crust over, leaving it undisturbed can curtail weed seed germination and prevent the penetration of some soil-inhabiting pests such as the pale western cutworm *Porosagrotis orthogonia*.

Yarwood (1968) believes that there are probably instances in which tillage reduces crop diseases, but states that most of the experimental evidence indicates that it favors disease. In some cereals, direct drilling (planting without seedbed preparation) seems to be more beneficial than conventional tillage in that it saves energy, reduces evaporation and erosion, and reduces the incidence of several pests.

### *Crop Rotation, Mixed Cropping and Fallow*

An unbroken sequence of growing the same crop in the same place often encourages the build-up of pest populations. This is particularly true for those pests such as nematodes and plant pathogens that infest the root zone, because they tend to persist readily from crop to crop. Since crops are often rotated for purposes of soil and nutrient conservation, the practice can provide bonus benefits in terms of pest reduction. When used mainly as a pest control measure, the crops rotated must be botanically unrelated. For example, there are a number of pests that damage almost all members of the family Cruciferae, so that rotating crops belonging to this group may not reduce the pest problem. The technique is, of course, little use against pests with a wide range of hosts, as is the case for some nematodes (e.g. beet eelworm) and insects (e.g. cabbage looper).

There are some pests that can be controlled effectively by crop rotation. The white-fringed beetles, which are unable to fly, have a higher fecundity when they feed on legumes such as peanuts and soybean than when they feed on grasses, including corn. Consequently, the larvae do not damage grasses to the same extent that they damage legumes, and the population declines during cultivation of a grass crop. Alternation of corn with a legume crop can result in excellent control. Similarly, the rotation of potatoes with alfalfa reduces wireworm damage, rotating oats and corn reduces corn rootworm damage, and winter grain mite can be controlled by not planting small grains in the same field for more than two years in succession (National Research Council, 1969).

Crop rotation is also effective against some, but not all, plant diseases. Outstanding results have been obtained in the reduction of *Ophiobolus* disease of wheat by using a wheat-oats cropping sequence. Alternating potatoes with soybean, and cotton with peas are effective means of reducing the incidence of potato scab and cotton root rot, respectively (National Research Council, 1969).

Pasture rotation can greatly aid in the control of livestock pests. It is particularly useful against pests such as the cattle tick, which have one primary host, as opposed to those such as the lone star ticks that attack a wide range of hosts. In the first case, keeping pastures free of cattle for a few months results in tick starvation, whereas in the case of the lone star tick, the pastures must be clear of all animals for several years to effect control.

Keeping pastures free of animals is similar to allowing crop land to lie fallow. Whereas this cultural technique is often implemented for water conservation and fertility improvement by allowing a period for nitrogen build-up, it also can be of benefit in pest control. Fallowing is sometimes carried out specifically for weed control. This method is usually used against perennial weeds and is particularly effective when the fallow fields are tilled periodically. As in the spelling of pastures, fallowing can result in the reduction of insect and disease pests because of a lack of suitable hosts for these pests.

Mixed crops or intercropping can be used as an alternative to crop rotation. The growing of a mixture of crops, as is the practice in some forms of subsistence agriculture, can reduce pest damage through the increased ecosystem diversity that reduces the food abundance for specific pests, and may serve to enhance and perpetuate natural enemy populations. In mechanized agriculture there are economic limitations that usually restrict this method to intercropping with two species. Probably the best results from application of this method have been achieved by interplanting strips of alfalfa in cotton fields for the control of lygus bugs (Stern, 1969). The lygus bugs prefer alfalfa to cotton and will concentrate in the strips of alfalfa leaving the cotton undamaged. The alfalfa can be harvested later for animal feed.

### *Timing of Planting and Harvest*

Organisms normally evolve life histories that synchronize their development with the availability of suitable environmental conditions, including the appropriate developmental stage of their hosts, such that their survival is assured. Since many of these natural relationships have been carried over to crop ecosystems, many agricultural pests attack crops in a specific stage of development. Sometimes it is possible to alter the timing of crop development by modifying normal cultural practice and thereby effect a substantial reduction in damage.

A rather simple way to alter the timing of crop development is to change the planting time. Sometimes it is advantageous to delay the planting date, and in other situations it may be better to advance it. Hessian fly adults usually live only for a few days following their emergence in the fall. During this brief period they mate and lay their eggs on young winter wheat plants. If planting is delayed so that most of the flies have expired before the emergence of the wheat, damaging infestations can be avoided. On the other hand, early planting of corn reduces infestations of the southwestern corn borer because the moths tend to lay fewer eggs on more mature plants. Likewise, early planting of corn in northern areas reduces damage by the corn earworm populations that overwinter in the south and move northward each year. In situations where crops are invaded from elsewhere rather than attacked by a resident population, some crop protection can be obtained by delaying the planting of a crop until major pest migrations are over. Damage to sugarbeets by curly top virus can be lessened by planting the crop after the spring migration of the beet leafhopper which vectors the disease.

Sometimes it is possible to reduce pest populations or the damage they cause by adjusting harvest time. Early harvest of the first two cuttings of alfalfa is useful as a means of controlling the alfalfa weevil larvae, which suffer from a shortage of suitable food and the more severe physical environment of a mowed field. Reduced losses result when wheat is harvested before attacks by the wheat stem sawfly that cause the plants to bend over.

### *Strip Harvesting and Trap Crops*

Strip harvesting is really only practical for use in forage crops that are harvested several times each season. It might also prove useful in the management of rapidly growing field crops that would lend themselves to cultivation in strips of different ages. This method seems to have been most successful as a means of reducing lygus bug damage in alfalfa. The alfalfa is harvested in alternate strips so that when one strip is cut, the strips to each side are half grown (Figure 24-1). This stabilizes the environment for both the lygus bugs and their natural enemies. In addition many of the lygus bug eggs laid in the uncut strips by adults that moved into them are removed with the next strip harvest (Stern, van den Bosch, and Leigh, 1964).

Trap crops consisting of small plantings of highly favored hosts are used

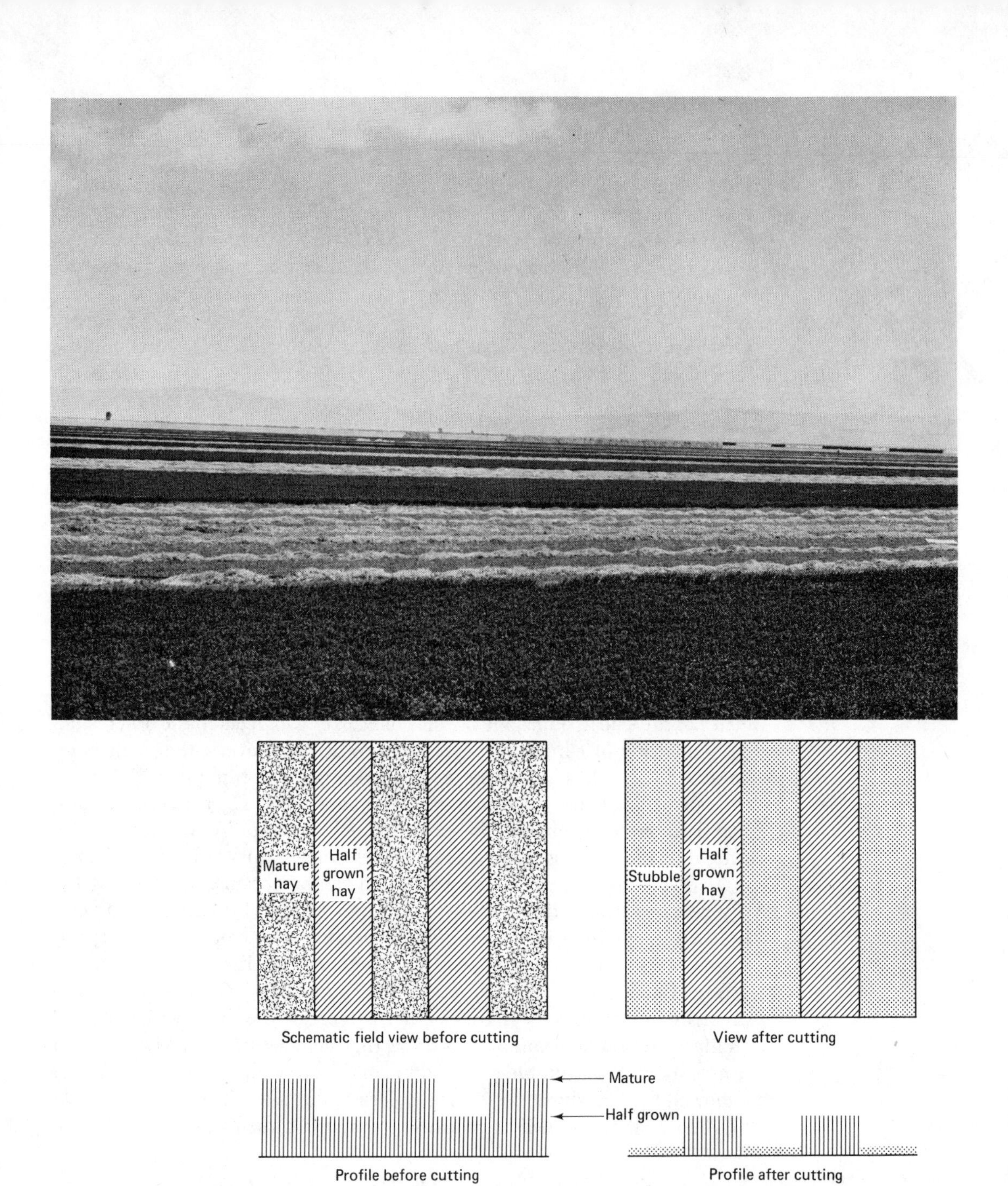


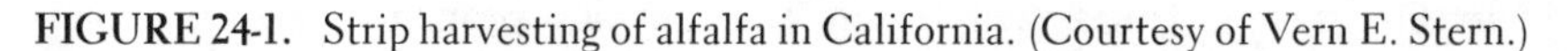

**FIGURE 24-1.** Strip harvesting of alfalfa in California. (Courtesy of Vern E. Stern.)

mainly to divert pests from the main crop or to draw them into a small area where they can be controlled without disrupting beneficial species in the main crop. Although lygus bugs attack cotton, they prefer alfalfa, so strips of alfalfa planted within cotton fields attract the bugs and reduce injury to the cotton plants. In Hawaii, corn, which is attractive to melon flies, is planted around fields of melons and squash; the corn is periodically treated with an insecticide to kill the flies without adversely affecting beneficial species in the main crop. This technique is not widely used as a primary method of insect control, but it can be combined effectively with other methods. Trap crops can provide an effective means of control against plant pathogens. Creeping legumes planted between rows of rubber trees serve as a trap crop for several root diseases. Interplantings of trap crops also are effective as a means of reducing nematode infections in cruciferous crops and peaches.

### *Sanitation*

Sanitation practices directed at reducing pest populations through the removal of suitable breeding material or sites in which pests pass dormant periods have broad application. However, sanitation as a pest control method must be considered on a case-by-case basis, since it can have a detrimental effect on populations of natural enemies.

The impact of a number of plant diseases can also be lessened through the practice of sanitation. The removal of wild plants and crop-plant volunteers from field margins can serve as a means of removing reservoirs of plant viruses that are transmitted to the crop by insects. Vegetated margins also can provide alternate hosts and a refuge for the plant disease vectors. For example, the incidence of aster yellows can be reduced by removing wild celery from the margins of nearby water courses.

Insect pests also can be controlled through sanitation. The destruction of prunings is effective against some orchard pests such as the peach tree borer *Conopia exitiosa* and lesser peach tree borer *Synanthedon pictipes.* The collection and utilization or destruction of dropped fruit reduces populations of some important direct pests, such as the plum curculeo *Conotrachelus nenuphar,* codling moth, and apple maggot *Rhagoletis pomenella.* The cutting and destruction of cotton plants after harvest is an important measure in the control of the pink bollworm, European corn borer, and sugar cane borer *Diatraea saccharalis* (National Research Council, 1969). Cleaning field margins of wild vegetation is a means of eliminating refuges that can be used by insect pests during the absence of the crop and of reducing pests that require alternate plant hosts.

Whereas most sanitation measures are directed at a specific pest and therefore can be carried out to produce mainly beneficial results, the clean field margin philosophy requires careful scrutiny because of certain benefits such vegetation can provide. Field margins provide a measure of diversity to areas of monoculture that can contribute to stabilizing the relationships between pests and their natural enemies. Many entomophagous parasites in the adult stage require pollen and nectar, which may not be available if the field margins are devoid of vegetation. The presence of certain flowering plants may actually attract beneficial species.

### *Water and Fertilizer Management*

The addition or withholding of water and fertilizer from crop ecosystems can have a variable impact on pest problems. The literature on the effect of different nutrients on the fecundity of plant-feeding insects is extensive. In situations in which certain nutrients adversely affect pest populations, some control can be obtained from altering their use. The growth response of some plants to water and nutrients helps to overcome early pest damage and make it less significant. Yarwood (1970) indicated that there are cases of increase in disease because of fertilization, and increased fertility can stimulate weed growth. Several examples of the effect of fertilizer on insect pests were given in Chapter 20. As another example, water and fertilizer delay the maturation of cotton and thereby contribute to late season build-up of pests such as boll weevil and bollworm.

As with fertilizer, the benefits from water management can be quite variable. In some cases, irrigation helps to reduce pests; in others, it is best to withhold water. Generally moist soil conditions favor weeds and plant diseases, but both flooding and drying can be useful in the control of various soil insects. The way in which irrigation water is applied may actually be of more significance. Overhead sprinkling tends to reduce some orchard pests and is quite effective as a means of reducing the populations of spider mites that are often protected by an accumulation of dust on the leaves. On the other hand, overhead irrigation encourages some wet-weather diseases such as halo blight of beans, black rot of melons, and angular leaf spot of cotton, and may increase rust mite populations on lemons.

## Host Resistance

The susceptibility or resistance of a host to attack by an insect represents the current status of the continuous process of coevolution of the interacting species. Resistant hosts occur naturally, and it would be surprising to learn that early man did not select at least some of his cultivars on the basis of some degree of freedom from attack by insect herbivores. Several early authors, including Theophrastus, commented on differences in the apparent susceptibility of crop plants to insects and diseases. However, the earliest reported recognition of plant resistance in an agricultural crop referred to the discovery in 1782 that "Underhill" wheat was resistant to Hessian fly attack. The first detailed report (1831) of a variety selected for its resistant quality followed the discovery that the Winter Majetin variety of apple was resistant to the woolly apple aphid. A few years later, it was found that some American varieties of grapes were resistant to the grape phylloxera (Homoptera: Phylloxeridae), whereas the European grapes were highly susceptible. The use of American root stock as a grafting base for the prized European grape varieties saved the European wine industry. The list of crop plant varieties resistant to insect attack has been growing ever since. In recent years, the search for ecologically sound alternatives to chemical pest control has accelerated the search for resistant plants and animals and has stimulated breeding programs. Reviews by Painter (1958), Beck (1965), Maxwell (1972), and Gallun and others (1975) provide ample evidence of both the complexity and usefulness of this approach to pest control.

Several workers have defined plant resistance, but probably the most widely accepted definition was provided by Painter (1951). He defined plant resistance as ". . . the relative amount of heritable qualities possessed by a plant which influences the ultimate degree of damage done by the insect. In practical agriculture it represents the ability of a certain variety to produce a larger crop of good quality than do ordinary varieties at the same level of insect population." Painter also segregated plant resistance mechanisms into three categories: (1) **preference** and **nonpreference,** (2) **antibiosis,** and (3) **tolerance.** All of these mechanisms can be under genetic control, and natural resistance may involve any combination of them.

### *Preference and Nonpreference*

Preference and nonpreference refers to those host characteristics that elicit responses from insects that lead them toward or away from a particular host as a source of food, for oviposition, for shelter, or for a combination of these. Recent research has demonstrated that the inherent responses of insects to inherited characteristics of plants are quite stable. The characteristics to which insects respond are variable. They include a variety of chemical constituents, surface and internal structural characters, color, and general conformation. Chemicals perceived as odor or taste seem to be the most important.

Some insects will neither feed nor oviposit on a nonpreferred host, whereas others will oviposit, but the resulting progeny will not feed. Nonpreference can be so strong that the affected insect will starve rather than feed on that particular host. Many, however, will avoid nonpreferred hosts if preferred hosts are available but will consume the nonpreferred host when an alternative food source is not available. When a female oviposits on a nonpreferred host, the young larvae sometimes take small nibbles but will not feed normally. Such is the case with the Colorado potato beetle on wild potato, and the European corn borer on some strains of corn. In such situations, it is difficult to determine whether the chemicals involved produce nonpreference or are actually toxic, feeding deterrents. At this level, nonpreference seems to overlap antibiosis. A nonpreference for certain host varieties also occurs among insects and ticks that attack domestic animals. Nonpreference here can result from the avoidance behavior used by the hosts to distract their parasites, but more predictable forms of nonpreference involve the chemistry of sebaceous gland secretions, and skin or coat characteristics. For example, Zebu cattle appear to be nonpreferred hosts for some ticks, horse flies, and mosquitoes because of their color and hide characteristics.

***Antibiosis*** Antibiosis refers to characteristics of the host that have adverse effects on a pest's survival, development, vigor, or fecundity. Some of the results include premature death, lowered reproduction, protracted development and exposure to causes of mortality, reduced vigor and dispersal capacity, and abnormal behavior. The mechanisms of antibiosis include the presence of toxins, growth regulators or reproductive inhibitors, the absence of nutrients such as essential vitamins, a low level of certain required nutrients, or a nutrient imbalance. Abnormal growth in response to insect damage, including the development of scar tissue, galls, or secondary periderm can also kill or injure attacking insects.

***Tolerance*** Tolerance is a form of resistance resulting from the ability of the host to repair insect damage or to tolerate the attack of a pest population that would cause economically significant damage to a susceptible host. In plants fed upon by chewing insects, resistance may result from an ability to bud adventitiously or to simply replace parts faster than they are severely damaged. For example, some varieties of corn can replace roots faster than they are rendered nonfunctional and consequently are tolerant to attack by corn rootworms, *Diabrotica* spp. (Coleoptera:Chrysomelidae), (Figure 24-2). The resistance to the wheat stem sawfly displayed by some varieties of wheat is generally attributed to the solidness of the stem. The solid stem acts as a deterrent to egg-laying and larval development. Furthermore, the damage that does occur can be tolerated because it does not weaken the stems sufficiently to cause lodging (bending over) and the related losses during harvest.

Some plants also display a capacity to tolerate pathogens. Such plants become infected but suffer little or no ill-effect. For example, the Huxley

**FIGURE 24-2.** Injury to the roots of corn plants caused by corn rootworms *Diabrotica spp.* The stalk on the right is of a susceptible variety whereas the four stalks on the left are of a resistant (tolerant) variety. At least part of the tolerance results from the plant's capacity to replace roots more quickly than they are destroyed. (Photograph from Guatemala—Iowa State Tropical Research Center.)

variety of strawberry can tolerate virus attack whereas other varieties are destroyed. The problem with developing tolerant varieties, as opposed to those that are resistant through antibiosis, is that the tolerant varieties form a reservoir of the disease or a refuge for an insect pest.

### *The Genetics of Resistance*

The inheritance of resistance to a specific pest is controlled by the same genetic principles that determine the transfer of other traits. It is usually classified according to whether it is controlled by a single gene (**monogenic**) or several genes (**polygenic**). Studies to determine the nature of the genetic control of resistance may use different approaches and consequently produce variable results for the same host. Although the genetic basis is not known for most resistant varieties that have been developed, there are a number of examples in which the genetic conditioning of resistance is well understood.

In addition to monogenic and polygenic resistance to specific insects and pathogens, there is substantial evidence that plants have a general resistance conferred by a collection of genetically controlled physiological and morphological characteristics. Van der Plank (1968) suggested that there is a form of polygenic resistance governed by genes that regulate ordinary processes rather than by special genes for resistance. This general resistance has developed as plants became increasingly specialized in response to new selective pressures in their environment. Baker and Cook (1974) state that ". . . the greater the genetic factors that condition resistance, the longer the evolutionary development of that resistance and the greater its spectrum of effectiveness . . ." They conclude, therefore, that polygenic resistance indicates antiquity, whereas monogenic resistance is a more recent evolutionary development. Polygenic resistance may begin as monogenic resistance which becomes enhanced by the accumulation of additional influential genes.

Monogenic resistance seems to be specific for a particular strain of pest with a specific gene for virulence that the plant encounters over part of its range. Such resistance may be effective over a wide range of environmental conditions. On the other hand, polygenic resistance provides protection against a number of variants of the pest, but does not provide total immunity or remain effective under some environmental conditions.

The single gene that conditions resistance may be dominant as is the cases of IR8 rice resistance to some leafhoppers, and other rice varieties resistant to rice blast. It can, however, be recessive as it is in peas resistant to powdery mildew or corn varieties resistant to western corn borer adults (Sifuantes and Painter, 1964). Although more complicated, the same is true of polygenic resistance. For example, the resistance of tobacco strains to bacterial wilt is polygenic and recessive, whereas the polygenic resistance of corn to the European corn borer is dominant (Penny and Dicke, 1966).

Within single species of plants the genetics of resistance to different pests may be highly variable. Pathak (1969) analyzed the inheritance patterns for resistance in rice to three insects. The resistance to the striped borer, *Chilo suppressalis*, is polygenic; for the plant hopper, *Nilaparavata lugens*, there are both dominant and recessive genes; and for the leafhopper, *Nephotettix impicticeps*, there is a single dominant gene.

### *The Importance of Pest Biotypes*

Genetic diversity exists not only in host organisms but is a characteristic of pest species as well. Consequently, when an insect or pathogen is subjected to the selective pressure of host resistance, variants of those insect or pathogen species that are less affected by the host resistance are better able to survive. These survivors then transfer the traits that counter the host's resistance through interbreeding. The result may be a population comprised of a new biotype. New biotypes also appear as a result of genetic changes caused by mutation. In areas where a previously resistant plant is exposed to a new virulent pest biotype, the plant becomes susceptible.

In insects and pathogens that reproduce without sexual recombination, there is no dilution of the genes that overcome the host resistance. If in addition the pest has several generations per year, a new biotype can predominate in a population in a very short time. Consequently it is difficult to determine how long a resistant variety will remain resistant. Generally, monogenic resistance is thought to be less persistent than polygenic resistance. When the resistance depends upon a gene-for-gene relationship in the host and its parasite, it can be overcome by a minor genetic adjustment in the pest species. For example, two varieties of wheat were known to be resistant to normal field populations of the greenbug, but a new biotype of the aphid developed that could damage both of these varieties. More recently, another biotype of the aphid appeared that is able to infest and reproduce on Piper sudangrass, a feat of which the other two biotypes are incapable (Wood, 1961 and Wood et al, 1969). Similarly, a developed strain of alfalfa that was resistant to the spotted alfalfa aphid *Therioaphis maculata*, was cultivated for only five years before a biotype of the aphid developed to which the alfalfa was susceptible. On the other hand, the breakdown in polygenic resistance has been rare. As an example, varieties of tobacco that display polygenic resistance to bacterial wilt have been planted in bacteria-infested soil for 20 years with no evidence that new virulent strains of the bacteria capable of infecting these varieties have developed.

Gallun, Starks, and Guthrie (1975) present an interesting situation involving eight known biotypes of Hessian fly that forms the basis for the following resumé, The Hessian fly biotypes and the genes of wheat that provide resistance to them are listed in Table 24-1. The wheat genes that condition

TABLE 24-1 Genes in Wheat That Condition Resistance to Races of Hessian Fly

| Hessian Fly Races | Genes in Wheat |
|---|---|
| GP | $H_1$ $H_2$, $H_3$, $H_5$, $H_6$, $H_7$ $H_8$ |
| A | $H_3$, $H_5$, $H_6$ |
| B | $H_5$, $H_6$ |
| C | $H_3$, $H_5$ |
| D | $H_5$ |
| E | $H_1$ $H_2$, $H_5$, $H_6$, $H_7$ $H_8$ |
| F | $H_1$ $H_2$, $H_3$, $H_5$, $H_7$ $H_8$ |
| G | $H_1$ $H_2$, $H_5$, $H_7$ $H_8$ |

the resistance are dominant, and each is specific for a comparable recessive gene for virulence in the Hessian fly. A wheat variety, therefore, remains resistant until a biotype of the insect develops with a set of genes to counter the set of genes conditioning resistance in the wheat. In other words, for every gene in the plant that conditions resistance, there must be a comparable gene for virulence in the insect, if the resistance is to be overcome. Consequently, those wheat varieties with a number of specific genes that condition resistance are least likely to become susceptible. However, the data in the table indicate that biotypes of Hessian fly have overcome resistance conditioned by a number of genes.

The discovery that in the Hessian fly the genes for virulence are recessive whereas those for avirulence are dominant has led to the suggestion that this could form the basis for genetic control of the pest. An avirulent strain (dominant) could be released in areas populated by a virulent strain (recessive). The progeny that would result from matings between the biotypes would be avirulent and would perish in the presence of a resistant variety of wheat.

### *Practical Considerations*

As long as a host remains resistant to an attack by a parasite species, its effect on the pest is specific, cumulative, and persistent and will usually tend to reduce the pest population. These features make host resistance a desirable alternative to pesticides, which are usually nonspecific, noncumulative, and temporary. Furthermore, the use of resistant crop varieties is not harmful to beneficial species or to the environment. Total resistance is not necessary. Any level of resistance that will suppress the pest population to a level below the economic threshold established for it meets the objective of pest control. Those pest individuals that remain in the crop ecosystem provide a source of hosts for natural control agents and therefore help to stabilize the pest population further.

Resistant varieties can be obtained by (1) selection of resistant individuals from heterogenous populations exposed to pest species, (2) crossing varieties carrying resistance with varieties that display other desirable characteristics, (3) hybridizing resistant wild species with susceptible cultivars, followed by selection, (4) backcrossing to the original parent to maintain desirable characteristics as a sequel to initial crossing of different varieties or species, (5) budding and grafting as a means of maintaining a resistant stock or scion, and (6) subjecting wild host plants to radiation or other mutagenic agents (National Research Council, 1968).

The development of resistant varieties can be time-consuming and there is, of course, no guarantee of success. Monogenic resistance is the easiest to obtain, as one might expect. Polygenic resistance is difficult to obtain and complicated by the fact that the increased number of genetic adjustments required increases the likelihood of inducing changes in other characteristics. The induction of resistance to one pest may inadvertently cause an increase in susceptibility to others. For example, progeny from a cross of Richland and Victoria oats were resistant to stem and crown rust and loose and covered smut. However, the hybrid variety was highly susceptible to Victoria blight.

Once a resistant variety has been developed, the benefits can be substantial even if the resistance breaks down several years later. The cost of this form of pest control to the grower is generally low, perhaps involving no more than a reasonable increase in the cost of seed. The benefits can be great, however. The cost of more conventional forms of pest control are eliminated, and increased yields often result. Estimates of crop loss caused by a specific pest are difficult to evaluate, so it is often difficult to determine the savings that can be attributed to host resistance. There also may be side benefits in the form of an over-all reduction in the pest population and a related decline in the damage to susceptible varieties grown in the area.

An evaluation of the programs to develop crop varieties resistant to the Hessian fly, wheat stem sawfly, and spotted alfalfa aphid revealed an estimated loss reduction of $308 million per year in the United States. On the basis of research and development costs of $9.3 million, the cost benefit ratio for a ten-year period would exceed 1 : 300 (Luginbill, 1969). The cultivation of crop varieties resistant to pathogens has been estimated to provide a monetary gain to American farmers in excess of $1 billion per year (National Research Council, 1968).

# Legislative or Regulatory Control

The implementation of special laws or government regulations implemented as a means of preventing the entry and establishment of foreign pests or the elimination or containment of pests within a defined area, is a major function of legislative or regulatory control. Additional regulations designed to prevent the spread of agricultural pests through commerce involving certain agricultural products and the restriction of the movement of infected plant and animal material comprise what is referred to as **quarantine.**

The purpose of quarantines is to exclude potential pests or prevent the spread of established pests with a well-defined, limited distribution. Such action cannot be left up to individuals or private concerns and therefore requires the establishment of specific regulations that can be enforced under the law. In the United States, the Plant Quarantine Act of 1912 gave the USDA the authority and responsibility for the enforcement and implementation of programs for the eradication or control of introduced pests. State agencies were made responsible for enforcing regulations to prevent the spread of pests within their geographic boundaries.

Under the provisions of the United States Quarantine Act, plants subject to restriction cannot be imported unless: (1) a special permit has been obtained prior to shipment and all necessary information has been provided, (2) the shipment can be inspected, (3) the importer accepts the responsibility for meeting any special requirements such as fumigation, (4) the importer agrees that hazardous materials can be seized, treated, or disposed of, and

(5) the products must remain in the custody of the Customs Department until released by a plant inspection officer.

Under the Quarantine Act, there are three kinds of action that can be undertaken with public support when a pest is inadvertently introduced in a quarantined area:

1. **Eradication.** The total elimination of the target species from the quarantined area may be possible. This type of action usually follows the discovery of an introduction if the infested area is small and can be well defined through an intensive survey. For example, both the Japanese beetle and the Oriental fruit fly were successfully eradicated following their discovery in San Diego County, California in recent years.
2. **Containment.** If eradication is not feasible because of the area infested or because no suitable control method is known, efforts are made to limit the spread of the pest. The gypsy moth is currently the subject of containment quarantine. The Secretary of Agriculture, after holding public hearings and providing adequate notice, can quarantine any portion of the United States. The pests that were subject to federal interstate quarantine in 1976 are listed in Table 24-2.
3. **Suppression.** The quarantine act provides special assistance in the control of periodic pest outbreaks that are too large for a single agency or landowner to undertake. Large-scale outbreaks of grasshoppers or forest pests may be controlled under this provision.

Most major nations now have quarantine laws and regulations to guard against the international spread of pests. Cooperation between many countries in the enforcement of quarantines resulted from a FAO conference and agreement reached in 1951. One aspect of this cooperation is the clearance or certification of plant material or animals destined for export to foreign countries.

In addition to the quarantine laws, there are additional laws specifically directed at weed control and plant disease protection. As mentioned earlier, the use of disease-free planting stock and of seeds relatively free of weeds is a sound cultural control procedure. This is aided by regulations requiring the inspection and certification of propagative material as free of insects and disease. Federal and state seed laws regulate the importation of seeds into the country as well as those shipped in interstate commerce.

## Autocidal Methods

Autocidal control can be defined as the utilization of an organism or some characteristic of it to destroy other members of the same species or to reduce its population through reduced reproductive capacity. The principal approaches involve sterilization and genetic manipulation.

### *Sterile Male Technique*

The sterile male approach to pest control was originated as a theory by E. F. Knipling more than a decade before it was established as a practical means of reducing a pest population. Knipling's calculations (Knipling, 1955) indicated that the use of an insecticide must remain quantitatively the same in each generation to provide the same proportionate mortality and, therefore, gives a diminishing return as the pest population density declines. On the other hand, the release of the same number of sterile males in each pest generation

TABLE 24-2 Pests Subject to Interstate Quarantine Within the United States in 1976

| | |
|---|---|
| Gypsy moth | *Porthetria dispar* |
| Brown-tail moth | *Nygmia phaeorrhoea* |
| Japanese beetle | *Popillia japonica* |
| Pink bollworm | *Pectinophora gossypiella* |
| Mexican fruit fly | *Anastrepha ludens* |
| White-fringed beetle | *Graphognathus spp.* |
| Imported fire ant | *Solenopsis saevissima* |
| Black stem rust | *Puccina graminis* |
| Witchweed | *Striga asiatica* |

would result in an increase in the ratio of sterile to nonsterile matings and therefore should become increasingly efficient as the native population declines. It is obvious from the simple mathematical models prepared by Knipling that it must be practical or feasible to swamp a wild population with sterile males so as to create a ratio of sterile to fertile matings that would initiate a downward trend in the population.

The initial project was directed against the screwworm fly, *Cochliomyia hominivorax*, a serious pest of cattle in the southern United States. Knipling noted that female screwworm flies mate only once, and recognized this as a desirable attribute. Furthermore, the fly survived over the winter at fairly low population densities along the climatically favorable southern portion of its range. If the program could be initiated before the populations increased and moved northward in the spring, control by the release of sterilized males seemed feasible. Indeed, it was feasible. The first totally successful field eradication occurred on the island of Curaçao in 1954 just 13 weeks after the sterile males were first released. Later, over a span of 18 months in 1958–1959, more than 2 billion flies were reared, sterilized with cobalt-60 radiation, and released in the southeastern United States. The result was the eradication of the screwworm from the southeast (Knipling, 1960; Baumhover, 1966). Sterile males are presently released annually along the United States-Mexico border as a means of reducing the losses caused by immigrant flies from Mexico.

Annual losses to the American cattle industry because of the screwworm prior to the sterile male program have been estimated at between $100 million and $140 million in spite of control attempts involving pesticides and judicious animal husbandry. These losses have been virtually eliminated as the result of an expenditure of about $10 million for the sterile male program.

The success of the sterile male technique as a control against the screwworm stimulated considerable interest in the approach as a means of solving other insect pest problems. However, there are a number of requirements that must be met before successful control can be anticipated:

1. Reproductive biology. If the female of the species mates only once, success of the program is more likely than if the female mates with several males. Monogamy is not essential, but if polyandry occurs, sterile males must produce active sperm that is competitive with normal sperm so that a high proportion of infertile eggs will result.
2. Sex separation. If it is not possible to separate the sexes by some efficient means, sterile males may be wasted in matings with released sterile females and thereby reduce the program's effectiveness.
3. Mass rearing. It must be possible to rear and sterilize large numbers of individuals in a short time to be able to release enough young, vigorous individuals to swamp the wild population.

4. Vigor. The reared individuals must be at least as vigorous as members of the wild population and, following sterilization, must have suitable survival characteristics and be able to compete with wild individuals for mates.
5. Biological relationship. If large numbers of individuals are to be released, they should not have any serious economic impact on the agroecosystem. For example, males of the screwworm, like many biting flies, do not feed on blood and are relatively harmless. On the other hand, this could not be said for a leaf-feeding beetle, the release of which would increase damage to the crop.
6. Population characteristics. The success of the approach as applied against the screwworm is predicated on the assumption that the wild population can be numerically swamped. If a large wild population exists, it may be necessary to reduce it substantially with chemicals prior to releasing the sterile males.

Several other control programs have been initiated as a practical means of pest suppression, and a number are in fairly advanced states of development. Most notable are the California programs directed against Mexican fruit flies entering from Mexico and the spread of the pink bollworm from the Imperial Valley into the San Joaquin Valley. Table 24-3, prepared by Knipling (1972), summarizes the status of the sterile male technique up to that time.

The technique of sterilizing members of a naturally occurring population provides an alternative to mass rearing, sterilization and release programs, and involves some important differences. Some method must be employed to bring wild individuals into contact with a chemosterilant to replace the mass sterilization with a radiation source in a laboratory. This can be accomplished by combining a chemosterilant with an attractant or arrestant distributed in the population's natural habitat. Although sterilized individuals will not leave fertile offspring and thereby contribute to the reduction of the population, they will persist in the population for their normal life span. As a result, they may contribute to crop losses on the one hand but serve as hosts for natural enemies on the other. Once perfected, this technique would be preferable to the release technique for species in which the released individuals may initially increase the damage.

The most promising chemosterilants are effective when ingested, although some are effective simply through tarsal contact. In either case, the chemical must be presented with a bait, attractant, or arrestant to assure contact by a large portion of the wild population. The preparation of these mixtures requires special precautions because of the danger of the chemosterilants. The powerful chemosterilants are not specific, so care must be exercised to minimize the danger to beneficial species.

### *Genetic Techniques*

The reduction of pest populations by genetic means would involve some alteration of the genetic make-up of the pest that would result in genetic sterility, a reduction in fecundity, a reduction in vigor, or the inability to complete development in a normal manner. Manipulations of the genetics of host species to make them resistant to attack by pests was discussed earlier.

Insect populations generally have a large gene pool, as often evidenced by numerous morphological, physiological, and behavioral races or strains. This high level of population variability serves effectively as a buffer against natural

**TABLE 24-3** **Insects Under Investigation, How Sterile Insects Will Be Used, and Status of Research or Practical Use**

| Insect | Proposed Manner of Use | Status of Research and Development |
|---|---|---|
| Screwworm | For suppressing populations on regional basis | In practical use |
| Mexican fruit fly | For preventing establishment of incipient infestations | In practical use |
| Pink bollworm | For preventing establishment of incipient infestations and to eliminate low level established populations | In practical use; additional improvements and pilot testing required |
| Oriental, Mediterranean, and melon fly | To eliminate low level populations and to prevent establishment of incipient populations | Effectiveness demonstrated in small island tests. Large pilot tests required |
| Codling moth | To maintain suppression after prior suppression of populations by cultural and chemical means | Effectiveness demonstrated in small orchard tests. Small pilot test under way |
| Boll weevil | To eliminate low level populations after prior suppression by chemical, cultural, and other means | Pilot tests planned |
| Bollworm and budworm | For area suppression of low level populations | Pilot tests required |
| Cabbage looper | For area suppression of low level populations | Pilot tests required |
| Fall armyworm | For area suppression of low level populations | Pilot tests required |
| Tobacco hornworm | For area suppression after prior suppression by cultural means | Pilot test required |
| Gypsy moth | For preventing spread and to eliminate incipient infestations | Pilot test required |
| Mosquitoes (important vector species) | To maintain suppression after prior suppression by sanitary and chemical means | One pilot test underway—others required |
| Tsetse flies | To eliminate low populations after prior suppression by chemicals and brush clearing | Pilot test planned |
| Horn fly | To eliminate low populations after animal spraying | Pilot test required |

*Source:* Knipling, 1972.

environmental changes in that it allows populations to adapt to new ecological situations. The fact that insects have a high reproductive potential and short life histories provides the means by which the production of variants in a population can change rapidly. One of the best examples of how this genetic plasticity can function was presented by the rapid development of insecticide resistance by a number of insect pests. Many entomologists conducting rearing programs under controlled conditions have also found changes in some aspect of the biology or behavior of their colony.

Although not yet at the stage of refinement required for practical application, the technique of genetic control has considerable potential. Genetic manipulations that could be useful in pest control include (1) creation of cytoplasmic incompatibility, (2) hybrid sterility, (3) introduction of lethal or harmful genes, and (4) developmental alteration.

In some insects there are strains which when cross-mated produce some sterile offspring. The sterility results from the fact that the sperm enters the egg and stimulates meiosis but does not fuse with the egg nucleus. Most of the eggs fail to hatch, although a small proportion develop parthenogenetically to produce females (see Laven, 1967 for more details). According to Woods (1974), this cytoplasmic incompatibility principle has been used successfully in Burma to eradicate a population of a mosquito *Culex pipiens fatigans* in an ecologically isolated area near Rangoon.

Hybrid sterility is simply the production of infertile hybrid offspring as a result of a cross between related species. The offspring are viable and may display hybrid vigor and, thus, compete with normal individuals for resources. Sometimes the females are normal and the males are sterile, but the over-all effect is population reduction. Employment of this method in control might be possible by rearing and then releasing a nonpest species into the habitat of a related pest species. Alternatively, it might be possible to bring together a pair of allopatric pest species that will mate but leave sterile progeny. This has been suggested as a way to reduce populations of the tsetse flies *Glossina swynnertoni* and *morsitans.*

The concept of inducing lethal or harmful genes into a population seems to contradict the process of natural selection. However, some workers have suggested that it might be possible to introduce genes that would be deleterious only at specific times or under particular circumstances. Certain genes could be propagated in a wild population if their deleterious effect is expressed only when particular environmental conditions prevail. Whitten (1971) suggested that a desirable situation would occur if a conditional lethal allele could be produced and be maintained in a homozygous condition. Hybridization with wild individuals would produce sterility, whereas the homozygotes would be destroyed only under certain conditions. For example, if the homozygote were susceptible to an insecticide to which the wild population was resistant, both populations eventually could be eliminated. As an alternative, Werhahn and Klassen (1971) suggest that desirable genes could be linked with those which confer insecticide resistance. Thus, releases of the genetically manipulated individuals could be made along with an insecticide application which would favor the carriers of the lethal gene.

In some insects that occupy a range that spans several distinctly different climatic zones, strains have evolved with life history patterns that enable them to exploit some particular portion of the range. For example, in areas with severe winters, the population may enter an obligatory dormancy (diapause) and therefore have only one generation per year. In a warmer part of the range, the population may be nondiapausing and complete several generations per year. If these developmental characteristics are under genetic control, it may be possible to reduce the populations by mixing genotypes. Such a manipulation has been suggested as a possibility for control of the European corn borer, *Ostrinia nubilalis,* in the United States and the field crickets, *Teleogryllus commodus* and *oceanicus,* in Australia (Woods, 1974). The genetic induction of diapause in populations in warm areas would reduce the number of generations per year. The introduction of genes for no diapause into populations in cool areas could increase winter mortality.

# Special Chemical Methods

There are a number of pest control methods, particularly useful against insects, that involve the use of chemicals, different from standard pesticides in either the way they are employed or their mode of action. These include feeding deterents, antimetabolites, repellents, attractants (including pheromones), and growth regulators.

***Feeding Deterrents*** Feeding deterrents are used to eliminate damage by preventing the feeding of a pest rather than by reducing the pest population, although strong deterrents can induce starvation among species with a narrow range of hosts. The only such substance tested on a large scale is 4-(dimethyltriazeno)-acetanilide. Treatment of foliage with this substance produced fair to excellent results against a variety of insect herbivores, including cabbage pests, the boll weevil, and cucumber beetles. However, the prevention of damage that resulted from feeding within the plant tissue by the corn ear worm and the codling moth was poor (see Wright, 1963). Perhaps combining the deterrent with an effective penetrant would prove useful.

A number of chemicals extracted from plants have been shown in the laboratory to be strong deterrents to feeding by such important pests as the European corn borer and the Colorado potato beetle (Smissman et al, 1957; Maxwell et al, 1965) but have yet to be proven in field trials.

Feeding deterrents present several advantages over conventional pesticides in that only the pest species is affected. In addition, the fact that the pests are not usually killed means that they persist as hosts for their natural enemies. The main disadvantages involve the fact that at least at present they are effective only against surface feeding pests, and since the pests are not killed, they can disperse to untreated areas. Furthermore, growth of the host plants following treatment provides a source of untreated food on which the pest population could concentrate and produce severe damage.

***Antimetabolites*** Antimetabolites are compounds that resemble essential nutrients but inhibit the utilization of metabolites. When these compounds are ingested with the food, the insect displays the symptoms of a dietary deficiency, which may result in death. The potential of such compounds as a practical control method is yet to be demonstrated and will probably be restricted to use against species with a limited range of foods or foods to which they have limited access.

***Repellents*** Repellents are compounds that cause an organism to react by moving away from the source. Generally, the distance from the source at which insects respond to repellents is relatively short. Insects tend to avoid treated surfaces rather than leave the area. For this reason, repellents are more important in providing protection for humans and domestic animals against irritating and disease-vectoring species than in providing protection for crops.

A long-lasting repellent can protect domestic animals from biting flies and reduce the annoyance and loss of blood that can result in decreased weight gain and milk production. Repellents also aid in the reduction of insect-transmitted diseases. However, of the hundreds of compounds that have been tested for use on domestic animals, only a few are recommended, and these have limited value, mainly because they are effective for only a day or so.

Although there are many naturally occurring and synthetic compounds that are repellent to plant-feeding insects, no practical methodology has been developed for their use in the protection of agricultural crops or forests. Since the pest population would not be reduced, repellents would act mainly to prevent invasion of a crop and would probably require treatment of a very large area to be effective. In special cases where part of a crop is more susceptible, as in a forest, for example, repellents could be used to disperse pests searching for susceptible hosts and thereby increase dispersal mortality.

### *Attractants*

Chemical messengers are of considerable importance in regulating insect behavior. Odors given off by host plants and animals and by other sources of food are used by insects in the location of feeding and breeding sites. In addition, insects produce a variety of odiferous compounds (pheromones), which are used in various kinds of communication, including species recognition, aggregation for mating and host colonization, defense, sex attraction, and mating stimulation (see Shorey, 1973). The powerful nature of these chemicals as behavioral modifiers has long been a source of interest to entomologists, but only in recent years has chemical technology advanced to the point where small quantities of such complex substances could be extracted, identified, and synthesized.

In a strict sense, attractants are stimuli that cause organisms to orient toward the source and therefore are important components of insect communication systems (see Chapter 13). As they bring individuals to a point in space, there are a variety of ways that they can be employed in insect control. Since control decisions are frequently made on the basis of pest population levels, attractants can be used as a tool to aid in the monitoring of population conditions. The specific nature of many insect attractants increases the efficiency of insect surveys directed against a limited number of pest species. Attractants are particularly useful for detecting the entry of pests into quarantined areas.

One of the largest survey programs based on the use of an attractant involves **disparlure,** a synthetic female sex pheromone of the gypsy moth *Porthetria dispar.* Approximately 50,000 small cardboard traps (Figure 24-3), combining a disparlure-saturated wick with a sticky coating on the inside of the cardboard, are placed throughout the infested area of New England to monitor the flying population of male moths (Jacobson, 1965). Attractant-baited traps are also used as an early warning system for the detection of a number of potentially serious pests, including the Mediterranean fruit fly, *Ceratilis capitata,* in Florida; the Mexican fruit fly, *Anastrepha ludens,* along the United States border with Mexico; and the Japanese beetle, *Popillia japonica,* and oriental fruit fly, *Dacus dorsalis,* in California. When properly set out in a systematic grid, such a detection program not only assists in the discovery of new invasions but also pinpoints the location so that an eradication program can be concentrated in a particular portion of the grid.

Some attractants, particularly those that stimulate population aggregation, can be used as a control tactic. Some pests can be drawn away from high-value or inaccessible resources and attracted to specific areas where they can be eliminated. For example, natural and synthetic bark beetle pheromones can be used to attract large numbers of beetles to trap trees or logs, where they may be permitted to establish their brood before the infested host material is removed or destroyed.

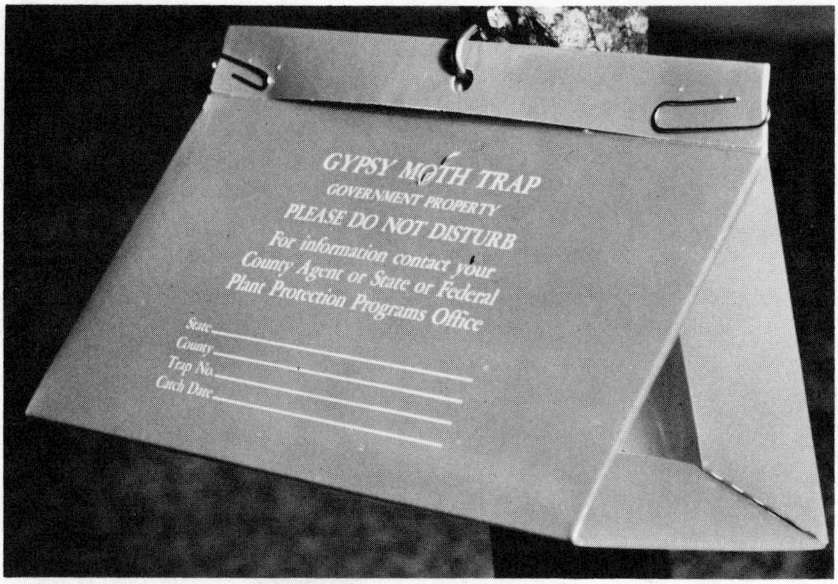

FIGURE 24-3. Pheromone trap of the type used to survey populations of pests such as the gypsy moth. A pheromone-saturated wick draws the insects into the sticky sleeve. (Photograph courtesy USDA.)

Attractants can also be used to draw pests into traps as a direct means of population reduction. This requires an efficient trap design as well as an effective attractant, which is released gradually over a period of time. The most simple arrangement involves a sticky surface on which the attracted insects become entangled (Figure 24-4). The efficacy of such approaches in suppressing pest populations is sometimes difficult to evaluate in that the number of individuals trapped may be unrelated to the local population prior to the release of the attractant. The placement of attractant sources throughout an area may change the behavior of dispersing individuals and actually draw additional individuals into the area. However, the intensive use of sex pheromone traps in an apple orchard substantially reduced the codling moth population and the level of fruit damage (MacLellan, 1976).

There are several other kinds of chemicals that act as arrestants or as feeding, mating, and ovipositional stimulants that can be included under the general heading of attractants. Some of these compounds can be used in combination with insecticides, sterilants, or pathogens as a means of improving the efficacy of the primary control agent. The Oriental fruit fly, *Dacus dorsalis*, was eradicated from the island of Rota by distributing 5 × 5 cm cardboard squares treated with an attractant and an insecticide. The attractant was so effective against the males that 15 bi-weekly treatments with the cards reduced the male population to a level at which most females were not mated and consequently left no fertile offspring.

As indicated earlier, many chemosterilants are too dangerous for use by general broadcasting and so must be combined with species-specific baits or arrestants. Similarly, attractants can be used to increase the effectiveness of pathogens. The attractant could serve to draw pest individuals into a portion of the environment where the pathogen has a better chance of survival and infection. Laboratory tests using this principle have demonstrated an in-

**FIGURE 24-4.** Hundreds of bark beetles caught on a large sticky trap, baited with a vial of attractant, used in a pilot control project. (Photograph courtesy of Gary Pitman, B.T.I.)

creased incidence of virus and sporozoan infection among corn earworms and boll weevils, respectively (National Research Council, 1969). Thuron Industries has developed a poison bait containing a fly attractant **Muscemone,** which is the first insecticide product registered by the EPA for insect control that has an attractant component.

One of the basic problems involved with the use of pheromones and other behavior modifiers in insect control results from population variability. Individuals respond to stimuli according to the relationship between the intensity of the stimulus and their response threshold. The threshold for response frequently varies according to physiological conditions and may change with time. For example, many insects have a migratory phase during which they are less responsive to hosts and mates than they are after the migratory drive has been satisfied. The behavioral make-up of a population may change in relation to how much migration has occurred or may vary according to conditions during development. Consequently, some populations may be more responsive to attractants than others, and control programs using these substances may produce variable results (see Atkins, 1966).

### *Insect Growth Regulators*

As with all organisms, the physiological processes related to reproduction, growth, molting, and metamorphosis in insects are under hormonal regulation. The understanding of hormones in the developmental physiology of insects is far from complete, but improved biochemical methods have produced major advances in recent years. There are basically three types of hormones known at this time: the brain hormone, which appears to mediate the production and interaction of the other hormones; ecdysones, or molting hormones, principally involved with the periodic casting off of the cuticle; and the juvenile hormone that governs the rate of maturation, including the development and maturation of the gonads. The chemical structure of the ecdysones and the juvenile hormone has been determined. Juvenile hormone and alpha-ecdysone have been synthesized. In addition, numerous compounds that exhibit the major biological activities of juvenile hormone (juvenoids) have been discovered.

The principal action of juvenile hormone in insect development is to determine the result of each molt by suppressing the expression of adult characteristics until the appropriate time (see Chapter 9). The treatment with juvenile hormone of developmental stages that normally progress to a more mature stage can inhibit normal development and actually prevent the completion of the life history. Juvenile hormonelike compounds also have gonadotropic effects and can be used experimentally to disrupt dormancy, normal ovarian development, and the apportionment of fat reserves between somatic and reproductive tissues. This has created considerable interest in the use of juvenile hormone mimics for pest control. The fact that many of the mimics seem to be somewhat specific in terms of the insect groups they affect, in contrast to natural juvenile hormones which affect all insects, creates some interesting possibilities. Specially designed compounds could be used to adversely affect the development of selected pests without harming beneficial species.

There are many problems that will have to be solved through research before these compounds can be fully evaluated under field conditions, but progress has been rapid. Several products have been tested against specific pests under field conditions with promising results. A California-based company, Zoecon, has developed one compound distributed under the trade name Altosid, which has been registered by the EPA for the control of flood water mosquitos (see Kosluchern, 1973). This was the first insect growth regulator approved for use by a regulatory agency. According to company reports, Altosid fed to chickens and cattle has also proved effective as a control for houseflies and hornflies that breed in the manure. More recently, a second Zoecon product, an insect growth regulator called Enstar-5E, has been approved for use in the control of aphids and white flies (Homoptera) in greenhouses.

Although not hormone mimics per se, another new group of compounds was recently reported in an undated Technical Information Bulletin of the Thompson-Hayward Chemical Company. These 1-(benzoyl)-3-(phenyl)-urea compounds are believed to interfere with the formation of normal insect cuticle. The newly formed cuticle of treated larvae apparently cannot withstand the increased turgor and/or muscle movement required during molting. As a result, treated larvae survive until ecdysis, at which time molting failure results in their death.

Bowers and others (1976) reported that a pair of plant compounds extracted from a common bedding plant *Ageratum* act as antijuvenile hormones and induced precocious metamorphosis in the milk weed bug. The researchers named the newly discovered compounds precocenes and suggest several ways in which they may be useful in pest control. Not only did the treated individuals pass through fewer instars and eat less, they were unable to reproduce.

## Miscellaneous Methods

There are a number of physical and mechanical methods of pest control involving the use of elements such as electromagnetic radiation, heat, barriers, and trapping devices. Generally, these are of less importance than the control methods already discussed, but some are effective in specific situations.

Low-energy radiation can be used as a means of creating a temperature increase and appears to be a satisfactory means of controlling pests of stored products, where the electrical properties of the pests are sufficiently different from the matrix that lethal heating of the pest tissue results without damage to the product. Heat produced by more conventional methods has been used to control insects and mites in stored commodities and wood products. Steam or dry heat can be used as a treatment for lumber, for delousing, and for soil sterilization.

Passive barriers of many kinds are a standard form of protection against a wide variety of pests. Common examples include netting to protect fruits against birds, fences to protect livestock against predators, and various kinds of bands to protect young trees against rodents and insects that attack the root collar. Sticky bands are often used to curtail the movement of crawling pests up and down the trunk of orchard trees.

Various mechanical methods are employed in pest control, the most noteworthy of which are the various types of machinery used for weed control and for the modification of insectan habitats. Not too many years ago, a horse-drawn device known as a hopper-dozer was used in the United States to scoop up large numbers of grasshoppers that frequently infested the fields.

## References Cited


ATKINS, M. D. 1966. Laboratory studies on the behaviour of the Douglas-fir beetle, *Dendroctonus pseudotsugae. Hopk. Can. Entomol.* 98: 953–991.

BAKER, K. F. and R. J. COOK. 1974. Biological control of plant pathogens. W. H. Freeman and Company, Publishers, San Francisco, Ca.

BAUMHOVER, A. H. 1966. Eradication of the screwworm fly—an agent of myiasis. *J. Am. Med. Assoc.* 196: 240–248.

BECK, S. D. 1965. Resistance of plants to insects. *Ann. Rev. Entomol.* 10: 207–232.

BOWERS, W. S., TOMIHISA OHTA, J. S. CLEERE and P. A. MARSELLA. 1976. Discovery of insect anti-juvenile hormones in plants. *Science* 193: 542–547.

COLE, J. S. and O. R. MATTHEWS. 1938. Tillage. Yearbook of Agriculture.

GALLUN, R. L., K. J. STARKS, and W. D. GUTHRIE. 1975. Plant resistance to insects attacking cereals. *Ann. Rev. Entomol.* 20: 337–357.

HAYNES, D. L. 1973. Population management of the cereal leaf beetle. Pages 232–240, *in* P. W. Geier, L. R. Clark, D. J. Anderson and H. A. Nix, editors.

Insects: Studies in population management. Ecol. Soc. Aust. (memoirs 1): Canberra.

Jacobson, M. 1965. Insect sex attractants. Interscience Publishers, New York, N. Y.

Knipling, E. F. 1955. Possibilities of insect control or eradication through the use of sexually sterile males. *J. Econ. Entomol.* 48: 459–462.

Knipling, E. F. 1960. The eradication of the screwworm fly. *Sci. Amer.* 203(4): 54–61.

Knipling, E. F. 1972. Sterilization and other genetic techniques. Pages 272–287, *in* Pest control strategies for the future. National Academy of Sciences, Washington, D. C.

Kosluchér, D. 1973. Altosid: A new method of mosquito control. *Proc. Louisiana Mosquito Control Assoc.* New Orleans.

Laven, H. 1967. Formal genetics of *Culex pipiens.* Pages 17–66, *in* J. W. Wright and R. Pal, editors. Genetics of insect vectors of disease. Elsevier, Amsterdam.

Luginbill, P. Jr. 1969. Developing resistant plants: The ideal method of controlling insects. *U. S. Dep. Agr. Prod. Res. Rep. 111.* 14 pp.

MacLellen, C. R. 1976. Suppression of codling moth (Lepidoptera: Olethreutidae) by sex pheromone trapping of males. *Can. Entomol.* 108: 1037–1040.

Maxwell, F. G. 1972. Host plant resistance to insects: Nutritional and pest management relationships. Pages 599–609, *in* J. G. Rodriguez, ed. Insect and mite nutrition. Amsterdam.

Maxwell, F. G., W. L. Parrott, J. N. Jenkins, and H. N. Lafever. 1965. A boll weevil feeding deterrent from the calyx of an alternate host, *Hibiscus syriacus. J. Econ. Entomol.* 58: 985–988.

National Research Council. 1968. Principles of plant and animal pest control. Vol. 1: Plant-disease development and control. National Academy of Sciences, Washington, D. C.

National Research Council. 1969. Principles of plant and animal pest control. Vol. 3: Insect-pest management and control. National Academy of Sciences, Washington, D. C.

Painter, R. H. 1951. Insect resistance in crop plants. Macmillan Publishing Co., Inc., New York, N. Y.

Painter, R. H. 1958. Resistance of plants to insects. *Ann. Rev. Entomol.* 3: 267–290.

Pathak, M. D. 1969. Stemborer and leafhopper-plant hopper resistance in rice varieties. *Entomol. Exp. Appl.* 12: 789–800.

Penny, L. H. and F. F. Dicke. 1966. Inheritance of resistance in corn to leaf feeding of the European corn borer. *Agron. J.* 48: 200–203.

Shorey, H. H. 1973. Behavioral responses to insect pheromones. *Ann. Rev. Entomol.* 18: 349–380.

Sifuentes, J. A. and R. H. Painter. 1964. Inheritance of resistance to western corn rootworm adults in field corn. *J. Econ. Entomol.* 57: 475–477.

Smissen, E. E., J. P. Lapidus, and S. D. Beck. 1957. Corn plant resistant factor. *J. Org. Chem.* 22: 220.

Stern, V. M. 1969. Interplanting alfalfa in cotton to control lygus bugs and other insect pests. *Proc. Tall. Timbers Conference.* 1: 55–69.

Stern, V. M., R. van den Bosch, and T. F. Leigh. 1964. Strip cutting alfalfa for lygus bug control. *Calif. Agric.* 18: 406.

Van der Plank, J. E. 1968. Disease resistance in plants. Academic Press, Inc., New York, N. Y.

Werhahn, C. F. and W. Klassen. 1971. Insect control methods involving the release of a relatively few laboratory-bred insects. *Can. Entomol.* 103: 1387–1396.

Whitten, M. J. 1971. Insect control by genetic manipulation of natural populations. *Science* 171: 682–684.

Wood, E. A. Jr. 1961. Biological studies of a new greenbug biotype. *J. Econ. Entomol.* 54: 1171–1173.

Wood, E. A. Jr., H. L. Chada, and P. N. Saxton. 1969. Reaction of small grains and grain sorghum to three greenbug biotypes. *Okla. State Univ. Agr. Res. Prog. Rep. 618.*

Woods, A. 1974. Pest control: A survey. John Wiley & Sons, Inc., New York, N. Y.

Wright, D. P. Jr. 1963. Antifeeding compounds for insect control. *Advan. Chem. Ser.* 41: 56–63.

Yarwood, C. E. 1968. Tillage and plant diseases. *BioScience* 18: 27–30.

Yarwood, C. E. 1970. Man-made plant diseases. *Science* 168: 218–220.

# Biological control—organism against organism

Biological control means different things to different people. In the broadest sense, it is the natural phenomenon of population regulation that results from the interaction of the biotic components of an ecosystem. Viewed in this way, biological control encompasses all of the biologically caused mortality, including that which results from competition between individuals of the same and different species, the effect of host defenses, host resistance, and the direct or indirect result of attack by organisms on a different trophic level. In a more applied sense, biological control involves the manipulation, conservation, and augmentation of specific kinds of organisms as a means of regulating the populations of undesirable species so as to reduce or prevent their impact on human well-being.

It is in this more narrow sense that biological control will be discussed here; host plant resistance and autocidal mechanisms such as induced lethal mutations were considered in Chapter 24. However, I will consider viruses as living organisms since they replicate in their hosts, and will include the use of proteinaceaus parasporal bodies formed by some bacteria even though they resemble chemical pesticides in their mode of action.

In general I will adhere to the wisdom of one of the foremost experts in the field, Dr. Paul DeBach, who defined biological control as "the action of parasites, predators and pathogens in maintaining another organism's density at a lower density than would occur in their absence." I will, therefore, consider herbivores as parasites and predators of plants in the biological control of weeds.

## The Ecological Basis

All of the organisms that make up natural ecosystems interact with each other and with their physical environment in such a way that each species evolves an average population, or characteristic level of abundance, around which its numbers fluctuate in relation to other events in the system that affect the relationship between natality and mortality. Although biologists do not normally think of plants as going through population changes because over a period of time their numbers do not change a great deal, it is useful in this context to at least liken changes in the abundance of certain kinds of vegetation to changes in the population of, say, a herbivore or a predator. When conditions are favorable for a particular species, its numbers usually increase (or in the case of a plant, there is above average vegetative growth). Sometimes improved environmental favorability is because of a period of better than average weather, the duration of which is entirely independent of the biological activity within the ecosystem. When such weather conditions continue for a long period, certain organisms increase abnormally and this sets in motion other phenomena, such as increased competition or increased attack by natural enemies, which gradually slow their rate of increase. Eventually, these interactions may bring about a reversal in the trend of population change of some species.

This regulation in the abundance of organisms is often called natural control and, over the long term, tends to maintain populations at their characteristic levels of abundance such that the energy flow through the particular ecosystem is more or less stabilized. Frequently this entire process is referred to as the **balance of nature.**

In the regulatory process, some environmental factors act upon populations of organisms in such a way that their intensity is not affected by the number of individuals present, whereas others vary in their effect in relation to population density. Factors of the first type are commonly said to be **density independent**, and those of the second type, **density dependent.**

Most living organisms interact with each other in a density-dependent way, and in most cases the relationship could be described as **reciprocal** in that the populations of interacting species tend to vary with each other. That is, when a host plant increases in abundance, the herbivores that feed upon it will increase in abundance. At the same time, parasites of the herbivore will increase and eventually be instrumental in causing a decline in the herbivore population, and so on.

In natural communities, the interplay of density-independent and density-dependent factors keeps most populations from attaining extraordinarily high numbers. In agroecosystems and some unstable natural systems, the environment often favors certain kinds of organisms, and some of the regulatory influences may be absent, leading to pest outbreaks. This is particularly true in ephemeral crop environments, because there is never time for well-forged relationships to develop between different trophic levels. In more permanent crop situations involving native species, the characteristic level of abundance of one or more organisms may be higher than desirable relative to man's management objectives. When exotic species are introduced into a new region in the absence of natural enemies, their natality may be high while mortality is substantially reduced, causing population outbreaks to occur. The objectives of applied biological control are to improve the biotic relationships in these situations so as to maintain undesirable species at population levels that are not injurious. Clearly, these situations are different in the difficulties and the control opportunities they present.

The biological control of native pests with native natural enemies poses problems based on long-term relationships. We have to assume that there has been coevolution between the species such that the natural enemies will avoid the problem of over-exploitation of their prey and, therefore, will not suppress the population equilibrium to any extent. At the same time, some crop environments favor an increase in the pest while impeding the efficiency of the natural enemies. In the case of native pests, it may be necessary to concentrate on improving the crop environment for natural enemies or to introduce new enemies that attack relatives of the pest in a different environment, thereby establishing a new relationship. On the basis of fundamental ecological relationships, it would seem easier to implement control against exotic species that are likely to become pests. This has indeed been the case, as most of the successful examples of biological control have resulted from the importation of the natural enemies of foreign pests that were collected from the pests' home ranges.

# Methodology

The methodology of biological control is a complex subject to which several authors have devoted entire books (DeBach, 1964; DeBach, 1974; Huffaker, 1971; and van den Bosch and Messenger, 1973, to mention a few). Biological control is very much an entomological subject inasmuch as the major areas of activity involve insects: control of insects and mites with insects and other animals, control of insects and mites with pathogens, control of weeds with insects, and the control of plant diseases through the biological control of insect vectors of the plant pathogens. In the next few pages I will present some of the fundamental aspects of biological control by using successful examples as a basis for the discussion.

# Insect Control with Parasites and Predators

***The Cottony Cushion Scale***

The biological control of the cottony cushion scale, *Icerya purchasi*, is a success story that has been told many times but bears repeating because of its historical as well as biological importance. The cottony cushion scale was unintentionally introduced into California from Australia in the 1860s and soon posed a serious threat to the state's citrus industry. C. V. Riley, the Chief of the Federal Division of Entomology, believed that the importation of natural enemies from Australia might provide a solution. Albert Koebele, in the guise of a government representative to the 1888 International Exposition at Melbourne, went to make an unofficial search for parasites and predators. In the meantime, an Australian, Frazer Crawford, had found a small fly, *Cryptochaetum iceryae*, parasitizing the pest near Adelaide and sent a small shipment of them to California. The flies were released near San Francisco in early 1888.

While in Australia, Koebel, with Crawford's help, found more *Cryptochaetum* but also discovered a small lady beetle, *Rodalia cardinalis*, commonly called vedalia, predating cottony cushion scale females. In addition to several thousand *Cryptochaetum*, Koebel shipped 129 live vedalia beetles to Los Angeles between November 30, 1888 and January 24, 1889. These and two later shipments that arrived in February and March were released on caged trees at three locations in the Los Angeles area. By June of 1889, more than 10,000 offspring of the original 129 vedalia had been sent to other parts of California. Within months of the release of the vedalia, citrus trees were cleared of the cottony cushion scale. At the same time, *Cryptochaetum* became established and spread, probably from the San Francisco area, to the entire coastal region of California.

From a few years after the introduction of the two control agents from Australia until recently when the system was disrupted by pesticides, the cottony cushion scale rarely attained pest status. Most of the credit for this success has been given to the predaceous vedalia, but DeBach (1974) believes that had *Cryptochaetum* alone been introduced, equally spectacular results would have been obtained.

The successful control of *Icerya purchasi* in California was followed by similar successes in other citrus-growing areas and clearly provided the impetus for an expansion of biological control. In addition to its historical significance, this example provides an introduction to one of the interesting questions in biological control. Do predators or parasites make the best biological control agents?

In more highly evolved parasitic relationships, the host is not usually killed, but in biological control, death of the prey or host individuals is implicit in the definition. Although we normally distinguish between predation and parasitism, they are in a sense different forms of the same symbiotic relationship.

Generally, predators are free-living forms in both the immature and adult stages and often attack the same species of prey throughout their entire life history, although there are a fair number that are predaceous only as immatures or as adults. In most cases predators search out and consume a number of prey individuals in the course of their lives. A lady beetle, for example, may consume several hundred aphids during its development. This is partially because they are active in all stages and are often larger than their prey. There is really no general statement that can be made about prey specificity because it varies widely from species to species, but many predators will attack several prey species and concentrate on the one that is most abundant.

On the other hand, insect parasites are usually free-living as adults and sometimes as first instar larvae but live more or less sedentary lives on or in their host for the rest of their life history. In most species, the adults either do not feed or feed on pollen and nectar rather than on hosts. Host location is a function of the adult female, and once parasitised, a single host provides sufficient food for the development of one or more immature parasites. Although there are exceptions, most insect parasites are rather host specific and consequently have life histories well synchronized with those of their host species so that they do not perish when their host is seasonally absent. Host specificity also requires searching efficiency. Earlier it was assumed that female parasites located the habitat of their host and then searched randomly for hosts, but at least some leave trail odors to avoid searching an area already traversed (Price, 1970)

On the surface, it would seem that predators would be better biological control agents simply because they consume more prey, but this alone would not result in effective prey regulation over a range of population densities. If a predator is unable to find prey when they are present in low numbers or switches to a more abundant prey species, its efficacy is reduced. Some predators, like *Rodalia cardinalis*, are effective control agents because they are specific, have a rapid rate of increase relative to their prey, and the prey are sedentary and gregarious. All insect predator-prey relationships do not display these attributes and most workers now believe that the host specificity, synchronized life histories, and superior searching ability that permits hosts to be located at low densities are attributes that tend to make parasites better biological control agents than predators.

However, it would be foolhardy to exclude predators from the search for useful natural enemies of pests, because a number have been highly effective, and as in the case of cottony cushion scale control, they can work in conjunction with one or more parasites. Furthermore, not all parasites are good control agents. Some are too specific and are unable to survive the absence of their host by attacking an alternate host. Those that require flowers as a source of adult food for egg maturation may disperse away from the host population, but this can be corrected through vegetation management if the requirements of the parasite are understood.

### *The Oriental Fruit Fly*

The Oriental fruit fly, *Dacus dorsalis*, is an Asiatic species which entered Hawaii about the end of World War II. It became a serious pest of ripe fruits of many kinds and posed a threat to the warm fruit-growing areas of the continental United States. Explorers were sent to a number of tropical areas in search of parasites and predators of the fruit fly. Among the variety of natural enemies shipped to Hawaii in 1947 and 1948 were three species of parasitic wasps of the genus *Opius* which at the time were thought to be only two species. One of these, *Opium longicaudatus*, increased rapidly. The second species released was thought to be *O. persulcatus* but later was found to be a mixture of *O. oophilus* and a new species, *O. vandenboschi*. *O. longicaudatus* was dominant at first but soon lost out to *O. vandenboschi*. Later, *O. vandenboschi* was displaced by *O. oophilus* (Clausen et al, 1965). *O. oophilus* alone has since provided effective control of the Oriental fruit fly.

This program, apart from being another successful case of biological control, illustrates that several parasites of the same pest can be released without having a detrimental effect on the over-all control. Although the three control agents were competing for the same host, the one with superior qualities displaced the others and became dominant. In this case, *O. vandenboschi* derived its advantage from attacking first instar larvae and thereby inhibiting the development of the eggs and larvae of *O. longicaudatus*, which favored older host larvae for oviposition. Likewise, *O. oophilus*, which oviposits in the eggs of the host, are already present as larvae by the time hosts are suitable for attack by *O. vandenboschi* (van den Bosch and Messenger, 1973).

A similar sequence of displacements occurred following the release over a number of years of parasites to control California red scale in California. As a result of competition between the parasites different species have become dominant in the climates to which they are best suited rather than because of the host stage attacked, as in the case of *Opius*. At the present time, *Aphytis lingnanensis* is dominant in coastal areas, where it has replaced *A. chrysomphali*. *A. melinus* later eliminated *A. lingnanensis* from the warmer interior areas and remains dominant there. Along the coast another parasite *Prospaltella perniciosi* complements *A. lingnanensis*, whereas *Comperiella bifasciata* complements *A. melinus* in the interior. Both of these complementary parasites, while not dominant, have been able to survive because they attack different host stages from *Aphytis* (DeBach, 1974).

Whereas a succession of parasites led to the establishment of control by a single dominant species quite quickly in the case of the Oriental fruit fly, the California red scale example illustrates that successful control may take a great deal of work and persistence. In the latter case, the first natural enemies were imported in 1889, but a stable situation did not develop until after the introduction of *A. melinus* in 1957.

### *The Winter Moth*

The geometrid moth *Operophtera brumata*, or winter moth as it is commonly called, is a defoliator of hardwood forest and ornamental trees that was introduced to eastern Canada from Europe during the 1930s. As a result of considerable damage to oaks and its gradual but persistent spread, a biological control program was initiated in 1954 with the introduction of six natural enemies from Europe. A tachinid fly, *Cyzenis albicans*, and an ichneumon wasp, *Agrypon flaveolatum*, became established and brought the pest under control in about six years.

This is a case in which there was no displacement. Instead, two species that are compatible and complement each other were able to bring about control. *C. albicans* is apparently very effective at high host densities, whereas the superior searching ability of *A. flaveolatum* makes it effective at low host densities. DeBach (1974) believes that this may have been reflected in the original collections in Europe that were made in areas with local outbreaks in which *C. albicans* was the predominant parasitoid, and indicates that the best control agent may not be the most common one.

The winter moth project is also an interesting example in that it counters some of the criticisms that have been leveled against biological control. Many of the early successes were obtained on islands or in areas with island characteristics (restricted native flora and fauna that offers little resistance to invading or introduced species) and in places with equitable climates. This led many workers to believe that biological control would not be successful in continental areas, but examples such as the Canadian winter moth project have shown this to be an erroneous assumption.

Many of the early successes of biological control were against gregarious, sedentary insects like scales and mealy bugs, and the current record indicates more success against pests of this type than against more active pests like caterpillars, aphids, and beetles. Certainly the scales and mealy bugs have many desirable attributes as targets for biological control, but we must also remember that many programs were patterned after the successful cottony cushion scale program. More recent successes such as that obtained against the winter moth and the walnut aphid in California (van den Bosch et al, 1964) indicate that with conscientious study there may be no limitations to the usefulness of this method.

A number of critics have suggested that often there has been only circumstantial evidence that the introduced natural enemies have actually caused the decline in pest populations and, unless data is collected to show it, biological control should not take the credit for pest reductions. This may be a legitimate criticism in that some introduced species tend initially to increase and spread rapidly and then stabilize; these events could confuse the interpretation of a biological control program initiated during a pest's stabilization phase. The winter moth project was extremely well analyzed and the use of life tables to develop a population model clearly showed that parasitism was the key factor in the control of the moth (Embree, 1971).

## Insect Control with Pathogens

Disease symptoms in insects were recorded by the ancient Chinese and the Greeks, but an experimental demonstration of micro-organisms as the causal agents did not occur until the nineteenth century. Interestingly, the first demonstrative investigations involved the study of diseases of two beneficial insects, the silkworm and the honeybee. An Italian named Agostino Bassi discovered in 1835 that a fungus, *Beauveria bassiana*, caused the white muscadine disease of silkworms. This was followed by Pasteur's famous discovery of a microsporidian as the cause of pebrine and flacherie of silkworms in 1870, and of *Bacillus alvei* as the cause of European foulbrood of honeybees by Cheshire and Cheyne in 1885. These initial discoveries led quickly to attempts to use micro-organisms to control insect pests. Now more than 1,200 species belonging to six major groups of micro-organisms have been found to be associated with insects, and most of them are pathogenic. I will consider briefly the three most important groups: bacteria, fungi, and viruses.

### Bacteria

The Japanese beetle, *Popillia japonica,* is a serious imported pest in parts of the eastern United States where the adults damage the foliage and fruits of more than 250 plant species and the larvae damage the roots of grass, nursery stock, and vegetables. The discovery and application of spore-forming bacteria, as a means of controlling the larvae of the Japanese beetle in the soil (White and Dutky, 1940) provided the first encouragement for the use of bacteria in insect control. *Bacillus popilliae* and *lentimorbus* that cause Types A and B milky disease of Japanese beetle can both be mass-produced and are sold as a spore dust for injection into the soil. Infected larvae that die in the soil become a source of contamination for other larvae feeding in the vicinity. Larval populations can be substantially reduced in this way, and the *Bacillus* spores persist in the soil to infect larvae from generation to generation. However, stabilization of larval populations at satisfactorily low levels takes several years, so some damage has to be sustained after the application of the disease preparation to the soil.

*B. popilliae* and *B. lentimorbus* are obligate parasites, which makes them extremely difficult to propagate on artificial media. The spores must be obtained by culturing the bacteria in host larvae, and thereby makes the dust preparation expensive. A related spore-forming bacteria, *Bacillus thuringiensis,* is a facultative pathogen that infects a variety of insects, including the larvae of lepidopterans, beetles, and flies. This bacteria can be cultured on artificial media and is therefore quite economical to produce.

*B. thuringiensis* produces proteinaceous parasporal crystals as well as spores. These crystals dissolve in the gut of a number of plant-feeding larvae and quickly cause gut paralysis, a cessation of feeding, and later a rupturing of the gut that permits the bacterial spores to enter the insect's tissues where they germinate and reproduce. Commercial preparations of *B. thuringiensis* (Biotrol, Dipel, Thuricide) containing both spores and crystals are registered with the USDA for use on a variety of crops (Table 25-1) as a biological insecticide at material costs competitive with chemical insecticides.

*B. thuringiensis* has several desirable attributes as a means of controlling a variety of foliage-feeding insect pests. The paralytic action of injested parasporal crystals results in a rapid cessation of feeding and the related crop damage; this is an improvement over the way the milky diseases must be used. However, the spores do not seem to persist in the crop environment to provide continuous control from generation to generation. Nevertheless, the inexpensive preparations can be applied as needed, much like a regular insecticide, once a pest population approaches its economic threshold. The rather specific nature of *B. thuringiensis* to kill a few groups of foliage-feeders and not to harm beneficial species is of great value in management programs designed to conserve natural enemies.

There have been a few reports that the repeated use of *B. thuringiensis* spray on the same land results in a residue of spores and crystals that is

TABLE 25-1 Some Registered Uses for *B. thuringiensis* Products in the U.S.A.[a]

| Pest | | Crop |
|---|---|---|
| | *Vegetable and field crops* | |
| Alfalfa caterpillar | *Colias eurytheme* | alfalfa |
| Artichoke plume moth | *Platyptilia carduidactyla* | artichokes |
| Bollworm | *Heliothis zea* | cotton |
| Cabbage looper | *Trichopulsia ni* | beans, broccoli, cabbage, cauliflower, celery, collards, cotton, cucumbers, kale, lettuce, melons, potatoes, spinach, tobacco |
| Diamondback moth | *Plutella maculipennis* | cabbage |
| European corn borer | *Ostrinia nubilalis* | sweet corn |
| Imported cabbageworm | *Pieris rapae* | broccoli, cabbage, cauliflower, collards, kale |
| Tobacco budworm | *Heliothis virescens* | tobacco |
| Tobacco hornworm | *Manduca sexta* | tobacco |
| Tomato hornworm | *Manduca quinquemaculata* | tomatoes |
| | *Fruit crops* | |
| Fruit-tree leaf roller | *Archips argyrospilus* | oranges |
| Orange dog | *Papilio cresphontes* | oranges |
| Grape leaf folder | *Desmia funeralis* | grapes |
| | *Forests, shade trees, ornamentals* | |
| California oakworm | *Phryganidia californica* | |
| Fall webworm | *Hyphantria cunea* | |
| Fall cankerworm | *Alsophila pometaria* | |
| Great Basin tent caterpillar | *Malacosoma fragile* | |
| Gypsy moth | *Lymantria* (*Porthetria*) *dispar* | |
| Linden looper | *Erannis tiliaria* | |
| Salt Marsh caterpillar | *Estigmene acrea* | |
| Spring cankerworm | *Paleacrita vernata* | |
| Winter moth | *Operophtera brumata* | |

[a]From information supplied, in part, through the kindness of International Minerals and Chemical Corp., Libertyville, Illinois and Nutrilite Products Inc., Buena Park, California, U.S.A. (Reproduced from Falcon, 1971 with permission of the author and Academic Press Inc.)

injurious to earthworms, but no evidence of parasite or predator mortality has been found. Mahr and Atkins (1976) fed infected caterpillars to several parasites and predators and found no indications that the bacteria caused any ill effects in the beneficial species. However, in an agroecosystem treated with *B. thuringiensis* there would be a decline in parasite and predator populations in response to the mortality induced in their food supply. In this respect *B. thuringiensis* sprays are much like insecticides.

***Fungi*** Entomogenous fungi belong to all of the four major taxonomic groups of true fungi, but only a few genera are frequently associated with insect disease outbreaks. Those most commonly used in insect control are *Beauvaria bassiana* (white muscardine disease) and *Metarrhizium anisopliae* (green muscardine disease), both of which are fungi imperfect. A natural collapse of chinch bug (*Blissus leucopterus*) populations caused by *B. bassiana* prompted the first, though unsuccessful, attempt to use fungi to control an agricultural pest. However, the project did provide some insight into the problems of using fungi in microbial control.

Most entomogenous fungi are internal pathogens. The infective unit is usually a spore which germinates on the surface of the host's integument. A special rootlike structure penetrates the integument by enzymatic action and mechanical force. Once the host tissue is invaded, the fungus can normally complete its life cycle, but the survival and germination of the spores is critical to the development of an epidemic. Successful penetration is usually followed by a ramification of the body by mycelia which continues until the infected insect is almost filled with fungus. Fruiting bodies then erupt through the integument and produce spores that are released into the external environment. During germination and at least the early stages of mycelial growth, an infected host appears to function normally.

Facultative fungi such as *Beauvaria* and *Metarrhizium* can be cultured on artificial media, thereby facilitating the production of spore preparations that can be used in biological control. As with most biological control agents, fungi can be used for either persistent or short term control.[1] A fungus can be introduced into an area where it becomes established and kills the host year after year similar to the way the Japanese beetle is controlled with milky disease. Alternatively, fungal spore preparations can be used as microbial insecticides similar to the way *Bacillus thuringiensis* is used.

Relatively few attempts have been made to colonize entomogenous fungi. Most projects have involved the redistribution of indigenous fungi or those associated with introduced pests, rather than the importation of foreign species. The best example of attempts to establish new fungal pathogens in disease-free areas involves the introduction of *Coelomomyces* against mosquito larvae, but so far the success has been limited.

On the other hand, the successful use of repeated applications of fungal spores as microbial insecticides have been reported (Müller-Kögler, 1965) as a means of achieving short-term reductions of pest populations. The major limiting factor in initiation of fungal disease in insect populations is the effect of the microclimate on spore survival and germination. The optimal temperature range for the growth of entomogenous fungi is fairly narrow, and relatively high humidity is needed by most fungi to germinate and successfully penetrate their host before they can produce the new spores needed to spread the disease. Spores are also known to be killed by sunlight. Consequently, the application of a spore preparation must coincide with both the presence of susceptible hosts and suitable environmental conditions. Best success can usually be obtained by applying the spores in the absence of sunlight such as on a warm evening following either rain or irrigation which provides the needed humidity.

Burges and Hussey (1971) outlined a number of areas in which further research is needed to optimize the use of fungi in insect control, including: (1) elucidation of aspects of disease induction including chemical constituents of the cuticle that inhibit or enhance spore germination and penetration, (2) the development of fungal strains suitable to microbial control, (3) development of methods for maintaining virulence in mass cultures, (4) standardization of microbial preparations, (5) further elucidation of environmental effects on the pathogens, and (6) determination of the effects of combinations of fungi, fungi with other pathogens, or fungi with sublethal doses of insecticides.

[1] Even parasites and predators can be used to inundate an area as a means of reducing a pest population temporarily without concern for permanent establishment of the control agent. For example, backswimmers (Hemiptera: Notonectidae) can be introduced into temporary ponds for the reduction of mosquito populations.

***Viruses*** The majority of insect pathogenic viruses are different from most viruses in that the virus particles or **virions** are enclosed in proteinaceous capsules or membranes. Such viruses are called **inclusion viruses,** as opposed to **noninclusion** viruses in which the virions are free within the cells of the host.

There are two main types of inclusion viruses, the **polyhedroses** and the **glanuloses.** The polyhedroses are characterized by a number of virus particles embedded in polyhedral-shaped protein matrix, whereas the glanuloses have single virus particles, each enclosed in a separate protein coat or capsule.

The polyhedroses are divided in nuclear and cytoplasmic forms, based on the site of virion multiplication within the cells of the host. Nuclear polyhedroses, most of which have been found in caterpillars and the larvae of sawflies, affect the epidermis, fat body and blood cells. Virus particles may infect the host through the mouth or the cuticle or be passed from one generation to the next on or within the egg. Once nuclear polyhedrosis viruses start to multiply in the host larval tissue, the larvae become sluggish but often crawl to high-branch tips where they die hanging by the prolegs at the tip of the abdomen. At this stage, the larval skin becomes fragile and ruptures easily, releasing a shower of virus particles onto the foliage below. Cytoplasmic polyhedroses cause diseases of caterpillars and the larvae of a few lacewings (Neuroptera). They mainly invade the cells of the alimentary canal, and infected larvae often become white and swollen. The virus is spread when polyhedral bodies are regurgitated or passed out with the feces.

Granulosis viruses are restricted to the larvae and pupae of Lepidoptera, and the fat body is the main tissue attacked. The virions first multiply in the nuclei but later continue to replicate in the cytoplasm. The disease eventually kills the insect, leaving it hanging as a fragile sack of virus similar to that which results from nuclear polyhedroses infection.

There are a few noninclusion viruses that attack specific insects, but with the exception of Tipula Iridescent Virus and Mosquito Iridescent Virus that might prove useful in mosquito control, most attention has been given to the inclusion viruses. The fact that virus particles encapsulated in a protein matrix maintain their infectivity for many years means that the inclusion viruses can be stored as concentrated preparations for later application with conventional pesticide spray equipment.

The production of large quantities of virus is often difficult because they normally must be obtained from infected hosts, thus involving an extensive rearing program. Although a supply of virus can be accumulated over a period of time, they often lose virulence in storage, so recently propagated virus gives the best control. Field trials involving a virus application have produced rather variable results, but this technique promises to be an effective means of insect pest control.

Work by a number of Canadian scientists has shown that a nuclear polyhedroses virus is highly effective against a variety of forest sawflies, and is particularly encouraging in that the virus has persisted in the environment and provides continuous regulation of the pest in some areas. However, one of the problems with viruses is that there are periods when they seem to have little effect on pest populations. Naturally occurring virus may remain latent in a pest population for several generations and then develop epizootics when the pest population comes under some form of stress. Just what causes viruses to behave in this manner is not understood, and more information in this area would improve pest control with viruses. Generally, short-term control can be gained by frequent applications of virus preparations so that there is an active innoculum in the pest environment for an extended period.

Several nuclear-polyhedrosis viruses are being produced on a large scale for possible use against a variety of pests, including cotton bollworm, tobacco budworm, corn ear worm, cabbage looper, forest tent caterpillar, and alfalfa butterfly. However, virus preparations can be used experimentally in the United States only if they are properly registered, as is the case with insecticides and bacterial preparations. So far, progress in the registration of viruses for insect control has been rather slow.

## The Biological Control of Weeds

The control of weeds by various biological agents can be an effective alternative to cultural and chemical techniques, especially in accessible areas or on land with low agricultural value. However, weeds are often useful plants elsewhere or are quite closely related to economically important species. Consequently, the use of a biological agent to reduce the numbers of a weed is an antithesis of normal insect or disease control. For this reason, there has been some reluctance to fully exploit the method, and there is no question that it must be employed with utmost care. Nevertheless, there have been some spectacular successes.

Inasmuch as plants form the foundation of all food chains, there is a wide range of organisms that feed upon them, but in natural communities, some balance has evolved between producer and consumer organisms. Consequently, the biological control of weeds, as with insects, has usually been employed against alien species or species that have escaped natural regulation. Potentially, at least, weeds can be controlled by a variety of vertebrates, snails, insects, mites, parasitic plants, and plant pathogens, including bacteria, fungi, viruses, and nematodes. So far, most success has been achieved using phytophagous insects.

Before an insect herbivore can be introduced into a new area for weed control, it must be determined beyond a doubt that it is specific to the weed species and will not switch its feeding habits to some other plant that is desirable. Insects vary considerably in their host plant specificity, and the degree to which they are tied to a particular host cannot be determined by observing it in its natural situation. For the purpose of weed control, an insect should be monophagous or at least restrict its feeding to a few related plants of no economic importance. To safely determine this requires rather exhaustive feeding tests under experimental conditions.

### *Klamath Weed*

The control of Klamath weed or St. Johnswort, *Hypericum perforatum*, by an insect herbivore provides one of the best examples of biological weed control.

A native of Europe and Asia, Klamath weed has invaded a number of subhumid areas of the world where it displaces desirable forage plants and interferes with the raising of livestock because of its poisonous nature. The weed was first discovered in northern California in the early 1900s and quickly spread over a considerable area of valuable rangeland. Although the weed could be controlled by chemical and cultural methods, this was generally too costly over large and inaccessible areas. A similar occurrence in Australia led to the search for insects that feed on the weed in parts of its natural range.

The search in Europe for potentially useful natural enemies of Klamath weed led to the discovery of three species of leaf-feeding beetles (Chrysomelidae), and after thorough testing they were shipped to Australia. In 1945, the beetles were imported from Australia to California for colonization. Within three years, two of the beetles (*Chrysolina* spp) had become established and were causing severe defoliation of the weed in the areas where they had been released. The beetles spread naturally and were also distributed artificially to other areas of California infested by the weed. By 1956, Klamath weed in California was effectively suppressed (Figure 25-1).

The widespread devastation of a host plant by an introduced insect, as has occurred in this case, often raises fears that the insects will respond to the lack of food by switching to another host. However, if the specificity of the herbivore has been properly tested prior to the introduction, there is little danger of this happening because the riciprocal relationship between food supply and population eventually results in a decline in the herbivore population. Some Klamath weed still exists in shaded locations where *Chrysolina* does not do particularly well, and this provides a small reservoir of food that maintains a low population of the insect. As the weed expands into sunnier sites, the insect population increases and suppresses it and good control of the weed still persists.

Outside of California, Klamath weed control by the leaf beetles has been less successful, but this seems to be a feature of biological control. Although there have been a number of complete successes, there also have been a number of partial successes and, of course, some failures. Climate is always an important consideration, but there also are numerous kinds of biological interference that can occur when new species are introduced into complex communities. Goeden and Louda (1977) in a review of biotic interference with insects imported for weed control indicate that this may be a very important aspect of the method.

A list compiled by the National Research Council (1968) shows that 25 serious weed pests have been controlled by insects to various degrees in a variety of insular and continental areas. The most effective insects in weed control belong to the orders Lepidoptera, Hymneoptera, Coleoptera, Hemiptera, and Diptera. Those that attack aboveground vegetative parts generally seem to be more effective than those that attack roots and seeds, although complete control of puncture vine on the island of Kauai was achieved with a seed-feeding weevil.

As yet, pathogenic organisms have not been seriously tried as a means of controlling weeds, although they have considerable potential. Accidentally introduced diseases such as chestnut blight, white pine blister rust, and Dutch elm disease have almost completely destroyed several desirable plants in certain areas, and the weed pamakani was controlled in Queensland following the inadvertent introduction of a fungus. As with insects, plant pathogens would have to be subjected to vigorous screening to determine their host specificity before being released for weed control. The complex interplay of factors such as climate, conditions needed for infection, size and location of the innoculum, and the availability of susceptible hosts may make diseases less reliable unless supplemented by laboratory-maintained cultures. The use of certain plant diseases vectored by insects could prove to be a successful approach in that the insects would act as effective dispersal agents and would aid in the infection process.

The most recent summary of the status of projects on the biological control

**FIGURE 25-1.** Biological control of Klamath weed by *Chrysolina quadrigemina* in California. **A.** High population of the weed in flower in 1946, **B.** Portion of the same field in 1949 after release of *Chrysolina* showing heavy cover of grass, **C.** Same field in 1966 showing full control of the weed.

of weeds in the United States and Canada was prepared by Goeden and others (1974). The list includes 78 species of weeds that are the subject of biological control, involving about 110 species of insects and a variety of plant pathogens. Projects involving a total of 34 kinds of weeds have progressed to the stage at which natural enemies have become established. The effect of the introduced insect herbivores on the weed population has been evaluated for 19 weed species. Table 25-2 lists the six weeds against which biological control has been completed, and the biological control agent involved.

TABLE 25-2 Successful Projects on the Biological Control of Weeds by Insects in the United States and Canada

| Target Weed | Biological Control Agent | Family | Order |
|---|---|---|---|
| Canadian thistle *Cirsium arvense* (L.) Scop. | *Altica carduorum* Guenee | Chrysomelidae | Coleoptera |
| Pamakani *Eupatorium adenophorum* Spr. | *Procecidochares utilis* Stone | Tephritidae | Diptera |
| St. Johnswort or Klamath weed *Hypericum perforatum* L. | *Chrysolina* spp | Chrysomelidae | Coleoptera |
| | *Agrillus hyperici* Cruetzer | Buprestidae | Coleoptera |
| | *Zeuxidiplosis giardi* (Kieffer) | Cecidomyidae | Diptera |
| Prickly pear *Opuntia littoralis* (Engelm.) Ckll. | *Cactoblastis cactorum* (Berg) | Pyralidae | Lepidoptera |
| Prickly pear *Opuntia oricola* Philbrick | *Dactylopius opuntiae* Lichtenstein | Coccidae | Homoptera |
| | *Chelinidea* spp | Coreidae | Hemiptera |
| Tansy ragwort *Senecio jacobaea* L. | *Hylemya seneciella* (Meade) | Arctiidae | Lepidoptera |

# Population Replacement

A somewhat different form of biological control potentially useful against insects and pathogens involves the replacement of an injurious population with an innocuous or less harmful one. The methodology may involve the gradual replacement of a pest by an introduced species or the eradication of a pest by some other means of control, followed by the introduction of a more desirable species to fill the void and prevent re-establishment of the original pest.

The concept of competitive displacement is based on Gause's principle, which states that two species with the same ecological requirements (ecological homologues) cannot coexist over an extended period of time. Eventually, the species with the highest rate of increase will survive. This does not imply that the pair of species must have identical requirements, but they must actively compete for some life requisite such as food, nesting sites, oviposition sites, and so on. There are a number of examples of species replacement that have occurred more or less naturally. The introduced pest *Pieris rapae* unfortunately replaced the native *P. oleracea* throughout much of North America. In this case, the species that took over is more troublesome.

Several possibilities for pest control exist. It may be possible to replace vectors of plant, animal, or human diseases by species that are nonvectors or species that attack other plants or animals so as to reduce their nuisance. Several examples have occurred accidentally. The antimalarial campaign in Sardinia reduced the population of the vector, *Anopheles labranchiae,* to the point where it was replaced by a normally more rare nonvector, *A. hispaniola.*

The chemical control of *A. funestus* in East Africa resulted in its replacement by *A. rivulorum* that attacks cattle instead of man. The use of a latent form of virus has been suggested as a possible means of prevention of the infection of plants by a more serious virus.

## General Considerations

Biological control presents the opportunity to suppress many pests of forestry and agriculture as well as medically important species with a minimal amount of ecological disruption. It can be particularly effective when applied as a colonization technique that reunites pest species, that have become established in new geographic areas, with their natural enemies. Once a biological control agent has become established, it becomes a part of the forest or crop ecosystem and serves in the regulation of the pest species in a continuous self-perpetuating manner. Often the need for future human intervention is eliminated or markedly reduced, as opposed to chemical control which is characterized by repeated intervention. Because biological control agents are fairly specific, they have little or no adverse effect on beneficial species, and there is little danger of pest resurgence or secondary pest outbreaks as a result of their activity.

The development costs of most biological controls are paid largely from public funds but, in many cases, are less than the research, development, and production costs of an insecticide. Sometimes the cost is very small relative to the savings that accrue once the program has been implemented. A striking example is provided by the citrophilus mealybug project which is estimated to have cost less than $2,600 when implemented in 1928 and to have resulted in savings to the citrus industry amounting to between $1,000,000 and $2,000,000 annually (Compere and Smith, 1932) with no additional input of funds. There are, of course, some costly failures, but no control tactic is a guarantee against losses.

The fact that once established, biological control provides ongoing damage suppression with little or no additional cost makes it a highly desirable approach for the protection of crop values that accrue slowly as in forestry or the protection of low-value resources such as grazing land. For example, the chemical control of Rhodesgrass scale, *Antonina graminis* (actually a mealybug), was economically unfeasible even though it caused serious damage to quality rangeland. The suppression of the pest to subeconomic levels through the introduction of the parasite *Neodusmetia sangwani* from India provided a highly satisfactory solution.

Inasmuch as biological control often results in substantial savings and frees growers from the annual expense of chemical control, it is surprising that more money for research and development is not available. Perhaps it is about time that growers be required to pay a small part of the savings they derive from biological control into a fund to support further research.

Not all agroecosystems are suitable for the establishment of natural enemies for ongoing pest regulation. As pointed out by Southwood and Way (1970), the best opportunities for mimicking nature occur in more permanent, complex ecosystems such as forests, orchards, plantations, vineyards, and other similar systems characterized by diversity, temporal continuity, reasonable climatic stability, and a minimal degree of isolation. Conversely, the use of colonized natural enemies is more difficult and less effective in short-term monocultures more or less isolated from other similar systems. Baker and Cook (1974) suggest that at least for the biological control of plant pathogens, the opportunity in monocultures is increased by large-scale agriculture, since the expenses of preparing the system for biological control can be borne more easily than is the case for a small grower.

However, biological control methods involving the use of mass-reared parasites and predators for temporary pest suppression by inundation can be used in small, short-lived agroecosystems. Insect pathogens, particularly *Bacillus thuringiensis*, are also useful in crops that do not lend themselves to natural enemy colonization. With the exception of self-perpetuation, this approach encompasses all of the desirable attributes of biological control.

Nevertheless, there has been some resistance to the use of biological control on the part of some growers who have grown accustomed to temporarily pest-free fields following the use of conventional chemical control. There seems to be some difficulty in accepting the fact that the objective of biological control is rarely eradication. The goal of merely suppressing the pest population below its economic threshold usually means that there is some evidence of pest presence, and some growers seem unable to accept this. Furthermore, permanent, self-perpetuating biological control cannot be achieved overnight, and during the establishment period, the project cannot be disrupted by the use of broad-spectrum pesticides. This may mean that some damage has to be tolerated for a year or so and requires that the growers be educated to the fact that the long-term results are worth waiting for. The use of biological insecticides during this period can be of considerable value and may also serve as a very persuasive tool.

## References Cited


Baker, K. F. and R. J. Cook. 1974. Biological control of plant pathogens. W. H. Freeman and Company, Publishers, San Francisco, Ca.

Burges, H. D. and N. W. Hussey. 1971. Microbial control of insects and mites. Academic Press, Inc., New York, N. Y.

Clausen, C. P., D. W. Clancy, and Q. C. Chock. 1965. Biological control of the oriental fruit fly (*Dacus dorsalis* Hendel) and other fruit flies in Hawaii. U. S. *Dept. Agric. Tech. Bull. 1322.*

Compere, H. and H. S. Smith. 1932. The control of the citrophilus mealybug, *Pseudococcus gahani*, by Australian parasites. *Higardia* 6: 585–618.

DeBach, Paul. 1964. Biological control of insect pests and weeds. Van Nostrand Reinhold Company, New York, N. Y.

DeBach, Paul. 1974. Biological control by natural enemies. Cambridge University Press, New York, N. Y.

Embree, D. G. 1971. The biological control of the winter moth in eastern Canada by introduced parasites. Pages 217–268, *in* C. B. Huffaker, ed. Biological control. Plenum Publishing Corporation, New York, N. Y.

Falcon, L. A. 1971. Use of bacteria for microbial control. Pages 67–96 *in* H. D. Burges and N. W. Hussey, editors. Microbial control of insects and mites. Academic Press, Inc., New York, N. Y.

Goeden, R. D., L. A. Andres, T. E. Freeman, P. Harris, R. L. Pienkowski, and C. R. Walker. 1974. Present status of projects on the biological

control of weeds with insects and plant pathogens in the United States and Canada. *Weed Sci.* 22: 490–495.

GOEDEN, RICHARD D. and SVATA M. LOUDA. 1977. Biotic interference with insects imported for weed control. *Ann. Rev. Entomol.* 22: 325–342.

HUFFAKER, C. B. (ed.) 1971. Biological control. Plenum Publishing Corporation, New York, N. Y.

MAHR, D. L. and M. D. ATKINS. 1976. Effect of *Bacillus thuringiensis* infected hosts on some entomophagous insects. (unpublished)

NATIONAL RESEARCH COUNCIL. 1968. Principles of plant and animal pest control. Vol. 2. Weed control. National Academy of Sciences, Washington, D. C.

PRICE, P. W. 1970. Trail odors: Recognition by insects parasitic on cocoons. *Science* 170: 546–547.

SOUTHWOOD, T. R. E. and M. J. WAY. 1970. Ecological background to pest management. Pages 6–29, *in* R. L. Rabb and F. E. Guthrie. Concepts of pest management. North Carolina State University, Raleigh, N. C.

VAN DEN BOSCH, R. and P. S. MESSENGER. 1973. Biological control. Thomas Y. Crowell Company, New York, N. Y.

VAN DEN BOSCH, R., E. I. SCHLINGER, J. C. HALL, and B. PUTTLER. 1964. Studies on succession, distribution, and phenology of imported parasites of *Therioaphis trifolii* (Monell) in Southern California. *Ecology* 45: 602–621.

WHITE, R. T. and S. R. DUTKY. 1940. Effect of the introduction of milky diseases on populations of Japanese beetle larvae. *J. Econ. Entomol.* 33: 306–309.

# Integrated pest management—an ecological approach

# 26

In the foregoing chapters I have presented some of the important methods of pest control along with some of the major principles underlying them. Some of the methods are time-tested, whereas others have great potential but are as yet unproven. Each of these approaches to pest control either has or will have an important place in pest suppression; each can be used alone and all are subject to failure if improperly used or abused.

For the most part the methods discussed have been used in an ad hoc manner, frequently with great success. But sometimes less than the desired results have been achieved, and far too often new pest problems have arisen out of an empirical approach to eliminating old pests. Chemical control stands out as the method that has produced enormous benefits on the one hand, and some catastrophes on the other. The catastrophes, in the form of ecological disruption, pest resurgence, secondary pest problems, and the induction of pesticide resistance (discussed in Chapter 23) have stimulated a re-examination of pest control methodology, particularly the wisdom of using any one tactic separately and to the exclusion of all others.

The development of pesticide-resistant strains of pests and the appearance of biotypes that overcome host resistance provide ample evidence of the genetic plasticity of insects. Pest control tactics are a strong selective pressure to which the target organisms are bound to respond. Experience has shown that we cannot expect any single form of pest control to be a permanent solution. The development of biotypes that do not respond to attractants, that circumvent male sterilization by becoming parthenogenic, that become resistant to primary natural enemies, or that adjust to certain cultural practices looms as a distinct possibility. We should never again make the mistake of assuming that pest resistance will not occur. Over the past few decades we have grown to realize that we must replace empirical control methods with an approach based on sound ecological and evolutionary principles. What is commonly referred to as **integrated control** or more appropriately, **integrated pest management,** is such an approach with the potential of providing the solution we are seeking.

The literature contains many definitions of integrated pest management. The early definitions stressed a blend of biological and chemical control, and this is still a popular view. However, some of the newer definitions, such as the one proposed by an FAO panel in 1967, stress that an integrated pest management strategy is a combination of all suitable techniques. The FAO experts defined integrated pest management as:

> a pest management system that in the context of the associated environment and the population dynamics of the pest species utilizes all suitable techniques and methods in as compatible a manner as possible and maintains the pest populations at levels below those causing economic injury.

Keeping in mind the need for systems based on sound ecological principles, an alternative definition would be, *the combination of as many suitable control methods as is practical into an ecologically harmonized system, designed to maintain pest populations at levels below those which would cause economically significant losses to agriculture and forestry or endanger human health.*

The definition does not exclude the use of pesticides, but it is implicit that the control methods used must be reasonably compatible if a harmonized system is to be the result. The value of chemicals in pest control is clearly established and recognized, but it is also clear that the use of chemicals should be substantially reduced. Furthermore, the multifaceted nature of integrated pest management reduces the intensity with which any single method acts upon a pest population, so there is less selective pressure leading to the development on new biotypes that are difficult to control.

The practice of integrated pest management is not new. Before the modern synthetic pesticides provided growers with the capability of reducing pest populations to near zero at very little cost, they used a combination of methods to reduce crop losses. In many cases pest control was an art, an art that was unfortunately pushed into the background by the proliferation of pesticides. The farmers who developed their own forms of integrated pest management did so largely on the basis of a trial-and-error approach without fully understanding the underlying ecological principles. It is rather interesting that many of these farm-grown programs involved cultural practices that made the crop environment less favorable to the pests and/or more favorable to their natural enemies.

Inasmuch as biological control was a well-established alternative to chemical control, it is not surprising that the first conceptualization of integration involved mainly a combination of these two approaches. Some programs still lean heavily toward biological control that is carefully augmented by the use of chemicals. But as the concept has evolved, various cultural methods, plant resistance, pheromones, and autocidal methods have all been considered as a part of integrated pest management programs. In some programs, chemical control is an important component; in others, chemicals are not used at all.

The old term *integrated control* is variously applied, but the concept has become so widely used and publicized that there seems no practical value in placing unnecessary restrictions on its meaning. For example, there is no need to distinguish between pest suppression and eradication when either is achieved by a multiplicity of control methods. Nevertheless, some workers consider eradication to be incompatible with the integrated control concept because the total elimination of a species reduces diversity. However, species designated for eradication are often introduced species. If they can be eliminated from an ecosystem in which they do not belong more efficiently by combining several control methods, then an integrated approach should be used.

In order to maximize the efficacy of an integrated pest management program, it is necessary to understand the whole crop or forest ecosystem of which the complex of pests is a part. It is necessary to recognize that several pests of varying importance may be present and to have adequate knowledge of each pest's population dynamics. The multiplicity of factors that influence a species population density has been described as its life system (see Figure 17-9), but usually such systems are imperfectly known. Consequently, integrated pest management programs still tend to be rather empirical.

Certainly the successful application of integrated pest management is dependent upon an understanding of the forest or agroecosystem and the biological relationships of the pests within it. Southwood and Way (1969) discussed the important characteristics of forest and crop ecosystems that influence pest population dynamics and also form a basis for the selection of suitable control methods. Two of the important considerations concern the size and delimitation of the ecosystem to be treated. As mentioned earlier, large monocultures form large targets for dispersing pests and natural enemies. On the other hand, large areas devoted to a single crop are less affected by the composition of adjacent ecosystems or the natural occupants of field margins, roadways, and other peripheral areas. In such cases, the flora and fauna of adjoining areas influence the ecology of the marginal portions of the crop considerably more than the more isolated central portions. Conversely, a small crop area is heavily influenced by adjacent ecosystems. Any integrated pest management program must take these features into account because accessible wild areas form refuges for both pests and beneficial organisms when the crop ecosystem is disrupted. Depending on the pest problem and the choice of control tactics, it may require a decision as to whether or not the control program should be extended into the marginal ecosystems.

Crop permanence is another feature that impacts the selection of pest control methods. In fairly long-lived situations such as vineyards, orchards, or groves, there is less need for organisms to disperse in and out, so the major pests are frequently less vagile than those of annual crops. Such agroecosystems tend to become more diverse and stable with time and provide ideal situations for the application of biological control. The permanence of the system permits the implementation of management procedures that conserve and augment natural enemies. Parasites, predators, and pathogens can be introduced with the objective of having them become permanently established in the ecosystem. On the other hand, the permanence of a crop may encourage the establishment of some pests such as root pathogens.

Southwood and Way graphically summarized the important characteristics of a range of agroecosystem types that have some bearing on the selection of control methods. The chart and some of the authors' generalizations are reproduced in Figure 26-1.

Ideally, when an integrated pest management program is formulated, the entire pest complex should be considered at once. Entomologists, plant pathologists, and weed control experts are generally adopting this view for the control of pest insects, pathogens, and plants. However, there are few cases where all types of pests are actually considered together. This should be the ultimate goal of any program designed for a specific crop ecosystem, even though it may complicate the situation considerably.

Generally, entomologists seem to have formalized the multiple pest approach to integrated pest management more than have other specialists. However, plant pathologists often attempt to control a variety of diseases at one time, and most forms of weed control are not particularly specific.

Chant (1964) categorized insect pests into four classes. Class I pests are those which appear perennially and normally attain population levels that would result in economic damage unless some control measure is implemented. These so-called **key pests** usually dominate the development of a control program; when chemical control is used alone, these insects are the principal target. Class II pests are species that are usually adequately controlled by natural enemies but which occasionally escape natural regulation and develop to a density at which economic damage results. Class III pests are those that normally do not cause significant damage even though they may occur in large numbers. However, changes in an agroecosystem, such as might

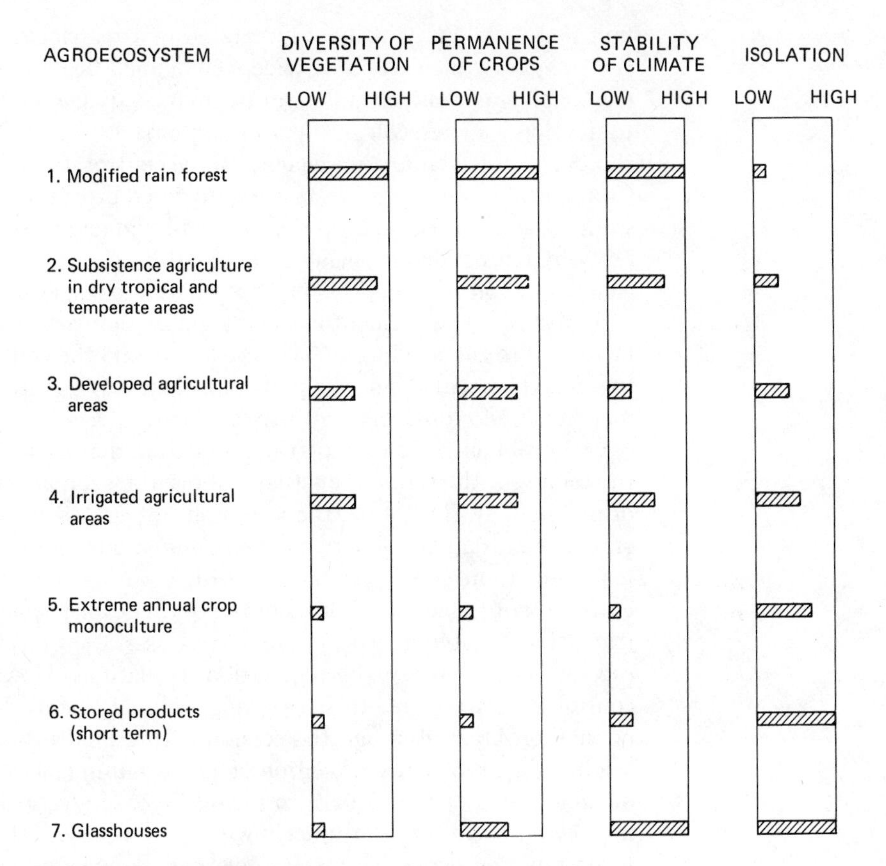


**FIGURE 26-1.** The characteristics of several major types of agroecosystems of importance in pest management. (Reproduced from Southwood and Way, 1970, courtesy of North Carolina University Press.) The following generalizations are possible:

1. The need and opportunities for manipulation as opposed to mimicking nature, increase through the series of agroecosystems 1–7.
2. The danger of upsetting natural biological control through the use of pesticides decreases through the series of agroecosystems from 1–5 and 6.
3. The opportunities for manipulated biological control in the isolated agroecosystems (4–7) are highest in greenhouses (7), moderate in irrigated areas (4) and lowest in monocultures (5) and stored products (6). They are directly correlated with the stability of the climate.
4. The role of "planting date" as a management procedure is greatest in those agroecosystems towards the bottom of the figure, for here the problems of invasion are greatest (because of degree of isolation), and the crops can be managed so that they pass through the susceptible stage simultaneously. This is a particularly powerful approach when the plant is only susceptible to or suffers damage from the insect for a brief period.

result from the use of a pesticide, can result in these species becoming troublesome. Class IV pests are nonresident species which may invade a crop ecosystem and cause economic damage. The invasion by Class IV pests is often unpredictable. When it occurs, direct control action that is disruptive to any ongoing integrated management plan may become necessary.

Just as chemical control programs were usually based on the most important pests, the same is true of integrated pest management. These then are the pests for which we must have well-defined economic thresholds and a reasonably good understanding of population dynamics. However, the need to develop a more ecologically sound approach to control may not allow a thorough study before some changes in the control strategy are implemented. Consequently, more often than not, integrated programs are initiated as pilot studies based on common sense and a fundamental understanding of the ecosystem function. In many respects, this is a practical approach, since changes in control practices tend to generate changes in the agroecosystem and subsequent adjustment in the integrated pest management program can be related to them. Thus, the agroecosystem and the control program evolve together. In essence then, the pest control activity goes beyond pest management to total crop ecosystem management.

Pest problems arise because the population density is either always above the economic threshold or fluctuates above the economic threshold regularly or occasionally. Therefore, the aim of an integrated pest management program depends upon the nature of the population suppression needed. In the first case, it may be necessary to reduce the over-all favorability of the environment to reduce the level of the pest's normal population equilibrium (characteristic abundance). In the latter cases, the objective of the control program is to dampen the population oscillations. However, even when a control program seems to be working well, situations develop in which a population breaks through the economic threshold unexpectedly. This can result from a temporary relaxation of a constraining factor or from an unpredictable influx of individuals from elsewhere. Under such circumstances, it may be necessary to react quickly with an insecticide. If an integrated pest management program has taken some time to develop to a satisfactory state, the degree to which the use of a pesticide may disrupt the program must be considered; it may be better to absorb the loss to the crop rather than sacrifice the integrated program.

If the use of chemicals appears to provide the best solution to a temporary problem or is necessary as a component of an integrated program, there are a number of ways they can be used to minimize their disruption of the normal functioning of the crop ecosystem. When a chemical application is necessary, a highly specific pesticide provides the least disruption, but usually no such compound is available. Consequently, steps must be taken to apply broader-spectrum compounds in such a way that the efficacy of the other control tactics in the program is not diminished. For instance, Bartlett (1964) discussed several ways to reduce the impact of pesticides on natural enemies:

(1) Preservation of natural enemies outside the treated area. Although it is better to conserve the natural enemies within the crop area, it is not always possible. The ultimate decision will depend upon the dispersal and searching characteristics of parasites on predators. For example, *Tetrastichus julis* is a good, gregarious parasite of the cereal leaf beetle, but it has poor powers of dispersal and cannot keep pace with shifting populations of the pest (Haynes, 1973). Such a parasite could fit nicely into an integrated pest management program involving insecticide use, because it can be protected, moved, and manipulated in a manner compatible with the application of insecticides.

Natural enemy populations can be maintained in environments adjacent to the crop or in parts of the crop that do not need to be sprayed. The natural vegetation of field margins, roadsides, and irrigation ditches can provide valuable refuges. A crop can also be spot-treated. DeBach and Landi (1959) applied a nonselective oil spray to alternate pairs of citrus trees at six-month intervals to reduce purple scale populations without eliminating the natural enemies. In Israel, only strips baited to concentrate adult Mediterranean fruit flies were treated with malathion in order to protect scale parasites formerly killed by blanket applications (Harpaz and Rosen, 1971). An effective, but less desirable, alternative is to culture natural enemies in insectaries and to release them in the crop once the danger of the pesticide has subsided.

(2) Selectivity based on differential susceptibility. Adult parasites are generally more susceptible to insecticides than are other stages. On the other hand, the nymphs or larvae of many herbivores are often more susceptible than adults. It may be possible to arrange the timing of spray applications to coincide with the presence of susceptible stages of the pest and less susceptible stages of their natural enemies.

(3) Selectivity based on mode of feeding. Since parasites and predators do not feed on the vegetative parts of plants, systemic poisons are preferable to contact poisons in integrated programs. The pest thereby acquires the poison with its food while the natural enemies are protected. However, natural enemies may be killed if they feed on hosts that have accumulated the toxic material.

(4) Selectivity based on life histories and habitats. It may be possible to time the application of a pesticide when the pest is in a stage of development not under attack by the more important natural enemies or when it is in a portion of the environment not frequented by the beneficial species.

(5) Selectivity based on characteristics, formulation, or application of the pesticide. If the choice of pesticide must be made from several nonselective compounds, a nonpersistent one would be preferred in order to reduce the risk of natural enemy exposure. Since the objective of integrated programs is to suppress a pest below the economic threshold rather than achieve in excess of 90% mortality, it is often possible to reduce the dosage and thereby reduce the danger to beneficial species. However, some parasites are so susceptible to poisons that they will be killed by dosage levels that permit the pest to survive, thereby aggravating the problem.

Although not much is known about the effect of the formulation of pesticides on beneficial species, some obvious possibilities exist. For example, seed dressings, granular formulations, or soil fumagents applied against pests in the soil environment would be generally less harmful to parasites and predators that are active above the soil.

Some groups of insecticides are generally nontoxic to parasites and predators. Petroleum oils, commonly used against phytophagous mites and scale insects, are not toxic to most parasites and predators. Biological insecticides such as *Bacillus thuringiensis* or special virus preparations are particularly useful in integrated pest management programs because they are specific to herbivorous species.

There is, however, one important aspect of the use of any insecticide in integrated pest management. Regardless of the care taken to reduce the direct impact of poisons on natural enemies, the desirable species will be affected by the reduction in the number of hosts or prey available. There may be a rather critical balance between an adequate or inadequate number of the pest species relative to the needs of the natural enemy population. If a pesticide application seriously depletes the food resource available to parasites and predators, they may die out or leave the crop environment.

Although existing pesticides can certainly be used in a manner more compatible with other control methods, there is still a great need for new chemical compounds that are more specific, safer, less persistent, and biodegradable. It is unfortunate that with the vast increase in the technological capability of the chemical industry over the past few decades that more desirable pesticides have not been developed. But the turning point could be approaching as the potential of some of the new control methods becomes more fully realized, and therefore more profitable.

Gonzalez (1970) likened developing an integrated pest management program to building a house. Sampling, the establishment of economic thresholds, and understanding natural mortality in agroecosystems form the foundation. The variety of control methods available, such as insecticides, biological control, host resistance, and autocidal methods form the studs. The integrated program forms the roof. How well the roof holds up depends upon the strength of the foundation and the number of studs that support it (Figure 26-2).

### *Examples of Integrated Control*

The formalization of integrated pest management as an approach to pest suppression in most crops is rather recent. In most cases, the need to implement controls less reliant on pesticides has precluded the expenditure of time and effort really necessary to compile a reasonably full understanding of the

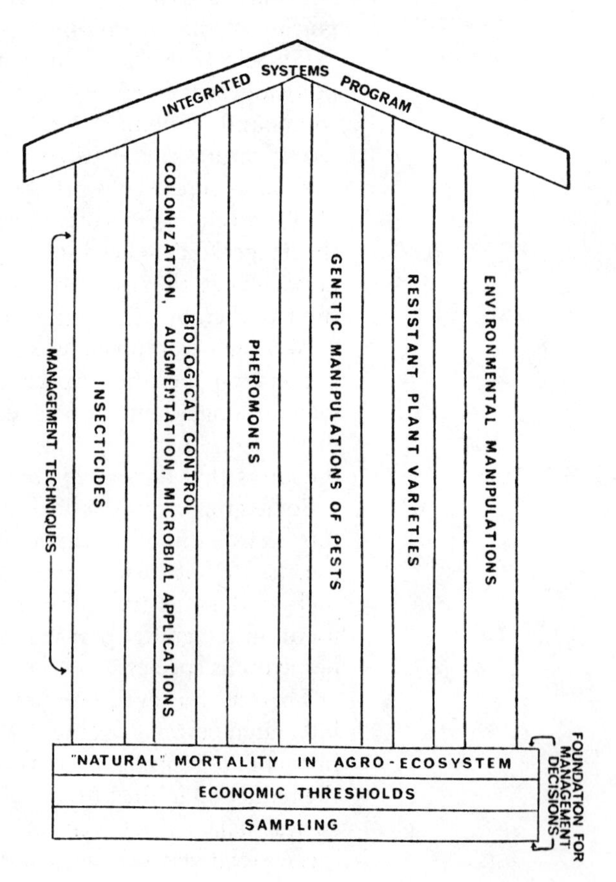


FIGURE 26-2. Diagrammatic representation of an integrated pest management program. (From Gonzalez, 1970, reproduced with permission of the author.)

population dynamics of the pest species. Consequently, current integrated pest management programs are being developed in a step-by-step manner. None can be considered as being in anything like a final format, and the cases given as examples should be viewed as representing only a stage in the process of dynamic development. The purpose of the examples is to illustrate how various tactics can be harmonized into a more ecologically sound approach to pest control.

*Cotton pest control in the Cañete Valley, Peru:* This story has been related many times, but it is worth repeating because it clearly shows the detrimental impact of empirical pest control with broad-spectrum insecticides and the recovery that can result from the implementation of a more ecologically sound program (Doutt and Smith, 1971).

The Cañete Valley is an irrigated coastal valley isolated from a number of similar agricultural areas by rugged topography and arid uplands bearing native vegetation. In the 1920s the farmers of the valley switched their emphasis from sugar cane to cotton. Prior to the advent of the broad-spectrum organochlorine insecticides, six prominent pests, including the bollworm, *Heliothis virescens,* the boll weevil, *Anthonomus vestitus,* and the leafworm, *Anomis texana,* were kept in check with arsenical and nicotine insecticides and manual removal. During the period from 1943 to 1948, the cotton yield in the valley ranged from 466 to 591 kilograms per hectar (415 to 526 pounds per acre). In 1949, the yield dropped to 366 kilograms per hectar (326 pounds per acre) because of an outbreak of an aphid and *Heliothis.* In the years that followed, heavy reliance was placed on the application of DDT, DHC, and Toxaphene to prevent pest outbreaks, and yields almost doubled.

The farmers became convinced that their yields were directly related to the amount of insecticide applied and began to blanket the fields with aerial applications, made possible by the removal of trees used by birds as nesting sites. As time passed, the pests developed resistance to the pesticides. Organophosphates such as parathion were substituted for the organochlorines, and the interval between insecticide applications was shortened. The almost total destruction of beneficial species resulted in outbreaks of the old pests, and previously innocuous species attained pest status (secondary pest problems). Finally, most of the pests developed resistance to the available pesticides, there were no nesting sites for insectivorous birds, and the isolated nature of the valley prevented the entry of natural enemies from elsewhere. By 1956, the valley's cotton yield dropped to 332 kilograms per hectar (296 pounds per acre) in spite of the massive use of insecticides.

The farmers then turned to their own research organization for assistance, and some new regulations were imposed. The use of organic insecticides was prohibited unless special dispensation was granted. Several cultural methods including uniform planting dates and fallow were employed. The acreage devoted to other crops such as corn and potatoes was increased, and flax was introduced, thereby increasing the ecological diversity and providing alternative hosts and habitats for parasites and predators. Natural enemies were brought in from other valleys and from abroad. Arsenicals and nicotine sulphate were reinstated as the predominant insecticides. As a result, the major problems subsided, and the secondary pests reverted to their previously innocuous status. In the following year, the cotton yield increased to 526 kilograms per hectar (468 pounds per acre) and since have been the highest in the history of the valley.

*Cotton pest control in the United States:* Cotton seems to be plagued by a variety of pests regardless of where it is grown, but there seems to be a rather marked difference in which pests are most important from place to place. In Texas, the key pests are the boll weevil and the cotton fleahopper; in California's Imperial Valley, they are the pink bollworm and the cotton leaf perforator; and in the San Joaquin Valley, it is a lygus bug. In all of these areas the traditional approach to pest control has been the repeated use of chemicals. In fact, in 1971, the last year for which data are available, 47% of the insecticides used by American farmers were applied to cotton (USDA, 1974). Most of the major cotton pests have now developed resistance to one or more of the poisons registered for use on the crop, and where insecticides are effective, they often cause secondary pest problems. Thus, we see in the cotton-growing areas of the United States a situation similar to that which developed in the Cañete Valley and a widespread effort to devise integrated pest management programs.

According to Adkisson (1972), chemical control of the boll weevil and the cotton fleahopper in Texas releases the bollworm and the tobacco budworm from regulation by natural enemies. Early attempts to control the key pests without disrupting the parasites and predators of the *Heliothis* species involved termination of the crop with defoliants and dessicants, late season insecticidal treatment of the boll weevil to reduce the diapausing population, and the plowing under of plant residues immediately after harvest.

The present program involves the same cultural and pesticide practices plus an attempt to avoid treatment of the fleahopper whenever possible and the use of early varieties that are forced to mature early by withholding water and fertilizer. This reduces the availability of host material available to the key pests some time before they are ready to enter dormancy. However, if cool, moist weather delays the maturation of the crop, pest populations can increase. If outbreaks occur that demand the use of chemicals, the present integrated pest management program is disrupted.

New short-season cotton varieties are now being developed that can be grown in dry areas without large doses of fertilizer, water, and pesticides. It is hoped these varieties will be resistant to some pests and will fruit so early and over such a short time that they will escape damage by others (Walker and Niles, 1971). Additional studies are underway to develop additional tactics for integration into the control program. These include the use of insect pathogens, pheromones, the sterile male technique, and the search for more selective insecticides.

In the San Joaquin Valley of California, the most important pest of cotton has been the lygus bug, *Lygus hesperus.* Three lepidopterans, the bollworm, cabbage looper, and beet army worm, become pests from time to time, particularly when their natural regulation is disrupted by insecticides directed against the lygus bug.

The present integrated control program outlined by van den Bosch and Messenger (1973) requires a more careful evaluation of the lygus bug population and a reduction of insecticidal treatment. Previously, sprays were used as a prophylaxis or at any time when the lygus bugs reached the economic threshold. Now insecticide treatment is only recommended when the population is found to exceed the economic threshold on two consecutive sampling dates three to five days apart during the period of flowering and budding. In some areas, lygus bug populations are substantially reduced by cultural practices. The strip harvesting of alfalfa (the favored host of *Lygus*) reduces the large-scale movement of the bugs into cotton following mowing. Interplanting strips of alfalfa as a trap crop in cotton fields also reduces the population on the cotton plants and improves natural control. However, as is so often the case in integrated pest management, additional changes in the present pest management strategy may be required as a result of the invasion of the valley in 1975 by the pink bollworm.

*Orchard pests:* Orchards provide a crop environment quite different from that of cotton. Tree fruits are a long-term crop and therefore provide a rather stable environment in which natural controls can flourish if not disrupted by chemical sprays. However, most of the key pests attack the fruits directly, and relatively low populations can cause serious losses. These are high-value crops, the marketability of which declines rapidly if the fruit is less than the highest quality, so the economic threshold for pests that do direct damage may be very low. In addition, tree fruits are plagued by a variety of diseases that may require numerous applications of sprays for their control. Any attempt to develop an integrated pest management plan is therefore complicated by the effect of these sprays on the resident parasites and predators.

Like cotton, tree fruits grown in different regions are often plagued by different pests, but the integrated pest management programs are rather similar. All of them involve efforts to reduce the use of insecticides against the key pests and thereby reduce secondary pest problems (see Hoyt and Burts, 1974).

In apple orchards, traditional pests such as coddling moth and apple scab have dominated the spray schedules, and various secondary pest problems, particularly those involving phytophagous mites, have developed. In Nova Scotia, coddling moth populations are usually low and satisfactorily controlled with a relatively specific botanical insecticide, ryania. This permits build-up of native parasites and predators, and in some orchards the need for sprays against coddling moth no longer exists. However, the apple maggot is the key pest, and since it is free from attack by predators, it must be controlled chemically. Lead arsenate is fairly specific, but may be banned, and this will make control of the pest within the integrated pest management program more difficult. The recent discovery of a pheromone which deters oviposition may alleviate the problem in the future (Prokopy, 1972).

The program for the control of peach pests in California (Hoyt and Caltagirone, 1971) is particularly interesting because it closely parallels the theoretical program suggested by Chant (1964) as a framework for his discussion of the strategy and tactics of insect control.

In California peach orchards, the Oriental fruit moth and the peach twig borer are pests in most orchards most of the time. Both of these insects attack fruits as well as twigs and cause economic damage at low population levels. Neither pest is controlled satisfactorily by natural enemies, so they have traditionally been controlled with insecticides applied on a regular schedule; these are the key or Class I pests.

The two-spotted spider mite, European red mite, and San Jose scale comprise the Class II pests. In the absence of disruption as a result of pesticides, these pests are normally held in check by natural enemies. However, when they escape regulation, they must be controlled chemically. The two mite pests can be particularly difficult to control because of the rapidity with which they develop resistance to acaricides. The peach silver mite is a typical Class III pest in that it is often present in very large numbers but apparently does not cause much in the way of economic damage; nevertheless, some growers spray for it on a regular basis. There are a few Class IV pests, such as the western peach tree borer, that become important from time to time and require a special insecticide application.

The predaceous mite *Metaseiulus occidentalis* provides effective control of the phytophagous mites as long as its population can be maintained at high levels. This requires a reduction in the use of pesticides and the preservation of the peach silver mite as a food source to support an expanding population of *Metaseiulus*.

The disruption caused by chemical sprays can be reduced by partially controlling the peach twig borer, San Jose scale, and phytophagous mites with dormant (winter) sprays such as petroleum oil, which are less harmful to beneficial species. Early season treatment of the Oriental fruit moth is eliminated in order to avoid disruption of the spring population expansion of beneficial species. Chemical control of the peach silver mite is no longer recommended in order to assure an adequate number of prey for a large population of *Metaseiulus*. This in turn reduces the need to apply acaricides against the two-spotted spider mite and European red mite.

So far, the developing integrated pest management program appears to be producing satisfactory results. However, the incorporation of additional tactics into the program may provide further benefits. The removal of twigs or fallen fruits infested by the Oriental fruit moth would be advantageous, if economically feasible. The Oriental fruit moth also has a pheromone that might be usable in control trapping. Attractant-baited traps could be used to reduce the population of adult moths, or a combination of the pheromone with chemosterilants could reduce the number of fertile females.

### *Evaluation and Prospect*

According to Wildbolz and Meier (1973), the development of integrated pest management programs in affluent countries like the United States is motivated by the biological and environmental consequences of the continuous use and misuse of pesticides. In some crops, cotton for example, the development of multiple resistance to previously useful pesticides has left no alternative but to search for new approaches to pest suppression. In other managed ecosystems such as grassland and forests, the economic thresholds for most pests are such that the relatively high per acre cost of repeated pesticide use is unwarranted. In these systems, alternative methods of pest control that are self-perpetuating and less expensive are more desirable; they are also easier to establish because of the temporal and spatial continuity of these agricultural systems.

Modern mechanized agriculture has become a standardized cultural approach, based on the economics of producing a quantity of uniformly high-quality produce, which consumers have grown to take for granted. Such an approach to food production and marketing is not particularly tolerant of pest damage, yet has led to the development of nondiversified ecosystems. These features hardly favor the development of integrated pest management.

There is also some psychological resistance to integrated control among growers. Many have grown accustomed to the pest-free fields that resulted from pesticide applications, guaranteed to produce pest mortality in excess of 90%. Furthermore, integrated pest management is more complex than chemical control. Consequently, growers feel a loss of independence when they have to seek outside assistance to implement their pest control. In some cases, the establishment of a pest management program takes several years before the full benefits are obtained, and growers, fearful of a loss of income because of pests, may disrupt the program by reverting to empirical methods with which they are more familiar. In agricultural areas consisting of a mosaic of private holdings, it is sometimes difficult to persuade all of the growers to cooperate. If one or two growers in such an area decide to rely on a previously established spray schedule, integrated pest management becomes difficult or impossible to implement on the remaining acreage. The best success is achieved where grower cooperatives or a government agency, such as Israel's Citrus Board, is in charge of the pest control program over a large area.

In spite of these difficulties, affluent countries have the research and development capability to provide the scientific and technical information necessary to the widespread implementation of integrated pest management programs. The changeover will not come about easily, because in many cases consumers will have to be re-educated, governments will have to make more funds available for research, and full cooperation between the growers and the purveyors of knowledge will have to prevail. The trend has begun, however. A recent article by Hall, Norgaard, and True (1975) indicates that some independent pest management consultants are meeting with success. The study showed that in California's San Joaquin Valley, cotton and citrus growers who employ private consultants use less pesticides and obtain higher per acre incomes than the growers who do not. These successes, it is hoped, will lead to a wider acceptance of more ecologically sound systems of pest control.

In developing economies, the problems are somewhat different. Many of the agricultural countries that have not yet realized their full potential lie within the tropics. Tropical pest outbreaks are often sporadic and so, discourage thorough investigation. When these outbreaks occur, they are usually combated with pesticides and little is learned about their cause or alternative means of prevention. This, combined with a tendency for administrators to equate crop protection with chemical control, and the fact that outside agencies gain prestige from spectacular short-term successes are creating a reliance on pesticides (Wood, 1973). The tropical environment would seem to favor integrated control, and it should be highly advantageous in economies that can ill-afford to spend money on imported chemicals for which other control methods can be substituted. Other problems in these countries are that they lack the scientific expertise to implement integrated pest management and their training facilities are inadequate. Often regional governments are too unstable to give any assurance that a program can be brought to fruition, especially if other projects present a better political profile. Yet, these areas will play an important role in providing the world's population with an adequate supply of food and fibre. If a conscientious effort is not made to support the development of integrated pest management in agriculturally emerging nations soon, we will find that the increasing reliance on pesticides has made the task considerably more difficult.

Integrated pest management seems to have advanced most rapidly as an approach to insect pest suppression. Unfortunately, we have not reached the stage where many workers are thinking in terms of the whole pest complex including plant diseases and weeds. This is understandable in view of the rather disciplinary nature of our training programs. However, the changeover to broader curricula that provide a breadth of training in agriculture, ecology, entomology, pathology, and other related areas has begun. This is an important development in that agroecosystems cannot be evaluated or managed in terms of a single class of pests. For example, clean field margins may be useful as a means of removing a source of weed seeds or eliminating the reservoir of a plant virus. But, as mentioned earlier, the wild vegetation of field margins and roadsides may provide the pollen and nectar required by parasites and create zones of diversity where natural relationships between pests and their natural enemies can flourish. If these areas are viewed only in terms of one kind of pest situation, any decision to alter them may have a detrimental effect on the over-all crop management program.

The consideration of the entire pest complex will, of course, make a complicated situation even more complicated, but it seems the only logical avenue to follow. Fortunately, computer science and modeling techniques are sufficiently advanced to be useful in sorting out the complexity. Computers can deal with the multiplicity of factors that must be considered and can store large quantities of data to provide an historical perspective. The development of models provides a means of testing the effects of various management strategies. Furthermore, computer simulation permits the exploration of a much larger number of strategies than would be possible by field experimentation. However, models must not be developed in a vacuum. There must be substantial input from entomologists, pathologists, and population ecologists who work in the field. At the present time, there are too few pest control specialists to provide the input to modeling projects and to carry the results of computerized gaming back to the crop systems for field testing.

We have come a long way, but we still have a long way to go. A strong commitment to research will add new and improved techniques for pest control. In the near future, we can expect major advances in the areas of pheromones, chemosterilants, hormones, and autocidal control methods. Nevertheless, we must not become blinded by any successes we obtain and must always bear in mind that the unilateral use of mortality agents invites a compensatory response from nature. The success of integrated pest management will depend heavily on all who use it having an open mind.

## References Cited


Adkisson, P. L. 1972. Integrated control of insect pests of cotton. *Proc. Tall Timbers Conf. on Ecol. Anim. Contr. by Habitat Manage.* 4: 175–189.

Bartlett, B. R. 1964. Integration of chemical and biological control. Pages 480–511 *in* P. DeBach, ed. Biological control of insect pests and weeds. Van Nostrand Reinhold Company, New York, N. Y.

Chant, D. A. 1964. Strategy and tactics of insect control. *Can. Entomol. 96: 182–201.*

DeBach, P. and J. Landi. 1959. Integrated chemical, biological control by strip treatment. *Calif. Citrograph.* 44: 324–352.

Doutt, R. L. and R. F. Smith. 1971. The pesticide syndrome—diagnosis and suggested prophylaxis. Pages 3–15 *in* C. B. Huffaker, ed. Biological control. Plenum Publishing Corporation, New York, N. Y.

Gonzalez, D. 1970. Sampling as a basis for management strategies. *Proc. Tall Timbers Conf. Ecol. Anim. Contr. Habitat Manage.* 2: 83–101.

Hall, D. C., R. B. Norgaard, and P. K. True. 1975. The performance of independent pest management. *California Agriculture* (October): 12–14.

Harpaz, I. and D. Rosen. 1971. Development of integrated control programs for crop pests in Israel. Pages 458–468 *in* C. B. Huffaker, ed. Biological Control. Plenum Publishing Corporation, New York, N. Y.

Haynes, D. L. 1973. Population management of the cereal leaf beetle. Pages 232–240 *in* P. W. Geier, L. R. Clark, D. J. Anderson, and H. A. Nix, editors. Insects: Studies in population management. Ecol. Soc. Aust. (Memoirs 1): Canberra.

Hoyt, S. C. and L. E. Caltagirone. 1971. The developing programs of integrated control of pests of apples in Washington and peaches in California. Pages 395–421 *in* C. B. Huffaker, ed. Biological Control. Plenum Publishing Corporation, New York, N. Y.

Hoyt, S. C. and E. C. Burts. 1974. Integrated control of fruit pests. *Ann. Rev. Entomol.* 19: 231–252.

Prokopy, R. J. 1972. Evidence for a marking pheromone deterring repeated oviposition in apple maggot flies. *Environ. Entomol.* 1: 326–332.

Southwood, T. R. E. and M. J. Way. 1970. Ecological background to pest management. Pages 6–29 *in* R. L. Rabb and F. E. Guthrie, editors. Concepts of pest management. North Carolina State University, Raleigh, N. C.

U.S.D.A. 1974. Farmer's use of pesticides in 1971. *Agr. Econ. Rep. No. 252. Econ. Res. Ser.*, 56 pp.

van den Bosch, R. and P. S. Messenger. 1973. Biological control. Thomas Y. Crowell Company, New York, N. Y.

Walker, J. K., Jr. and G. A. Niles. 1971. Population dynamics of the boll weevil and modified cotton types. *Texas Agr. Exp. Sta. Bull. 1109*, 14 p.

Wildbolz, T. and W. Meier. 1973. Integrated control: critical assessment of case histories in affluent economies. Pages 221–231 *in* P. W. Geier, L. R. Clark, D. J. Anderson, and H. A. Nix, editors. Insects: Studies in population management. Ecol. Soc. Aust. (Memoirs 1): Canberra.

Wood, B. J. 1973. Integrated control: critical assessment of case histories in developing economies. Pages 196–220 *in* P. W. Geier, L. R. Clark, D. J. Anderson, and H. A. Nix, editors. Ecol. Soc. Aust. (Memoirs 1): Canberra.

# Appendix
# The orders of insects

The classification of the Class Insecta has been in a constant state of ferment, even in terms of what should be included or excluded, as well as at the subclass and ordinal levels of organization. As a result of the size and complexity of the group, combined with the ever-increasing acquisition of new information, different groupings become recognized by different systematists. Consequently, it is virtually impossible to present a system of classification that will find general acceptance. There is no other group of organisms for which such a variety of ordinal groupings or nomenclatural variations have been proposed.

In the system of classification to be presented here, a few of the more important variations will be discussed briefly to provide a prospective of the complexity of the subject. In addition, some of the more widely used alternative names and groupings will be presented in parentheses, followed by the most widely accepted common names.

Since the main objective of this appendix is to provide the beginning student with a practical way of identifying most of the major groups, I have chosen to use more orders than preferred by many authors. This provides more opportunity for both the morphological and biological distinction between the members of some otherwise rather unruly assemblages. To clarify some of the interrelationships, I have then grouped the orders into reasonably well-defined complexes.

One rather important area of controversy in the classification of insects involves whether or not all of the six-legged arthropods should be placed in the Class Insecta. For many years, the Subclass Apterygota contained four orders of primitively wingless, insectlike, hexapod arthropods. Many now consider that there is sufficient morphological and biological evidence to place the Protura, Collembola, and Entotrophi (Diplura) outside the insects as a distinctly different assemblage of animals, collectively referred to as the **Entognathus Hexapods,** because their mouth parts are housed within folds of the head capsule. These small arthropods form an important group from an evolutionary as well as ecological point of view and are recognized to be very much part of the purview of entomologists. However, students should realize the distinctive nature of these organisms.

## General Classification

Entognathus Hexapods
- Class and Order Protura—telsontails
- Class and Order Collembola—springtails
- Class and Order Entotrophi (Diplura)—diplurans

Ectognathus Hexapods
- Class Insecta
  - Subclass Apterygota—primitively wingless insects
    - Order Thysanura—bristletails, firebrats, and silverfish

Subclass Pterygota—winged insects
Paleopteran Complex
  Order Ephemeroptera (Ephemerida)—mayflies
  Order Odonata—dragonflies and damselflies
Orthopteroid Complex
  Order Blattodea (Blattaria; Dictyoptera in part)—roaches
  Order Isoptera—termites
  Order Mantodea (Dictyoptera in part)—praying mantids
  Order Plecoptera—stoneflies
  Order Zoraptera—zorapterans
  Order Dermaptera—earwigs
  Order Embioptera (Embiidina)—webspinners
  Order Grylloblattodea—grylloblattids
  Order Orthoptera—grasshoppers, katydids, and crickets
  Order Phasmatodea (Phasmida)—stick insects
Hemipteroid Complex
  Order Psocoptera (Corrodentia)—psocids or booklice
  Order Thysanoptera—thrips
  Order Homoptera (Hemiptera in part)—cicadas, hoppers, aphids, whiteflies, scales
  Order Hemiptera—bugs
  Order Mallophaga (Phthiraptera in part)—chewing lice
  Order Anoplura (Phthiraptera in part)—sucking lice
Endopterygote Complex
  Order Megaloptera (Neuroptera in part)—alderflies and dobsonflies
  Order Raphidioptera (Neuroptera in part)—snakeflies
  Order Neuroptera—lacewings, mantispids, and antlions
  Order Coleoptera—beetles
  Order Strepsiptera (Coleoptera in part)—twisted-winged parasites
  Order Mecoptera—scorpionflies
  Order Trichoptera—caddisflies
  Order Lepidoptera—butterflies, moths, and skippers
  Order Diptera—flies
  Order Siphonaptera—fleas
  Order Hymenoptera—sawflies, parasitic wasps, ants, wasps, and bees

## Entognathus Hexapods

***Class and Order Protura***

The telsontails are very small (less than 2 mm), elongate, mostly unpigmented hexapods that lack both eyes and antennae. The forelegs are much more well developed than the other two pair and are held forward as tactile organs (Figure A-1). Proturans are inhabitants of soil rich in organic matter and often occur in quite large numbers in decaying plant debris. The larvae are essentially like the adults with the exception that they lack gonads and

FIGURE A-1. A typical proturan.

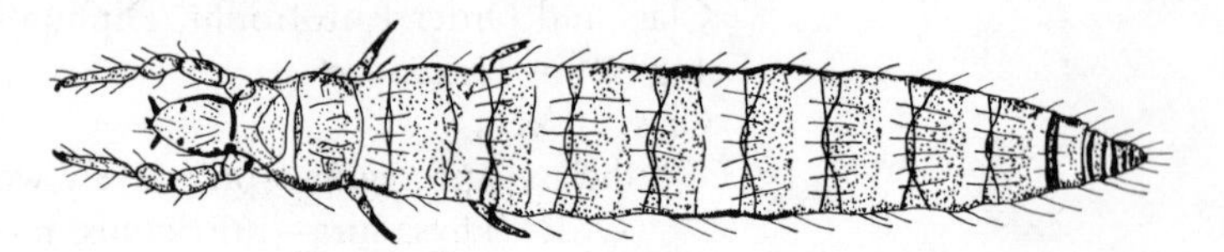

FIGURE A-2. Collembola, springtails. A. Suborder Symphypleona, Family Sminthuridae. B. Suborder Arthropleona, Family Isotomidae. (Redrawn from Rankin, 1970, in *Insects of Australia.*)

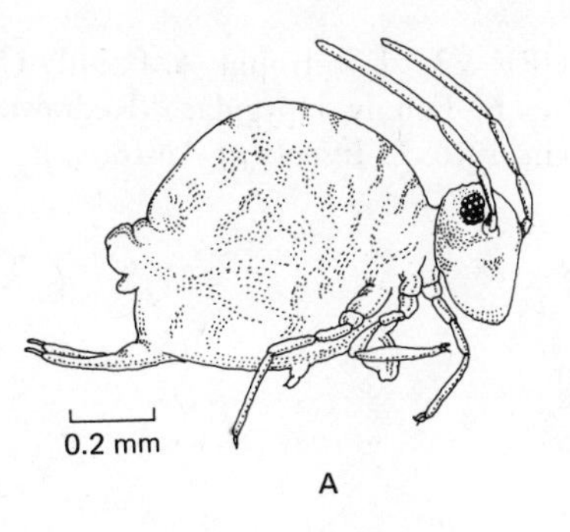


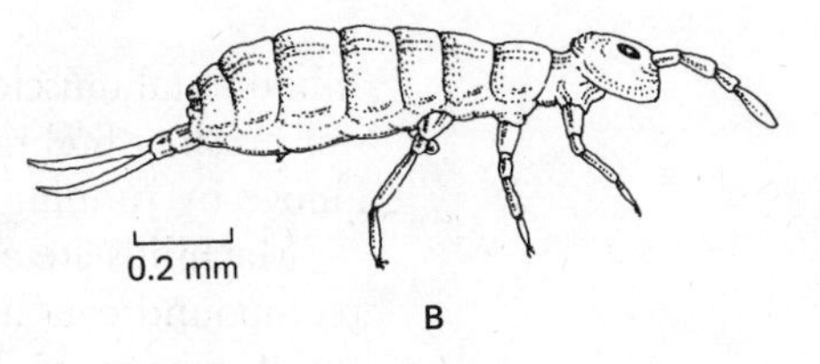


have nine instead of twelve abdominal segments. An abdominal segment is added with each larval molt, but no other structural changes occur; they are thus said to develop anamorphically.

***Class and Order Collembola***

The springtails are small insectlike creatures with both simple eyes and antennae. Although the members of this group are variable in form (Figure A-2), they are characterized by the presence of an adhesive device on the first abdominal segment and a special jumping apparatus on the third and fourth abdominal segments (see Figure 7-11). The immatures develop by anamorphosis and so resemble the adults. Collembolans are widely distributed wherever the environment can support animal life. They inhabit the surface layer of the soil and abound in areas rich in plant litter, where they feed on a variety of things, including lichens, fungus spores, pollen, and plant debris.

***Class and Order Entotrophi***

Diplurans are small- to medium-sized, elongate, mainly unpigmented hexapods that lack eyes but have well-developed beadlike antennae. The abdomen is 10-segmented and terminates in a pair of well-developed cerci that are either multisegmented or forceplike (Figure A-3). The larvae develop anamorphically. Like the other entognathus hexapods, diplurans are largely soil inhabitants where they feed as scavengers and predators on other small arthropods.

## Ectognathus Hexapods

***Class Insecta Subclass Apterygota***

ORDER THYSANURA. The Thysanura form the sole living order of true insects that is primitively wingless (never having possessed wings). Within the order, there are two quite distinct groups that some recognize as separate orders. The machilids or bristletails are subcylindrical and have special longitudinal

FIGURE A-3. Entotrophi. **A.** Family Campodeidae. **B.** Family Japygidae. (Redrawn from Rankin, 1970, in *Insects of Australia.*)

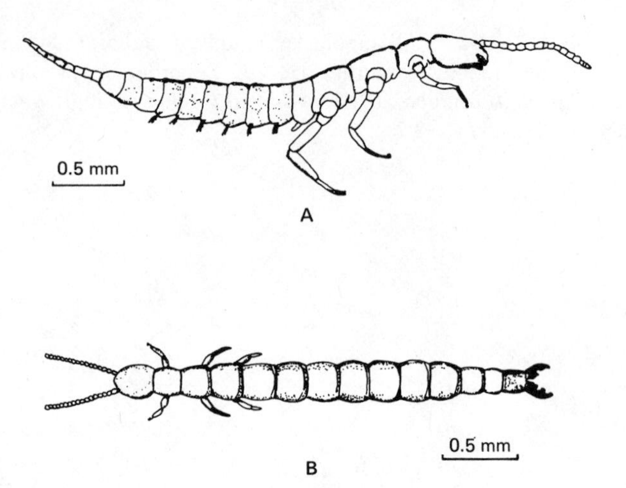


abdominal muscles that form the basis of a jumping mechanism, whereas the silverfish or firebrats are more dorsoventrally flattened, lack such muscles, and move by running.

Machilids are clothed in pigmented scales and have ocelli, well-developed compound eyes and a caudal filament that is considerably longer than the multisegmented cerci (Figure A-4*A*). They are mainly inhabitants of the

FIGURE A-4. Thysanura. **A.** Family Machilidae. **B.** Family Lepismatidae. (Redrawn from Quick, 1970, in *Insects of Australia.*)

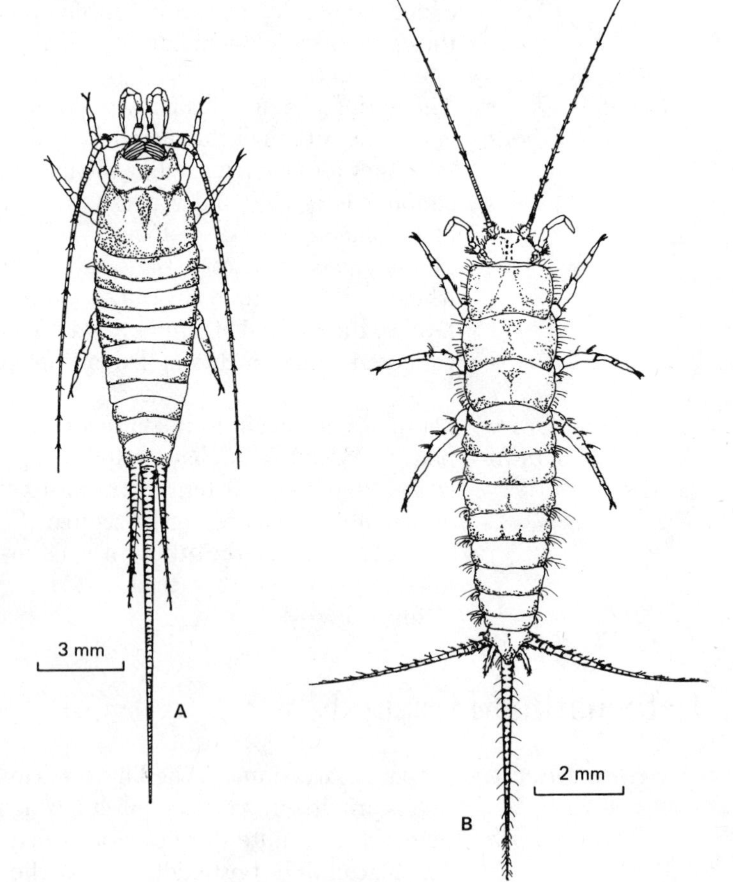

soil-litter interface and small natural crevices where they feed on lichens, algae, and plant debris.

The silverfish often have a mixture of scales and hairs on their body surface, have either reduced or absent compound eyes, and usually lack ocelli; the caudal filament is usually about the same length as the cerci (Figure A-4*B*). Like the machilids, silverfish are mainly vegetarian and are found in the litter, under bark or in the nests of other animals. Some species have become pests of human inhabitations where they feed on a variety of materials with a high starch content such as cereals.

### *Subclass Pterygota Palaeopteran Complex*

Almost all of the known fossil insects, like their modern descendants, bore wings, but many of the primitive insects that thrived during Carboniferous times rested with their wings either held more or less vertically above the body or extended outward at right angles to the body. These insects had humeral and axillary plates at the base of the wings fused with the wing veins and consequently were unable to flex the wings so as to bring them parallel to the body. These ancient groups have therefore been grouped in the **Palaeoptera** (*palaeo* = ancient, *aptera* = wings). All of the more advanced insects have discrete axillary sclerites, thus enabling them to flex the wings at their base and fold them flat over or beside the body. These forms have been grouped in the **Neoptera** (*neo* = new, *aptera* = wings). Two orders, the Ephemeroptera and Odonata, are the only living representatives of the Palaeoptera.

ORDER EPHEMEROPTERA. The mayflies comprise a relatively small order (about 2,000 species), the adults of which are delicate, weak-flying, short-lived insects, found in close proximity to the fresh water in which the immature stages pass their life. The adults are characterized by their slender, elongate bodies, large, protruding compound eyes, two or three long caudal filaments, and small hind wings (Figure A-5*A*). The immatures (Figure A-5*B*) are similar in general form to the adults but have a pair of articulated tracheal gills that arise from the lateral margins of from four to seven abdominal segments. The nymphs of most species are scavengers or herbivores for which diatoms and other microscopic organisms are an important dietary component.

An interesting feature of the development of mayflies is the occurrence of an incomplete adult stage called a **subimago.** The subimago is not sexually mature and does not have fully developed wings but can fly; this is the only order in which there is a subadult with functional wings. A subimaginal molt results in the emergence of a full-fledged adult.

ORDER ODONATA. The approximately 5,000 living species of dragonflies and damselflies are all elongate species that closely resemble the ancestral forms that thrived in the swamp forests of the Carboniferous. The dragonflies (Suborder Anisoptera) rest with their wings extended laterally (Figure A-6), whereas the damselflies (Suborder Zygoptera) fold them above their back without flexion as the mayflies do (Figure A-7). Both groups have highly movable heads with large compound eyes and short bristlelike antennae. The thorax is well-developed and tilted backward to a considerable degree; the abdomen is long and slender. The dragonflies are accomplished fliers and prey on a variety of insects while in flight; damselflies tend to fly poorly. Odonate nymphs are aquatic predators that use their specialized prehensile labium (see

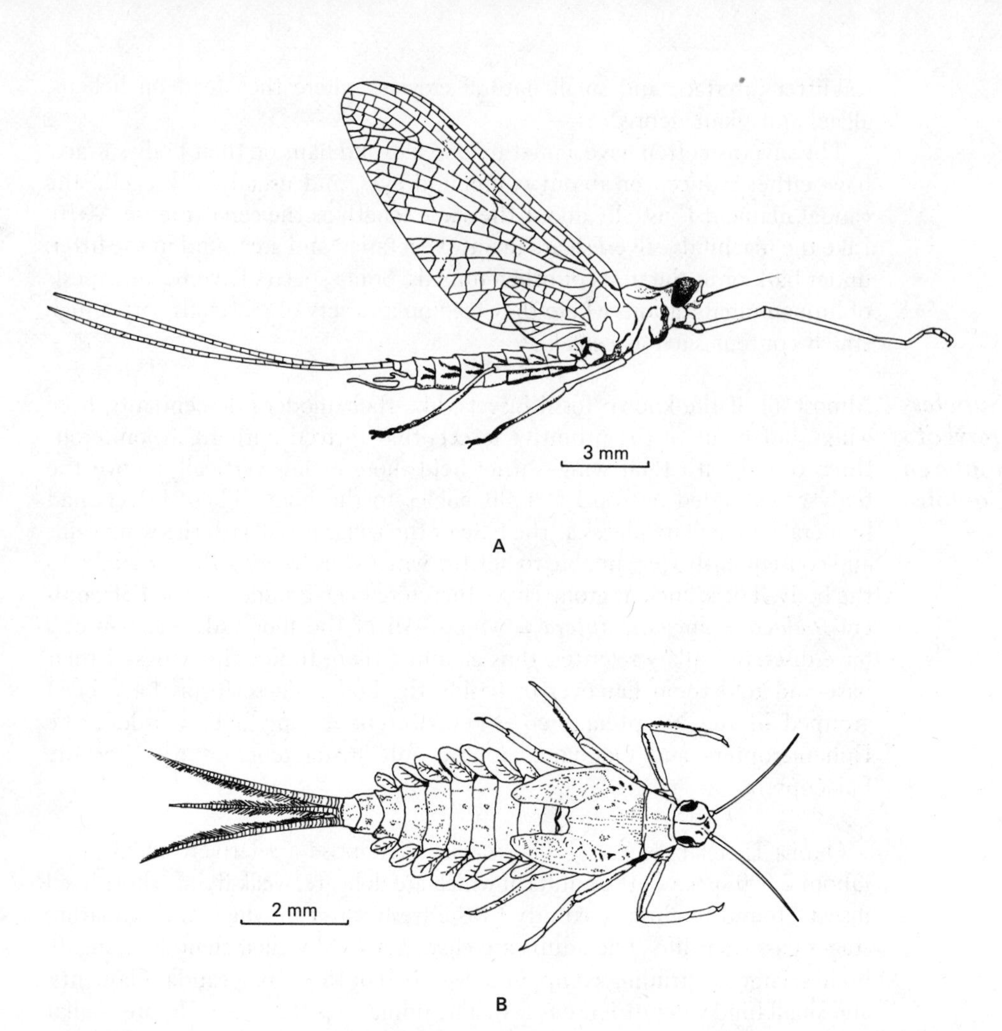


**FIGURE A-5.** Ephemeroptera, mayflies. **A.** Typical adult. **B.** Typical aquatic nymph.

**FIGURE A-6.** Odonata; Suborder Anisoptera; dragonfly about normal size.

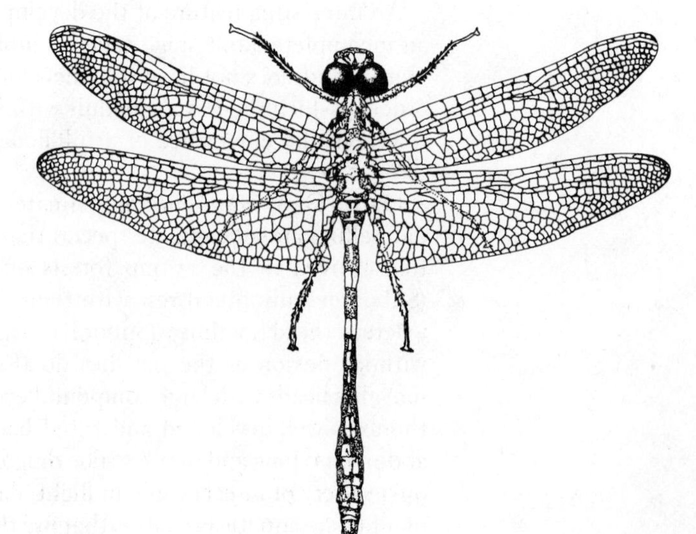

FIGURE A-7. Odonata; Suborder Zygoptera; damselfly.

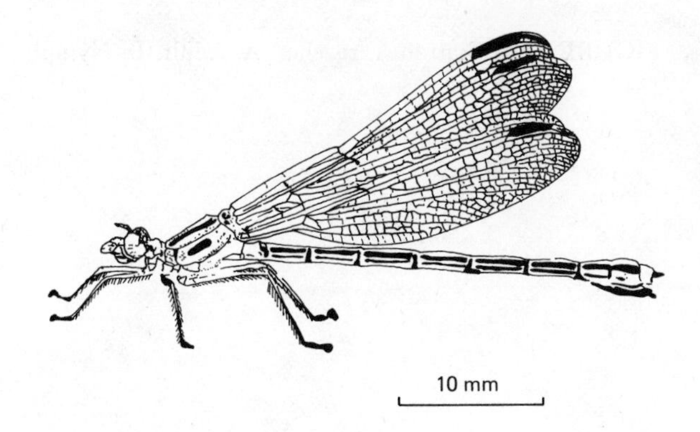


Figure 6-9) to capture prey. Dragonfly nymphs (Figure A-8*A*) tend to be rather robust and lack featherlike terminal abdominal gills. Damselfly nymphs (Figure A-8*B*) are more elongate and have three terminal tracheal gills arising from the tip of the abdomen.

FIGURE A-8. Odonata nymphs. **A.** Suborder Anisoptera. **B.** Suborder Zygoptera.

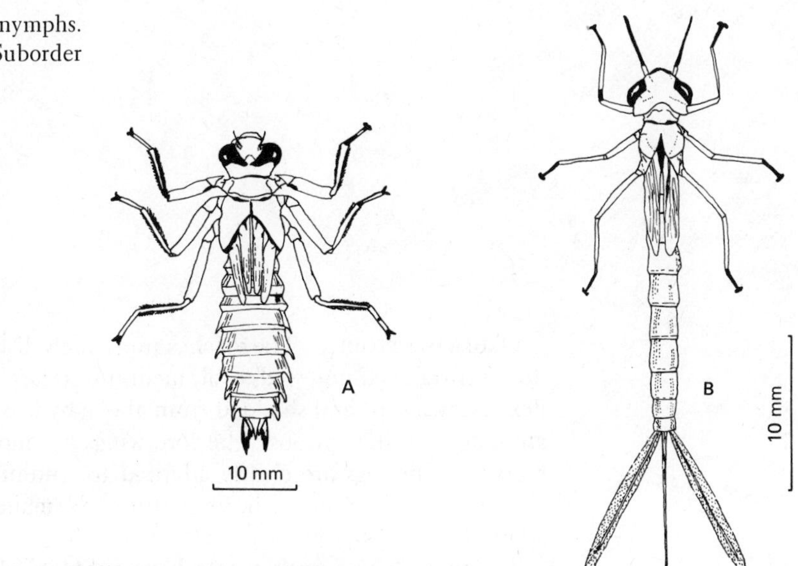


## *Orthopteroid Complex*

The orthopteroid complex consists of a fairly distinct group of orders that share a number of rather generalized characteristics. Although most representatives of this complex have generalized wings with numerous cross-veins, and hind wings with a large anal fan, they all have a wing-flexing mechanism. In addition, they have a number of other primitive characteristics, including chewing mouth parts, the retention of cerci, a large number of Malphigian tubes, and a ventral nerve cord with several discrete abdominal ganglia. The immatures generally resemble the adults and display gradual and relatively slight metamorphosis (incomplete metamorphosis) during which the wings develop externally (exopterygote). In this complex, the orders Embioptera and Zoraptera display some more advanced conditions, particularly more simplified wing venation and reduced anal fans, but they are not clearly related to the more advanced orders that comprise the hemipteroid complex.

FIGURE A-9. Blattodea, roaches. A. Adult. B. Nymph.

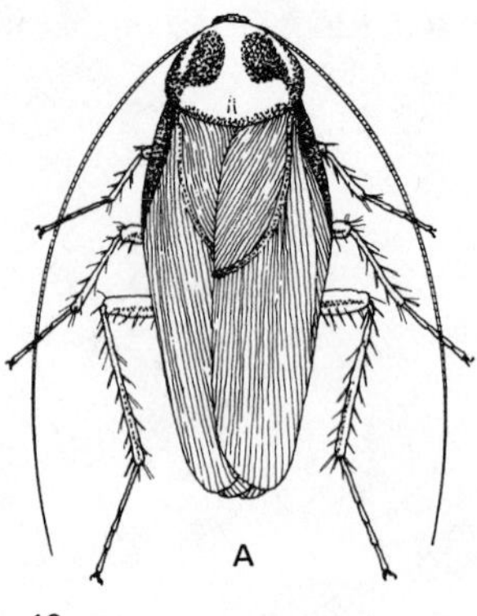


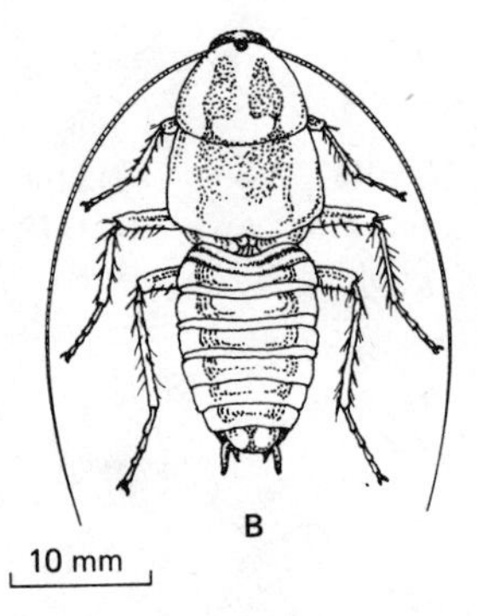


ORDER BLATTODEA. The roaches are a widely distributed group of somewhat dorsoventrally compressed, oval, medium- to large-sized insects. The head is flexed downward and shielded from above by the large pronotum; antennae are long and filamentous. The fore wings are modified into parchmentlike tegmina. The legs are clearly adapted for running and are equipped with numerous stout spines. The immature are smaller wingless versions of the adult (Figure A-9).

The roaches are largely tropical and subtropical omniverous insects, but a few of the approximately 4,000 known species have spread by world commerce to become more or less domesticated in temperate regions. Most are nocturnal, terrestrial species that rarely utilize their ability to fly. Many are gregarious and occur in groups consisting of both adults and nymphs of various ages. Their habit of establishing themselves in man's dwellings, where they contaminate foodstuffs with their excreta, has resulted in a few species being recognized as household pests.

ORDER ISOPTERA. The termites form a polymorphic group of roughly 2,000 species, all of which live in social units composed of a limited number of reproductives and numerous wingless workers, soldiers, and immatures. Termites are relatively soft-bodied. The alates have two pair of wings, similar in form and usually longer than the body, which are shed after their colony-founding flight. The nonreproductives are wingless and vary morphologically, depending on their role in the colony; the body form of the nymphs resembles

FIGURE A-10. Isoptera, termites. A. Winged reproductive. B. Worker. C. Soldier.

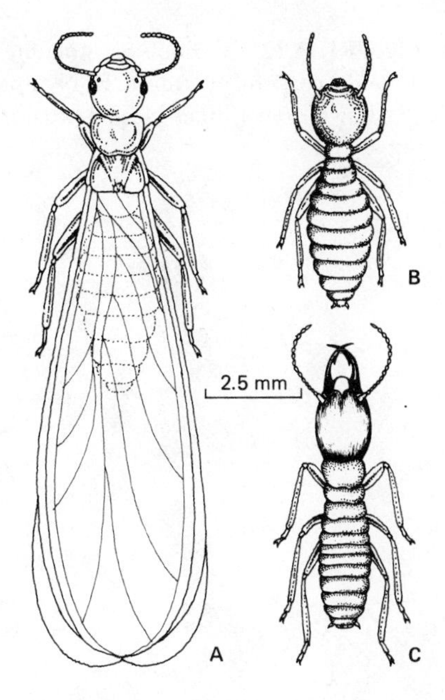


the adult (Figure A-10). Some colonies contain subordinate reproductives that never have wings.

Most termites feed on materials high in cellulose which is broken down by symbiotic protozoans harbored in the digestive tract. A number of wood-feeding termites are well known as serious structural pests, but many play a significant role in the building of tropical soils. A few species culture fungi on special comblike structures in the heart of their nests or termitaria. The social biology of the termites is discussed in more detail in Chapter 16.

ORDER MANTODEA. The mantids are slow-moving terrestrial predators of moderate to large size. They form a group of about 1,800 known species, characterized by the specialization of the front legs to form a pair of raptorial organs used in the capture of their prey. In addition, the prothorax is elongated so that the forelegs that arise anteriorly are widely separated from the two pair of walking legs (Figure A-11). The nymphs closely resemble the adults except for the absence of wings.

Mantids are solitary, often cryptically colored, predators that normally frequent shrubs and trees in preference to ground vegetation. They spend much of the time motionless with the forelegs raised ready to capture insects or spiders that come within range. The eggs are laid in groups of up to several

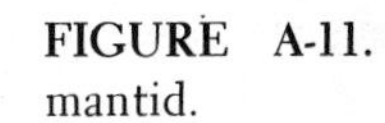
FIGURE A-11. Mantodea, mantid.

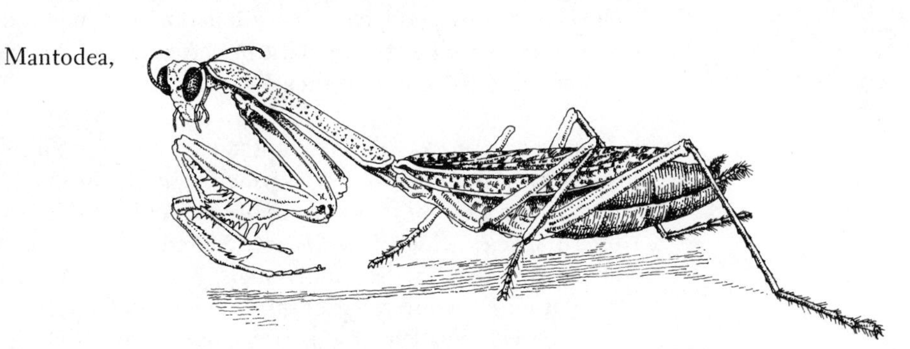

FIGURE A-12. Plecoptera, stoneflies, *Isoperla confusa*. **A.** Adult. **B.** Nymph. (Redrawn from Illinois Natural History Survey.)

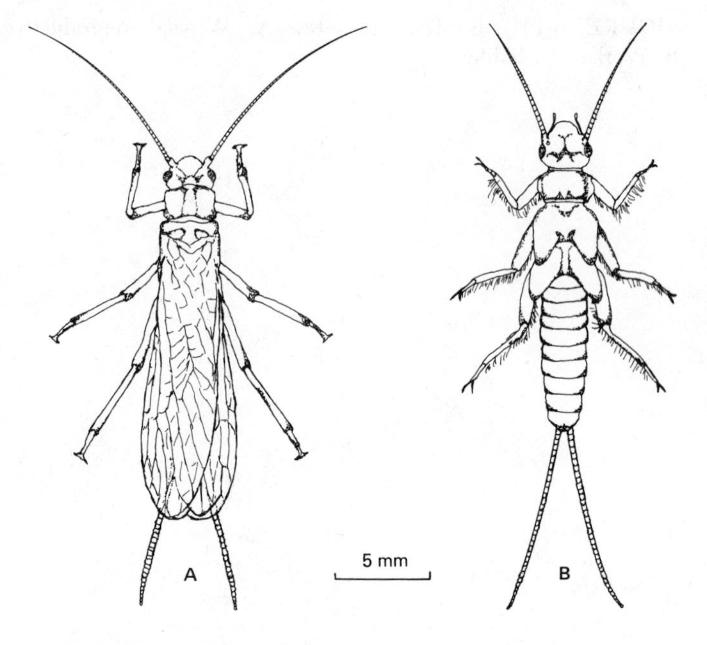


hundred and are then covered by a frothy material that hardens into an ootheca. Egg parasitism is high and the young nymphs are cannibalistic, so the population density of mantids is usually low. This, combined with their general predatory nature and territorialism, reduces their usefulness as regulators of specific pest populations.

ORDER PLECOPTERA. The roughly 1,000 known species of stoneflies comprise the only order in the orthopteroid complex in which a large portion of the life history is spent in an aquatic environment. The adults display some highly generalized morphological characteristics that are little changed from their Permian ancestors. They are elongate, soft-bodied insects with clearly separated thoracic segments, large, somewhat fluted membranous wings, long filiform antennae, and prominent, multisegmented cerci (Figure A-12*A*). The aquatic nymphs (Figure A-12*B*) generally resemble the adult but may have external gills located variously over the body; remnants of these gills are often retained in the adult.

Adult stoneflies are found only in close proximity to fresh water from which they emerge throughout the year, depending on the species and location. They are weak fliers and are often found crawling about feeding on algae and lichens near the water's edge. The nymphs usually live beneath stones in well-aerated water and feed on various kinds of aquatic plants; a few are carnivorous and take other aquatic insects as food. The nymphs may take up to three years to complete their development.

ORDER ZORAPTERA. The zorapterans are very small, rarely encountered insects with both winged and wingless adults (Figure A-13). The order contains but a single family comprised of only 22 known species. Most are tropical species which live under the bark of trees or in rotten wood in colonies of a few hundred individuals, but there is no evidence of parental care or other social interrelationships. The biology is rather poorly known, and although other small arthropods form part of their diet, it is not clear if they are predators or scavengers.

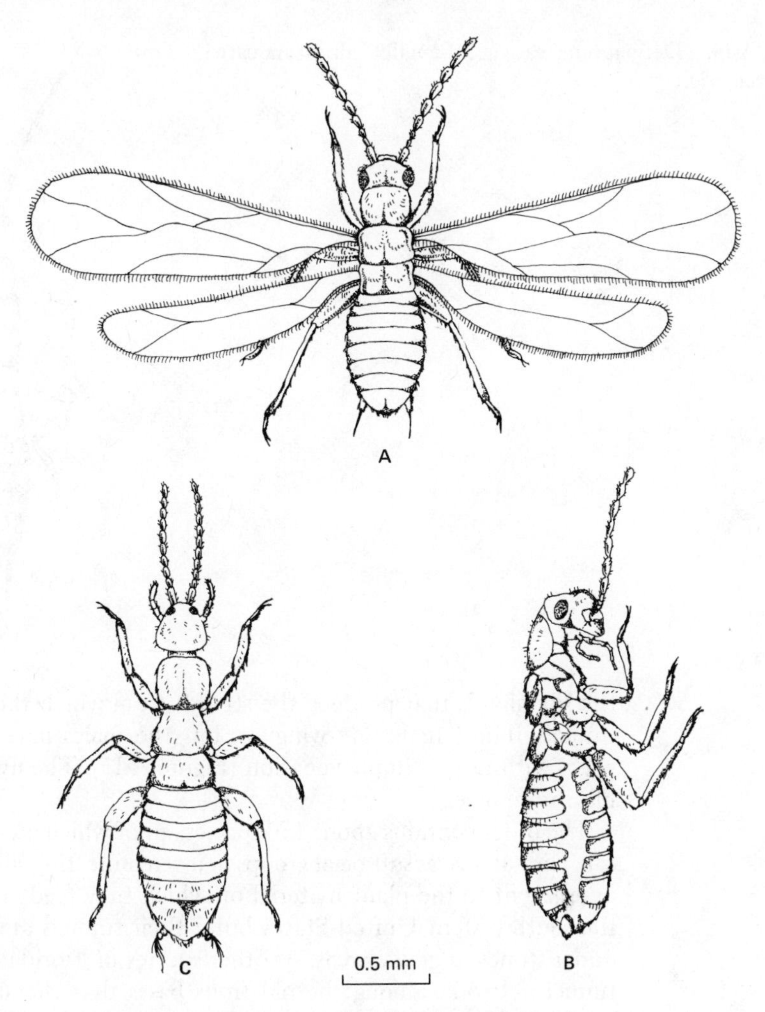


**FIGURE A-13.** Zoraptera, *Zoroptypus hubbardi*. **A.** Winged female. **B.** Dewinged female. **C.** Wingless female. (Redrawn from Caudell.)

ORDER DERMAPTERA. Earwigs are medium sized, elongate, somewhat flattened, well-sclerotized insects, generally characterized by a pair of terminal abdominal forceps formed by the single-segmented cerci (Figure A-14). There are both winged and wingless forms among the approximately 1,200 described species. When wings are present, the anterior pair are modified to form short leathery pads beneath which the large fanlike hind wings are almost completely concealed. Like many members of the orthopteroid complex, earwigs do not fly a great deal.

The adults and nymphs are very similar in appearance, and both tend to be active at night and secretive during the day. They are positively thigmotactic (respond to close contact), resting under stones or in other natural crevices. Earwigs are rather omnivorous and occasionally become floricultural pests in home gardens, feeding on the nectaries of flowers and causing a premature dropping of the petals. One African family has become ectoparasitic on rodents.



ORDER EMBIOPTERA. Web spinners are medium sized, slender insects with short, stout legs. The first segment of the foretarsi are enlarged and contain

**FIGURE A-14.** Dermaptera, earwigs. Female European earwig, *Forficula auricularia.*

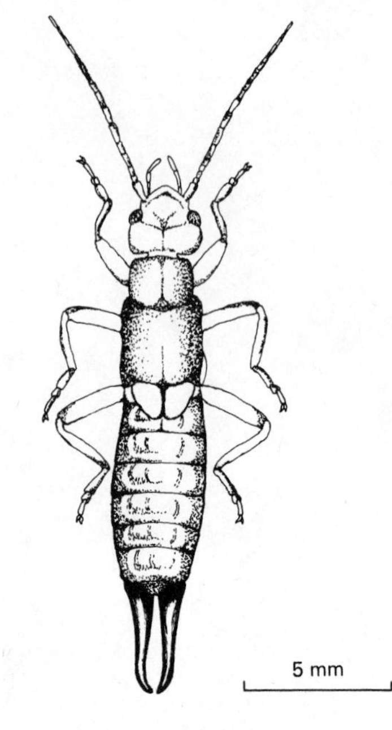


the silk glands that produce the strands from which they construct tubular tunnels. The females are wingless, but the males have two pair of similar wings with rather simple venation (Figure A-15). The nymphs closely resemble the female.

The order contains about 150 species, all of which are tropical or subtropical. They all live as subsocial groups in interconnecting silken tunnels, spun on or adjacent to the plant material on which they feed. The species found in the southwestern United States build their tunnels amongst dried grass or under stones in grassy areas. Another species in Florida constructs extensive tunnel networks among the old frons bases that characterize the trunk of some palm trees.

**FIGURE A-15.** Embioptera, web-spinners. **A.** Winged male. **B.** Wingless female.

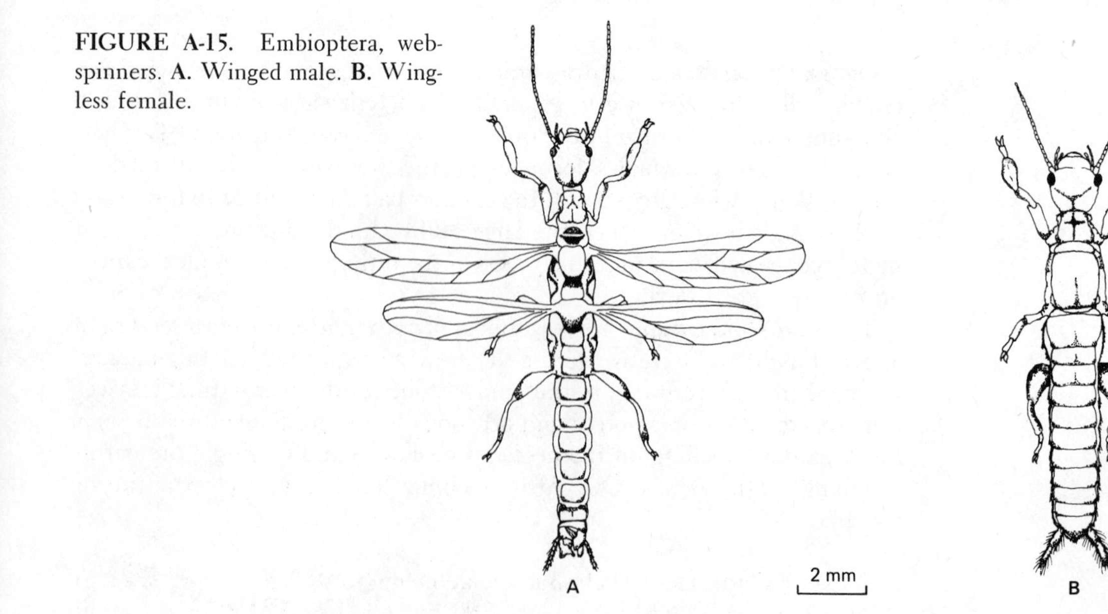

The embiids are highly adapted for life in their tunnels, being able to move backwards as quickly as they can forwards. In the males, the wings have become flexible to facilitate their backward motion, but when needed for flight they are temporarily stiffened by blood pressure in an elongate sac adjacent to the major longitudinal wing vein.

ORDER GRYLLOBLATTODEA. The grylloblattids form a small, rarely seen order, consisting of a single family that contains 10 to 12 species belonging to three genera. All are somewhat flattened, elongate insects with a prominent head, small eyes, elongate antennae, fairly long, multisegmented cerci, and a stout projecting ovipositor (Figure A-16).

All known species live in cold, wet habitats such as ice caves, crevices among the rocks of glacial morains, or rotting wood in areas of almost perpetual snow, where they feed as scavengers. The optimal temperature for their development ranges only a few degrees above freezing, and the temperature of a collector's hand is said to cause death. Embryonic development may take up to a year, with up to more than five additional years being required to attain the adult stage.

ORDER ORTHOPTERA. The Orthoptera proper is by far the largest and most diverse order in this complex, consisting of roughly 20,000 species belonging to seven common families and a number of lesser known ones. Orthopterans are all medium- to large-sized insects, many of which have elongated hind legs with some enlargement of the femora. The pronotum is usually large and extends downward to produce a protective collar. Most have large functional wings, although short-winged and wingless forms are widespread. When wings are present, the anterior pair are leathery or parchmentlike and provide protection to the membranous hind wings. All females bear an external ovipositor and in most cases the cerci are obvious. Members of the seven common families are illustrated in Figure A-17.

**FIGURE A-16.** Grylloblattodea, *Grylloblatta campodeiformis* female. (Redrawn from Nanninga, 1970, in *Insects of Australia.*)

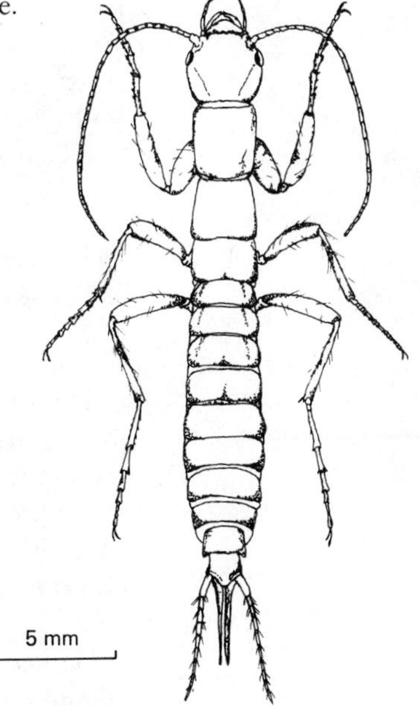

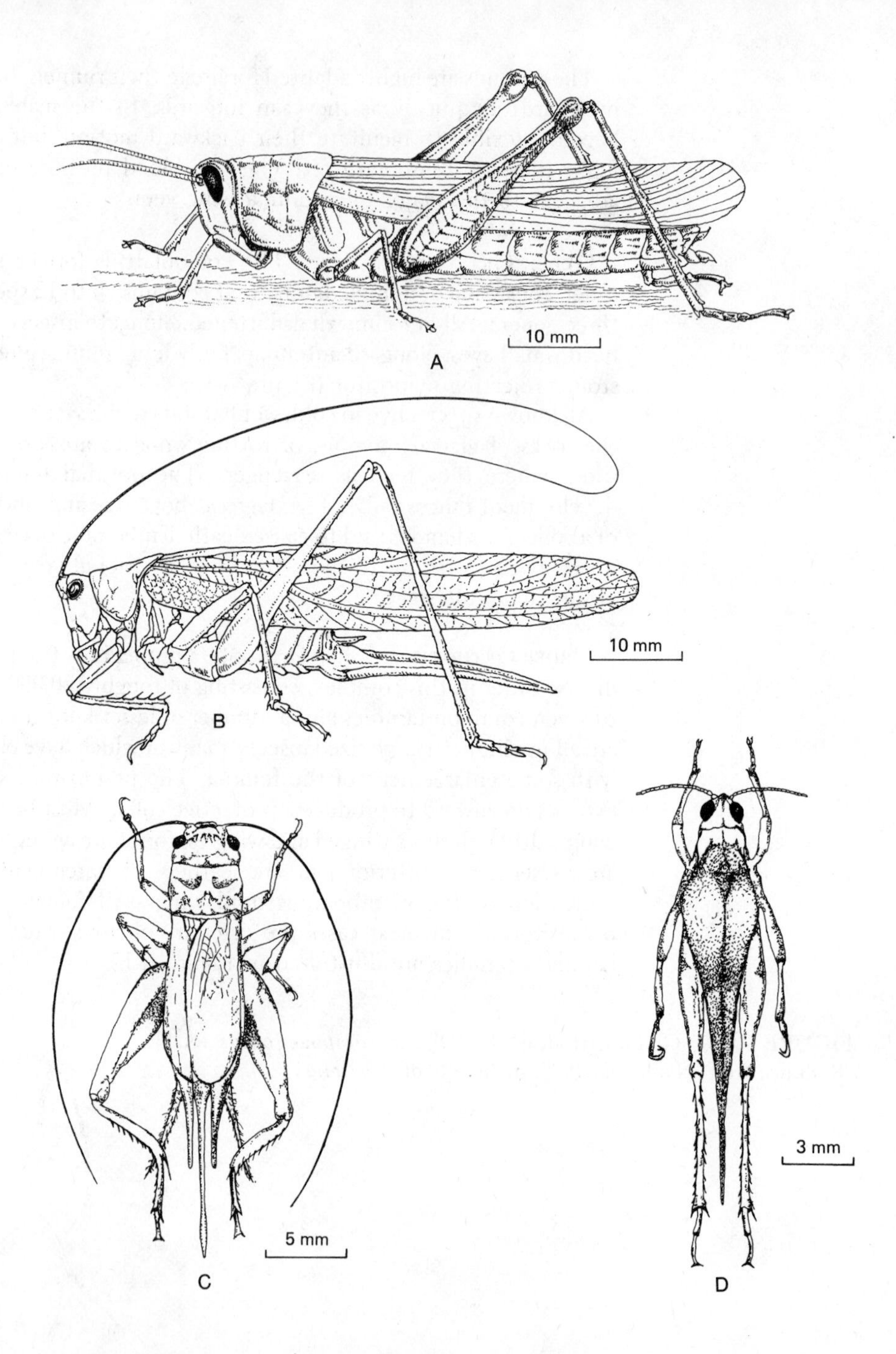


The Orthoptera display considerable ecological diversity. Most are phytophagous and, as a consequence, the order contains a number of important economic pests, but some katydids are predators on other orthopterans. The group probably evolved to live in the open or among shrubby vegetation for which they seem particularly well adapted. However, several groups have become adapted to a subterranean life.

Order Phasmatodea. The phasmids are large- to very large-winged, brachypterous or apterous, slow-moving creatures, often referred to as stick

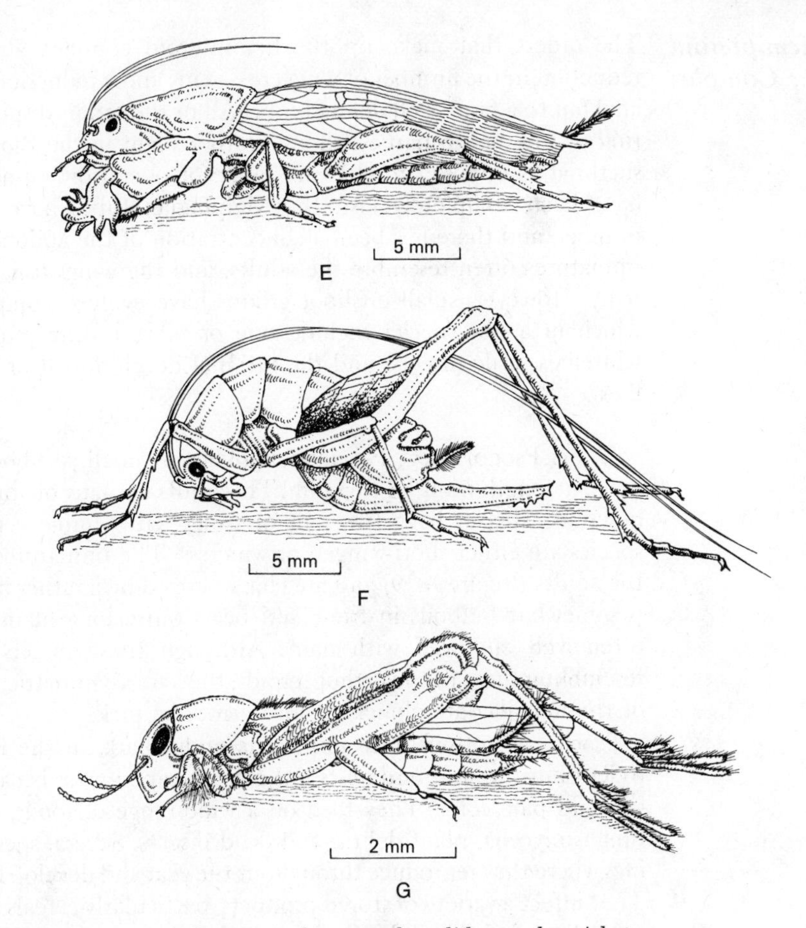


FIGURE A-17. Orthoptera, grasshoppers, locusts, katydids, and crickets. A. Acrididae (grasshoppers and locusts). B. Tettigoniidae (katydids and long-horned grasshoppers). C. Gryllidae (crickets). D. Tetrigidae (pigmy grasshoppers). E. Gryllotalpidae (mole crickets). F. Gryllacrididae (wingless long-horned grasshoppers). G. Tridactylidae (pigmy mole crickets).

insects because many are elongate and twiglike (Figure A-18). However, some are flattened and leaflike or rather robust and well endowed with stout spines (see Figure 4-6); all the North American species are wingless and sticklike.

All of the roughly 2,500 known species are leaf-feeders, most commonly found feeding on trees. Although solitary, they do occur in populations large enough to cause serious defoliation and are recurring pests of eucalyptus woods in Australia. The eggs, which are often highly ornamented (see Figure 9-1), are laid singly and drop to the ground where they pass the winter. The nymphs closely resemble the adults.

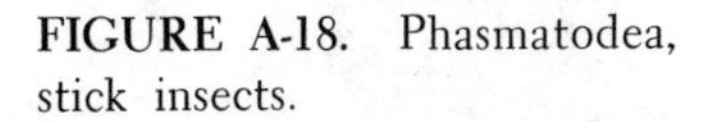
FIGURE A-18. Phasmatodea, stick insects.

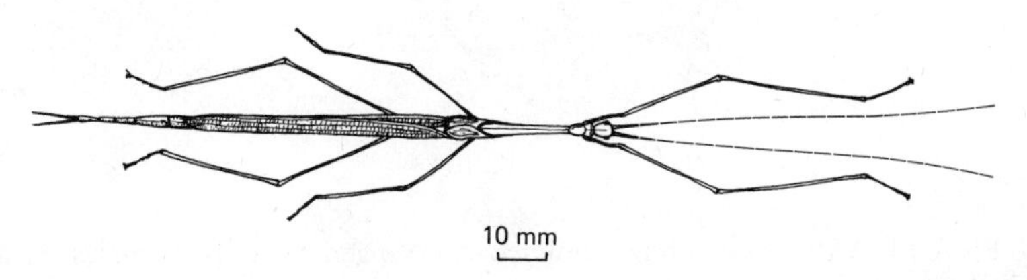

**Hemipteroid Complex** The orders that make up the hemipteroid complex all display an early reduction in the number of wing cross-veins and a reduction in the size of the anal fan to a fairly well-defined lobe. All of the orders display some modification of the mouth parts, although rather slight in the Psocoptera, toward a suctorial as opposed to chewing mode of operation. In addition, the cerci have been completely lost, the Malphigian tubes have been reduced in number, and there has been a concentration of the abdominal ganglia. The immatures often resemble the adults, and the wings usually develop externally. However, small offshoot groups have evolved complex life histories, which in a few species include one or more resting pupal stages (thrips, whiteflies, scale insects) and the internal development of the wings (whiteflies).

ORDER PSOCOPTERA. Psocids or booklice are small, soft-bodied insects which range in length from 1.5 to 5 mm. The adults of many of approximately 1,800 described species have two pair of wings with simple venation, but some species are either short-winged or wingless. The immatures closely resemble the adults (Figure A-19) and are characterized by a rather broad head, which is somewhat bulbous in front and bears quite long filamentous antennae, often well endowed with hairs. Although these insects have mandibles resembling those of the orthopteroids, they are asymmetrical, and the lacinea of the maxillae are elongated into a rodlike pick.

Psocids are often quite numerous under bark, in the leaf litter, and in rodent burrows and bird nests but often escape notice because of their small size and pale color. They feed on a wide range of foods, including pollen, fungus mycelia, plant debris, and dead insects. Several species invade buildings where they reproduce throughout the year and develop large populations. They infect a variety of stored products, particularly cereals, which they cause to be discarded. They also feed on starch-based paste and sizing used in books, thereby causing considerable damage in libraries, especially where older books and documents are stored.

ORDER THYSANOPTERA. Thrips are elongate insects between 2 and 3 mm long with asymmetrical piercing mouth parts in a short conelike projection from the ventral hind margin of the head. Many of the 4,000 known species are wingless, but when wings are present, they consist of a narrow, middle membranous area with greatly reduced venation and a marginal fringe of long

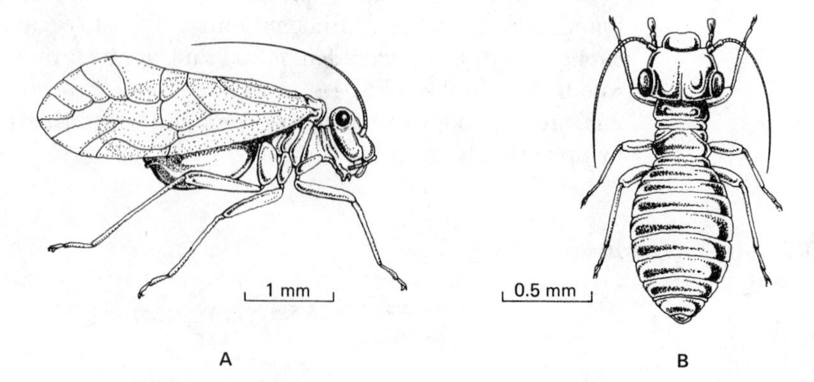


FIGURE A-19. Psocoptera, booklice. A. Winged male. B. Wingless female.

**FIGURE A-20.** Thysanoptera, thrips.

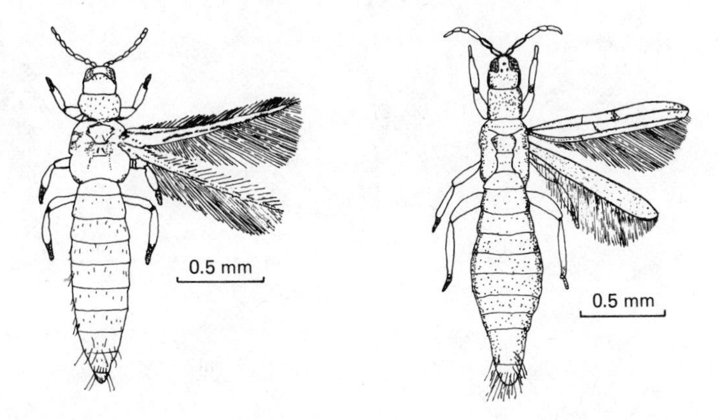


hairs (Figure A-20). The immatures are similar to wingless adults, but the last one or two instars are inactive and are called pupal stages.

Most thrips are general to host-specific plant juice-feeders, but a few are predaceous on mites and small insects. The phytophagous forms feed on both leaves and floral parts but are known to bite man when searching for suitable hosts. The short mouth parts restrict feeding to tender, actively growing tissues, which can be seriously distorted by the mechanical damage and salivary secretions. In addition to feeding lacerations, damage may also be caused by the ovipositor. Thrips frequently occur in large populations and cause economic damage to a variety of agricultural and floral crops.

ORDER HOMOPTERA. This order (often recognized as a suborder of the Hemiptera) contains more than 32,000 species of plant sap-feeding insects. They exhibit considerable morphological and biological variation but are clearly related through the form of their piercing-sucking mouth parts (see Figure 6-1) that appear to arise from the posterior margin of the head. There are fully winged, short-winged, and wingless forms. When wings are present, they are typically held in a rooflike manner over the body, and the structure of the fore wings is uniformly membranous or slightly thickened; male scale insects have only mesothoracic wings. Representatives of some common families are illustrated in Figure A-21.

Most homopterans undergo gradual metamorphosis and have immatures of similar morphology to the adults. However, some groups have very complex life histories, and in the whiteflies and scale insects, development resembles complete metamorphosis.

Many homopterans are serious pests of cultivated plants because of the feeding damage they cause and because they serve as vectors of a number of important plant pathogens. In some agricultural regions, the aphids collectively are the most important group of crop pests.

ORDER HEMIPTERA. The true bugs are a morphologically diverse group of approximately 24,000 described species, unified by several consistently shared characteristics. All have piercing-sucking mouth parts similar to those of the Homoptera (see Figure 6-1), except they clearly arise from the front of the head which is deflected posteriorly. Most species have two pair of wings, the front pair of which are modified into hemelytra with a leathery basal portion and apical membranous area. When folded, the wings lie flat over the body. Some species are wingless and in some short-winged forms, the membranous

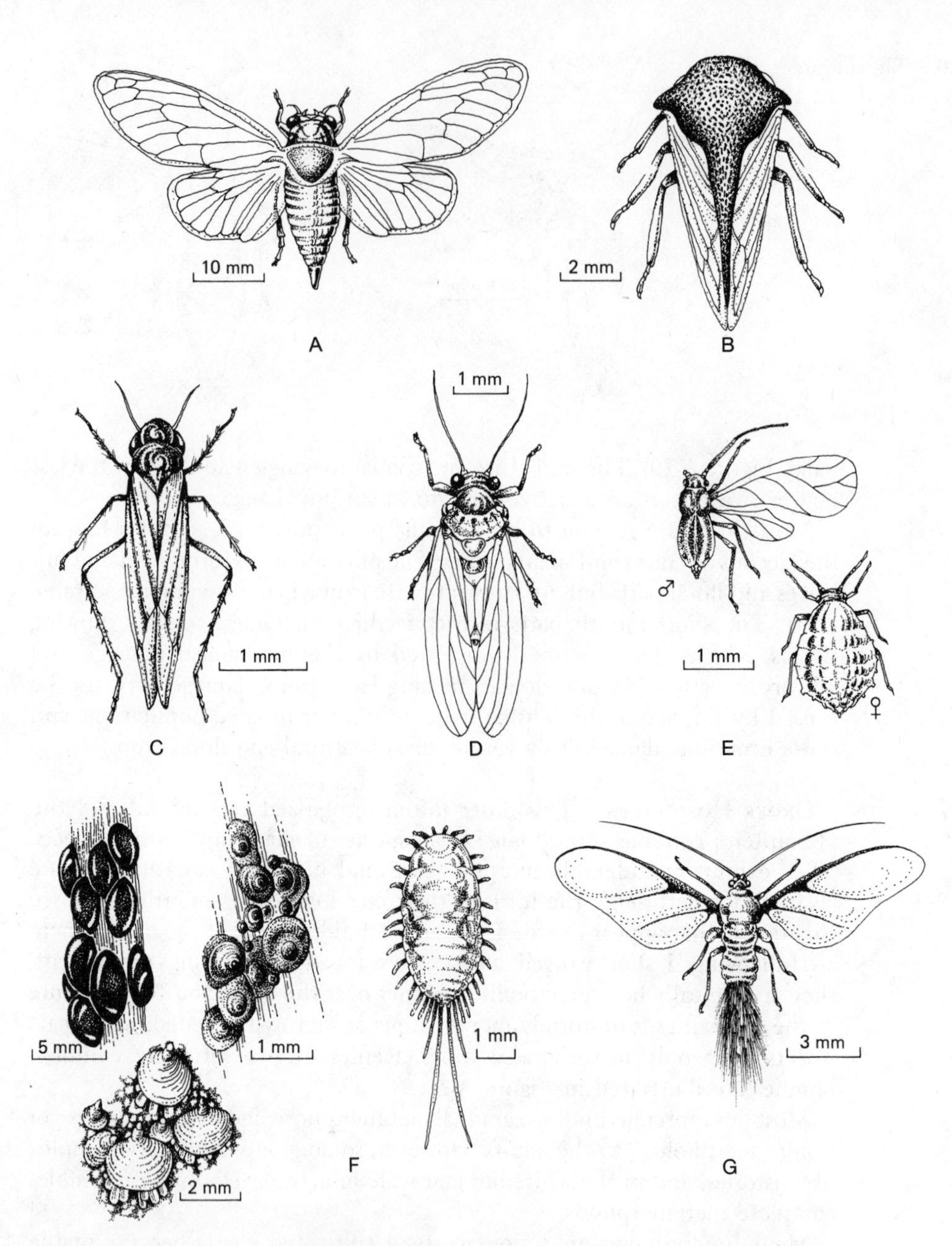


**FIGURE A-21.** Common families of Homoptera. **A.** Cicadidae (cicada). **B.** Membracidae (tree hopper). **C.** Cicadellidae (leafhopper). **D.** Psyllidae (psyllid). **E.** Aphididae (aphid). **F.** Coccoidea (female scale insects). **G.** Coccoidea (male scale insect).

area of the fore wings is lacking. Hemipterans all display gradual metamorphosis and the nymphs resemble the adults rather closely. The life cycle is uncomplicated. Representatives of some of the more common families are illustrated in Figure A-22.

The Hemiptera are widely distributed and although primarily a terrestrial group, a few families are well adapted to an aquatic life style. The group is composed primarily of plant sap-feeders and some are serious pests of culti-

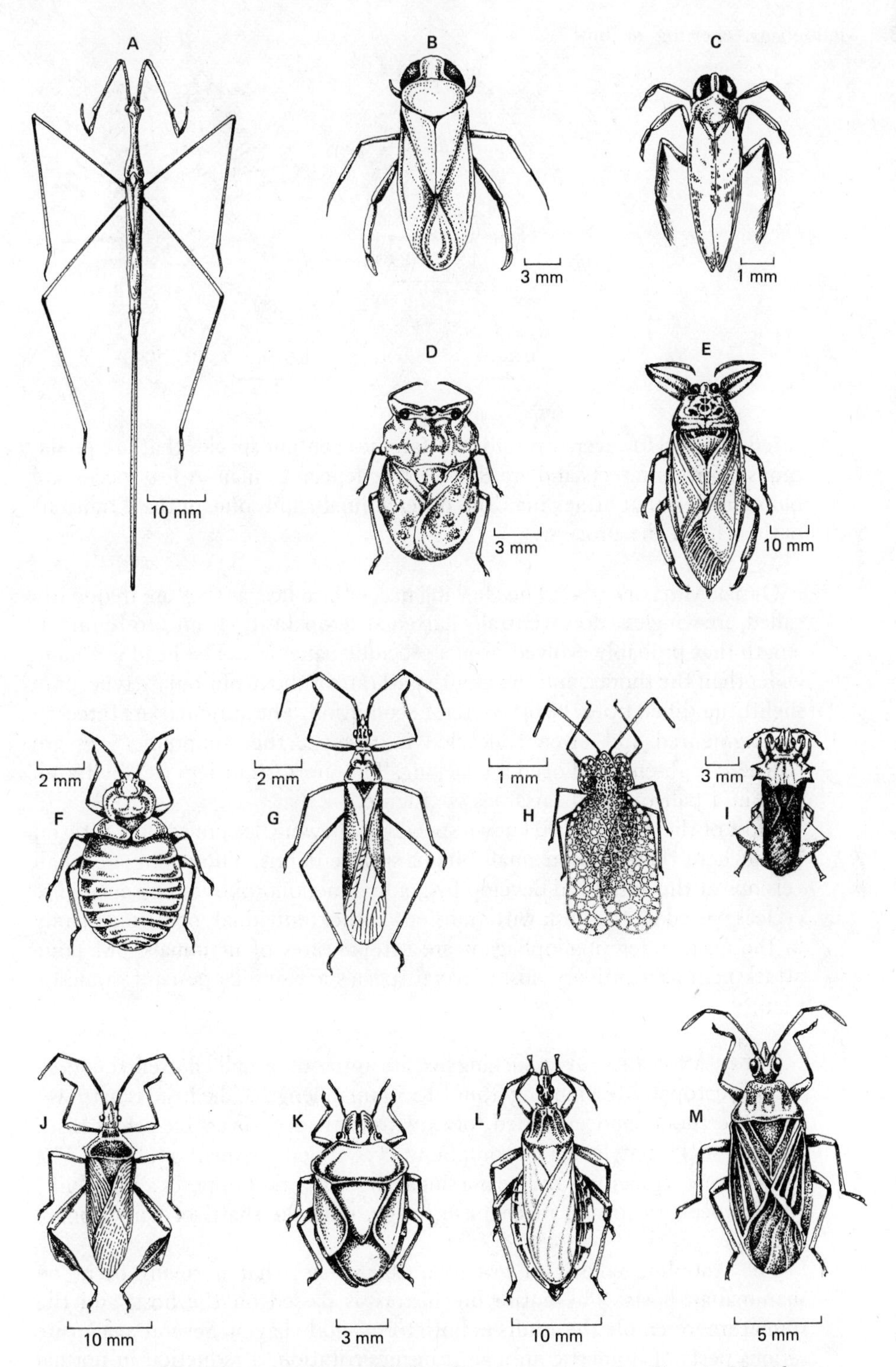


**FIGURE A-22.** Hemiptera, bugs. Adults of some common families. **A.** Nepidae (water scorpion). **B.** Corixidae (water boatman). **C.** Notonectidae (backswimmer). **D.** Gelastocoridae (toad bug). **E.** Belostomatidae (giant water bug). **F.** Cimicidae (bed bug). **G.** Nabidae (damsel bug). **H.** Tingidae (lace bug). **I.** Phymatidae (ambush bug). **J.** Coreidae (leaf-footed bug). **K.** Pentatomidae (stink bug). **L.** Reduviidae (assassin bug) **M.** Lygaeidae (Lygaeid bugs).

FIGURE A-23. Mallophaga, chewing or bird lice.

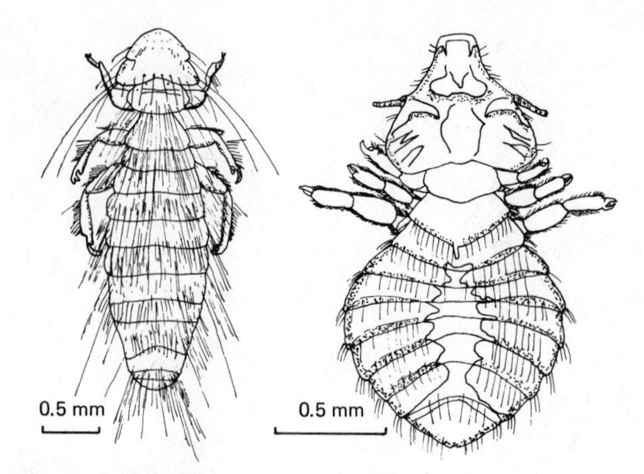


vated plants. However, a number of families contain species that are predaceous on other insects and are therefore beneficial to man. A few species are blood-feeders that attack man and other animals, and some of these transmit pathogens in the process.

ORDER MALLOPHAGA. The chewing lice, or bird lice, as they are frequently called, are wingless, dorsoventrally flattened ectoparasites from 2 to 10 mm in length that probably evolved from a psocidlike ancestor. The head is usually wider than the thorax, and the mouth parts are of the mandibulate type, only slightly modified from the orthopteroid condition. The antennae are three- to five-segmented and often concealed in recesses; the compound eyes are reduced or absent, and oceli are lacking. The short, stout legs usually terminate in a pair of small tarsal claws (Figure A-23).

Most of the nearly 3,000 known species of chewing lice are ectoparasitic on birds where they feed on small bits of surface debris. The young are small versions of the adult and develop by gradual metamorphosis. The entire life cycle is passed on the host, with transfer to other individuals occurring mainly in the nest. A few mallophagans are ectoparasites of mammals, but none attack man as a primary host. Several species are serious pests of domestic poultry.

ORDER ANOPLURA. The sucking lice are apterous, usually flattened, blood-sucking ectoparasites, ranging from 1 to 5 mm in length. The head is narrower than the thorax and extended forward to enclose an eversible set of three piercing-sucking stylets. The compound eyes are either small or absent. The tarsi are one-segmented and bear a single large claw that opposes a thumblike tibial process to form a mechanism for grasping the shafts of hairs (Figure A-24A).

The Anoplura consist of less than 300 species that normally occur on mammalian hosts. The entire life history is passed on the host, and the immatures resemble the adults in both form and behavior. Several species are serious pests of domestic animals, causing irritation, a reduction in normal weight-gain, and a general loss of vigor. Two species, the crab louse, *Phthirius pubis*, and the body louse, *Pediculus humanus*, are ectoparasites of man (Figures A-24*B*, *C*).

***Endopterygote Complex***

All of the orders in the endopterygote complex are characterized by complete metamorphosis during which the wings develop internally. The immatures are

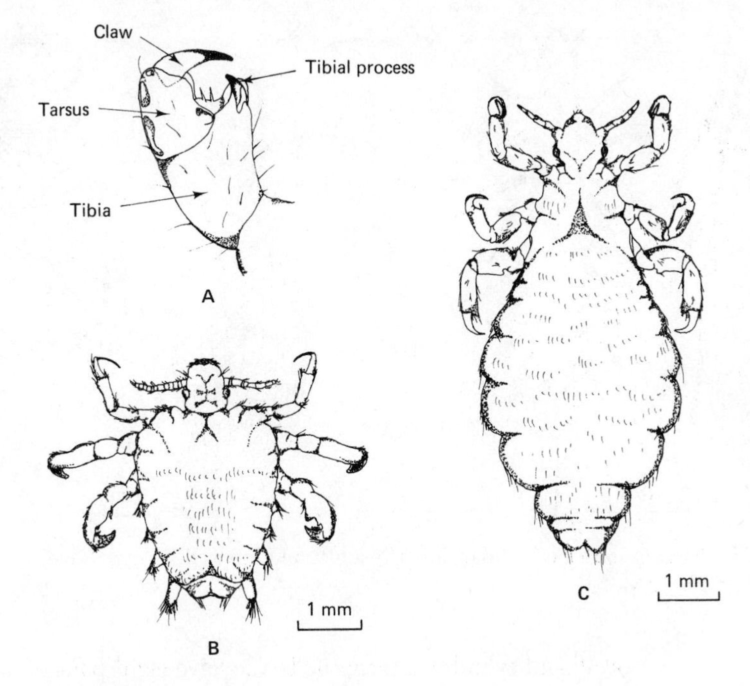


**FIGURE A-24.** Anoplura, sucking lice. **A.** Grasping tarsal claw. **B.** The crab louse *Phthirius pubis.* **C.** Human louse, *Pediculus humanus.*

transformed to adults through a pupal stage during which there is considerable tissue reorganization and muscle histolysis. The complex is morphologically diverse, ranging from rather primitive members such as the Megaloptera, which have retained some of the primitive characteristics of the orthopteroids, to highly specialized forms such as the fleas.

All the endopterygote orders seem to be derivable from a megalopteran-type ancestor. Each major order has made some adaptive morphological breakthrough that allowed it to exploit a new situation and diversify in a different direction. Yet, the primitive members of this complex all show some affinity with the Megaloptera. Although the adults of orders such as the Coleoptera, Mecoptera, Trichoptera, Lepidoptera and Hymenoptera are very different, the larvae of the primitive members of each of these groups are clearly related; all have reduced antennae, compound eyes replaced by groups of lateral ocelli, reduced body sclerotization, unsegmented tarsi, and a single tarsal claw.

Of all the orders, the Coleoptera share the fewest characters with other members of the complex, perhaps as a result of their early divergence and long separate evolution. Their hard exoskeleton and the modification of the fore wings into elytra seem to have enabled them to expand into a number of unoccupied niches early in their evolution without undergoing many major structural modifications.

ORDER MEGALOPTERA. The dobsonflies and alderflies are rather large, soft-bodied insects with subequal, membranous wings with numerous cross veins; long, many-segmented antennae; unmodified chewing mouth parts, except for some elongation of the mandibles in the dobsonflies, and modified cerci. The two groups that comprise the order can be distinguished readily in both the adult and larval stages. The dobsonfly adults (family Corydalidae) have three

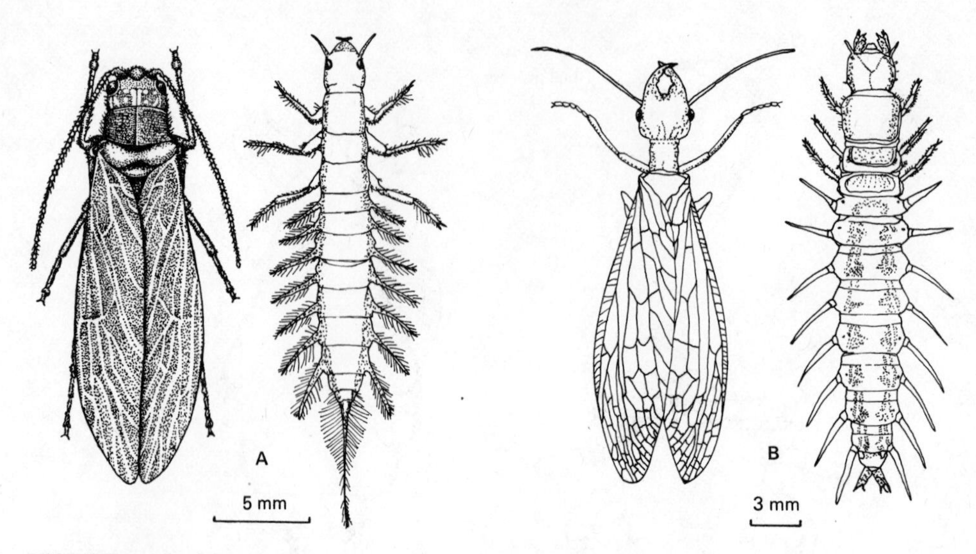


FIGURE A-25. Megaloptera. **A.** Sialidae, alderfly adult and larva, **B.** Corydalidae, dobsonfly adult and larva.

ocelli and cylindrical tarsi; the larvae have eight pairs of filamentous abdominal gills, no caudal filament, and a pair of anal prolegs (Figure A-25*A*). The alderfly adults (family Sialidae) lack ocelli and have bi-lobed, fourth tarsal segments; the larvae have seven pair of featherlike abdominal gills, a caudal filament, and no anal prolegs (Figure A-25*B*). Combined, there are only about 300 described species of megalopterans.

Adult megalopterns are terrestrial and usually found near fresh water. The larvae are aquatic predators of other aquatic insects and are usually found singly under stones in clean running streams. They can be fiercely cannibalistic, so they are usually found in rather low numbers.

Order Raphidioptera. There are approximately 100 known species of snake flies, divided into two families and separated on the basis of the presence or absence of ocelli. They are all delicate, medium-sized insects with subequal membranous wings held rooflike over the body at rest. Though closely related to megalopterans, they are readily identified by their elongate prothorax, which gives the impression that the head is mounted on a long serpentine neck (Figure A-26*A*). The larvae, illustrated in Figure A-26*B*, are elongate and soft-bodied, except for the well-sclerotized head and prothorax. The legs are relatively long and there are not abdominal projections or appendages.

Both the adults and larvae are terrestrial and predaceous. The adults are usually found among vegetation where they prey on small weak insects such as aphids. The larvae occur most commonly under loose bark where they feed on a variety of soft-bodied insects.

Order Neuroptera. The Neuroptera is a fairly diverse order containing about 4,500 known species. The adults are small to large soft-bodied insects, usually with two pair of membranous wings with numerous veins and cross veins, long antennae, chewing mouth parts and prominent, compound eyes. Beyond these similarities, the body shape varies considerably, as illustrated by

**FIGURE A-26.** Raphidioptera, snakeflies. **A.** Adult. **B.** Larva.

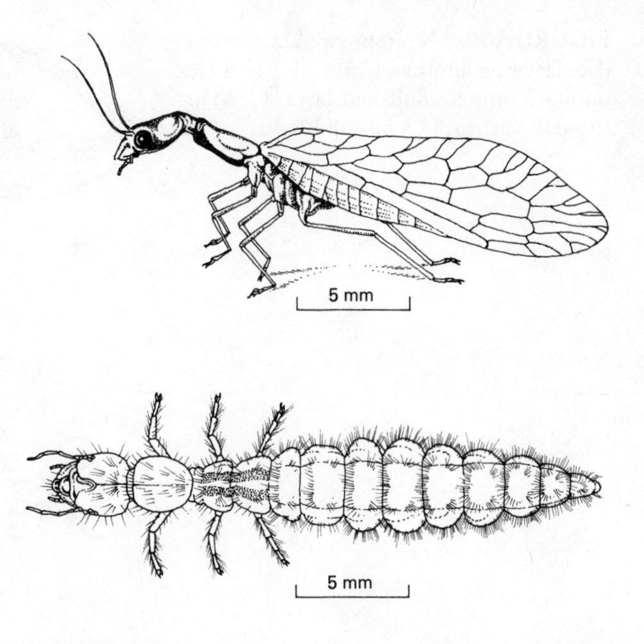


the three major groups shown in Figure A-27. The larvae are variable, as illustrated, but are characterized by their modified mouth parts. The mandibles fit together with the maxillae to form a pair of sucking jaws.

Both adults and larvae are predaceous. The lacewings are found wherever there are numerous aphids and other small, soft-bodied prey; several species are important in the natural control of pestiferous aphids. The mantispids are quite rare and like the mantids which they resemble are sedentary predators; the larvae feed in wasp nests and on spider egg masses. The larvae of ant lions bury themselves at the bottom of pits that serve as effective fall-traps for the terrestrial insects such as ants upon which they feed.

Order Coleoptera. The beetles form by far the largest order of insects. Well over 300,000 species have been described, and many new species are discovered each year. In spite of this enormous diversification, the order is well unified by the modification of the fore wings into sclerotized shields called elytra, and the retention of chewing mouth parts quite similar to those of the orthopteroid orders. The beetles have been variously grouped into in excess of 150 families of which roughly one third could be considered common in most populated parts of the world. Many of these are readily identified on the basis of their highly characteristic body conformation. Representatives of a few such families are illustrated in Figure A-28.

The larvae of beetles are highly variable but generally fall into one of the four groups illustrated in Figure A-29. They all have a well-defined, sclerotized head capsule and mouth parts similar to the adults. Antennae are always present. The ocelli and thoracic legs, however, are variously developed and may be absent. Some species pass through two or more distinctly different larval forms in the course of their development.

The Coleoptera form a group with great ecological diversity, but in many families the larvae and adults share similar situations and resources. Many are herbivorous and as such are well known as pests of forest and agricultural ecosystems. On the other hand, a number of families are predaceous or

**FIGURE A-27.** Neuroptera. **A.** Chrysopidae, lacewing adult and larva. **B.** Myrmeleontidae, antlion adult and larva. **C.** Mantispidae, mantispid adult and larva.

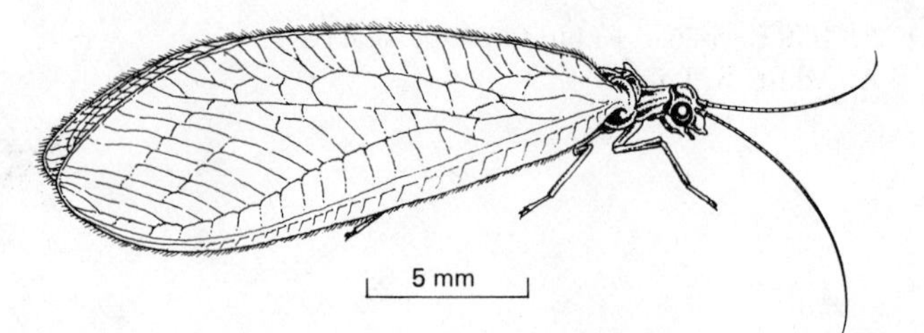


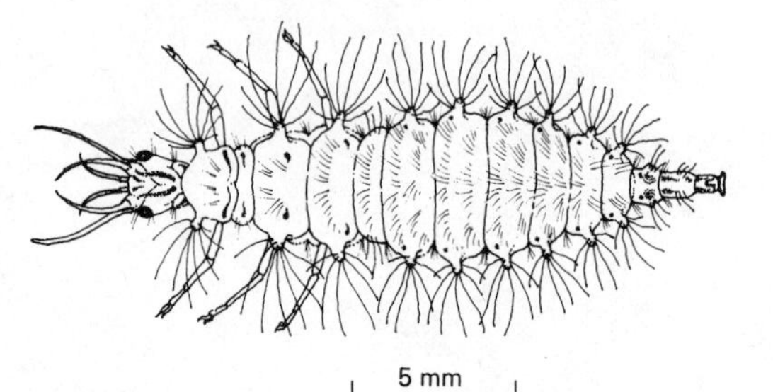


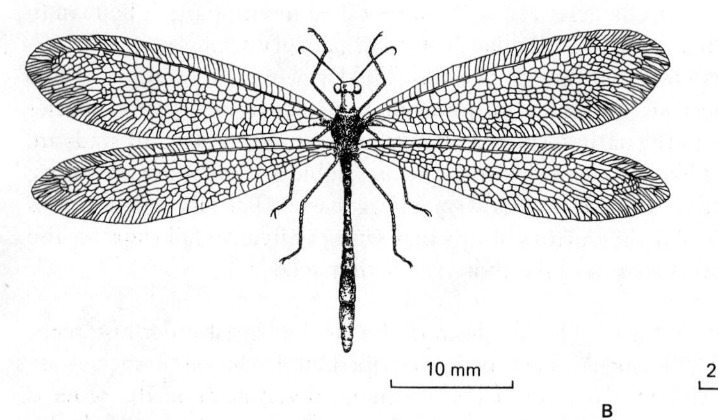


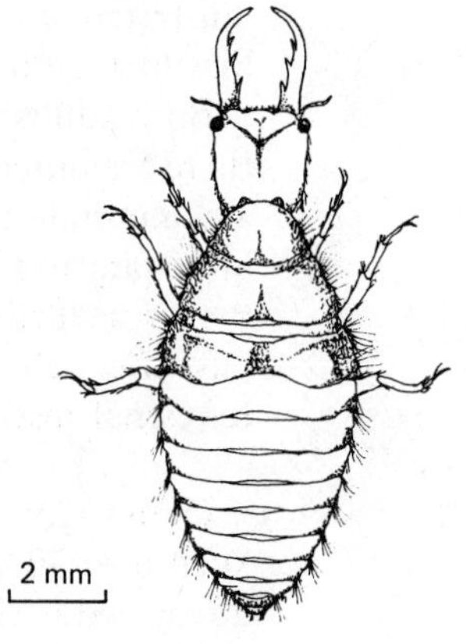


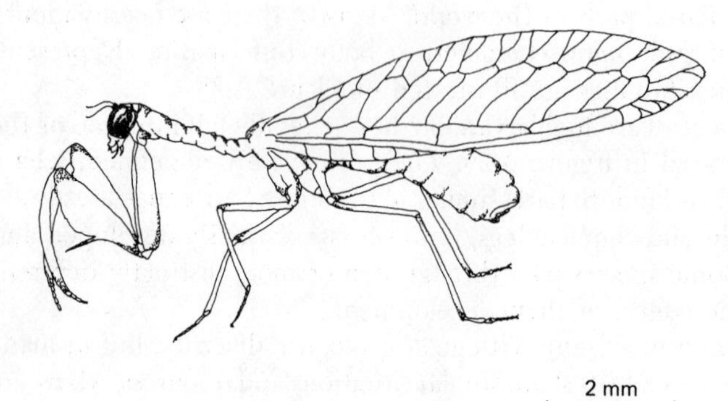


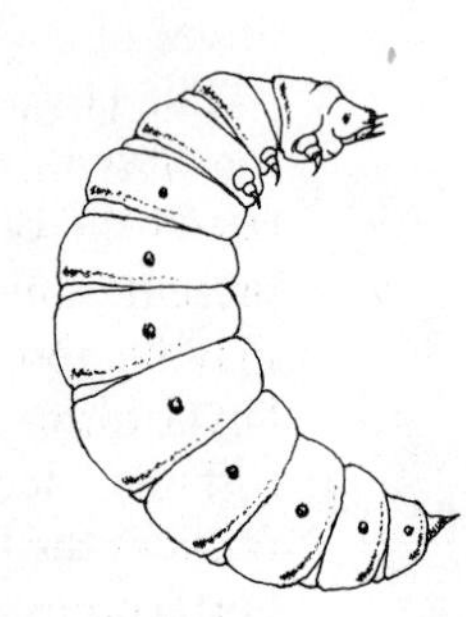

parasitic and play an important role in the natural regulation of other insects. Many different families feed on dead plant and animal material and are as a consequence important in the recycling of nutrients bound up in all forms of organic debris. A few families even have representatives that have become parasitic on mammals. There may in fact be no group of animals that consumes a wider range of foods than do the beetles.

ORDER STREPSIPTERA. The small twisted-winged parasites are a highly specialized and rarely encountered group consisting of about 300 described species. The females are larviform parasitoids, but the males are free-living insects with functional wings. They are readily characterized by their flabellate antennae, small elytralike fore wings and large hind wings with greatly reduced venation (Figure A-30).

Strepsipterans are endoparasites of a number of insects, especially homopterans and aculeate Hymenoptera. The first instar larva (triungulin) actively gains access to a host. The remaining instars are legless and develop as internal parasitoids. Pupation occurs in the host; upon emergence, the male leaves to seek and mate with the females that remain in their host except for the combined head and thorax, which protrudes from the host's body. The stages of this hypermetamorphic development are illustrated in Figure 9-9.

ORDER MECOPTERA. The scorpionflies are small- to medium-sized, soft-bodied insects with chewing mouth parts extended ventrally into snoutlike structures. The body is quite slender as are the legs, which have enlarged and elongated coxae. They have two pair of rather slender, membranous wings with complex venation; in a few species, the wings are short and aborted. Males of the family Panorpidae have a bulbous recurved genital structure at the tip of the abdomen that resembles the sting of a scorpion (Figure A-31).

The adults are usually found among dense vegetation where they feed on small insects, pollen, nectar, and tender vegetation. The larvae are found in moist, decaying vegetable matter or moss and feed on organic debris.

ORDER TRICHOPTERA. Caddisfly adults are mothlike insects, ranging from 1.5 to 40 mm in length. They have prominent compound eyes, long filamentous antennae, and reduced chewing mouth parts except for the well-developed palps. Most of the roughly 5,000 described species have two pair of large wings with an almost full complement of longitudinal veins but few cross veins. The body, wings, and slender legs are usually clothed with hairs (Figure A-32).

The larvae (Figure A-33) are variable in form but usually have a well-sclerotized head capsule and thorax and a soft abdomen. They have well-developed chewing mouth parts, one-segmented antennae, and a single pair of lateral ocelli. The thoracic legs are well developed, but they lack ventral abdominal prolegs; the last segment bears a pair of terminal prolegs that terminate in a stout claw used to anchor the larva to the substrate or in its case. Some have filamentous, lateral, abdominal gills.

Caddisfly adults are mainly crepuscular or nocturnal and are most commonly found near water. The aquatic larvae live in both lakes and streams with cold, clean water, where they are often one of the most important groups in the food chain. They are quite active and feed mainly as grazers on the micro-organisms that encrust organic debris. A number build silken traps that filter drifting material from the current. A few are predaceous on the larvae of

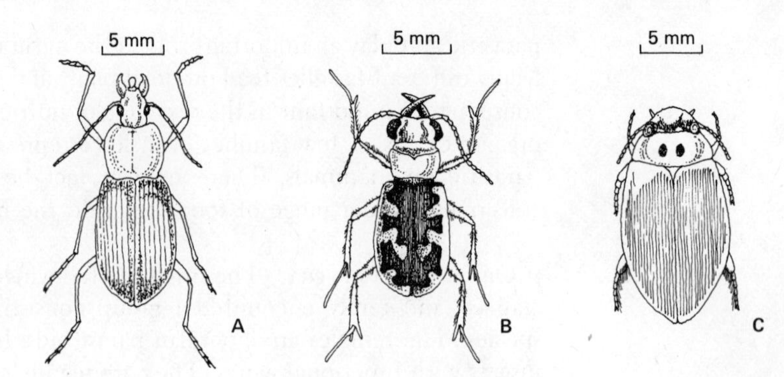
5 mm
5 mm
5 mm
A
B
C

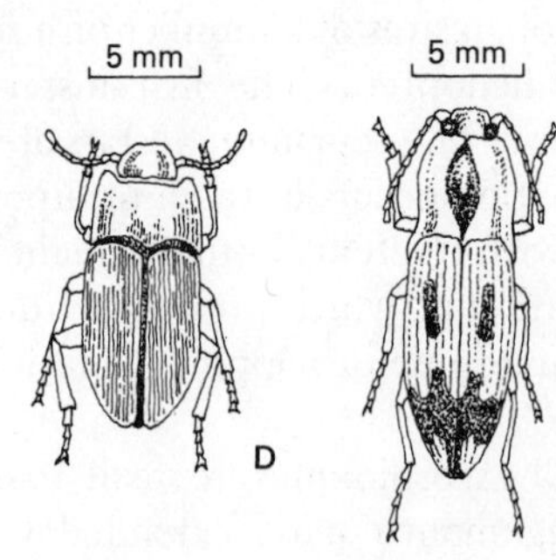
5 mm
5 mm
D
E

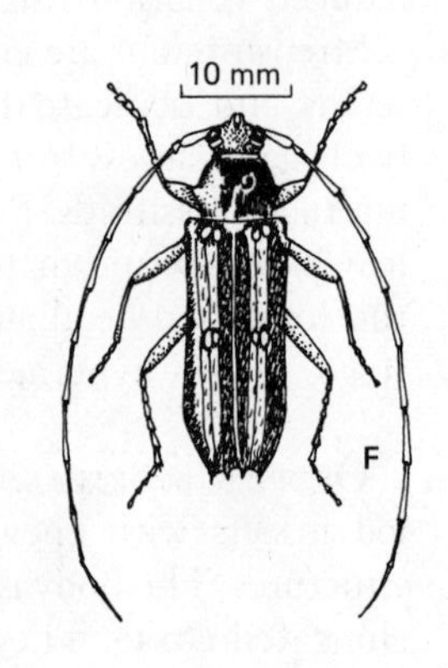
10 mm
F

5 mm
G

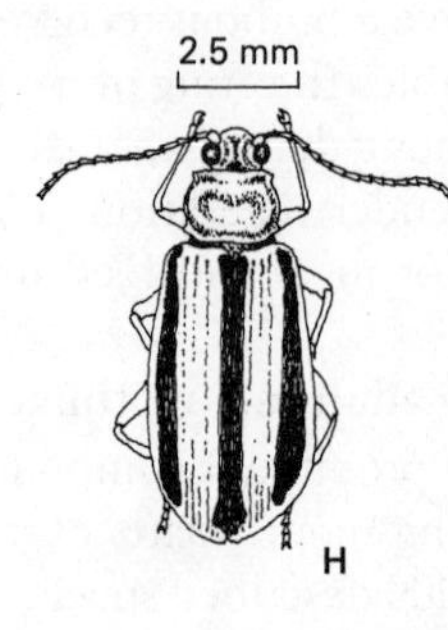
2.5 mm
H

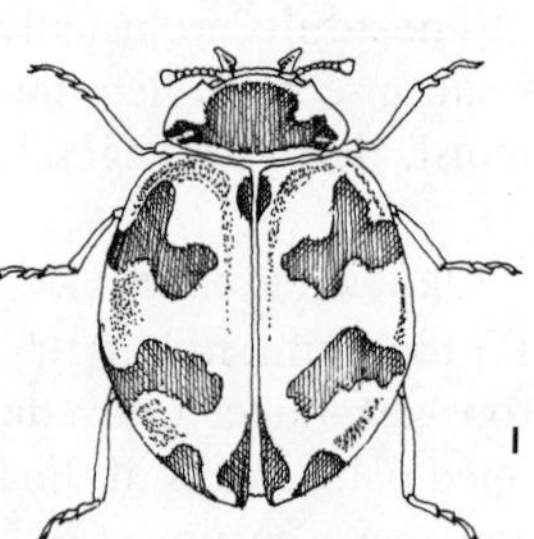
1 mm
I

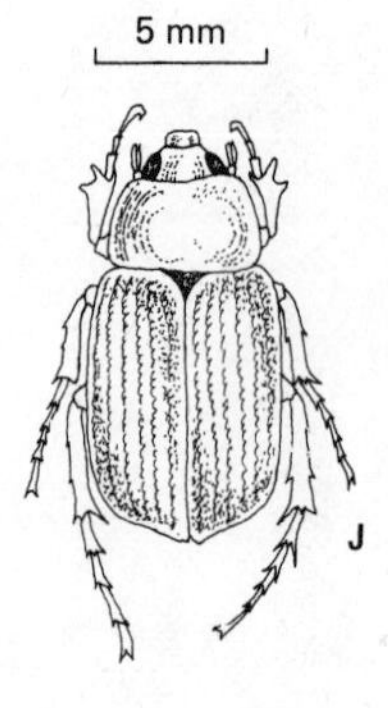
5 mm
J

K

5 mm

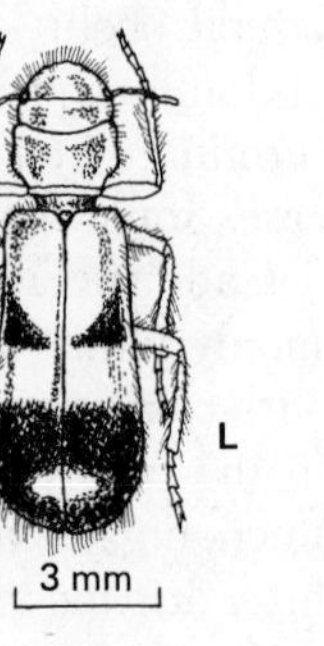
L
3 mm

M
3 mm

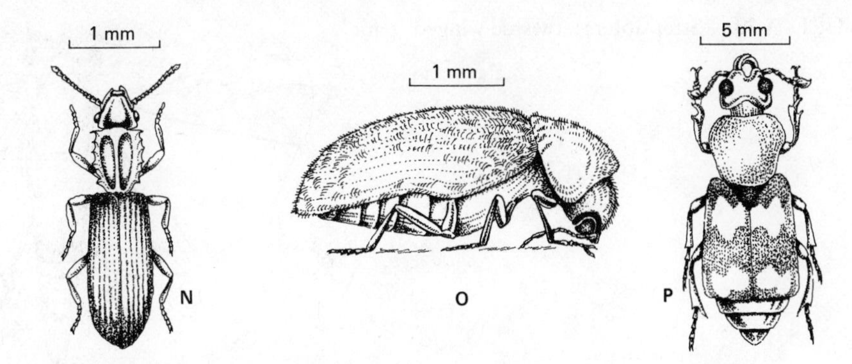


FIGURE A-28. Coleoptera, beetles. Representatives of some common families. **A.** Carabidae (ground beetle). **B.** Cicindelidae (tiger beetle). **C.** Dytiscidae (diving water beetle) **D.** Tenebrionidae (darkling beetle). **E.** Elateridae (click beetle). **F.** Cerambycidae (long-horned beetle). **G.** Buprestidae (metalic wood-boring beetle). **H.** Chrysomelidae (leaf beetle). **I.** Coccinellidae (lady beetle). **J.** Scarabaeidae (scarab beetle). **K.** Staphylinidae (rove beetle). **L.** Cleridae (checkered beetle). **M.** Curculionidia (weevil). **N.** Cucujidae (flat bark beetle). **O.** Anobiidae (drugstore beetle). **P.** Silphidae (carrion beetle).

other aquatic insects. The larvae of most families construct cases that protect the body and later serve as a pupation site (see Figure 15-10).

Order Lepidoptera. The Lepidoptera consists of approximately 120,000 species, commonly called butterflies, moths, and skippers. Although ranging in size from 3 to 250 mm and varying greatly in form and color, most lepidopterans are readily distinguished by a dense covering of overlapping scales and the modification of the mouth parts into a specialized sucking proboscis. Most have large compound eyes, well-developed antennae, and long legs. Though not always the case, butterflies usually have knobbed

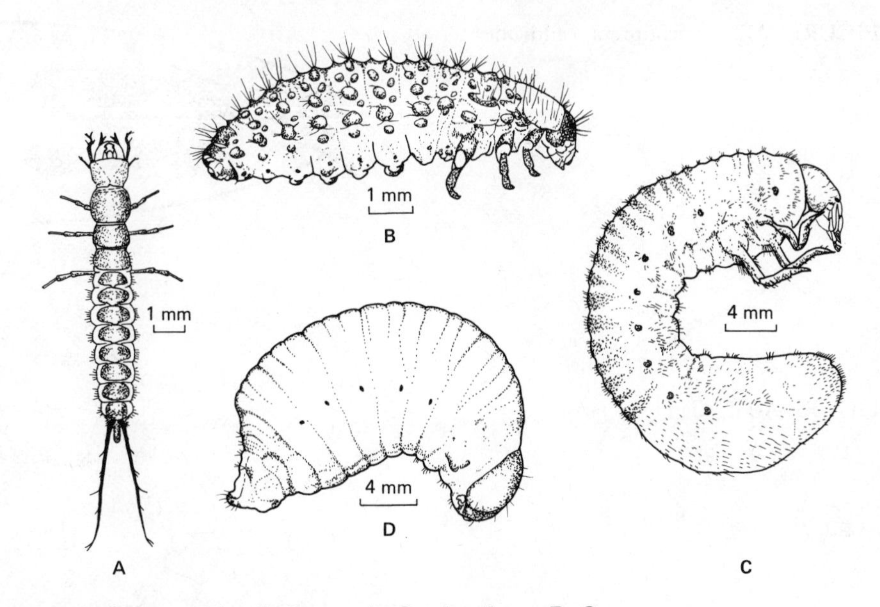


FIGURE A-29. Four principal types of beetle larvae. **A.** Carabaeiform. **B.** Campodeiform. **C.** Scarabaeiform. **D.** Apodus.

**FIGURE A-30.** Strepsiptera, twisted-winged parasites.

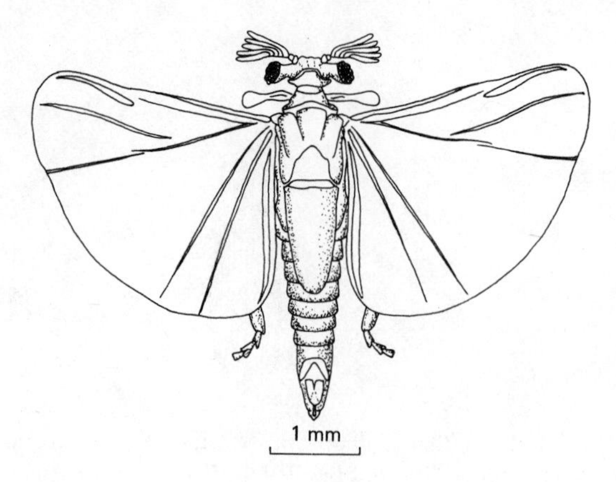


**FIGURE A-31.** Mecoptera, scorpionflies, adult male.

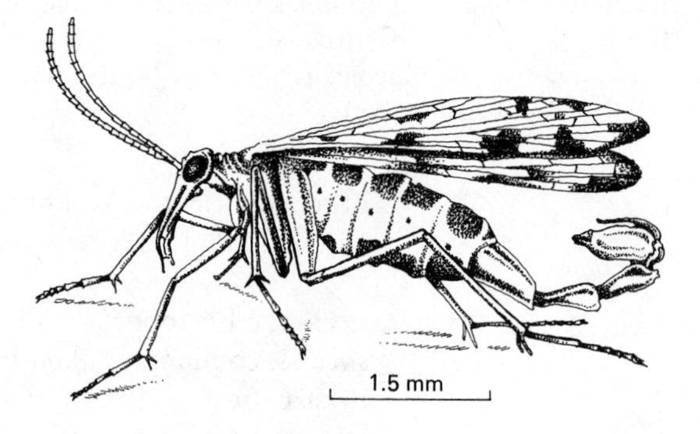


**FIGURE A-32.** Trichoptera, caddisflies.

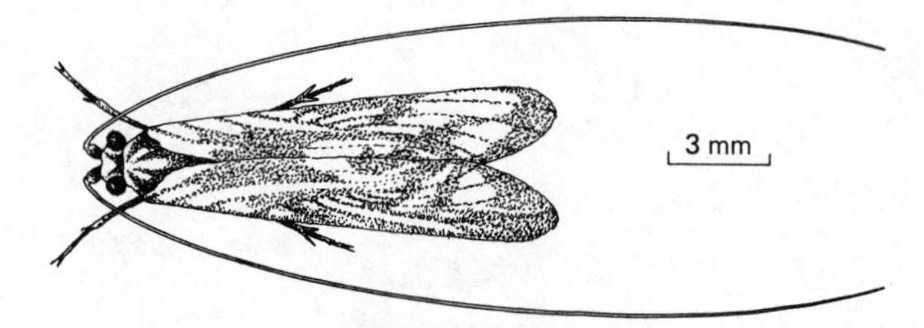


**FIGURE A-33.** Caddisfly larva.

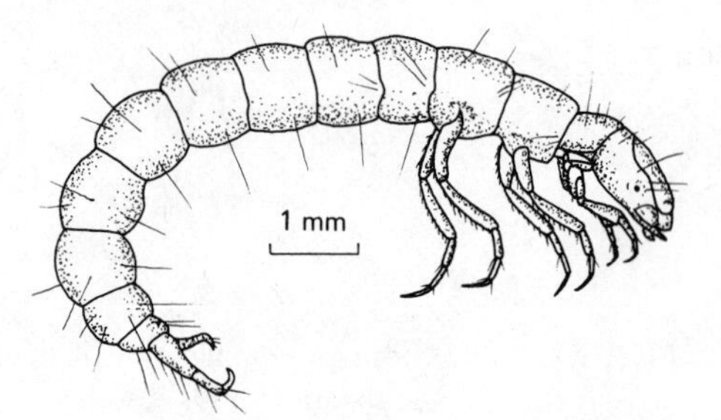

antennae, are generally of medium to large size, and have brightly colored wings that are not folded over the body at rest. Moths, on the other hand, may be very small, do not have knobbed antennae, are often dull colored, and usually fold their wings against the body. Some of the variation displayed by the adults of this group is shown in Figure A-34.

The larvae are commonly called caterpillars (Figure A-35). They have a well-sclerotized head capsule, clearly defined thorax, and ten-segmented abdomen. The mouth parts are of the chewing type and all have well-developed mandibles. Most have a group of six ocelli just behind the base of the three-segmented antennae. Each thoracic segment usually bears a pair of five-segmented legs terminating in a single claw. The abdomen has from one to five pair of ventral prolegs on segments three through six, and a pair of anal prolegs on segment 10; the prolegs bear a number of small hooks or crochets.

Most adult lepidopterans are nectar feeders, but some do not feed at all, and a few primitive forms are predaceous. As a group they are important as pollinators of flowering plants. The larvae are mainly phytophagous and are the most important group of forest and agricultural pests, causing large-scale, direct and indirect damage to wide varieties of crops and stored products.

ORDER DIPTERA. The flies form another large and diverse order containing close to 90,000 known species. Although a few are wingless, the adults are characterized by a single pair of membranous mesothoracic wings with somewhat reduced venation; the hind wings are reduced to clublike balancing organs called halteres. The head is usually well defined and bears a pair of antennae of variable form. The mouth parts are basically suctorial but are highly adapted for cutting, piercing, sponging, and rasping (see Figures 6-5 to 6-7). The thorax is characterized by a pronounced development of the wing-bearing segment. Some of the more common families of flies are illustrated in Figure A-36 to provide an indication of the structural diversity of the order.

The larvae are as variable in form as the adults. They are most commonly cylindrical and soft-bodied, and true, segmented, thoracic legs are never present. Larvae of families in the Suborder Nematocera have a definite head capsule, antennae, and mandible that move in a horizontal plane. In the more primitive members of the Suborder Brachycera, the head capsule is reduced and the mandibles move vertically. In the higher Brachycera, there is no defined head and a pair of mouth hooks replace the more normal mouth parts. Several larval types are illustrated in Figure A-37.

Adult flies abound in a wide variety of situations. Frequently, they occur in such large numbers as to constitute a nuisance. A number of species require a blood meal before they can produce viable eggs. Such species can cause considerable irritation to man and his domestic animals, and some are important as obligate vectors of major human pathogens. A number of scavenging species mechanically transmit the micro-organisms that contaminate their food and breeding places. The larvae range from terrestrial to aquatic and from saprophagous to parasitic. As a consequence, many species are highly beneficial, whereas others are serious pests.

ORDER SIPHONAPTERA. Adult fleas are small, highly specialized, wingless, ectoparasitic insects of rather recent origin (Figure A-38). All of the roughly 1,200 known species are similar in that they are laterally compressed and have

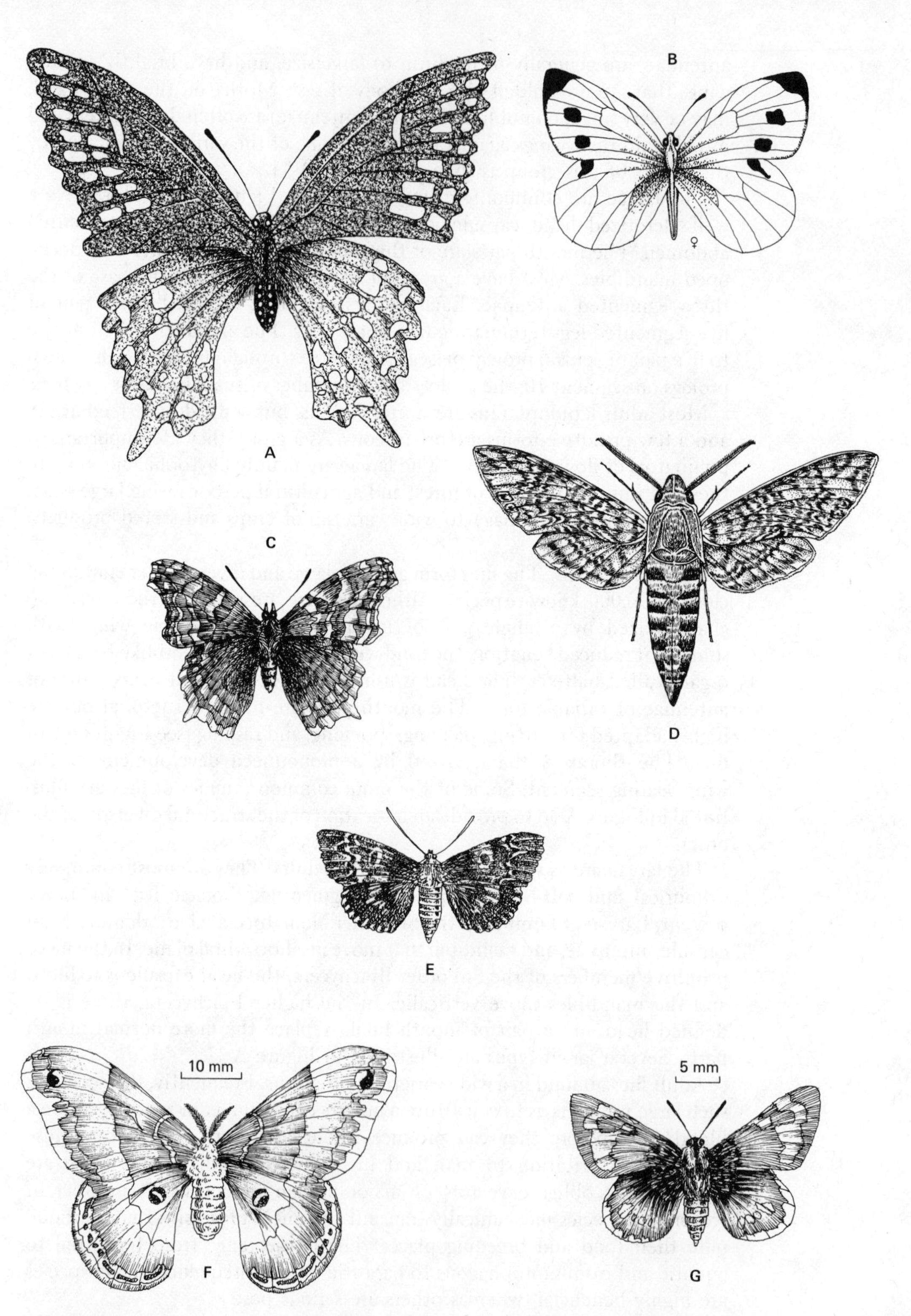


**FIGURE A-34.** Lepidoptera, butterflies, skippers, and moths. **A.** Papillionidae (swallowtail). **B.** Pieridae (cabbage). **C.** Nymphalidae (angle wing). **D.** Sphingidae (sphinx moth). **E.** Phalaenidae (noctuid moth). **F.** Saturniidae (promethea moth). **G.** Hesperiidae (skipper).

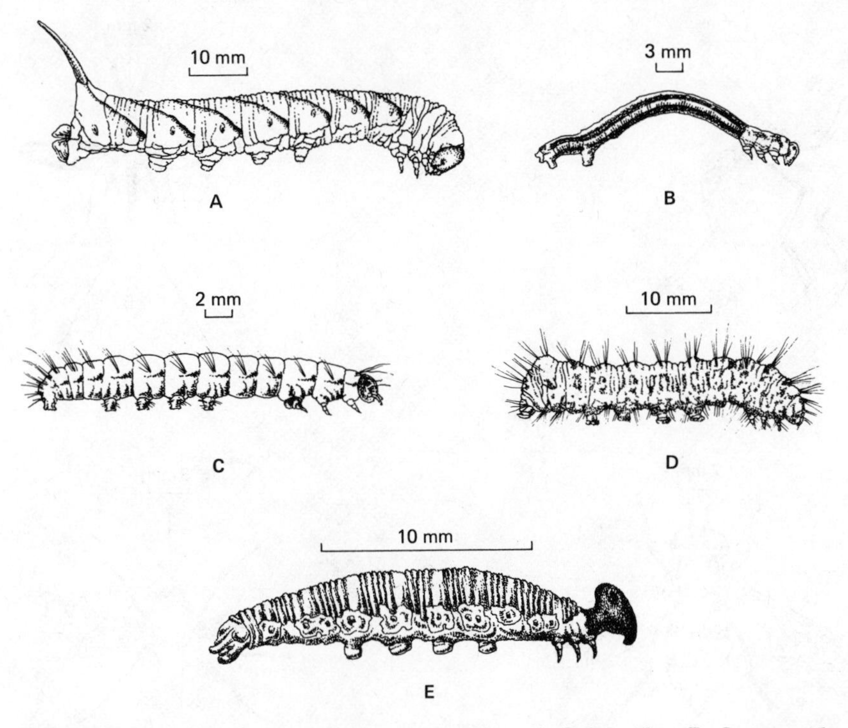


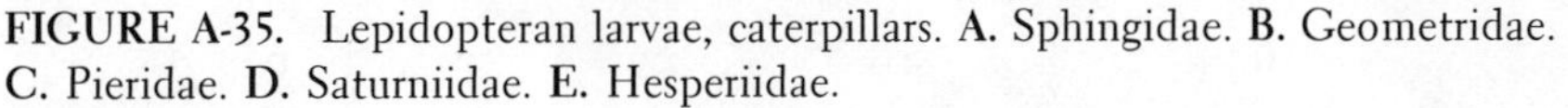

**FIGURE A-35.** Lepidopteran larvae, caterpillars. **A.** Sphingidae. **B.** Geometridae. **C.** Pieridae. **D.** Saturniidae. **E.** Hesperiidae.

unusual piercing-sucking mouth parts, no compound eyes, lateral or no ocelli, and three-segmented antennae that lie in lateral grooves. The body and legs often bear stout backward-pointing spines. The legs are usually long and well developed with the hind pair greatly enlarged for jumping. The larvae (Figure A-39) are whitish, wormlike creatures with a fairly well-developed head capsule, serrated mandibles, one-segmented antennae, and no ocelli. They have no legs and no clear differentiation of the thorax; all body segments bear a number of rigid setae.

The adults are blood-feeding ectoparasites on birds and mammals but spend much of their adult life off their host. The eggs may be laid on or off the host but hatch in the nest or on the ground. The larvae live in the soil or nest debris where they develop as scavengers. Fleas can be serious domestic pests but are mainly important as vectors of disease-causing micro-organisms.

ORDER HYMENOPTERA. Hymenopterans come in a wide variety of shapes and sizes. Some are delicate and only one tenth of a mm in length, whereas others are robust and exceed 5 cm. The adults all have chewing mandibles, but other components of the mouth parts may be adapted for lapping or sucking. The wings are usually well developed and membranous with a reduction in the longitudinal veins but a moderate number of cross veins. The hind wings are usually much smaller than the fore wings to which they are coupled by a row of hooks called a humulus. Some females have a well-developed ovipositor (modified to form a sting in some) making them unique

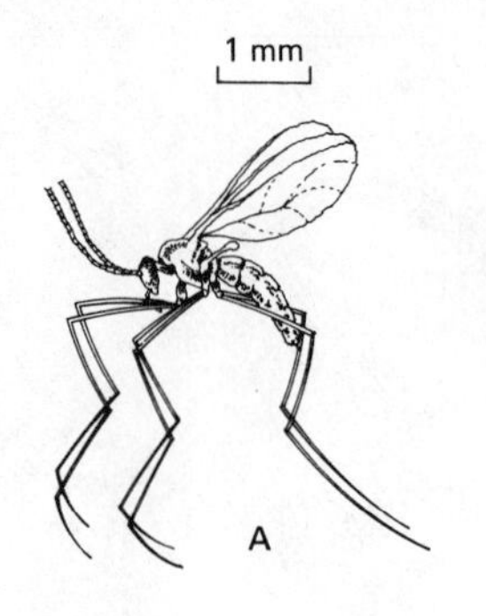


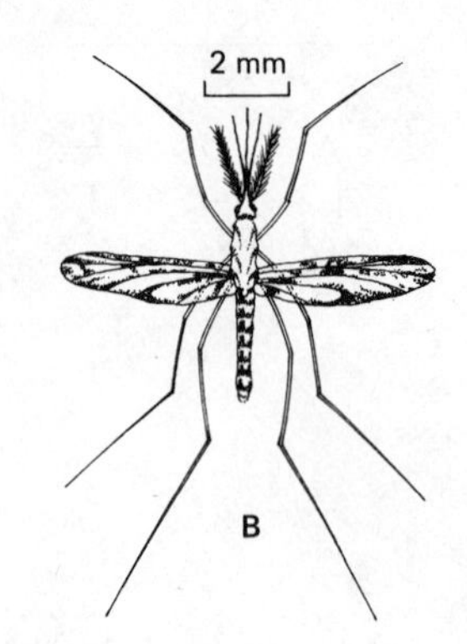


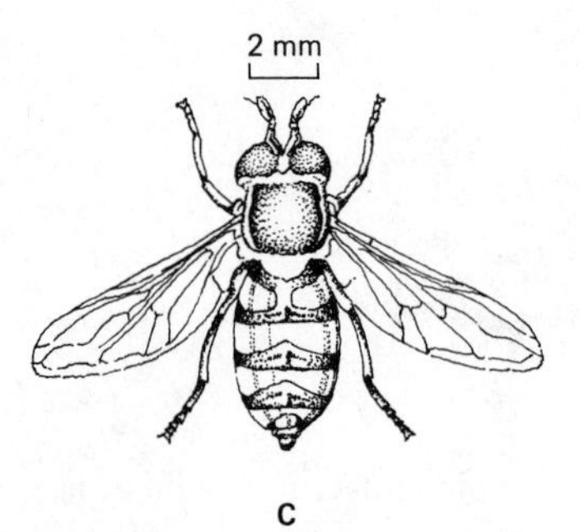


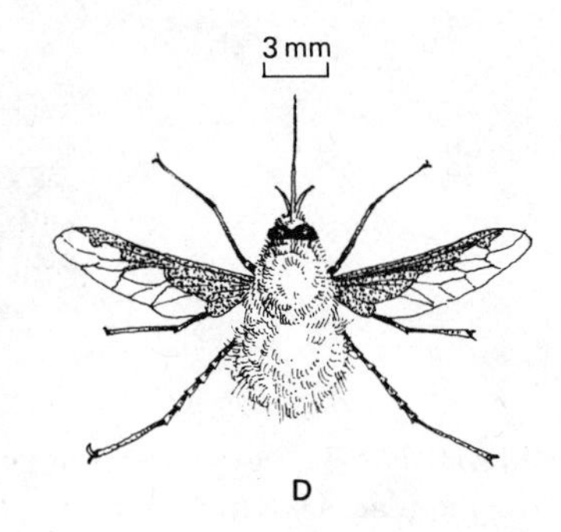


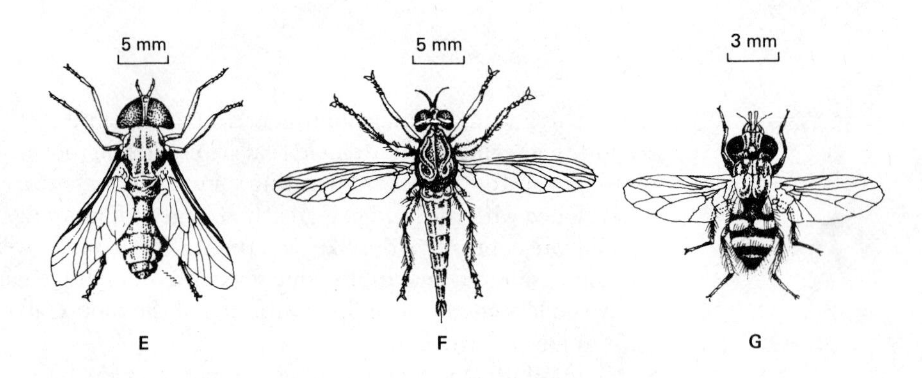


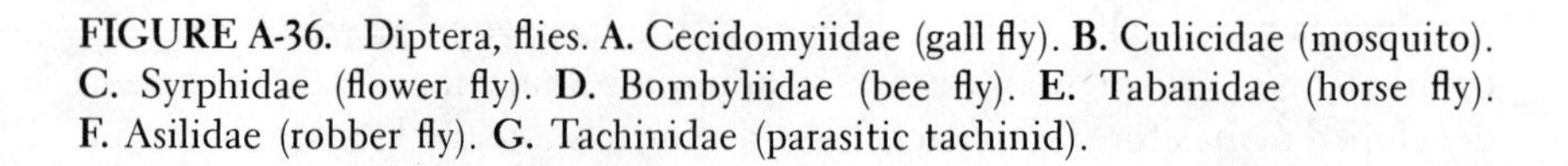
**FIGURE A-36.** Diptera, flies. **A.** Cecidomyiidae (gall fly). **B.** Culicidae (mosquito). **C.** Syrphidae (flower fly). **D.** Bombyliidae (bee fly). **E.** Tabanidae (horse fly). **F.** Asilidae (robber fly). **G.** Tachinidae (parasitic tachinid).

FIGURE A-37. Fly larvae. **A.** Culicidae (mosquito). **B.** Chironomidae (midge). **C.** Syrphidae (rat-tailed maggot). **D.** Syrphidae (predaceous flower fly). **E.** Calliphoridae (screw worm maggot).

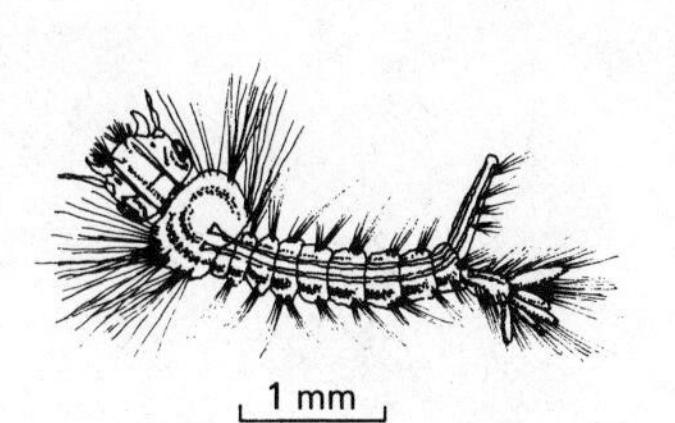


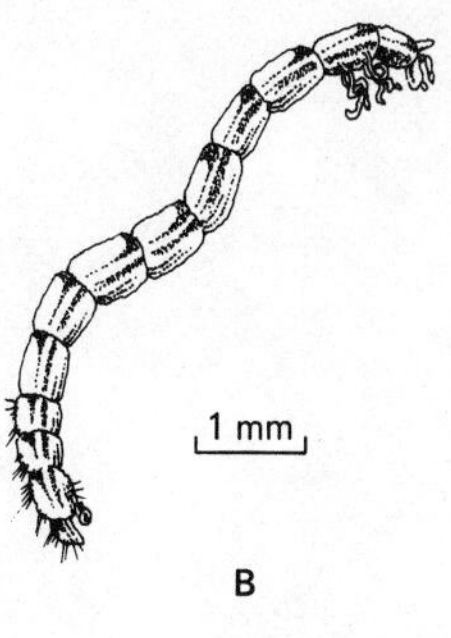


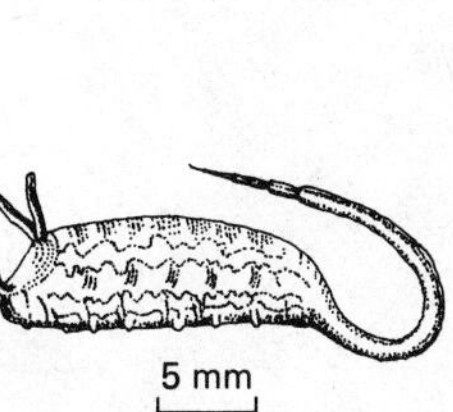

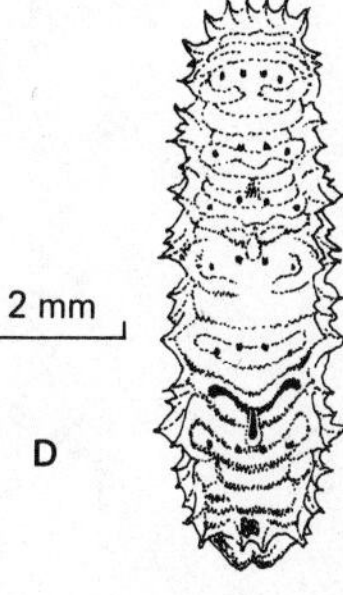


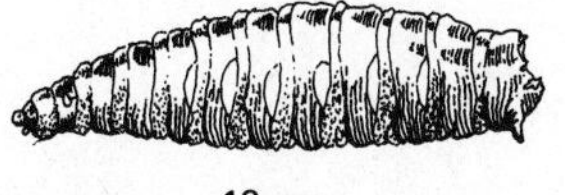


FIGURE A-38. Siphonaptera, fleas.

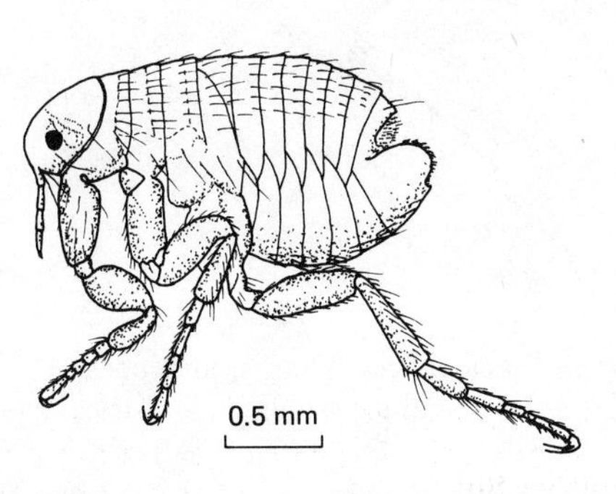


FIGURE A-39. Flea larva.

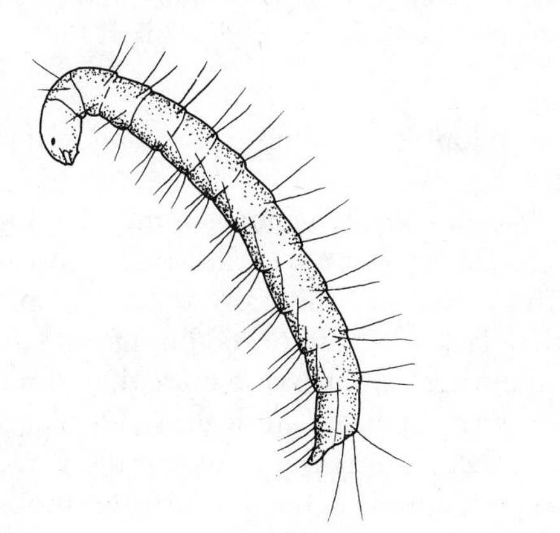

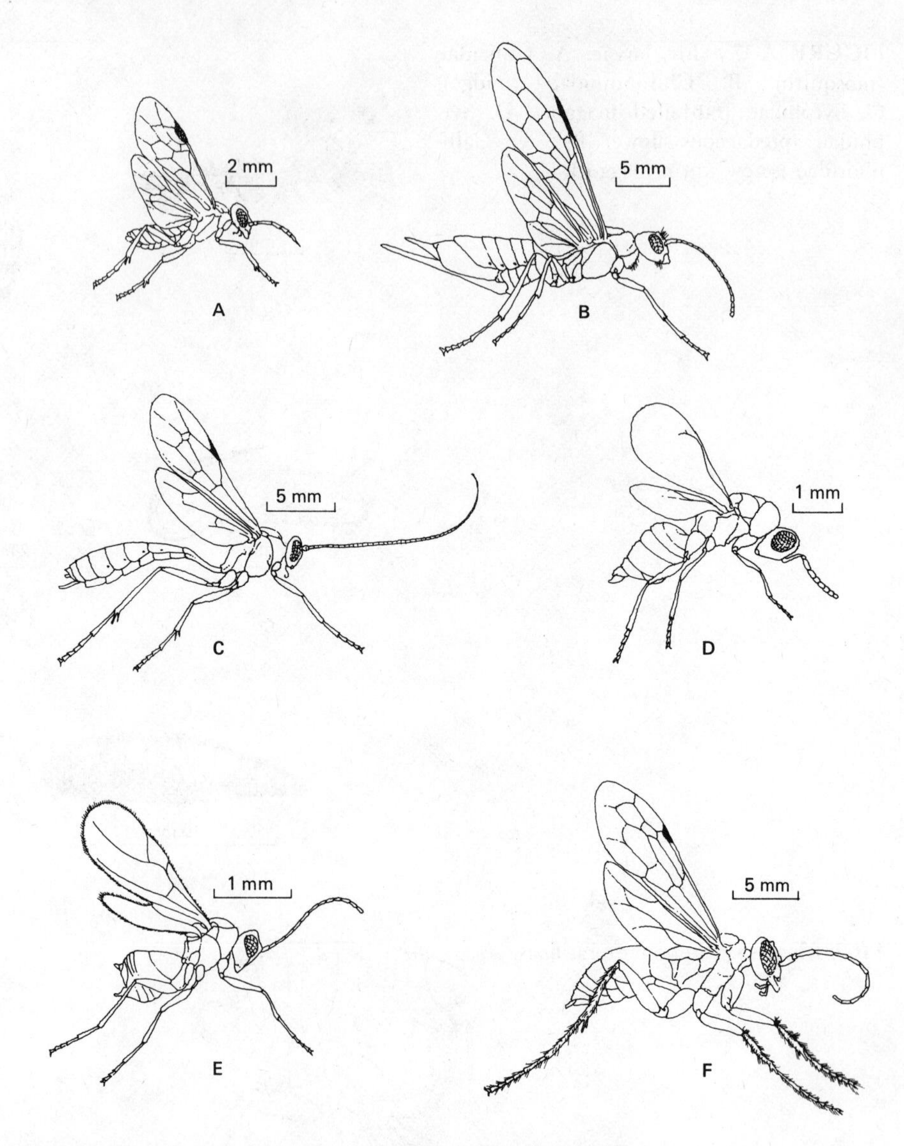


**FIGURE A-40.** Hymenoptera, sawflies, wasps, ants, and bees. **A.** Tenthridinidae (sawfly). **B.** Siricidae (horntail). **C.** Ichneumonidae (parasitic ichneuman wasp). **D.** Eurytomidae (chalcid parasite). **E.** Cynipidae (gall wasp). **F.** Pompilidae (pompilid wasp). **G.** Scoliidae (parasite of white grubs). **H.** Vespidae (polistes wasp). **I.** Sphecidae (spider wasp). **J.** Formicidae (ant). **K.** Halictidae (alkali bee).

among the endoptygote orders. Some representatives are illustrated in Figure A-40.

The larvae of the Hymenoptera are also highly variable. Those of the primitive Suborder Symphyta are similar to caterpillars but can be distinguished from them on the basis of having six to eight pair of abdominal prolegs that lack crochets (wood-boring and leaf-mining species may lack abdominal prolegs), and have no more than three pair of lateral ocelli. Larvae of the Suborder Apocrita are legless and maggotlike. Their mandibles are usually well defined but the head capsule is reduced and in some parasitic forms may be partially retracted into the prothorax.

The Symphyta are all herbivores, and several families are well known as pests. Some sawflies cause serious defoliation of forest trees and others are agricultural pests. The larvae of some wood wasps do considerable damage to commercial timber by tunneling in the wood and introducing rot-causing fungi. The Apocrita, on the other hand, are largely beneficial, especially in their roles as insect parasites and pollinators. Many of the higher Hymenoptera have evolved a social way of life, which is discussed in Chapter 16.

# Glossary

**Abdomen:** The posterior of the three main body divisions of an insect.
**Aculea** (pl. aculeae): With a sting (Hymenoptera).
**Adaptive color change:** A fairly rapid change of color in response to a change in the color characteristics of the environment.
**Adaptive radiation:** The diversification through evolution of organisms of common origin in response to a variety of new environmental opportunities.
**Aestivation:** Dormancy during a warm or dry season; usually the summer.
**Agroecosystem:** A crop ecosystem managed by man to maximize the production of an agricultural commodity.
**Allele:** Any of several alternative gene forms at a given chromosomal locus.
**Allometric growth:** The abnormal development of one part of the body.
**Alopatric:** Having ranges that do not overlap.
**Anal:** Pertaining to the last abdominal segment which bears the anus.
**Anal lobe:** A lobe in the posterior basal part of the wing.
**Anaphylaxis:** Hypersensitivity to a foreign protein or poison (such as a venom) marked by an intense systemic reaction.
**Antenna** (pl. antennae): One of a pair of segmented appendages located on the head above the mouth parts; usually sensory in function (Figure 3-4).
**Antennal club:** The enlarged distal segments of a clubbed antenna (Figure 10-5).
**Anterior:** Front.
**Anus:** The posterior opening of the alimentary tract (Figure 3-8).
**Aphrodisiac:** A chemical substance that stimulates sexual desire.
**Apical:** At the end, tip, or outermost part.
**Apiculture:** The keeping of honeybees.
**Apodeme:** An invagination of the body wall that forms a rigid process which serves for muscle attachment and for strengthening the body wall (Figure 3-2).
**Apophysis** (pl. apophyses): A tubercular or elongate process of the body wall, either external or internal.
**Aposematic coloration:** Conspicuous or warning coloration.
**Apterous:** Wingless.
**Apterygote:** Belonging to the subclass Apterygota.
**Aquatic:** Living in water.
**Arbovirus:** A virus transmitted by an arthropod vector.
**Articulation:** A joint, as between two segments or structures.
**Auditory:** The perception of sound.
**Autecology:** The ecology of a single species.
**Axon:** A fiber of a nerve cell that transmits nerve impulses away from the cell body.
**Basement membrane:** A noncellular membrane underlying the epithelial cells of the body wall (Figure 5-1).
**Batesian mimicry:** A form of similarity in the coloration of two or more species, one of which is unpalatable or distasteful, the others of which are edible.

**Biomass:** The total dry weight or volume of all the organisms in a given area.
**Biotype:** A variety or a population of a species that displays a distinct and recognizable characteristic that is at variance with other populations of the same species.
**Biramous:** With two branches.
**Bisexual:** With males and females.
**Book gill:** The leaflike gills of a horseshoe crab.
**Book lung:** A respiratory cavity containing a series of leaflike folds in spiders.
**Brachypterous:** With short wings that do not cover the abdomen.
**Brood:** The individuals which hatch from the eggs laid by one mother, individuals which hatch at about the same time and normally mature at about the same time.
**Campaniform sensillum:** A sense organ consisting of a dome-shaped cuticular area into which the sensory cell process is inserted (Figure 10-1).
**Campodeiform larva:** A larva shaped like the thysanuran *Campodea*, i.e., elongate, flattened, with well-developed legs and antennae.
**Cannibalistic:** Feeding on other individuals of the same species.
**Cantharidin:** An irritating chemical constituent of the cuticle of meloid beetles that has the properties of an aphrodisiac.
**Capitate:** With an apical knoblike enlargement.
**Carnivorous:** Feeding on the flesh of other animals.
**Caste:** A form or type of individual of a social insect.
**Caterpillar:** The larva of a butterfly, skipper, or moth.
**Caudal:** Pertaining to the tail or the posterior part of the body.
**Cell:** A unit mass of protoplasm, surrounded by a cell membrane, and containing one or more nuclei, or nuclear material.
**Cephalic:** On or attached to the head.
**Cephalothorax:** A body region consisting of head and thoracic segments.
**Cercus** (pl. cerci): One of a pair of appendages at the end of the abdomen (Figure 3-8).
**Cervical:** Pertaining to the neck or cervix.
**Cervical sclerite:** A sclerite located in the lateral part of the cervix, between the head and the prothorax (Figure 3-4).
**Cervix:** The neck, a membranous region between the head and prothorax.
**Chelicera** (pl. chelicerae): An anterior appendage of arachnids.
**Cheliped:** A leg terminating in an enlarged pincerlike structure, as in many crustaceans (Figure 2-6).
**Chitin:** A nitrogenous polysaccharide that occurs in the cuticle of arthropods.
**Chordotonal organ:** A sense organ, the cellular elements of which form an elongate structure, attached at both ends to the body wall.
**Chrysalis** (pl. chrysalids or chrysalides): The pupa of a butterfly (Figure 9-8).
**Circulative virus:** A virus that circulates through and passes some part of its life cycle within its vector.
**Clypeus:** A sclerite on the lower part of the face, between the frons and the labrum (Figure 3-4).
**Coevolution:** The evolution of a specific relationship between two or more distinctly different species.
**Collophore:** A tubelike structure located on the ventral side of the first abdominal segment of springtails (Figure 7-11).
**Commensalism:** A living together of two or more species, none of which is injured thereby, and at least one of which is benefited.

**Competitive displacement:** The replacement of an indigenous species by a recent arrival with similar ecological requirements.

**Compound eye:** An eye composed of many individual elements or ommatidia, each of which is represented externally by a facet (Figure 10-7).

**Condyle:** A knoblike process forming an articulation (Figure 3-4).

**Counter shading:** A form of coloration that tends to reduce the three-dimensional appearance of an animal by suppressing highlights.

**Coxa** (pl. coxae): The basal segment of the leg (Figure 3-7).

**Crawler:** The active first instar of some homopterans.

**Crop:** The dilated portion of the foregut just behind the esophagus (Figure 3-10).

**Cultivar:** A cultivated plant.

**Cuticle:** The noncellular part of the integument (Figure 5-1).

**Deciduous:** Having a part or parts that may fall off or be shed.

**Delusory parasitosis:** A psychological illness in which the inflicted persons imagine themselves to be infected with living parasites.

**Density-dependent factor:** An environmental factor that causes a level of mortality that varies with the number of individuals in the population.

**Density-independent factory:** An environmental factor that causes a level of mortality independent of the number of individuals in the population.

**Detritus food chain:** The passage of nutrients and energy through an unstructured assemblage of decomposer organisms.

**Diapause:** A period of arrested development or suspended animation.

**Diecious:** Having the male and female sex organs in different individuals, any one individual being either male or female.

**Disruptive coloration:** A color pattern that breaks up the outline of the body such that it does not stand out against a variable background.

**Distal:** Near or toward the free end of an appendage; that part of a segment or appendage farthest from the body.

**Diurnal:** Active during the daytime.

**Dormancy:** A state of quiescence or inactivity.

**Dorsal:** Top or uppermost; pertaining to the back or upper side.

**Dorsoventral:** From top to bottom, or from the upper to the lower surface.

**Ecdysis** (pl. ecdyses): Molting; the process of shedding the skin.

**Ecdysone:** The hormone that induces molting in arthropods (Figure 9-11).

**Ecological homologues:** Species that have the same fundamental ecological requirements and therefore cannot occupy the same area indefinitely without one species displacing the other.

**Ectoparasite:** A parasite that lives on the outside of its host.

**Ejaculatory duct:** The terminal portion of the male sperm duct (Figure 3-17).

**Elytron** (pl. elytra): A thickened, leathery, or horny front wing as in Coleoptera.

**Emergence:** The act of the adult insect leaving the pupal case or the last nymphal skin.

**Empodium** (pl. empodia): A padlike or bristlelike structure at the apex of the last tarsal segment, between the claws (Figure 7-4).

**Encephalitides:** A group of diseases of the central nervous system caused by mosquito-borne viruses.

**Endemic:** Restricted to or native to a particular locality; a population in equilibrium with its environment (not an epidemic).

**Endoparasite:** A parasite that lives inside its host.

**Endopterygote:** Having the wings developing internally; with complete metamorphosis.

**Entomophagous:** To feed upon insects.

**Entomophobia:** A pronounced fear of insects.

**Enzyme:** A protein that acts as a catalyst in the chemical reactions of living organisms.

**Epipharynx:** A mouth part structure attached to the inner surface of the labrum or clypeus (Figure 6-5).

**Esophagus:** The narrow portion of the alimentary canal immediately posterior to the pharynx (Figure 3-10).

**Eusocial:** The highest level of social development.

**Evagination:** An outpocketing, or saclike structure on the outside.

**Eversible:** Capable of being everted or turned outward.

**Exopterygote:** With the wings developing on the outside of the body, as in insects with simple metamorphosis.

**Exoskeleton:** A skeleton or supporting structure on the outside of the body.

**Exuviae** (always used in the plural): The cast skin of an arthropod.

**Facet:** The external surface of an individual compound eye unit of ommatidium (Figure 10-7).

**Family:** A subdivision of an order, suborder, or superfamily, and containing a group of related genera, tribes, or subfamilies. Family names end in *-idae.*

**Feces:** Indigestible wastes discharged from the digestive tract.

**Femur** (pl. femora): The third leg segment, between the trochanter and the tibia (Figure 3-7).

**Filiform:** Hairlike or threadlike.

**Filter chamber:** A bladderlike structure in which hind-gut loops back on the mid-gut in some homopterans (Figure 6-9).

**Food chain:** An assemblage of organisms, including producers, consumers, and decomposers, through which energy and materials may move in a community.

**Foramen magnum:** The opening on the posterior side of the head, through which pass the internal structures that extend from the head to the thorax (Figure 3-4).

**Fore-gut:** The anterior portion of the alimentary tract, from the mouth to the mid-gut (Figure 3-10).

**Fossorial:** Fitted for or with the habit of digging (Figure 7-12).

**Frass:** Wood fragments made by a wood-boring insect, usually mixed with feces.

**Frons:** The head sclerite bounded by the frontal and epistomal sutures and including the median ocellus (Figure 3-4).

**Furcula:** The forked springing apparatus of Collembola (Figure 7-11).

**Gall:** An abnormal growth of plant tissues, caused by the stimulus of an animal or another plant (Figure 15-4).

**Gamete:** A sexual reproductive cell that must usually fuse with another before development occurs; an egg or sperm.

**Ganglion** (pl. ganglia): A knotlike enlargement of a nerve, containing a coordinating mass of nerve cells (Figure 3-14).

**Gastric mill:** A grinding structure within the fore-gut of some orthopteroid insects (Figure 3-11).

**Gene:** A unit of inheritance.

**Gene pool:** The sum total of all the genes of all the individuals in a population.

**Generation:** From any given stage in the life cycle to the same stage in the offspring.

**Genitalia:** The sexual organs and associated structures; the external sexual organs.

**Genotype:** The particular combination of genes present in the cells of an individual.

**Genus** (pl. genera): A group of closely related species; the first name in a binomial or trinomial scientific name. Genus names are Latinized, capitalized, and when printed are italicized.

**Gill:** Respiratory evaginations of the body wall or hind-gut, in an aquatic animal (Figure 8-5).

**Glossa** (pl. glossae): One of a pair of lobes at the apex of the labium between the paraglossae (Figure 3-5).

**Gonopore:** The external opening of the reproductive organs.

**Gregarious:** Living in groups.

**Gustation:** The act of sensation of tasting.

**Haltere** (or halter): A small knobbed structure on each side of the metathorax in place of the hind wings in Diptera (Figure 7-27).

**Hamuli** (sing. hamulus): Minute hooks; a series of minute hooks on the anterior margin of the hind wing, with which the front and hind wings are attached together in Hymenoptera.

**Haustellate:** Formed for sucking, the mandibles wanting or not suited for chewing.

**Haustellum:** A part of the beak.

**Head:** The anterior body region, which bears the eyes, antennae, and mouth parts.

**Hemelytron** (pl. hemelytra): The front wing of Hemiptera.

**Hemimetabolous:** having a simple metamorphosis; lacking a pupal stage.

**Hemocoel:** The body cavity which is normally filled with blood or hemolymph.

**Hemorrhagic fever:** Fever resulting from the infection of blood cells by a parasite.

**Herbivorous:** Feeding on plants.

**Hermaphroditic:** One individual with both male and female sex organs.

**Heterogamy:** Alternation of bisexual reproduction with parthenogenetic reproduction.

**Heterozygote:** Having two different alleles of a given gene.

**Hibernation:** Dormancy during the winter.

**Hind-gut:** The posterior portion of the alimentary tract, between the mid-gut and anus (Figure 3-10).

**Holometabolous:** With complete metamorphosis; having a pupal stage.

**Homologous:** Organs or structures that are similar in their position and developmental origin, but not necessarily similar in appearance or function.

**Homozygote:** Having two similar alleles of a given gene.

**Honeydew:** Liquid discharged from the anus of certain Homoptera.

**Hormone:** A chemical secreted in one part of the body that regulates the function of some other part of the body.

**Host:** The organism in or on which a parasite lives; the plant on which an insect feeds.
**Hybrid:** A cross between two species, or a cross between two genetic types.
**Hypermetamorphosis:** A type of complete metamorphosis in which the different larval instars represent two or more different types of larvae.
**Hyperparasite:** A parasite whose host is another parasite.
**Hypognathus:** With the head vertical and the mouth parts located ventrally (Figure 3-3).
**Hypopharynx:** A mouth part structure arising on the upper surface of the labium (Figure 3-4).
**Imago** (pl. imagoes or imagines): The adult or reproductive stage of an insect.
**Instar:** The form of an insect between successive molts, the first instar being the stage between hatching and the first molt.
**Instinctive behavior:** Unlearned stereotyped behavior, in which the nerve pathways involved are hereditary.
**Integument:** The outer covering of the body including the epithelium and the cuticle (Figure 5-1).
***Inter-*:** Between; e.g., interspecific, between two or more different species.
***Intra-*:** Within; e.g., intraspecific, within a single species.
**Invagination:** An infolding or inpocketing.
**Johnston's organ:** A sense organ similar to a chordotonal organ, located in the second antennal segment of most insects (Figure 10-3).
**Joint:** An articulation of two successive segments or parts.
**Juvabione:** A substance isolated from trees of the genus *Abies* and found to have juvenile hormone characteristics.
**Juvenile hormone:** The hormone secreted by the corpora allata that maintains the expression of juvenile characteristics and suppresses adult characteristics (Figure 9-11).
**Knock-down insecticide:** A poison that knocks down flying insects rapidly but may not kill them.
**Labellum:** The expanded tip of the labium (Figure 6-7).
**Labial:** Of or pertaining to the labium.
**Labial palpus** (pl. labial palpi): One of a pair of small feelerlike structures arising from the labium (Figure 3-5).
**Labial suture:** The suture on the labium between the postmentum and the prementum.
**Labium:** One of the mouth part structures; the lower lip (Figure 3-5).
**Labrum:** The upper lip, lying just below the clypeus (Figure 3-5).
**Labrum-epipharynx:** A mouth part representing the labrum and epipharynx (Figure 6-5).
**Lac:** A resinous substance secreted by some scale insects.
**Lacinia** (pl. laciniae): The inner lobe of the maxilla, borne by the stipes (Figure 3-5).
**Lamella** (pl. lamellae): A leaflike plate.
**Lamellate:** Composed of a series of leaflike plates.
**Larva** (pl. larvae): The immature stages between the egg and pupa in holometabolous insects or egg and adult in hemimetabolous insects.
**Latency:** Dormancy, or the period within a host during which no symptoms of infection are evident even though a pathogen is present.
**Lateral:** On or pertaining to the side.

**Leishmaniasis:** Any disease caused by flagellate protozoans of the genus *Leishmania.*

**Lipid:** Any of a variety of compounds such as fats, oils, waxes, and steroids that are insoluble in water but soluble in alcohols and ethers.

**Longitudinal:** Lengthwise of the body or of an appendage.

**Luminescent:** Producing light.

**Maggot:** A vermiform larva; a legless larva without a well-developed head capsule (Figure 9-7*E*).

**Malaria:** An infectious, fever-causing disease caused by protozoans of the genus *Plasmodium*, which invade the red blood corpuscles after transmission to a human most commonly by an *Anopheles* mosquito.

**Malpighian tubule:** An excretory diverticulum of the digestive tract of insects (Figure 3-10).

**Mandible:** Jaw; one of the anterior pair of the paired mouth part structures of insects (Figure 3-5).

**Mandibulate:** With jaws suited for chewing; mouth parts derived from mandibles.

**Maxilla** (pl. maxillae): One of the paired mouth part structures immediately posterior to the mandibles (Figure 3-5).

**Maxillary palpus** (pl. maxillary palpi): A small feelerlike structure arising from the maxilla (Figure 3-5).

**Meiosis:** Nuclear division in which the number of chromosomes is reduced by half.

**Mentum:** The distal part of the labium, which bears the palpi (Figure 3-5).

**Mesonotum:** The dorsal sclerite of the mesothorax.

**Mesosternum:** The sternum, or ventral sclerite, of the mesothorax.

**Mesothorax:** The middle or second segment of the thorax.

**Metamorphosis:** Change in form during development.

**Metathorax:** The third or posterior segment of the thorax.

**Mid-gut:** The mesenteron, or middle portion of the alimentary tract.

**Mitosis:** Nuclear division characterized by a complex movement of chromosomes, which results in the formation of new nuclei having the same number of chromosomes as the original nucleus.

**Molt:** A process of shedding the cuticle, ecdysis.

**Moniliform:** Beadlike, with rounded segments.

**Monoculture:** A single species or single varietal planting.

**Monogenic:** A characteristic (such as resistance) controlled by a single gene.

**Morphology:** The study of form or structure.

**Motor neuron:** A neuron leading from the central nervous system to an effector organ (Figure 11-1).

**Müllerian mimicry:** A number of unpalatable or distasteful species that share a similar color pattern.

**Mutation:** Any relatively stable heritable change in the genetic material.

**Mycangium** (pl. mycangia): An invaginated cuticular pouch in which fungal spores are carried (Figure 6-10).

**Myiasis:** The infection of animal tissue by the larvae of certain flies.

**Naiad:** An aquatic, gill-breathing nymph.

**Nasutus** (pl. nasuti): An individual of a termite caste in which the head narrows anteriorly into a snoutlike projection (Figure 14-6).

**Natural selection:** An increase in the frequency of occurrence of some genes or gene combinations and a decrease in the frequency of others because of the selective action of the environment.

**Nectar:** A sugary liquid secreted by the flowers of many plants.

**Nectary:** The floral gland which secretes nectar.

**Nerve:** A bundle of neuron fibers.

**Neuron:** A nerve cell (Figure 11-1).

**Nocturnal:** Active at night.

**Nonpersistent virus:** A virus that does not undergo part of its development within the body of its vector; a stylet-borne virus.

**Notum** (pl. nota): The dorsal surface of a body segment (usually used when speaking of the thoracic segments).

**Nymph:** An immature stage (following hatching) of an insect that does not have a pupal stage; the immature stages of Acarina that have eight legs.

**Occipital suture:** A transverse suture in the posterior part of the head which separates the vertex from the occiput dorsally and the genae from the postgenae laterally (Figure 3-4).

**Occiput:** The dorsal posterior part of the head, between the occipital and postoccipital sutures (Figure 3-4).

**Ocellus** (pl. ocelli): A simple eye of an insect or other arthropod (Figure 10-12).

**Olfaction:** The sense of smell.

**Ommatidium** (pl. ommatidia): A single unit or visual section of a compound eye (Figure 10-6).

**Omnivore:** An animal that eats a variety of foods including both plant and animal matter.

**Onchocerciasis:** Infection by filarial worms of the genus *Onchocerca* transmitted by some species of black flies.

**Ootheca** (pl. oothecae): The covering or case of an egg mass.

**Operculum** (pl. opercula): A lid or cover.

**Opisthognathus:** Having the front of the head deflected backward so the mouth parts project to the rear (Figure 3-3).

**Oral:** Pertaining to the mouth.

**Order:** A subdivision of a class or subclass, containing a group of related families.

**Osmeterium** (pl. osmeteria): A fleshy, tubular, eversible, usually Y-shaped scent gland at the anterior end of certain caterpillars.

**Ostium** (pl. ostia): A slitlike opening of the insect heart (Figure 3-9).

**Ovary:** The egg-producing organ of the female (Figure 3-16).

**Oviduct:** The tube leading away from the ovary through which the eggs pass (Figure 3-16).

**Oviparous:** Egg laying.

**Oviposit:** To lay or deposit eggs.

**Ovipositor:** The egg-laying apparatus; the external genitalia of the female (Figures 9-3, 9-4).

**Ovoviviparous:** The production of living young immediately following the hatching of the eggs within the female.

**Paedogenesis:** The production of eggs or young by an immature or larval stage of an animal.

**Palpus** (pl. palpi): A segmented process borne by the maxillae or labium (Figure 3-5).

**Paraglossa** (pl. paraglossae): One of a pair of lobes at the apex of the labium, laterad to the glossae (Figure 3-5).

**Parasite:** An animal that lives in or on the body of another living animal (its host), at least during a part of its life cycle and at the host's expense.

**Parasitic:** Living as a parasite.

**Parasitoid:** A parasite that usually kills its host.

**Parthenogenesis:** Reproducing by eggs which develop without being fertilized.

**Pathogen:** A disease-causing organism.

**Pathogen reservoir:** A host or group of hosts harboring a disease-causing organism.

**Pectinate:** With branches or processes like the teeth of a comb.

**Pedicel:** The second segment of the antenna; the stem of the abdomen, between the thorax and the gaster in ants.

**Persistent virus:** A virus that persists within the body of a vector; circulative virus.

**Pharynx:** The anterior part of the fore-gut, between the mouth and the esophagus (Figure 3-10).

**Phenotype:** The physical manifestation of a genetic trait.

**Pheromone:** A special chemical substance which, secreted by one individual, influences the behavior of other individuals of the same species.

**Phoresy:** The transport of one organism by another.

**Photoperiodism:** A response by an organism to the duration or pattern of light and dark periods.

**Photosynthesis:** The utilization of solar energy in the primary synthesis of organic materials.

**Phragma** (pl. phragmata): A platelike apodeme of the dorsal wall of the thorax (Figure 3-6).

**Phylum** (pl. phyla): One of the major divisions of the animal kingdom.

**Phytophagous:** Feeding on plants.

**Phytotoxicity:** Poisonous or chemically injurious to plants.

**Plastron:** A film of air over part of the body of some aquatic insects; like a bubble, it serves as a physical gill.

**Pleural:** Pertaining to the pleura, or lateral sclerites of the body; lateral.

**Pleuron** (pl. pleura): The lateral area of a thoracic segment (Figure 3-6).

**Plumose:** Featherlike.

**Pollen grain:** A microgamete of a seed plant.

**Pollinium:** A mass of pollen grains characteristic of the orchid family; often bearing a stalk with an adhesive disc that facilitates its transfer by insects (Figure 18-4).

**Polyembryony:** An egg developing into two or more embryos.

**Polygenic:** A characteristic, such as resistance, conferred by more than one gene.

**Polymorphism:** The simultaneous occurrence of more than one phenotype in a population.

**Population:** A group of individuals of the same species living together in a limited and defined space.

**Posterior:** Hind or rear.

**Preapical:** Situated just before the apex.

**Predaceous:** Feeding as a predator.

**Predator:** An animal that attacks and feeds on other animals, usually smaller or less powerful than itself.

**Pregenital:** Anterior to the genital segments of the abdomen.

**Prepupa:** A quiescent stage of the last larval instar immediately preceding pupation; the third instar of a thrips.

**Primary infection:** The initial introduction of a disease into a crop by insect vectors invading from elsewhere.

**Proboscis:** The extended beaklike mouth parts of sucking insects.

**Prognathous:** Having the head horizonal and the mouth parts projecting forward (Figure 3-3).

**Proleg:** One of the fleshy abdominal legs of certain insect larvae (Figure 7-7).

**Prothorax:** The anterior of the three thoracic segments.

**Proximal:** Nearer to the body or to the base of an appendage.

**Pterygote:** Winged; a member of the subclass Pterygota.

**Ptilinum:** A bladderlike organ that when inflated extends from the front of the head of the adult of some species of flies and ruptures the puparium thereby facilitating emergence.

**Pubescent:** Downy, covered with short fine hairs.

**Pulvillus** (pl. pulvilli): A pad or lobe beneath each claw (Figure 7-4).

**Pupa** (pl. pupae): The stage between the larva and the adult in insects with complete metamorphosis, a nonfeeding, usually inactive stage (Figure 9-8).

**Puparium** (pl. puparia): The thickened, hardened last larval skin within which some flies pass the pupal stage (Figure 5-4).

**Pupate:** Transform to a pupa.

**Raptorial:** Fitted for grasping prey; raptorial front legs.

**Recessive characteristic:** A characteristic governed by an allele that does not express itself in the presence of another allele of the same gene. Recessive traits are only expressed in homozygous individuals.

**Rectum:** The posterior region of the hind-gut (Figure 3-10).

**Reflex:** A functional unit of the nervous system involving the entire pathway from a receptor cell to an effector (Figure 11-1).

**Respiration:** The release of energy by the oxidation of fuel molecules within a cell.

**Retina:** The receptive apparatus of an eye (Figure 10-6).

**Rhabdom:** A rodlike structure formed of the inner surfaces of adjacent sensory cells in the ommatidium of a compound eye (Figure 10-6).

**Rickettsia:** An obligate intracellular parasite of arthropods. Somewhat intermediate between bacteria and viruses, many are pathogenic of man and other vertebrates.

**Sap:** Water and dissolved materials moving through the conductive tissues of plants.

**Saprophagous:** Feeding on decaying matter.

**Scape:** The basal segment of the antenna (Figure 10-5).

**Scavenger:** An animal that feeds on dead plants or animals, on decaying materials, or on animal wastes.

**Scientific name:** A Latinized name, internationally recognized, of a species or subspecies. The scientific name of a species consists of the generic and a trivial name, and that of a subspecies consists of the generic and two trivial names. Scientific names are always written in italics.

**Sclerite:** A hardened integumentary plate bounded by sutures or membranous areas.

**Sclerotin:** A proteinaceous material that contributes to the hardness of the cuticle (Figure 5-2).

**Sclerotized:** Hardened by the presence of sclerotin.

**Scutellum:** A sclerite of a thoracic notum; the mesoscutellum, appearing as a triangular sclerite behind the pronotum of some insects (Figure 3-6).

**Secondary infection:** The infection of healthy plants within a crop after the disease has entered the crop from outside (primary infection).

**Segment:** A subdivision of the body or of an appendage, between joints or articulations.

**Selective pressure:** The main force that induces a genetic change through natural selection.

**Semiaquatic:** Living in wet places, or partially in water.

**Sensory neuron:** A neuron leading from a receptor cell to the central nervous system (Figure 11-1).

**Sericulture:** The rearing of silkworms for silk production.

**Serrate:** Toothed along one edge like a saw.

**Sessile:** Attached or fastened, incapable of moving from place to place.

**Seta** (pl. setae): A bristle or hairlike cuticular process (Figure 5-3).

**Setaceous:** Bristlelike.

**Species:** A group of individuals or populations that are similar in structure and physiology and are capable of interbreeding and producing fertile offspring; the largest group of individuals within which effective gene flow occurs or could occur.

**Spine:** A thornlike outgrowth of the cuticle (Figure 5-3).

**Spiracle:** An external opening of the tracheal system; a breathing pore (Figure 8-1).

**Stadium** (pl. stadia): The period between molts in a developing insect.

**Stemma** (pl. stammata): The simple eye of a holometabolous larva; a lateral ocellus (Figure 10-12).

**Sternite:** A sclerite on the ventral side of the body; the ventral sclerite of an abdominal segment (Figure 3-8).

**Stipes** (pl. stipites): The second segment or division of a maxilla, which bears the palpus, the galea, and the lacinia (Figure 3-5).

**Stomodaeum:** The fore-gut (Figure 3-10).

**Stridulate:** To make a noise (chirp or creak) by rubbing two structures or surfaces together.

**Stylet:** A needlelike structure.

**Stylet-borne virus:** A nonpersistent virus carried on the mouth parts of its vector.

**Subclass:** A major subdivision of a class, containing a group of related orders.

**Subesophageal ganglion:** The knotlike swelling at the anterior end of the ventral nerve cord, just below the esophagus (Figure 3-14).

**Subfamily:** A major subdivision of a family, containing a group of related genera. Subfamily names end in *-inae.*

**Subimago:** The first of the two winged instars of a Mayfly after it emerges from the water.

**Submentum:** The basal part of the labium (Figure 3-5).

**Suborder:** A major subdivision of an order, containing a group of related families.

**Subspecies:** A subdivision of a species, usually a geographic race. The different subspecies of a species are ordinarily not sharply differentiated, intergrade with one another, and are capable of interbreeding.

**Suctorial:** Suited for sucking.

**Suture:** An external linelike groove in the cuticle or a narrow membranous area between sclerites; the line of juncture between the elytra.

**Swimmeret:** An abdominal appendage that functions as a swimming organ in crustaceans (Figure 2-6).

**Symbiont:** An organism living in symbiosis with another organism.

**Symbiosis:** Two species living together in a more or less intimate association.

**Sympatric:** Having overlapping ranges.

**Synanthropy:** A close association with man.

**Synecology:** The study of groups of interacting species and their environment.

**Tarsal claw:** A claw at the apex of the tarsus (Figure 7-4).

**Tarsus** (pl. tarsi): That part of the leg beyond the tibia, consisting of one or more segments or subdivisions (Figure 3-7).

**Taxonomy:** Classification into categories of varying rank, based on similarities and differences, and the describing and naming of these categories.

**Telson:** The last abdominal segment of crustaceans (Figure 2-6).

**Tenaculum:** A minute structure on the sternum of the third abdominal segment which serves as a clasp for the furcula of collembolans (Figure 7-11).

**Tentorial pits:** Pitlike depressions on the surface of the head that mark the points of invagination of the arms of the tentorium (Figure 3-4).

**Tentorium:** The endoskeleton of the head, usually consisting of two pairs of apodemes.

**Tergite:** A tergal sclerite; the dorsal surface of an abdominal segment (Figure 3-8).

**Tergum** (pl. terga): The dorsal surface of any body segment.

**Terrestrial:** Living on land.

**Testis** (pl. testes): The sex organ in the male that produces sperm (Figure 3-17).

**Thorax:** The body region between the head and abdomen which bears the legs and wings (Figure 3-1).

**Tibia** (pl. tibiae): The fourth segment of the leg, between the femur and the tarsus (Figure 3-7).

**Trachea** (pl. tracheae): A tube of the respiratory system, lined with taenidia, ending externally at a spiracle, and terminating internally with tracheoles (Figure 8-1).

**Tracheoles:** The fine terminal branches of the respiratory tubes which lack taenidia (Figure 8-1).

**Trench fever:** A nonfatal disease of humans caused by a rickettsia transmitted by the body louse.

**Tribe:** A subdivision of a subfamily, containing a group of closely related genera. Names of tribes end in *-ini.*

**Trochanter:** The second segment of the leg, between the coxa and the femur (Figure 3-7).

**Trochantin:** A small sclerite in the thoracic wall immediately anterior to the base of the coxa (Figure 7-1).

**Tympanum** (pl. tympana): A vibrating membrane; an auditory membrane or eardrum (Figure 13-2).

**Typhus fever:** An often fatal disease of humans caused by a rickettsia transmitted by the body louse.

**Vagina:** The terminal portion of the female reproductive system, which opens to the outside.

**Vector:** An organism that transmits a pathogen from one host to another.
**Vein:** A thickened line in the wing of an insect.
**Ventral:** Lower or underneath; pertaining to the under side of the body.
**Vertex:** The top of the head, between the eyes and anterior to the occipital suture.
**Vesicle:** A sac, bladder, or cyst.
**Vestigial:** Small, poorly developed, degenerate, nonfunctional.
**Viviparous:** Giving birth to living young, rather than egg-laying.

# General index

**Boldface** numerals indicate major treatment; *italics* indicate illustrations; "t" indicates table.

# Taxonomic index